47

Topics in Heterocyclic Chemistry

Aims and Scope

The series Topics in Heterocyclic Chemistry presents critical reviews on present and future trends in the research of heterocyclic compounds. Overall the scope is to cover topics dealing with all areas within heterocyclic chemistry, both experimental and theoretical, of interest to the general heterocyclic chemistry community.

The series consists of topic related volumes edited by renowned editors with contributions of experts in the field.

More information about this series at http://www.springer.com/series/7081

Joëlle Prunet
Editor

Synthesis of Heterocycles by Metathesis Reactions

With contributions by

M. Bassetti · J.F. Bower · P. Compain · A. D'Annibale · A. Gradillas · P.R. Hanson · D. Hazelard · S. Javed · J.H. Jun · A. Letort · D.M. Lindsay · C.N. Ndi · J. Pérez-Castells · J. Prunet · N.J. Race · D.T.S. Rijkers · G. Vincent

Springer

Editor
Joëlle Prunet
School of Chemistry
University of Glasgow
Glasgow
United Kingdom

ISSN 1861-9282 ISSN 1861-9290 (electronic)
Topics in Heterocyclic Chemistry
ISBN 978-3-319-39939-3 ISBN 978-3-319-39941-6 (eBook)
DOI 10.1007/978-3-319-39941-6

Library of Congress Control Number: 2016957352

Printed on acid-free paper

This Springer imprint is published by Springer Nature
The registered company is Springer International Publishing AG Switzerland

Foreword

One would be hard pressed to find a set of chemical transformations that can deliver heterocyclic structures with a wider scope than catalytic olefin metathesis. It is therefore of little surprise that in countless academic laboratories and in many programs in industry, whether it is to develop a potent therapeutic, a tantalizing fragrance, or a polymeric structure that has a particular set of properties, catalytic olefin and/or alkyne metathesis is the perennial favorite. In a great number of cases, the targeted molecules are cyclic structures that contain one or more heteroatoms, which render the present collection of articles of exceptional value.

First, there is a concise overview of some of the more important fundamental aspects of metathesis reactions that involve alkenyl and/or alkynyl substrates. This is the most welcome beginning because it will appeal especially to the younger scientists for whom this book will serve as their introduction to different varieties of metathesis reactions. The hardened chemist with many years of experience may skip this initial treatment but hardly what comes afterward.

We then have the article by Bower and Race who expertly demonstrate that a myriad of heteroaromatic structures – the "bread and butter" of medicinal chemistry – can be accessed by a variety of olefin metathesis transformations or one might consider an enyne as the optimal starting point. The chapter by Lindsey reveals how catalytic ring-closing metathesis (RCM) can allow one assemble, with remarkable efficiency, beautiful rows of five-, six-, seven-, and nine-membered ring ethers that are referred to as gambieric acids A–D or ciguatoxin CTX3C. Then there is kendomycin and neopeltolide – smaller targets no doubt but ones that bring to the table their own unique flavor of difficulties and challenges.

The ensuing chapters by Bassetti and D'Annibale and then by Compain and Hazelard offer different views of a somewhat different landscape, one that deals less with total synthesis and more with how one might prepare most efficiently some of the more commonly occurring medium-sized lactones and nitrogen-containing rings. The special significance of cyclic amines and related derivatives is underscored by Vincent in the subsequent overview, showing us how one might go about designing pathways that lead to lactam structures. These are routes that could include a ring-closing alkene metathesis, a ring-closing enyne metathesis, a ring rearrangement metathesis, or a cross metathesis (CM) reaction.

And then there are all those beautiful large-ring heterocyclic structures that can be prepared by a metathesis-type process. There are two well-written chapters that cover this crucial topic. The first is by Rijkers who treats us with an account of the emerging field of macrocyclic peptide synthesis; here we can read about how alkene metathesis is being utilized in the design of amino acid-based therapeutic candidates. What follows is an article by Gradillas and Pérez-Castells vis-à-vis the state of the art in preparation of complex macrocycles through alkene or alkyne metathesis. These contributors make it clear that metathesis-based catalytic strategies have allowed chemists to reach their final targets through remarkably brief sequences.

The last part of the story, penned by Hanson, Jun, Javed, and Ndi takes us on a tour of cyclic structures that contain some of the less encountered – but equally important – heteroatoms: phosphorous, sulfur, silicon, boron, and selenium. We see that catalytic metathesis strategies may be utilized to access some intriguing structures with plenty of untapped potential.

What I find most appealing about this book, other than the fact that it contains countless pieces of useful information, is the exciting number of yet unaddressed issues that would render olefin, enyne, or alkyne metathesis more powerful still. We have come a long way no doubt, but we have yet a considerable distance to go. What lies ahead is likely to be a more arduous road; the problems that remain are compelling but also probably far more demanding to solve.

Merkert Chemistry Center, Boston College, USA Amir H. Hoveyda

Preface

This volume is dedicated to the syntheses of heterocycles that involve a metathesis reaction as a key step, but not necessarily as the ring-forming step. The different chapters are not comprehensive reviews, but outline recent progress toward the synthesis of specific heterocycles while giving the reader insights of the factors that govern the formation of these heterocycles.

The introduction entitled "Metathesis Reactions: General Considerations," written by Aurélien Letort and myself, presents the different kinds of alkene, enyne, and alkyne metathesis reactions and the associated mechanisms.

In the first chapter "Synthesis of Heteroaromatic Compounds by Alkene and Enyne Metathesis," Nicholas J. Race and John F. Bower describe the synthesis of aromatic heterocycles such as furans and pyrroles, pyridones and pyridines, and pyridazinones and pyridazines, using ring-closing metathesis (RCM) or cross metathesis (CM) reactions.

The chapter entitled "The Synthesis of Cyclic Ether-Containing Natural and Non-natural Products by Metathesis Reactions," written by David Lindsay, focuses on the use of metathesis in total synthesis, since the methodological aspects of this topic have been very recently reviewed in another book of this series.

Mauro Bassetti and Andrea D'Annibale, in the chapter entitled "Metathetic Synthesis of Common and Medium-Sized Lactones: The State of the Art," cover the synthesis of 5- to 11-membered lactones, focusing on the most recent publications. For each ring size, they distinguish the acrylate approach, where one of the alkenes participating in the RCM reaction is conjugated to the ester, from the other possible ring formations.

The fourth chapter entitled "Synthesis of Amine-Containing Heterocycles by Metathesis Reactions: Recent Advances and Opportunities" by Philippe Compain and Damien Hazelard highlights the difficulties of metathesis reactions in the presence of amines and details the steric and electronic parameters that prevent the deleterious coordination of the amino group to the catalyst metal center.

Guillaume Vincent describes in the fifth chapter the "Synthesis of Lactams by Metathesis Reactions." This chapter is structured according to the metathesis reaction employed, RCM, CM, ring-closing enyne metathesis (RCEYM), or ring-rearrangement metathesis (RRM) and discusses the various parameters that govern the metathesis reactions.

The following chapter, written by Dirk T. S. Rijkers, reviews recent advances on the "Synthesis of Cyclic Peptides and Peptidomimetics by Metathesis Reactions" and shows the prominent role of metathesis reactions in, inter alia, alternatives for disulfide bridge formation, stabilization of peptide secondary structures, and α-helix stabilization.

In the seventh chapter entitled "Synthesis of Macrocycles other than Peptides by Metathesis," Ana Gradillas and Javier Pérez-Castells discuss the conditions that may favor the formation of these macrocycles. They then present recent syntheses of natural macrocyclic heterocycles according to a biogenetic classification, as well as nonnatural products including macromolecules for supramolecular chemistry.

The last chapter "Synthesis of P-, S-, Si-, B-, and Se-Heterocycles via Ring-Closing Metathesis," written by Jung Ho Jun, Salim Javed, Cornelius N. Ndi, and Paul R. Hanson, deals with the synthesis of less common heterocycles. These heterocycles are synthesized as a goal per se, or the heterocyclic element is part of a temporary tether that facilitates difficult intermolecular olefin formations as well as directs their stereoselective outcome.

I would like to thank all the authors who have participated in this book for their exceptional work. I am also grateful to Janine Cossy for having given me the opportunity to edit this volume, and I hope it will be a useful support for the chemists involved in heterocyclic synthesis.

Glasgow, Scotland Joëlle Prunet

Contents

Abbreviations

Ac	Acetyl
AChE	Acetyl cholinesterase
Abu	α-Aminobutyric acid
ADMET	Acyclic diene metathesis polymerization
Aib	α-Aminoisobutyric acid
Ala	Alanine
Alg	Allylglycine
All	Allyl
Amp	(2*R*,3*R*)-2-amino-3-methyl-4-pentenoic acid
AngIV	Angiotensin IV
anhyd	Anhydrous
Ar	Aryl
Arg	Arginine
Asn	Asparagine
Asp	Aspartic acid
ATPH	Aluminum tris(2,6-diphenyl)phenoxide
9-BBN	9-Borabicyclo[3.3.1]nonane
BHT	3,5-Di-*tert*-4-butylhydroxytoluene
Bn	Benzyl
Boc	*tert*-Butoxycarbonyl
BOP	(Benzotriazol-1-yloxy)tris(dimethylamino)phosphonium hexafluorophosphate
BQ	Benzoquinone
Bu	Butyl
s-Bu	*sec*-Butyl
t-Bu or ^{t}Bu	*tert*-Butyl
Bz	Benzoyl
cat	Catalyst
CBS	Corey–Bakshi–Shibata
Cbz	Benzyloxycarbonyl
CD	Circular dichroism

CDP	Cyclodepolymerization
CIP	Cahn–Ingold–Prelog
CM	Cross metathesis
Conv	Conversion
COSY	Correlation spectroscopy
COX	Cyclooxygenase
m-CPBA	3-Chloroperbenzoic acid
CSA	Camphorsulfonic acid
CSI	Chlorosulfonyl isocyanate
CTAB	Cetyltrimethylammonium bromide
Cy	Cyclohexyl
Cys	Cysteine (C)
dba	Dibenzylideneacetone
DBU	1,8-Diazabicyclo[5.4.0]undec-7-ene
DCC	*N,N*-dicyclohexylcarbodiimide
DCE	1,2-Dichloroethane
DCM	Dichloromethane
DDQ	2,3-Dichloro-5,6-dicyano-1,4-benzoquinone
de	Diastereomeric excess
DEAD	Diethyl azodicarboxylate
Dha	Dehydroalanine
Dhb	*Z*-dehydrobutyrine
DIAD	Diisopropyl azodicarboxylate
DIBAL	Diisobutylaluminium hydride
DIPEA	Diisopropylethylamine
DMA	Dimethylacetamide
DMAD	Dimethyl acetylenedicarboxylate
DMAP	4-(Dimethylamino)pyridine
DMB	2,4- or 3,4-dimethoxybenzyl
DME	1,2-Dimethoxyethane
DMF	Dimethylformamide
DMSO	Dimethyl sulfoxide
dppp	1,3-Bis(diphenylphosphino)propane
dr	Diastereomer ratio
EAS	Electrophilic aromatic substitution
ECE	Endothelin-converting enzyme
EDCI	*N*-(3-dimethylaminopropyl)-*N′*-ethylcarbodiimide hydrochloride
ee	Enantiomeric excess
EM	Effective molarity
EOM	Ethoxymethyl
equiv or eq	Equivalent(s)
ESI	Electrospray ionization
Et	Ethyl

Fmoc	9-Fluorenylmethoxycarbonyl
G0	Grubbs cinnamylidene catalyst
G1	Grubbs catalyst 1st generation
G2	Grubbs catalyst 2nd generation
Gln	Glutamine (Q)
Glu	Glutamic acid (E)
Gly	Glycine (G)
GN	Grubbs–Nolan catalyst
Hag	Homoallylglycine
HATU	1-[Bis(dimethylamino)methylene]1*H*-1,2,3,triazolo[4,5-*b*] pyridinium 3-oxid hexafluorophosphate
HBS	Hydrogen-bond surrogate
HDAC	Histone deacetylase
HG1	Hoveyda–Grubbs catalyst 1st generation
HG2	Hoveyda–Grubbs catalyst 2nd generation
His	Histidine (H)
HPLC	High-pressure liquid chromatography
HOAt	1-Hydroxy-7-azabenzotriazole
HONSu	*N*-hydroxysuccinimide
HSL	Hormone-sensitive lipase
Hyp	Hydroxyproline
Ile	Isoleucine (I)
Im	Imidazole
IMPDH	Inosine 5′-monophosphate dehydrogenase
IRAP	Insulin-regulated aminopeptidase
ITC	Isothermal titration calorimetry
KHMDS	Potassium bis(trimethylsilyl)amide
LAAM	Linked amino acid mimetics
LAH	Lithium aluminum hydride
LCES	MS Liquid chromatography electrospray mass spectrometry
LDA	Lithium diisopropylamide
Leu	Leucine (L)
LHMDS	Lithium bis(trimethylsilyl)amide
5-LO	5-Lipoxygenase
LPA	Lysophosphatidic acid
Lys	Lysine (K)
M1	Umicore indenylidene catalyst 1st generation
M2	Umicore indenylidene catalyst 2nd generation
MC	Melanocortin
MCR	Multicomponent reaction
Me	Methyl
MEM	(2-Methoxyethoxy)methyl
Mes	Mesityl (2,4,6-trimethylphenyl)
Met	Methionine (M)

MIC	Minimum inhibitory concentration
MOM	Methoxymethyl
MS	Molecular sieves
Ms	Mesyl (methanesulfonyl)
MW/μW	Microwaves
o-NBSH	*ortho*-Nitrobenzenesulfonylhydrazide
n.d.	Not determined
NHC	*N*-heterocyclic carbene
Nle	Norleucine
NMM	*N*-methylmorpholine
NMO	*N*-methylmorpholine-*N*-oxide
NMR	Nuclear magnetic resonance
NOE	Nuclear Overhauser effect
nr	No reaction
NSAID	Non-steroidal anti-inflammatory drug
Nu	Nucleophile
Oas	*O*-allylserine
OLEC	Olefin-ester cyclization
OM	Olefin metathesis
Orn	Ornithine
PA	Phosphatidic acid
Pbf	2,2,4,6,7-Pentamethyldihydrobenzofuran-5-sulfonyl
PEG	Polyethylene glycol
PG	Protecting group
Pgl	Pentenylglycine
Ph	Phenyl
Phe	Phenylalanine (F)
phth	Phthalate
PMB	4-Methoxybenzyl
Pmc	2,2,5,7,8-Pentamethylchroman
PMP	4-Methoxyphenyl
PPII	Polyproline II
Pr or *n*-Pr	Propyl
i-Pr or ^{i}Pr	Isopropyl
PRD	Proline-rich motif-recognizing domain
Pre	Prenylglycine
PRM	Proline-rich motifs
Pro	Proline (P)
quant.	Quantitative
RCAM	Ring-closing alkyne metathesis
RCEYM	Ring-closing enyne metathesis
RCM	Ring-closing metathesis
Ref.	Reference
RMSD	Root mean square deviation

ROESY	Rotating-frame nuclear Overhauser effect
RO	Ring opening
RO/CM	Ring-opening/cross metathesis
ROM	Ring-opening metathesis
ROMP	Ring-opening metathesis polymerization
RRCM	Relay ring-closing metathesis
RRM	Ring-rearrangement metathesis
rt	Room temperature
S	Schrock catalyst
Sac	*S*-allylcysteine
SAR	Structure–activity relationship
SDS	Sodium dodecyl sulfate
SEM	2-(Trimethylsilyl)ethoxymethyl
Ser	Serine (S)
SPPS	Solid-phase peptide synthesis
TBAF	Tetrabutylammonium fluoride
TBDMS or TBS	*tert*-Butyldimethylsilyl
TBDPS	*tert*-Butyldiphenylsilyl
TCB	2,4,6-Trichlorobenzoylchloride
TCE	1,1,2-Trichloroethane
TES	Triethylsilyl
TFA	Trifluoroacetic acid
THF	Tetrahydrofuran
THP	Tetrahydropyranyl
TIPS	Triisopropylsilyl
T.M.	Target molecule
TMEDA	*N,N,N',N'*-tetramethylethylenediamine
TMS	Trimethylsilyl
TOCSY	Total correlation spectroscopy
TOF	Time of flight
Tol	4-Methylphenyl
tol	Toluene
Trp	Tryptophan (W)
Trt	Triphenylmethyl (trityl)
Ts	Tosyl (4-toluenesulfonyl)
TST-RCM	Temporary silicon-tethered ring-closing metathesis
TTF	Tetrathiafulvalene
Tyr	Tyrosine (Y)
hTyr	Homotyrosine
uPA	Urokinase-type plasminogen activator
Val	Valine (V)

Catalysts

The structures of the most common alkene metathesis catalysts mentioned in this book are shown here.

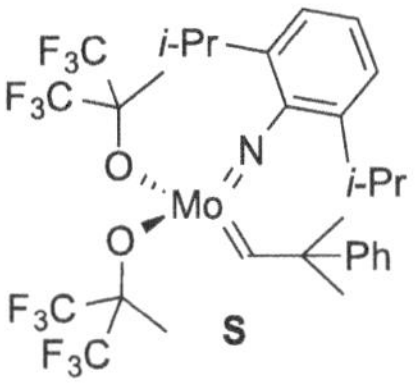

S

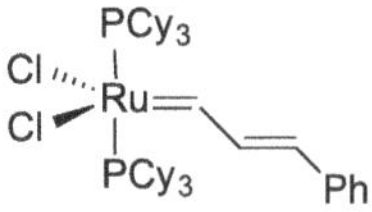

G0

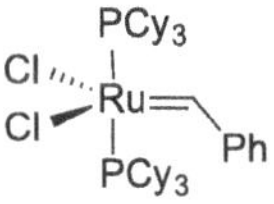

G1

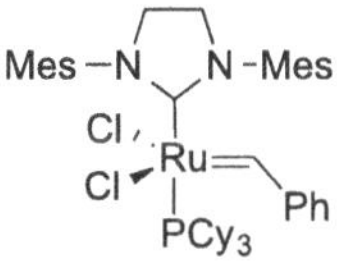

G2

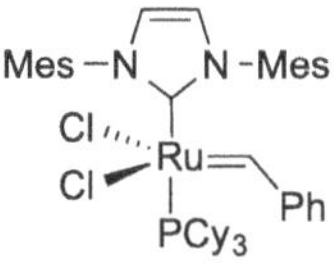

GN

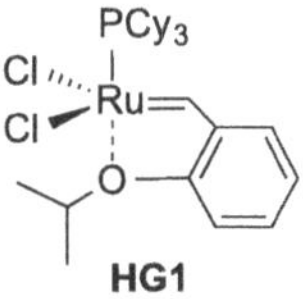

HG1

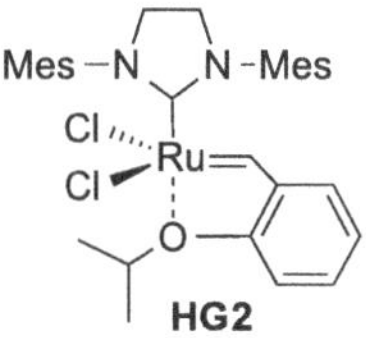

HG2

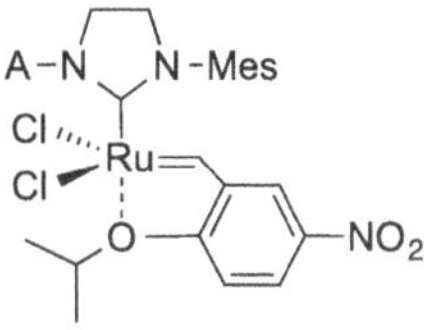

Grela

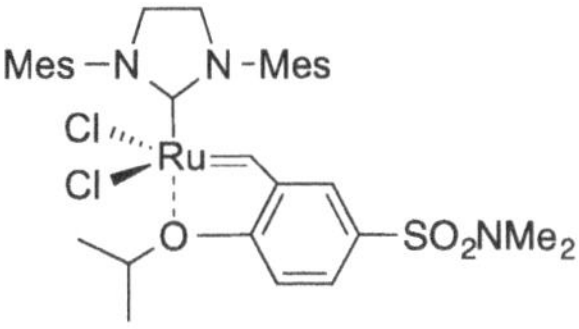

Zhan-1B

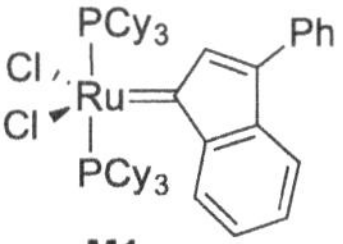

M1

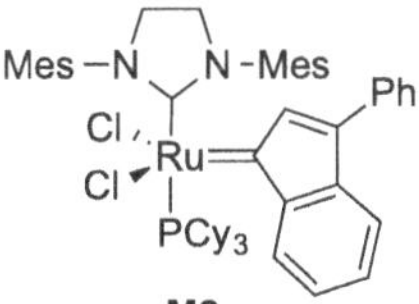

M2

Introduction to Volume: Metathesis Reactions: General Considerations

Aurélien Letort and Joëlle Prunet

Abstract The different types of metathesis reactions are outlined, with an emphasis on those used for the synthesis of heterocycles. The mechanism of alkene metathesis is shown, and ring-closing and cross metathesis reactions are discussed. Ring-closing ene-yne metathesis is also presented, with some insight into the mechanism of this reaction, and tandem processes are evoked. This introduction ends with ring-closing alkyne metathesis and presentation of its mechanism.

Contents

1 Background and Different Types of Alkene Metathesis

Alkene metathesis was discovered during early studies on olefin polymerisation. Later on, the mechanism proposed by Hérisson and Chauvin involving alkylidene complexes led to the development of well-defined, highly active catalysts tolerant towards a wide range of functional groups by Schrock et al. and Grubbs et al. These advancements contributed to make metathesis one of the most powerful synthetic tools in organic chemistry in the last two decades [1–3], and in 2005, the Nobel Prize was awarded to the latter three chemists.

A. Letort • J. Prunet (✉)
WestCHEM, School of Chemistry, University of Glasgow, Joseph Black Building, University Avenue, Glasgow G12 8QQ, UK
e-mail: joelle.prunet@glasgow.ac.uk

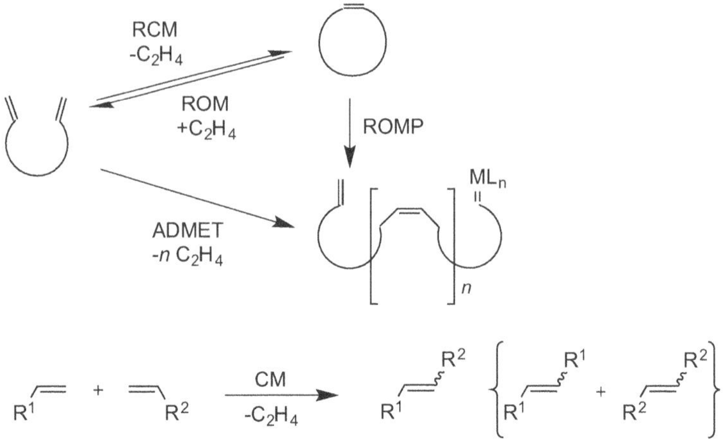

Scheme 1 Different types of alkene metathesis

Olefin metathesis, from the Greek word "μεταθεσις" meaning "transposition", refers to the cleavage and reformation of double bonds catalysed by alkylidene complexes. Five main types of olefin metathesis have been defined: ring-closure metathesis (RCM) and the reverse reaction ring-opening metathesis (ROM); ring-opening metathesis polymerisation (ROMP) and acyclic diene metathesis polymerisation (ADMET), which are beyond the scope of this book; and cross metathesis (CM) (Scheme 1). These processes are induced by metallacarbene catalysts. The structures of the most common alkene metathesis catalysts mentioned in this book can be found after the list of abbreviations.

The mechanism of alkene metathesis [4, 5] consists in a sequence of [2+2]-cycloadditions/cycloreversions proceeding via metallacyclobutane intermediates (Scheme 2). Each step of the mechanism is reversible, so an equilibrium mixture of olefins is obtained. In the case of RCM, the reaction is driven by entropy, as it produces two molecules from one; if one of them is volatile (e.g. ethylene), the equilibrium is generally displaced towards the cyclised product.

Alkene cross metathesis (CM) [6] emerged later than the ring-closing process. The first efficient CM reactions were performed with molybdenum catalysts [7]. The formation of undesired products such as homodimers and issues of stereoselectivity often plagued the reaction (Scheme 1). These problems were solved by the development of the more active "second-generation" ruthenium catalysts such as **G2** and **HG2** [8]. A general model to predict the outcome of this reaction was established by Grubbs et al. [9], based on the relative abilities of the olefins to undergo homodimerisation by CM.

Alkene metathesis converts two alkenes into two new ones, so cascade processes are possible [10]. ROM followed by CM is not relevant to this book, but other cascade processes such as RCM/RCM including relay ring-closing metathesis

Scheme 2 Mechanism of alkene ring-closing metathesis (RCM)

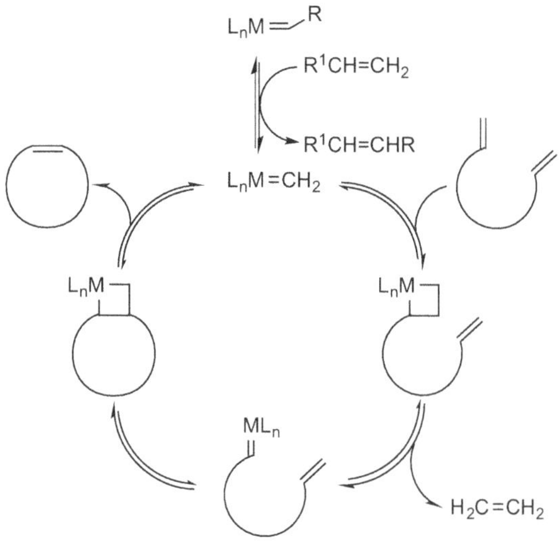

(RRCM) for the synthesis of highly hindered olefins [11] and RCM/ROM (or ROM/RCM) also known as ring-rearrangement metathesis (RRM) have been employed for the synthesis of heterocycles.

When chiral catalysts are used, enantioselective RCM can be performed by desymmetrisation of prochiral olefins [12, 13]. The structures of the chiral catalysts are shown in the different chapters.

2 Ene-yne Metathesis

Ene-yne (or enyne) metathesis was first reported by Katz et al. and further developed by Mori et al. [14]. The catalysts used for this reaction are the same as for alkene metathesis.

The different possible mechanistic pathways are shown in Scheme 3. Ene-yne RCM is not driven by entropy but by the enthalpic stability of the conjugated diene product. In contrast to olefin metathesis, ene-yne metathesis is an irreversible process because 1,3-butadienes are less active than alkenes or alkynes towards the catalysts. Ene-yne ring-closing metathesis is a remarkably versatile strategy, but it presents an issue of *endo* versus *exo* selectivity [15]. The general tendency is the following one: *exo* selectivity is favoured for small- and medium-sized cycles and *endo* selectivity for macrocycles (the ene-then-yne *endo* pathway 2b, only possible for macrocycles, is not shown in Scheme 3).

No definite proof of a mechanism beginning with the yne or the ene moiety has been established yet. The yne-then-ene mechanism (pathway 1) was first proposed,

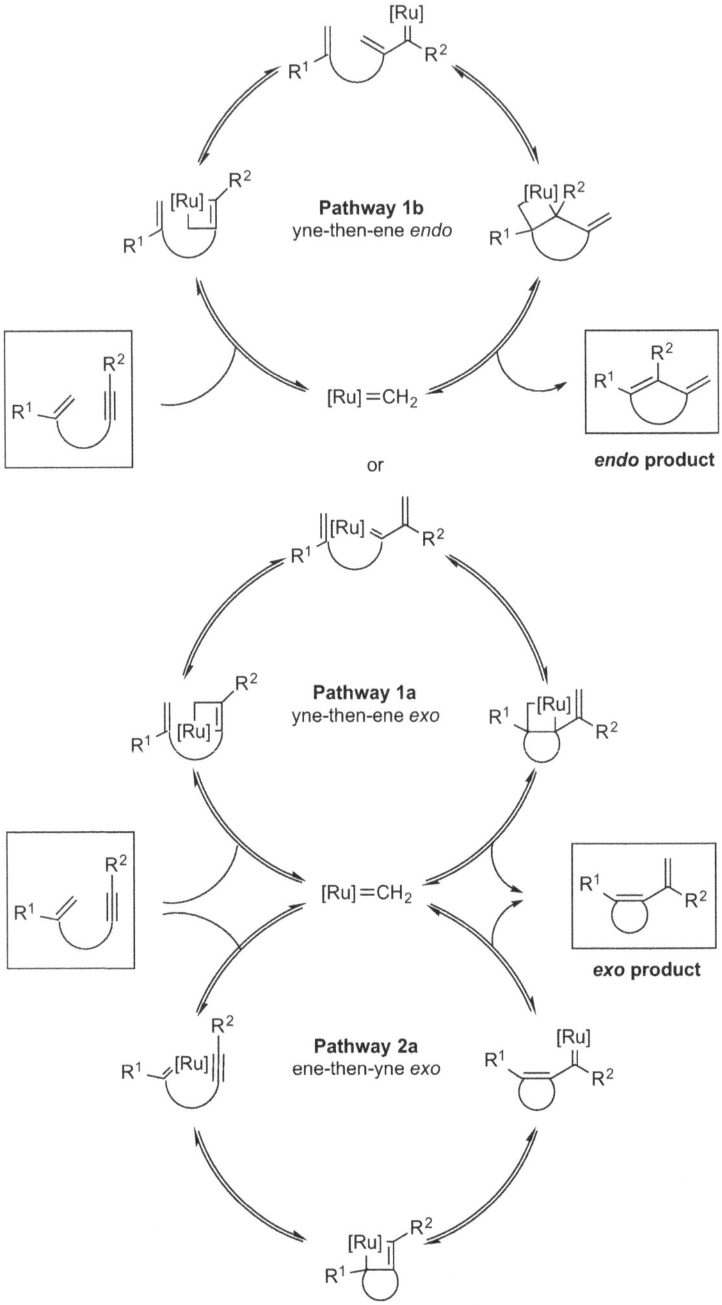

Scheme 3 Mechanism of ene-yne ring-closing metathesis (ene-yne RCM)

which explained the origin of recurrent side products as well as the formation of *endo* products for small cycles (formation that is not explainable by an ene-then-yne pathway) [16]. However, recent studies including isotopic labelling experiments gave results that tend to be explained by an alternative and more complicated ene-then-yne mechanism [17]. Moreover, carrying out the reaction under an ethylene atmosphere has a drastic influence on the reaction outcome: it seems to "temporarily protect" the catalyst from unproductive resting states of the ruthenium methylidene by helping the release of the catalyst from the conjugated carbene and therefore increases the reaction rate; this effect can only be explained by the ene-then-yne pathway. A safe conclusion seems to be that both mechanisms can take place and that the reaction course depends on the substitution pattern of the alkene and alkyne involved in the metathesis reaction.

Ene-yne metathesis has been proven to be a useful method for the synthesis of highly versatile 1,3-butadienes building blocks that can be used in numerous tandem processes such as ene-yne-ene RCM/CM, ene-yne-ene RCM/RCM, ene-yne ROM/RCM, ene-yne RCM/Diels-Alder, etc.

3 Alkyne Metathesis

Alkyne metathesis [18] was discovered roughly at the same time as its alkene counterpart, but it was not before the onset of alkyne RCM that it was exploited in organic synthesis [19]. Because of the inherent ring strain of small- and medium-size cycloalkynes, alkyne RCM is almost exclusively employed for the synthesis of macrocycles. Since macrocyclic *E*-selective alkene RCM is not known and catalysts for *Z*-selective RCM were only recently designed [20–23], the use of alkyne RCM was developed by Fürstner et al. for the stereoselective synthesis of macrocyclic alkenes: further hydrogenation of the cycloalkyne gives the *Z*-alkene, and hydrosilylation/desilylation leads to the *E*-olefin.

Catalysts are alkylidyne complexes or precursors of alkylidynes. The different catalysts are shown throughout the chapters.

The mechanism of alkyne metathesis is similar to that of alkene metathesis and is presented in Scheme 4. Since alkyne RCM does not work well with terminal alkynes, methyl-substituted alkynes are often used, and release of 2-butyne displaces the equilibrium towards the cyclic product.

Alkyne metathesis is orthogonal to alkene metathesis: the alkylidyne catalysts do not affect alkenes, and alkynes can resist to some alkylidene complexes such as the Grubbs first-generation catalyst [18].

Scheme 4 Mechanism of alkyne ring-closing metathesis (alkyne RCM)

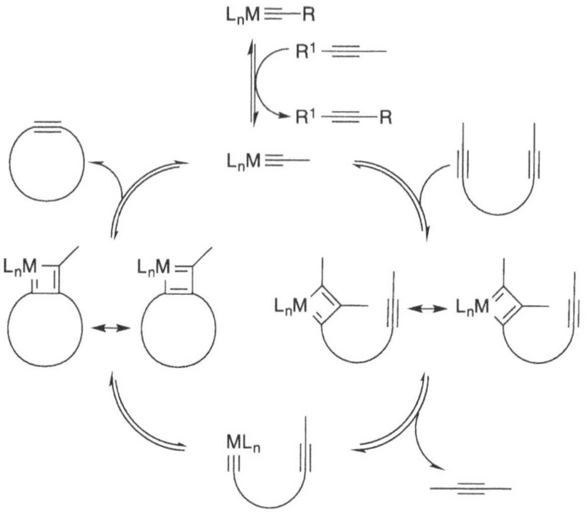

4 Conclusion

Metathesis reactions have been widely used for the construction of heterocycles, due to the relatively low affinity of the metathesis catalysts (especially the ruthenium complexes) for heteroelements. Numerous syntheses of aromatic heterocycles, ethers, lactones, amines and lactams, but also nitrogen-, phosphorus- and silicon-containing heterocycles, have been recently reported.

References

1. Grubbs RH (ed) (2003) Handbook of metathesis. Wiley-VCH, Weinheim
2. Cossy J, Arseniyadis S, Meyer C (eds) (2010) Metathesis in natural product synthesis. Wiley-VCH, Weinheim
3. Prunet J (2011) Progress in metathesis through natural product synthesis. Eur J Org Chem 3634–3647
4. Hérisson JL, Chauvin Y (1971) Catalyse de transformation des oléfines par les complexes du tungstène. II. Télomérisation des oléfines cycliques en présence d'oléfines acycliques. Makromol Chem 141:161–176
5. Nelson DJ, Manzini S, Urbina-Blanco CA, Nolan SP (2014) Key processes in ruthenium-catalysed olefin metathesis. Chem Commun. doi:10.1039/c4cc02515f
6. Prunet J, Grimaud L (2010) Cross-metathesis in natural products synthesis. In: Cossy J, Arseniyadis S, Meyer C (eds) Metathesis in natural product synthesis. Wiley-VCH, Weinheim
7. Schrock RR, Hoveyda AH (2003) Molybdenum and tungsten imido alkylidene complexes as efficient olefin-metathesis catalysts. Angew Chem Int Ed 42:4592–4633
8. Vougioukalakis G, Grubbs RH (2010) Ruthenium-based heterocyclic carbene-coordinated olefin metathesis catalysts. Chem Rev 110:1746–1787

9. Chatterjee AK, Choi TL, Sanders DP, Grubbs RH (2003) A general model for selectivity in olefin cross metathesis. J Am Chem Soc 125:11360–11370
10. Porta M, Blechert S (2010) Cascade metathesis in natural products synthesis. In: Cossy J, Arseniyadis S, Meyer C (eds) Metathesis in natural product synthesis. Wiley-VCH, Weinheim
11. Hoye TR, Jeon J (2010) Metathesis involving a relay and applications in natural products synthesis. In: Cossy J, Arseniyadis S, Meyer C (eds) Metathesis in natural product synthesis. Wiley-VCH, Weinheim
12. Hoveyda AH, Malcolmson SJ, Meek SJ, Zhugralin AR (2010) Catalytic enantioselective olefin metathesis and natural products synthesis. In: Cossy J, Arseniyadis S, Meyer C (eds) Metathesis in natural product synthesis. Wiley-VCH, Weinheim
13. Hoveyda AH (2014) Evolution of catalytic stereoselective olefin metathesis: from ancillary transformation to purveyor of stereochemical identity. J Org Chem 79:4763–4792
14. Mori M (2010) Synthesis of natural products and related compounds using ene-yne metathesis. In: Cossy J, Arseniyadis S, Meyer C (eds) Metathesis in natural product synthesis. Wiley-VCH, Weinheim
15. Hansen EC, Lee D (2006) Search for solutions to the reactivity and selectivity problems in enyne metathesis. Acc Chem Res 39:509–519
16. Kitamura T, Sato Y, Mori M (2002) Effects of substituents on the multiple bonds on ring-closing metathesis of enynes. Adv Synth Catal 344:678–693
17. Lloyd-Jones G, Margue RG, de Vries JG (2005) Rate enhancement by ethylene in the Ru-catalyzed ring-closing metathesis of enynes: evidence for an "Ene-then-Yne" pathway that diverts through a second catalytic cycle. Angew Chem Int Ed 44:7442–7447
18. Fürstner A (2013) Alkyne metathesis on the rise. Angew Chem Int Ed 52:2794–2819
19. Davies PW (2010) Ring-closing alkyne metathesis in natural products synthesis. In: Cossy J, Arseniyadis S, Meyer C (eds) Metathesis in natural product synthesis. Wiley-VCH, Weinheim
20. Yu M, Wang C, Kyle AF, Jakubec PJ, Dixon DJ, Schrock RR, Hoveyda AH (2011) Synthesis of macrocyclic natural products by catalyst-controlled stereoselective ring-closing metathesis. Nature 479:88–93
21. Wang C, Haeffner F, Schrock RR, Hoveyda AH (2013) Molybdenum-based complexes with two aryloxides and a pentafluoroimido ligand: catalysts for efficient *Z*-selective synthesis of a macrocyclic trisubstituted alkene by ring-closing olefin metathesis. Angew Chem Int Ed 52:1939–1943
22. Keitz BK, Endo K, Patel PR, Herbert MB, Grubbs RH (2012) Improved ruthenium catalysts for *Z*-selective olefin metathesis. J Am Chem Soc 134:693–699
23. Bronner SM, Herbert MB, Patel PR, Marx VM, Grubbs RH (2014) Ru-based *Z*-selective metathesis catalysts with modified cyclometalated carbene ligands. Chem Sci. doi:10.1039/C4SC01541J

Top Heterocycl Chem (2017) 47: 1–32
DOI: 10.1007/7081_2015_145

Published online: 16 June 2015

Synthesis of Heteroaromatic Compounds by Alkene and Enyne Metathesis

Nicholas J. Race and John F. Bower

Abstract The development of robust and functional group tolerant metathesis catalysts has resulted in a rapid expansion of synthetic methodologies that employ this technology. In this chapter, alkene and enyne metathesis-based methodologies for the de novo construction of heteroaromatic compounds are discussed. These protocols capitalise upon the ability to generate complex olefinic products from the combination of two simpler precursors. Advancement to the heteroaromatic target can then be achieved either directly or after modification of the immediate metathesis product. Strategies that rely upon ring-closing alkene metathesis, ring-closing enyne metathesis and alkene cross-metathesis are outlined, along with applications in target-directed synthesis.

Keywords Metathesis • Heteroaromatic • De novo • Furan • Pyrrole • Pyridine • Indole

Contents

N.J. Race and J.F. Bower (✉)
School of Chemistry, University of Bristol, Bristol BS8 1TS, UK
e-mail: john.bower@bris.ac.uk

1 Introduction

Heteroaromatic compounds continue to play an important role in drug discovery, agrochemicals and materials chemistry [1, 2]. Two distinct approaches are available for the preparation of substituted derivatives: (a) functionalise a pre-existing heteroaromatic framework and (b) undertake de novo construction of the target heteroarene. Increasingly sophisticated methods have been developed for the modification of heteroarenes, resulting in a rapid departure from protocols that require prefunctionalisation to those that exploit the innate reactivity of heteroarene C–H bonds [3, 4]. This field is necessarily dependent upon substrates provided by de novo synthesis and, consequently, a natural synergy emerges between the two approaches. However, progress in de novo heteroarene construction has lagged, with core methods relying upon classical condensation chemistries. Further refinement can be achieved by more efficient access to condensation precursors or by the identification of new methods that achieve heteroaromatisation. Methodologies that are based upon robust and operationally simple catalysis platforms are likely to be especially valuable. In this chapter, the rapidly expanding area of metathesis-based heteroaromatic methodologies is outlined [5–8]. This approach capitalises upon the combination of user-friendly ruthenium-based catalysts and readily available alkene substrates to provide regiocontrolled access to a wide range of heteroarene targets (Scheme 1). The discussion is focussed upon the most general metathesis–heteroaromatisation methodologies and outlines the range of substrates available,

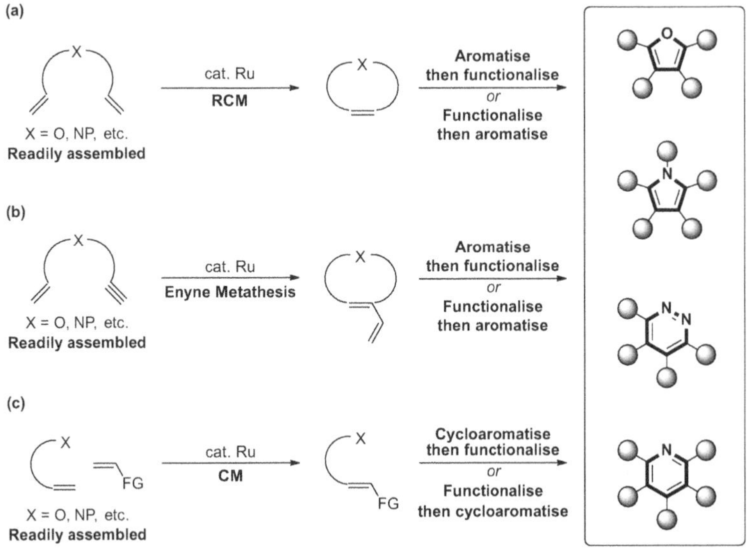

Scheme 1 Metathesis-based strategies to access heteroarenes

as well as applications in target-directed settings. Metathesis-based approaches to carbocyclic and boracyclic aromatics have been reviewed recently and are not covered here [7]. Progress in the wider area of transition metal-catalysed de novo heteroaromatic synthesis has also been the subject of a recent and comprehensive review [9].

2 Synthesis of Heteroaromatic Compounds by Ring-Closing Alkene Metathesis (RCM)

2.1 *Synthesis of 5-Membered Heteroaromatic Compounds by RCM*

Research in this area has focussed on the use of metathesis for the synthesis of furans, pyrroles and benzofused variants (e.g. indoles). Methodologies for the provision of furans and pyrroles have employed metathesis to form predominantly the C3–C4 linkage (Scheme 2). Strategies where RCM forms cyclic enol ethers, and therefore the C2–C3 linkage of furans, have received less attention, presumably due to the more challenging nature of precursor synthesis. Following the RCM event, aromatisation can be achieved by either (a) oxidation, (b) elimination of a leaving group or (c) isomerisation.

2.1.1 Synthesis of Furans and Pyrroles by RCM–Oxidation Sequences

The groups of Xiao and Wilson reported entries to pyrroles wherein RCM of *N*,*N*-diallylamines was followed by in situ oxidation of the intermediate pyrroline [10, 11]. Under microwave conditions, a series of *N*-alkylated diallylamines **1** underwent efficient ring closure when exposed to **G2** to afford mixtures of pyrrolines **2** and pyrroles (e.g. **3a**) (Scheme 3a). Extension to *N*-arylated systems provided pyrroles (e.g. **3b**,**c**) as the sole product using **G1** as the catalyst. In both approaches, the precise nature of the oxidant was not established and only monosubstituted alkenes were tolerated. Variants of this approach have employed discrete oxidants or oxidation catalysts, either at the outset or after RCM is complete. For example, Stevens and co-workers reported that a synergistic combination of **G2** (10 mol%) and $RuCl_3$ (2 mol%) can effect ring closure of a series of *N*-alkylated or *N*-arylated diallylamines to the corresponding pyrroles in 30–74% yield, with varying amounts of intermediate pyrrolines remaining [12]. In an effort to enhance the efficiency of

Scheme 2 RCM strategies to access furans and pyrroles

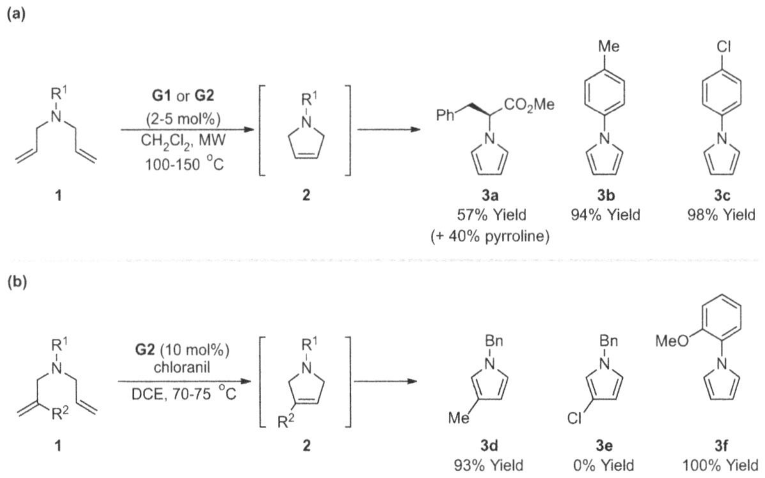

Scheme 3 RCM–oxidation approaches to pyrroles. (**a**) RCM–oxidation under "oxidant free" conditions; (**b**) RCM–oxidation using chloranil

the oxidation step, the Stevens group subsequently reported that $RuCl_3$ could be replaced by stoichiometric chloranil, and this provided the target pyrroles in high yield (>90%) and in shorter time periods (Scheme 3b) [13]. In these cases, internal alkylation was tolerated on one of the alkenes and this provided access to C-3-alkylated targets (e.g. **3d**). C-3 chlorinated adduct **3e** was not observed and only dimeric metathesis products were isolated; this reflects a general difficulty in promoting efficient metathesis of halide-substituted alkenes [14–16]. Extensions of this approach provide 2-phosphonopyrroles [17]. Related methodologies have employed oxidants such as Pd/C [18], *t*-BuOOH [19], $FeCl_3/O_2$ or $CuCl_2/O_2$ [20] to promote aromatisation.

Oxidation of dihydrofurans to furans is more challenging than for pyrrolines to pyrroles. In the context of RCM-based furan syntheses, approaches employing NiO_2 [21] or benzoquinones [22] have been used to achieve modest yields of the targets. The most extensive studies have been conducted by Schmidt and co-workers and they have shown that *t*-BuOOH is effective for oxidative RCM sequences to afford a series of C-2 substituted furans in moderate to good yield for the two-step (one-pot) protocol (Scheme 4) [19].

2,5-Disubstituted furans can be accessed via Heck arylation of the corresponding 2,3-dihydrofurans [23, 24]. Here, RCM of diallylether precursors **4** using **G1** afforded the expected 3,4-dihydrofurans **5** in high yield (Scheme 5). Addition of ethyl vinyl ether then converted **G1** into an Ru-H catalyst [25] that effected isomerisation to the corresponding 2,3-dihydrofurans **6**. Heck arylation, using a series of electron-rich or neutral arenediazonium salts [24], was combined with in situ oxidation by chloranil to afford the 2,5-disubstituted furan targets in good yield. A notable feature of this approach is the modification of the Ru-metathesis catalyst such that it can promote a second and mechanistically distinct process.

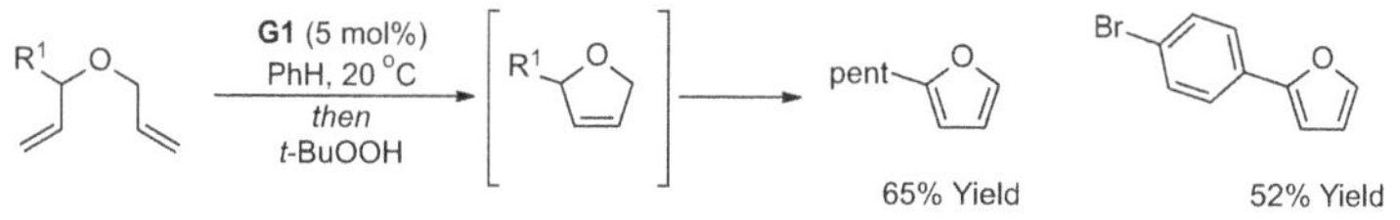

Scheme 4 RCM–oxidation approach to C-2-substituted furans

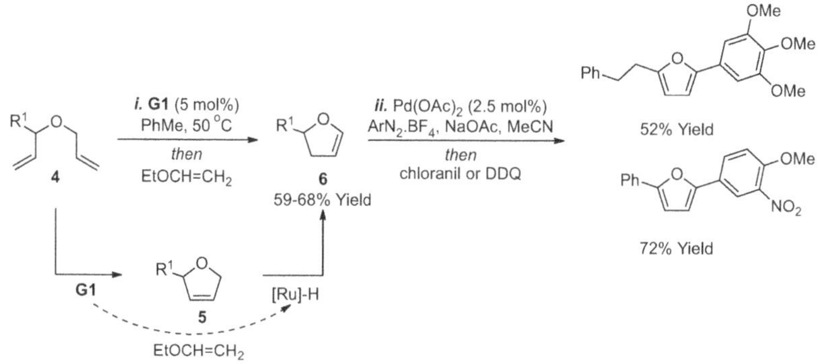

Scheme 5 RCM–isomerisation–Heck–oxidation approach to 2,5-disubstituted furans

2.1.2 Synthesis of Furans and Pyrroles by RCM–Elimination Sequences

It is well appreciated that many furans and pyrroles are unstable towards oxidants, and this, in turn, limits the scope of the approaches outlined in Sect. 2.1.1. With this in mind, metathesis-based strategies that employ precursors at the correct oxidation level for advancement to the final target are appealing. One way to achieve this is to incorporate suitable leaving groups on the substrate, such that post-metathesis heteroaromatisation can be achieved via elimination. The feasibility of this approach was demonstrated by Harrity and co-workers, who observed that acid-catalysed elimination of unsaturated cyclic ketals generated furans [26]. This observation was exploited by the Donohoe group, who developed a series of furan methodologies that employ acid-promoted elimination of alcohol leaving groups [27–29]. In one case, a series of acetal substrates **7** were prepared by Pd(II)-catalysed hydroetherification of methoxyallene (Scheme 6a) [27, 29, 30]. These underwent efficient RCM using **G2** to afford, ultimately, the target mono- or 2,3-disubstituted furans upon TFA-promoted elimination of MeOH (from **8**). The protocol shows good scope and even enables the bidirectional synthesis of sensitive difuran derivative **9**. The methodology was used as a key step in an efficient synthesis of (-)-(*Z*)-deoxypukalide (Scheme 6b). Here, acetal exchange provided precursor **10**, which underwent clean RCM and acid-promoted aromatisation to provide furan **11** in 85% yield [31].

A related approach provides access to pyrrole derivatives [27, 29]. In these cases the substrates **12** were accessed via Pd(II)-catalysed hydroamination of methoxyallene [32], and the acid-promoted aromatisation step was telescoped

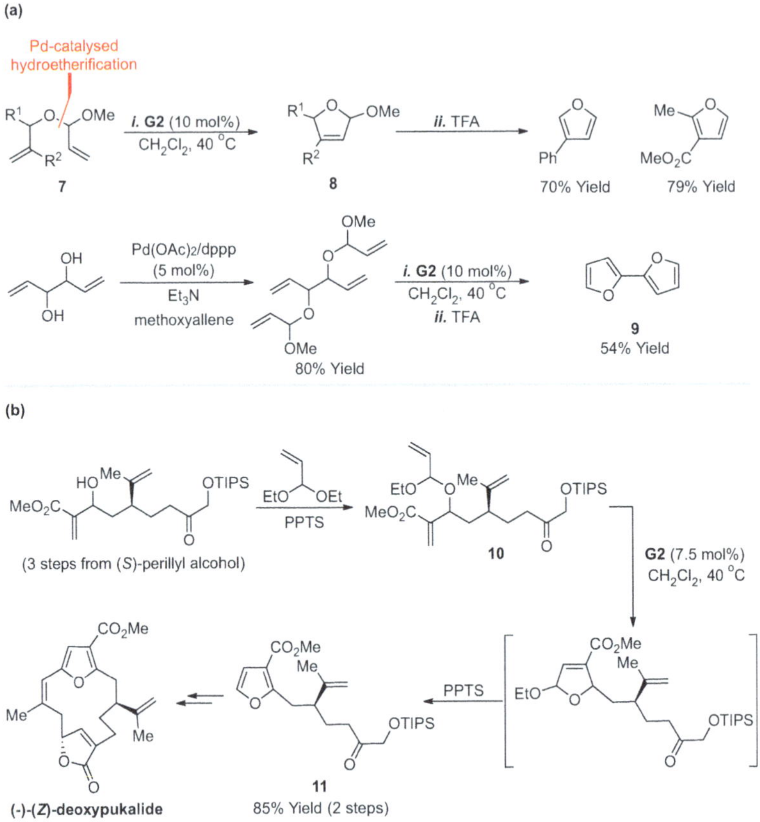

Scheme 6 RCM–elimination approach to furans and application to (-)-(*Z*)-deoxypukalide. (**a**) Furans via RCM–elimination; (**b**) application to (−)-(Z)-deoxypukalide

with the RCM step (Scheme 7a). An alternative option involves Pd(0)-catalysed aminoarylation of methoxyallene [33, 34], which in turn, provides access to C-3-arylated pyrroles (Scheme 7b).

The RCM–elimination approaches outlined so far do not provide access to 2,5-disubstituted derivatives, but this can be achieved by altering the position of the leaving group (Scheme 8) [28]. Precursors **13** were prepared by sequential indium-mediated aldehyde allylation [35] and esterification. Takai–Utimoto olefination [36, 37] then provided enol ether metathesis substrates **14**. RCM–elimination was then conducted as described earlier to provide the 2,5-disubstituted furan targets **15b-e** in moderate to good yield (38–75%). In certain cases, the efficiencies of **G2** vs **HG2** were compared, with the latter providing modestly improved yields. In the case of **15a**, ring closure was not achievable, presumably due to the increased steric demands of the methyl-substituted alkene.

The approach outlined in Scheme 8 is not applicable to the synthesis of analogous pyrroles. However, for these targets, another option is to employ an *N*-based

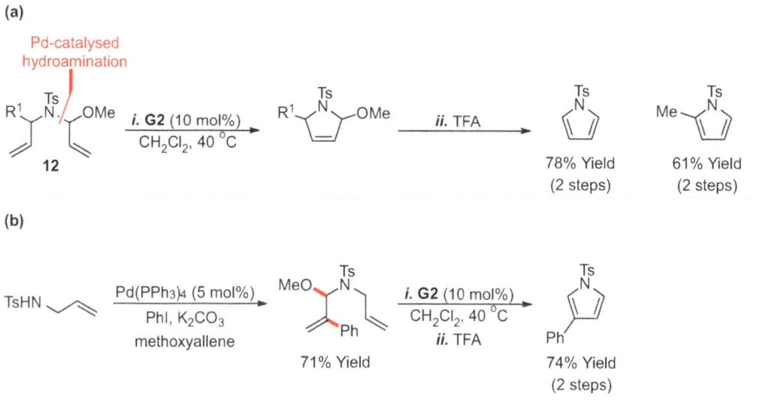

Scheme 7 RCM–elimination approaches to pyrroles. (**a**) Hydroamination-RCM–elimination approach to pyrroles; (**b**) aminoarylation-RCM–elimination approach to pyrroles

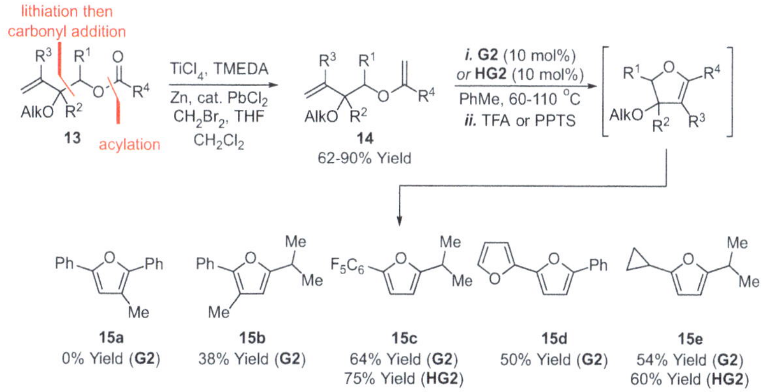

Scheme 8 RCM–elimination approach to 2,5-disubstituted furans

leaving group. Lamaty and co-workers took advantage of easily prepared aza-Baylis–Hillman adducts **16**, which upon *N*-allylation (to **17**) underwent efficient RCM with **G2** to afford the corresponding pyrrolines **18** (Scheme 9) [38]. The methyl ester moiety then facilitated base-induced sulfinate elimination to provide the targets (e.g. **19a–c**). In this approach, a series of 2,3-disubstituted pyrroles can be accessed, but the presence of an ester at C-3 and an arene as R^1 is required. Both functionalities are crucial for substrate synthesis and the former provides acidification for the final elimination step. This methodology was later adapted to the synthesis of pyrrolo-[3,2-*c*]quinolines [39, 40].

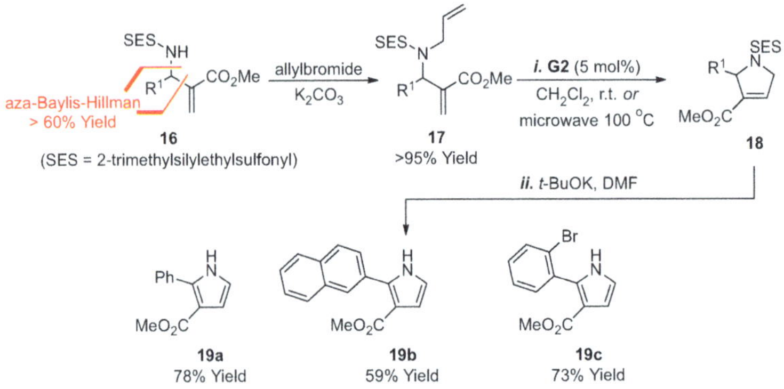

Scheme 9 Aza-Baylis–Hillman-enabled RCM approach to pyrroles

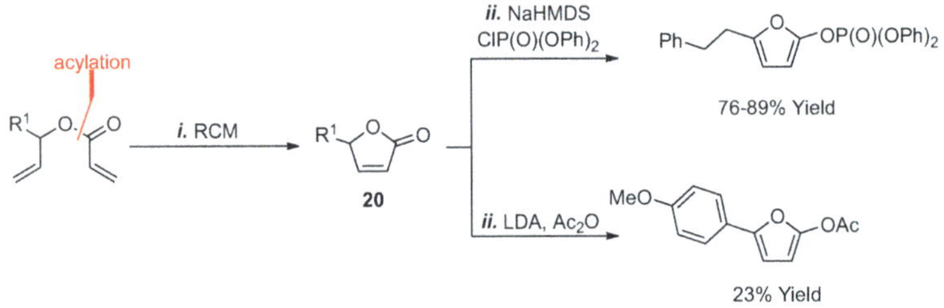

Scheme 10 RCM–enolisation approach to furans

2.1.3 Synthesis of Furans and Pyrroles by RCM–Isomerisation Sequences

A final option for achieving aromatisation is to exploit strategies that rely upon isomerisation. A simple access to furans is via the corresponding butenolides, which are formed easily by RCM [41]. This has been exploited by the Schmidt group, who showed that the extended enolate, formed by deprotonation of **20** with NaHMDS or LDA, can be *O*-phosphorylated or *O*-acetylated to provide the target furans in moderate to good yield (Scheme 10) [42]. Here, heteroaromaticity is achieved via transfer of the unsaturation from the carbonyl group to the ring system of the furan product. The resulting adducts are useful as substrates for Diels–Alder reactions [42].

RCM–isomerisation approaches to pyrroles are rare (Scheme 11). Pérez-Castells and co-workers used **G1** to effect RCM of indole derivative **21a** [43]. The initial product **22a** underwent isomerisation to afford tricyclic pyrrole derivative **23a** in 65% yield. This approach, however, is not general and **22b** underwent isomerisation to afford the alternative enamine product **23b** in 65% yield.

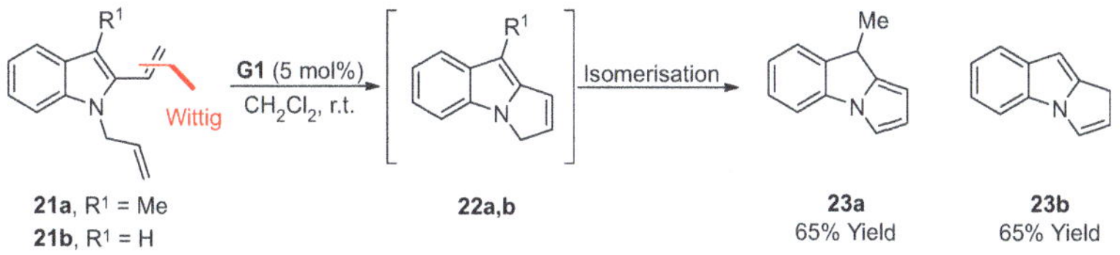

Scheme 11 RCM–isomerisation approach to a pyrrole

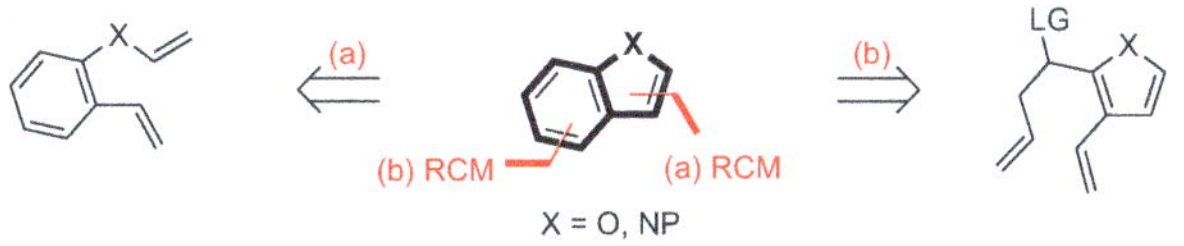

Scheme 12 RCM strategies to access benzofurans and indoles

2.1.4 Synthesis of Benzofused 5-Ring Heteroaromatic Systems by RCM

RCM has also been employed as a key step in the synthesis of benzofurans and indoles. For these targets, the metathesis step can be used to construct either (a) the C2–C3 linkage or (b) the benzenoid portion of the target (Scheme 12). For both approaches, the key issue is gaining efficient synthetic access to the metathesis precursor. Note that RCM to afford the C2–C3 linkage accesses directly the heteroaromatic target without the need for additional steps (e.g. oxidation, elimination, etc.) as required for pyrroles and furans.

In 1994, Grubbs and co-workers reported that the Schrock Mo-based catalyst **S1** would effect RCM of substrates **25** to afford the corresponding C-2 substituted benzofurans **26a,b** (Scheme 13a) [44]. At the time of this research, Ru-based systems were not sufficiently developed to promote the metathesis step. The precursors **25** were prepared in high yield (>70%) by Takai–Utimoto olefination [36, 37] of the corresponding ester **24**. This approach was employed in an efficient synthesis of *Sophora* compound I **30** (Scheme 13b). *O*-Allylation of phenol **27** was followed by Claisen rearrangement and isomerisation to styrene **28**. Esterification and olefination then provided metathesis precursor **29**. This underwent smooth RCM, and following hydrogenative debenzylation, target **30** was isolated in good yield. van Otterlo and co-workers have reported an alternative entry to metathesis precursors **32** via Ru-hydride-catalysed double isomerisation [45–48] of diallylated phenols **31** (Scheme 13c) [49, 50]. RCM, using **G2**, provided a series of benzofurans (e.g. **33a,b**) that possess substitution on the benzenoid portion. Other approaches have directly accessed precursors related to **32** and enable the introduction of esters or phosphates at C-2 [51, 52].

An analogous approach allows construction of the C2–C3 linkage of indoles. Nishida and co-workers reported the synthesis of precursors **35** by Ru-H-catalysed isomerisation of the corresponding allylic sulfonamides **34** (Scheme 14a) [53–55]. The Ru-hydride catalyst, in this case, was generated through the combination of **G2** and vinyloxytrimethylsilane [54]. RCM using **G2** then generated the

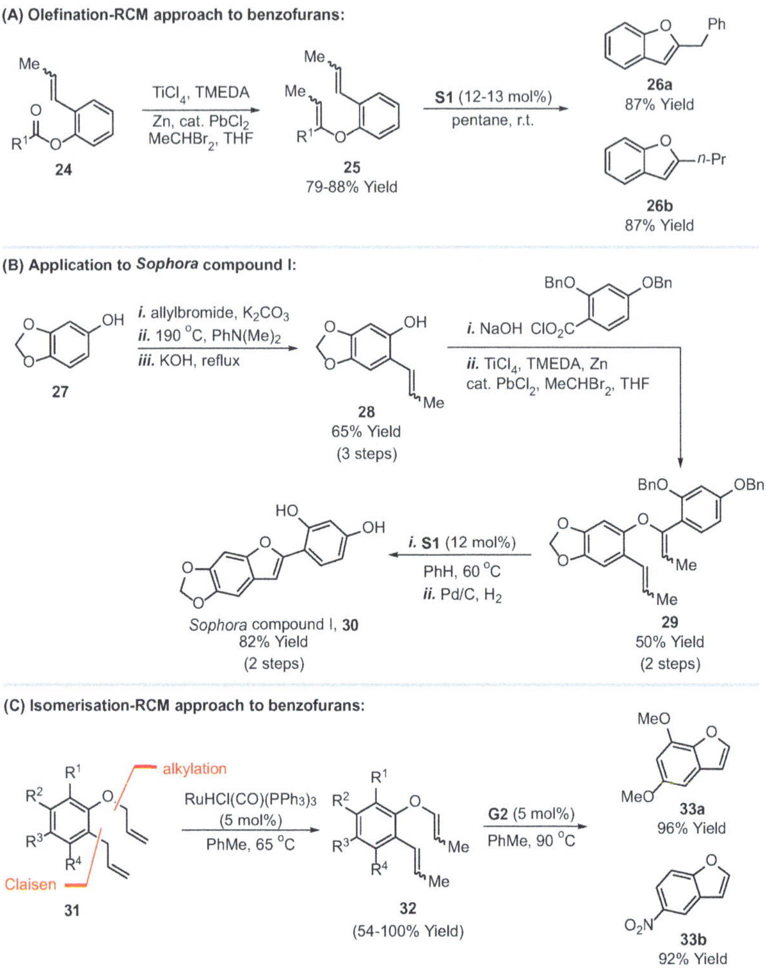

Scheme 13 RCM approaches to benzofurans. (**a**) Olefination–RCM approach to benzofurans; (**b**) application to *Sophora* compound I; (**c**) isomerisation–RCM approach to benzofurans

target indoles (e.g. **36a/b**), which bear substitution on the benzenoid portion, in high yield. A variety of *N*-protecting groups (e.g. Ts, Cbz) are compatible with the approach [53]. Bennasar and co-workers have developed an approach that enables the direct synthesis of C-2 alkylated indoles (Scheme 14b) [56]. Here, Petasis olefination [57] of **37** was used to generate enamide precursors **38** and RCM, using **G2**, occurred in moderate to high yield to provide the indole products (e.g. **39a/b**).

Approaches to indoles that use RCM to construct the benzenoid portion have also been reported. Yoshida, Yanagisawa and co-workers have shown that RCM of homoallylic alcohols **40** using **G2** generates bicycles **41**, which undergo facile acid-promoted elimination to the target indoles (e.g. **42a/b**) (Scheme 15a)

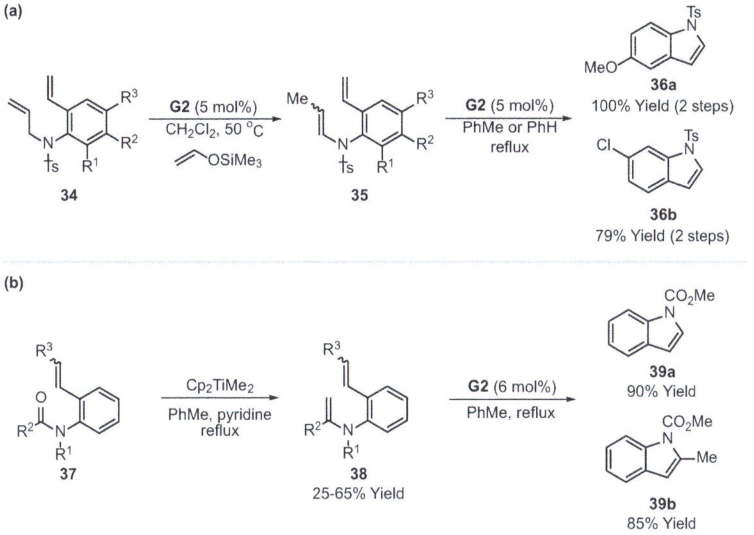

Scheme 14 RCM approaches to indoles via formation of the heteroaromatic portion. (**a**) Isomerisation–RCM approach to indoles; (**b**) olefination–RCM approach to indoles

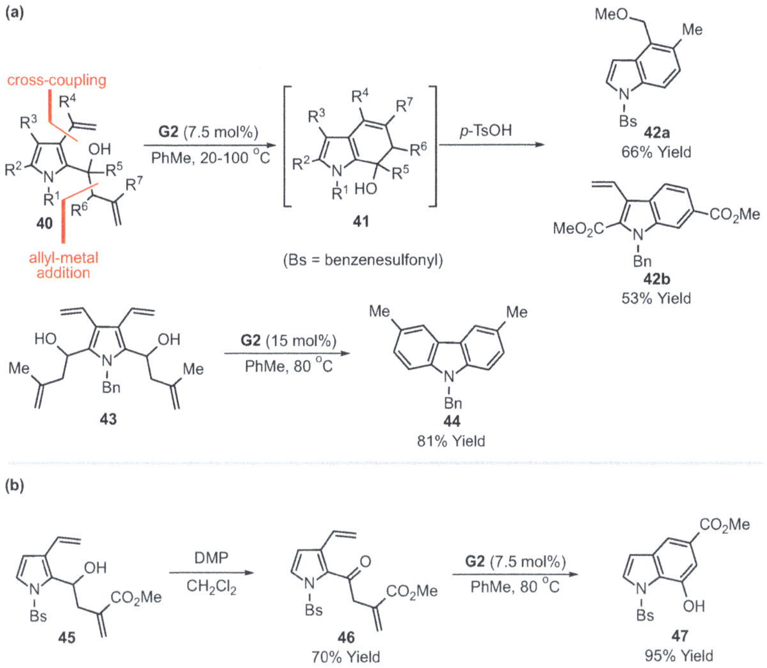

Scheme 15 RCM approaches to indoles via formation of the benzenoid portion. (**a**) RCM–elimination approach to indoles; (**b**) oxidation–RCM approach to indoles

[58]. The methodology enables access to a wide range of substitution patterns and tolerates *N*-benzyl and *N*-sulfonyl protecting groups. Bidirectional RCM–elimination provided substituted carbazoles (e.g. **44**); here dehydration occurred spontaneously after RCM. Hydroxyindoles (e.g. **47**) can be accessed by Dess–Martin oxidation of the homoallylic alcohol precursors (e.g. **45** to **46**). RCM was accompanied by spontaneous isomerisation to the target **47** (Scheme 15b). Note that the option of adjusting the oxidation level of the metathesis precursor enables access to indole products at a higher oxidation level (cf. **47** vs **42a/b**). Overall, these sequences provide rare and versatile methodologies for the de novo construction of the benzenoid portion of indoles.

2.2 *Synthesis of 6-Membered Heteroaromatic Compounds by RCM*

2.2.1 Synthesis of Pyridones and Pyridines by RCM–Oxidation Sequences

For 6-ring monocyclic heteroaromatics, RCM–oxidation approaches are limited to the synthesis of pyridones and pyridines. In 2004, Nan and co-workers reported an efficient RCM–oxidation sequence that enabled conversion of acrylamides **48** to a series of pyridones (e.g. **49a,b**) in 51–84% yield (Scheme 16a) [59]. In these processes, DDQ was the oxidant of choice. The starting materials were accessed readily via acylation of the corresponding allylic amine and, although exemplification was limited to C-3 amino derivatives, the modularity of the approach potentially enables the introduction of a wide range of substituents at C-3 and C-6 of the heteroarene target. The direct synthesis of more densely substituted products was not demonstrated but should be possible. Note, however, that Ru-based metathesis catalysts prefer at least one sterically accessible electron neutral/rich alkene for efficient initiation [60]. O'Brien and co-workers used a related approach en route to racemic cytisine (Scheme 16b) [61]. Here, RCM was followed by oxidation using Pd/C and cyclohexene to generate the target pyridone **51** in 68% overall yield for the two steps. This latter step also accomplished *N*-debenzylation thereby indicating that the cyclohexene serves as both an oxidant (**50** to pyridone) *and* reductant (*N*-debenzylation to **51**) in the two-step process.

Yoshida, Imamoto, Yanagisawa and co-workers have reported an RCM–oxidation approach to pyridines (Scheme 17) [62]. *N*-Benzylated precursors **52** are available from either the corresponding allylic amine or amino acid and underwent efficient RCM using **G2** to generate the corresponding cyclic enones **53** in 35–92% yield. Notably, even tetrasubstituted alkenes could be formed, albeit with modest efficiency. Aromatisation was achieved by a sequence of DDQ oxidation (to the *N*-benzyl pyridinium salt **54**) and *N*-benzyl reduction to furnish a series of 3-hydroxypyridines (e.g. **55a-c**) in 68–81% yield for the two-step procedure. The approach allows substituent introduction at the 2-, 3-, 4- and 5-positions of the heteroarene and further manipulations of the hydroxy group can be envisaged.

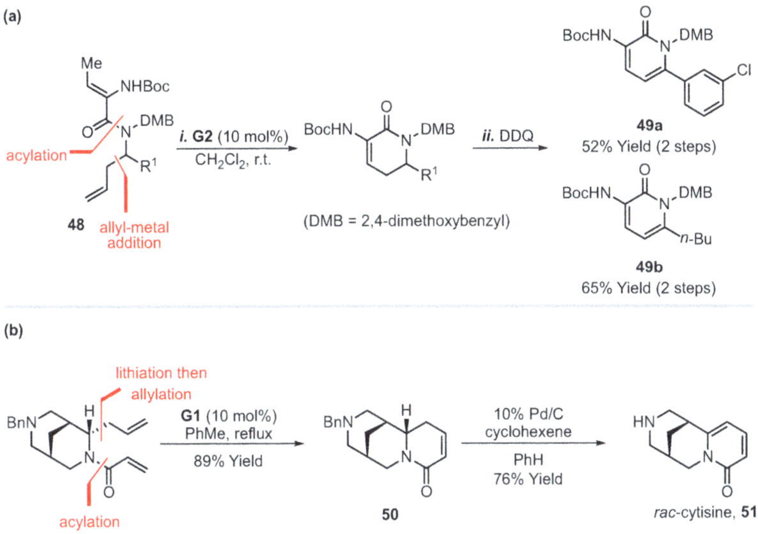

Scheme 16 Pyridones via RCM–oxidation and application en route to cytisine. (**a**) Pyridones via RCM–oxidation; (**b**) RCM–oxidation to a pyridone en route to cytisine

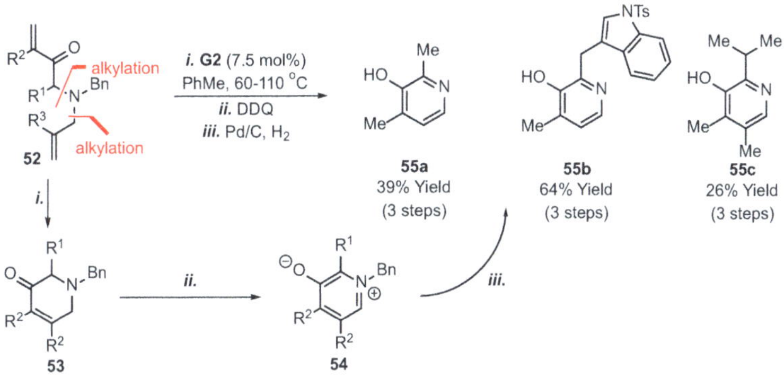

Scheme 17 RCM–oxidation–debenzylation sequence to 3-hydroxypyridines

2.2.2 Synthesis of Pyridones, Pyridines, Pyridazinones and Pyridazines by RCM–Elimination Sequences

In the strategies described above, the requirement for *N*-protection raises the question of whether the protecting group that can serve a dual role as a leaving group to facilitate RCM–elimination approaches to pyridines. This idea was delineated by the Donohoe group who showed that *N*-alkoxy acrylamides **56** undergo smooth ring closure using **HG2** to provide cyclic acrylamides **57** (Scheme 18a) [63–65]. These systems underwent base-induced E1cB elimination of BnOH or MeOH to provide NH pyridones (e.g. **58a–d**). Conversion to a range of pyridyl triflates was achieved in

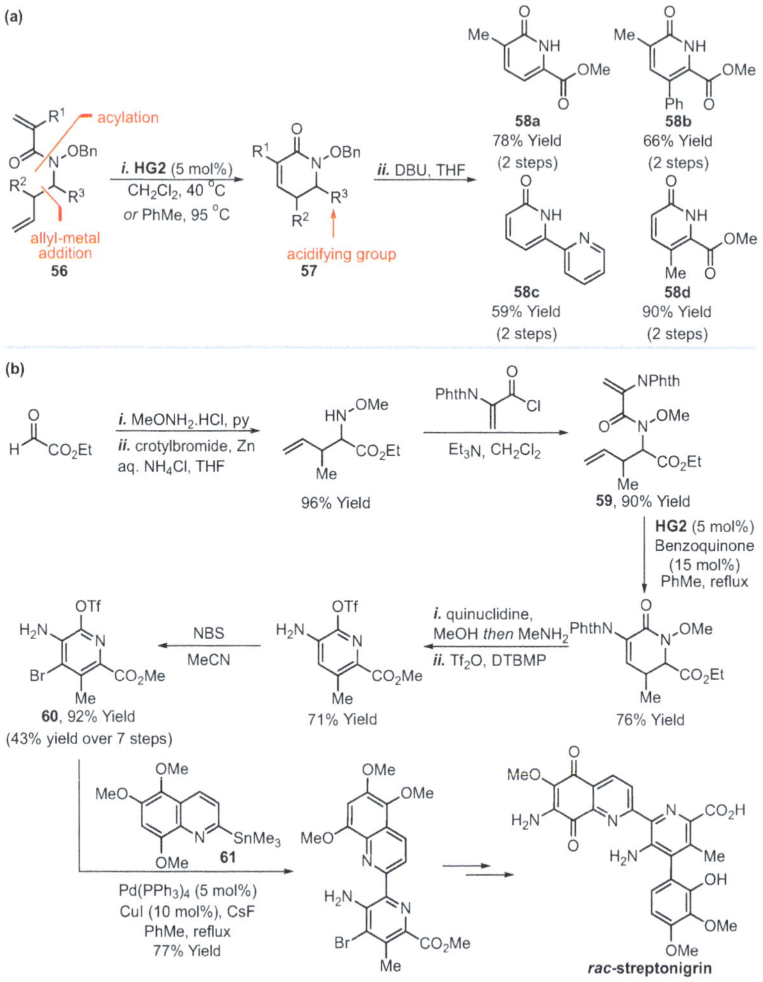

Scheme 18 RCM–elimination approach to pyridones/pyridines and application to streptonigrin (DTBMP = 2,6-di-*tert*-butyl-4-methylpyridine). (**a**) RCM–elimination approach to pyridones; (**b**) application to streptonigrin

high yield (62–94%) by treatment with Comins' reagent [66] under basic conditions (KHMDS). The protocol has good scope with respect to the acrylamide moiety, but 1,1-disubstitution is not tolerated on the allyl fragment, presumably because the increased steric demands hamper metathesis initiation. The methodology was applied to the synthesis of the challenging pyridyl subunit **60** of streptonigrin (Scheme 18b) [67, 68]. Metathesis precursor **59** was synthesised using a three-step sequence that is representative of the wider methodology. In this case, RCM was challenging and required the addition of benzoquinone, an additive that negates the effects of Ru-hydride by-products [69]. Conversion to the pyridyl triflate **60** was followed by

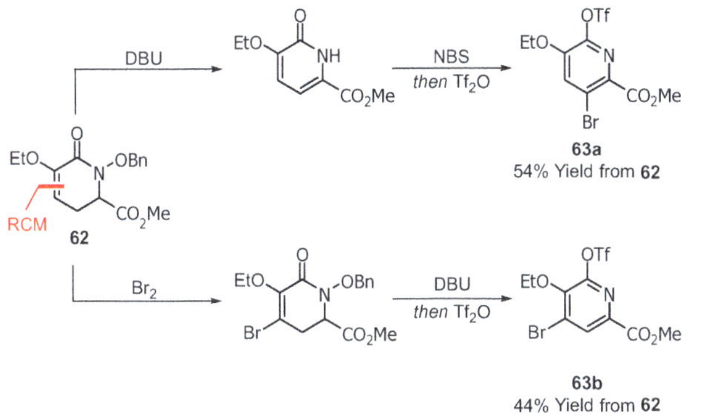

Scheme 19 Sequence-specific regiodivergent pyridine synthesis

efficient Stille cross-coupling with quinoline derivative **61**, thereby highlighting the value of the triflate moiety. In this approach the fully substituted pyridine **60** is available in 43% yield over 7 steps from commercial reagents.

A benefit of de novo heteroarene syntheses is that synthetic manipulations can be conducted prior to aromatisation, thereby allowing the exploitation of nonaromatic reactivity and/or selectivity. In the Donohoe approach this is highlighted by the regiocontrolled preparation of brominated pyridines **63a** and **63b** (Scheme 19) [65]. A sequence of aromatisation (by elimination) and then bromination provided isomer **63a**. Conversely, inversion of this sequence provided the alternate regioisomer **63b**.

In the approaches outlined in Scheme 18, elimination requires an acidifying group at R^3 to promote E1cB elimination. This, in turn, provides a limitation with respect to substituents on the eventual pyridine target. Further flexibility can be achieved by varying the position of the acidifying group. The groups of Donohoe and Yoshida, Imamoto and Yanagisawa developed concurrently approaches that rely upon elimination of sulfinate groups from metathesis products **65** (Scheme 20) [62, 65, 70]. This provides access to 3-hydroxypyridines, which can be further modified via *O*-triflation and cross-coupling. Using this approach, it is possible to achieve substitution at all positions on the final pyridine. However, tetrasubstituted alkenes cannot be generated easily and, in cases where **64** possesses bulky substituents at R^3 or R^4 (e.g. aryl groups), metathesis initiation was challenging. A relay-metathesis approach [71] can overcome both of these issues, but substrate synthesis requires extra synthetic effort [65, 70]. Donohoe and co-workers provided an alternative option where the oxidation state of the enone was adjusted to an allylic alcohol (as in **67**) [65, 70]. Catalyst initiation at this site is relatively easy and ring closure onto the more hindered alkene is facile (because it is intramolecular). The RCM products were subjected to alcohol oxidation, elimination and triflation to provide pyridyl triflates **68a/b**, which possess aryl groups at C-2 or C-4.

Donohoe and co-workers extended the sulfinate elimination strategy to provide an entry to substituted pyridazines (Scheme 21) [65, 70]. Metathesis precursors **69** were prepared from *N*-tosyl hydrazine in good yield by a sequence of sulfonamide

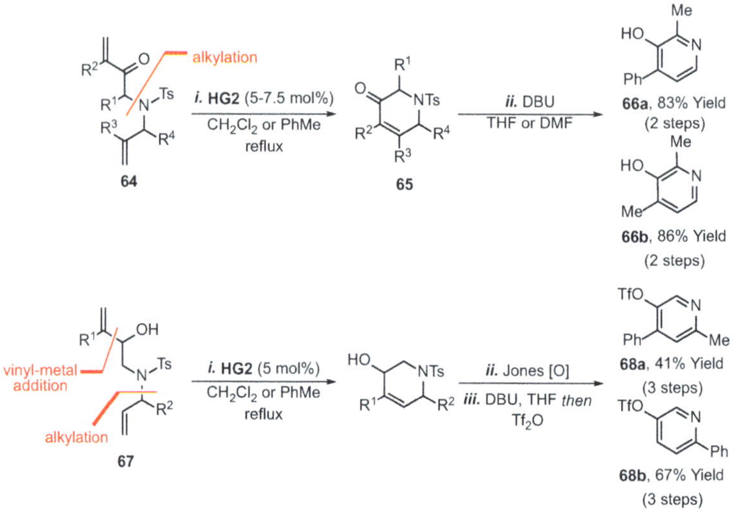

Scheme 20 Pyridines via RCM–elimination of *N*-tosyl substrates

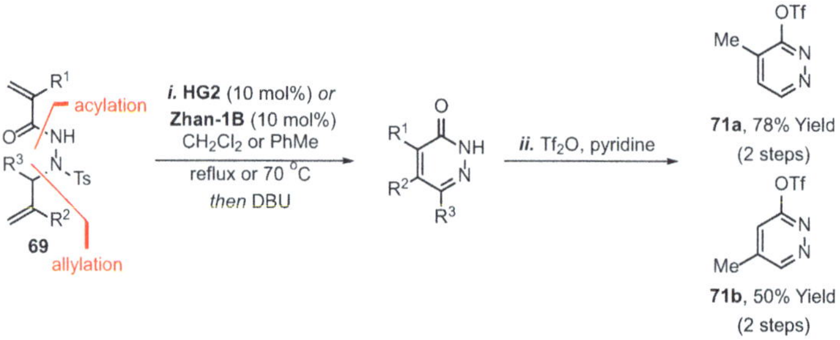

Scheme 21 Pyridazinones and pyridazines via RCM–elimination of *N*-tosyl substrates

allylation (under Mitsunobu or basic conditions) and *N*-acylation. RCM using **HG2** provided cyclic systems **70** which were subjected to in situ elimination (DBU) to provide the corresponding pyridazinones **71**. *O*-Triflation (Tf_2O) provided pyridazinyl triflates (e.g. **71a,b**) in good yield (60–86%). Using this approach, alkyl substituents can be introduced at all positions in the final target. For **71b**, RCM required initiation on a challenging 1,1-disubstituted alkene and this necessitated the use of the more active **Zhan-1B** catalyst system in this case [72].

2.2.3 Synthesis of Benzofused 6-Ring Heteroaromatic Systems by RCM

RCM methodologies that access monocyclic 6-ring heteroarenes are well developed. However, related strategies that access benzofused 6-ring systems have received less attention, perhaps due to the more challenging nature of substrate

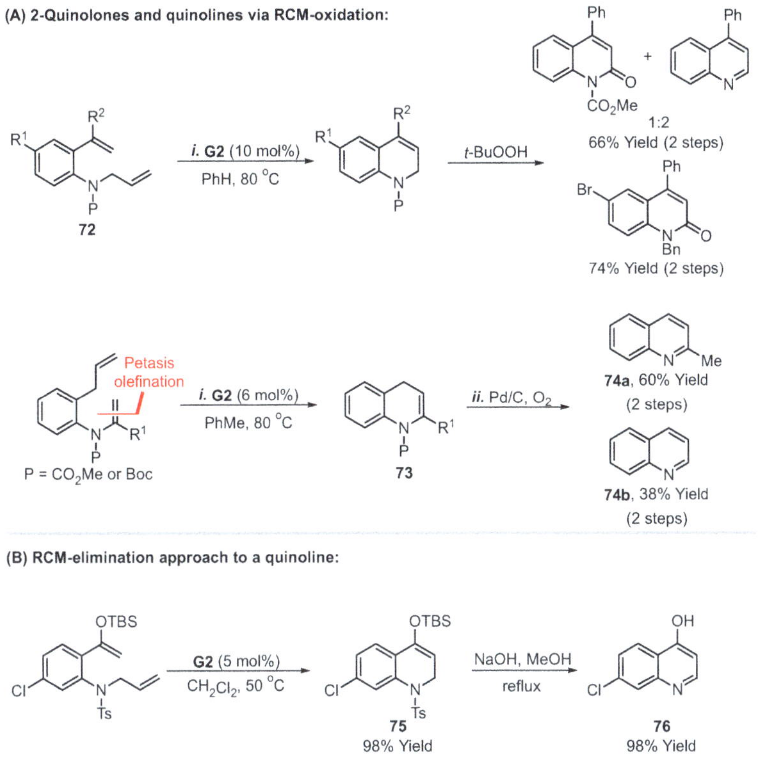

Scheme 22 RCM–oxidation and RCM–elimination approaches to quinolines and 2-quinolones. (**a**) 2-Quinolones and quinolines via RCM–oxidation; (**b**) RCM–elimination approach to quinoline

synthesis. The majority of efforts have focussed upon the synthesis of quinolones and quinolines and employ aromatisation strategies that are analogous to those described earlier for pyridines. For example, Arisawa, Shuto and co-workers have shown that aniline derivatives **72** undergo sequential ring closure with **G2** and oxidation (with *t*-BuOOH) to afford either 2-quinolones or quinolines (Scheme 22a) [73]. The outcome is often unpredictable but *N*-benzylated derivatives selectively afford the former products. In related work, Nakagawa and co-workers demonstrated that air oxidation was effective for an RCM–oxidation approach to 4-methylquinoline [74, 75]. Bennasar and co-workers pursued an alternative disconnection wherein RCM generated cyclic enamides **73** that underwent Pd/C-catalysed oxidation to quinoline targets **74a,b** [56, 76]. An *N*-sulfinate elimination strategy was used by the group of Nakagawa and Nishida to achieve aromatisation of RCM product **75** to quinoline **76** in high yield; further exemplification was not disclosed (Scheme 22b) [75].

The methodologies outlined in Scheme 22 are limited in scope and require further development to enhance their applicability. RCM-based protocols that access quinolizinones and quinolizinium salts offer greater flexibility. The Spring

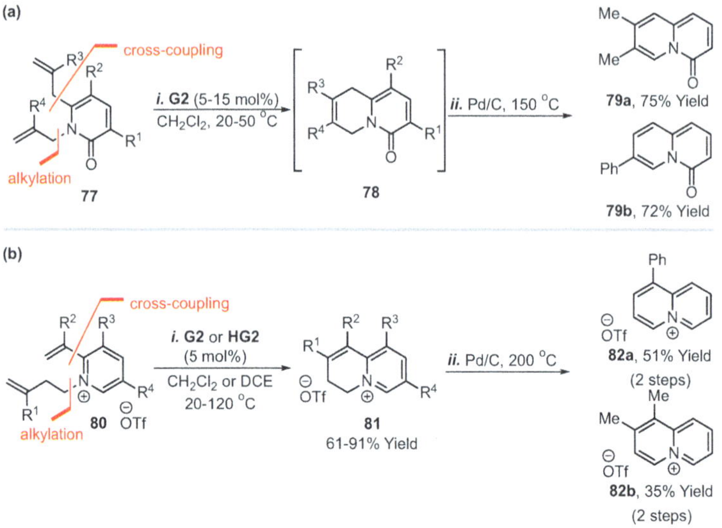

Scheme 23 RCM approaches to quinolizin-4-ones and quinolizinium cations. (**a**) Quinolizin-4-ones by RCM–oxidation; (**b**) quinolizinium cations by RCM–oxidation

group has reported an RCM–oxidation sequence that generates the benzenoid portion of quinolizin-4-ones (Scheme 23a) [77]. The substrates **77** were accessed by functionalisation of 6-bromo-2-pyridone (and related species) and underwent efficient RCM, using **G2**, to intermediates **78**. Oxidation to the targets (e.g. **79a**,**b**) was promoted by Pd/C at high temperature. Product **79a** demonstrates that tetrasubstituted alkenes can be generated efficiently during the RCM step. In earlier work, Cuadro, Vaquero and co-workers showed that pyridinium salts **80** cyclise efficiently using **G2** or **HG2** to provide intermediates **81**, which afford quinolizinium cations upon Pd/C-catalysed dehydrogenation (Scheme 23b) [78]. Complementary approaches were also investigated and the methodology offers wide scope for these challenging targets. Again, note that the RCM step could generate tetrasubstituted alkenes (e.g. to afford **82b**), although, as highlighted in earlier methodologies, vinyl halides were not tolerated. The approach was subsequently extended to provide benzofused analogues [79].

3 Synthesis of Heteroaromatic Compounds by Ring-Closing Enyne Metathesis

As outlined above, ring-closing alkene metathesis (RCM) has become a valuable and predictable tool for designing de novo heteroaromatic methodologies. By contrast, protocols that exploit ring-closing enyne metathesis are rare and are limited to a few systematic studies. Elegant but isolated examples have been

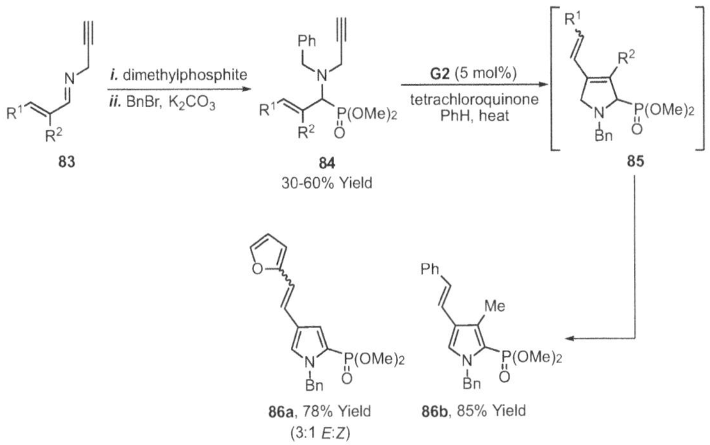

Scheme 24 Enyne metathesis–oxidation approach to 2-phosphono pyrroles

employed in total synthesis [80], but general and flexible methodologies are not well developed; the most promising are highlighted here.

Stevens and co-workers reported a pyrrole methodology based upon ring closure of substrates **84** (Scheme 24) [81]. These were prepared by a two-step sequence from imines **83** and underwent efficient cyclisation to **85** using **G2**. In the optimised protocol, in situ oxidation using tetrachloroquinone afforded directly the 2-phosphono pyrrole targets (e.g. **86a**,**b**) in moderate to high yield. Other reports have noted the formation of pyrroles or furans as by-products of enyne metathesis reactions [82–84].

Enyne metathesis can also be used to access indoles as demonstrated by the group of Pérez-Castells (Scheme 25a) [85]. Here, aniline derivative **87** underwent ring closure to vinyl indole **88** using **G2** under high dilution conditions. This was accompanied by substantial quantities of dimer **89**, the amount of which increased when the reaction was run under more concentrated conditions. Other related reports have noted indoles as side products of enyne metathesis processes [86]. The most efficient enyne metathesis methodology for indoles reported to date is by Yoshida and Yanagisawa and co-workers, where metathesis was used to construct the benzenoid portion of the targets (Scheme 25b) [87]. Ring closure of precursors **90**, using **G2**, was conducted under an ethylene atmosphere [88, 89] and this resulted in the efficient generation of a variety of vinyl indoles (e.g. **91a**,**b**) after acid-catalysed dehydration. Bidirectional application of this approach gave rise to complex polycyclic derivatives, such as **92**. As outlined in Scheme 15B, oxidation of the homoallylic alcohol precursor to the corresponding β,γ-enone (e.g. **93**) provides related hydroxyindoles (e.g. **94**) via an enyne metathesis–isomerisation sequence.

General methodologies for the construction of 6-membered ring heteroaromatics via enyne metathesis–aromatisation sequences have not been reported. An isolated example by Pérez-Castells and co-workers, which formed a complex, polycyclic system, is noteworthy [83]. Additionally, enyne metathesis has been used to build nonaromatic rings that are fused to heteroaromatic cations [90].

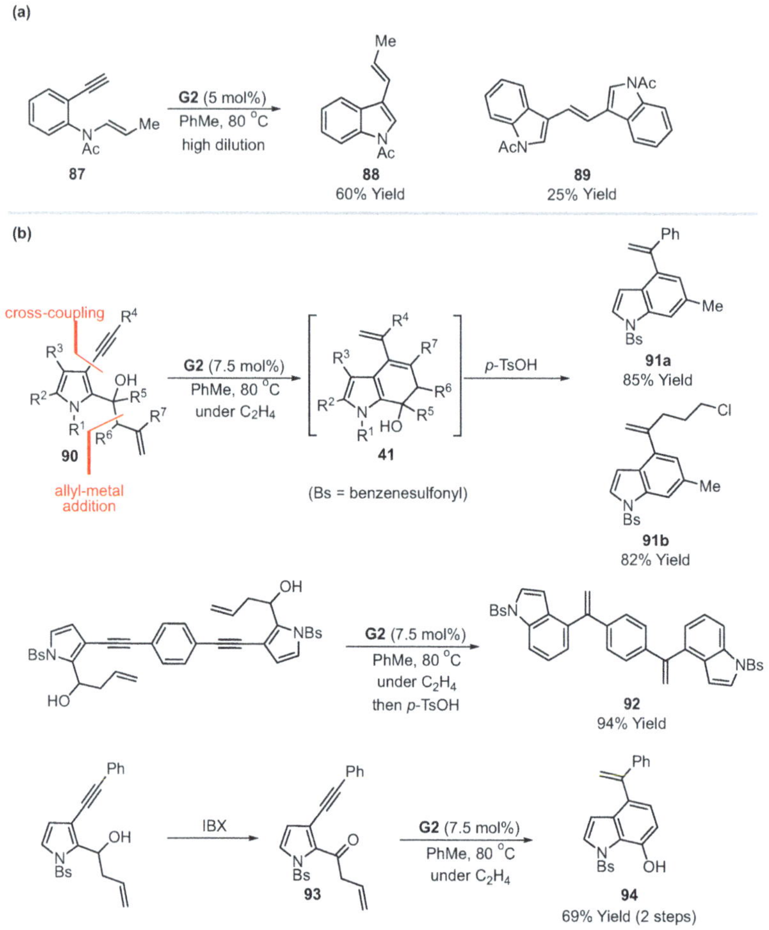

Scheme 25 Enyne metathesis approaches to indoles. (**a**) Enyne metathesis approach to the heteroaromatic portion of indoles; (**b**) enyne metathesis–elimination/isomerisation approaches to indoles

4 Synthesis of Heteroaromatic Compounds by Alkene Cross-Metathesis (CM)

For intramolecular metathesis-based approaches to heteroarenes, a key limitation in many cases is the length and generality of the synthetic sequence leading to the metathesis precursor. Additionally, linear synthetic sequences do not lend themselves readily to convergent and, therefore, diversifiable heteroaromatic methodologies. While synthetic chemists have been quick to embrace the synthetic power of RCM, methodologies based upon intermolecular metathesis between two different alkene partners (i.e. the CM reaction) have been slow to emerge. Challenges in this area include the efficiency of catalyst systems for inherently more challenging

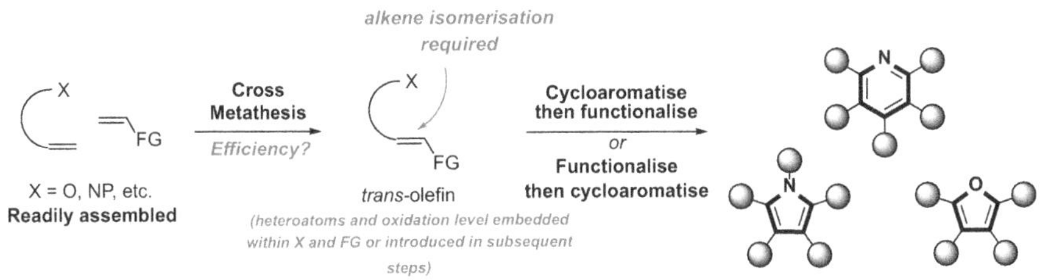

Scheme 26 CM as an approach to heteroaromatic targets

intermolecular processes and the identification of alkene partners that exhibit high cross-selectivity [91]. Within the context of heteroaromatic methodologies, a further consideration is that, until recently [92–94], catalyst systems that provide *Z*-alkenes (as required for the eventual target) were not available. Nevertheless, the power of this approach is evident. In principle, CM enables the convergent union of two simple building blocks to provide a new and more heavily functionalised alkene, which can be modified or converted directly to the heteroaromatic target (Scheme 26). Furthermore, the heteroatom can be introduced before or after the metathesis event. This second option removes the requirement to protect, for example, basic amines, which is a key consideration in RCM approaches. Because the CM process usually generate *E*-alkenes and good reactivity requires monosubstituted partners, the most profitable CM-based methodologies combine further functionalisation of the CM product with alkene isomerisation.

4.1 Synthesis of Furans and Pyrroles by CM

The first CM-based (hetero)aromatic methodologies were reported by Donohoe and co-workers in the context of furan syntheses [95, 96]. These approaches rely upon CM between readily available allylic alcohols and enones to provide γ-hydroxyenone products **95** (Scheme 27). This process is efficient using **HG2**, but requires a two- to fivefold excess of the enone to achieve optimal yields. Conversion of the γ-hydroxyenone CM product to 2,5-disubstituted furans (e.g. **96a-d**) was readily achieved via acid-catalysed alkene isomerisation and cycloaromatisation. Optimised conditions, which employed catalytic PPTS in combination with **HG2**, provided the targets directly and in moderate to good yields (Scheme 27a). Because, at the present time, CM to form intermediates **95** is only viable with monosubstituted alkenes, more highly functionalised furans could not be accessed directly. However, Heck arylation of intermediates **95**, under conditions developed by Fu and co-workers [97], afforded efficiently 2,3,5-trisubstituted furan derivatives (e.g. **97a,b**) (Scheme 27b). In this process, following *syn*-arylpalladation of the enone, bond rotation occurs to allow *syn*-β-hydride elimination. The net result is an inversion of alkene geometry which triggers ketal formation and dehydration to the furan. Note that the Heck arylation is highly regioselective (for the β-position of the enone), and this programmes where the aryl group is installed

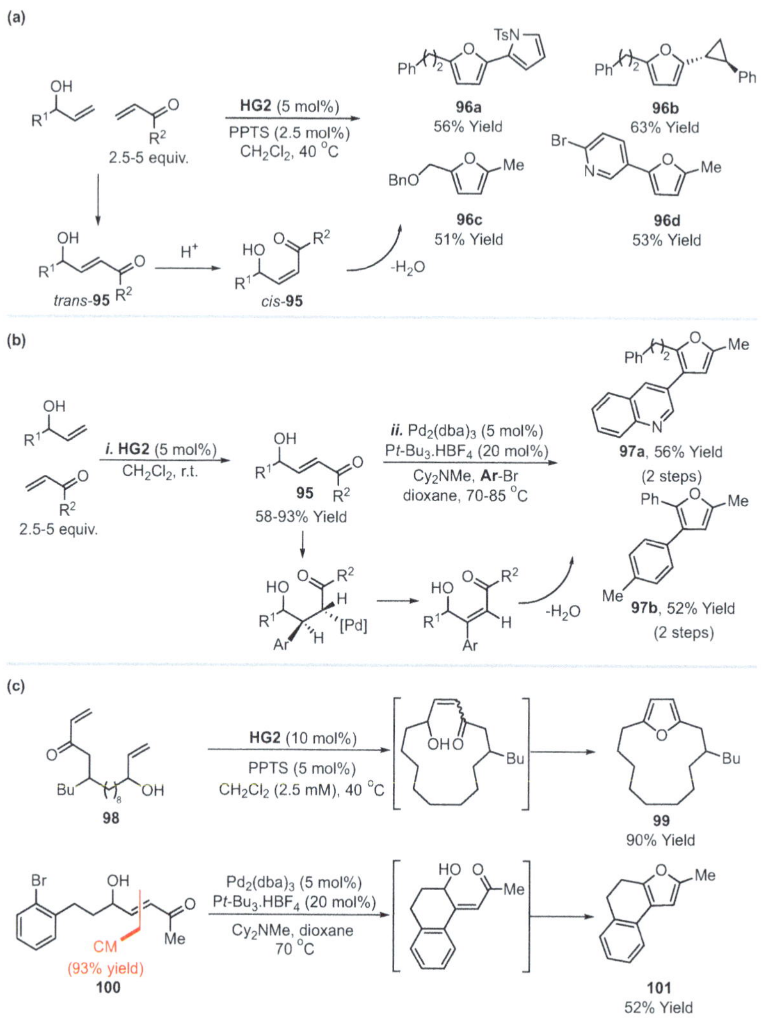

Scheme 27 CM-based approaches to di- and trisubstituted furans. (**a**) CM–isomerisation–condensation approach to furans; (**b**) CM–Heck–condensation approach to furans; (**c**) intramolecular variants

on the furan product. Intramolecular variants of both processes provided access to polycyclic systems. For example, macrocycloaromatisation of **98** generated macrocyclic furan **99** in excellent yield. Alternatively, intramolecular Heck arylation of CM product **100** resulted in aromatisation to tricyclic furan **101** (Scheme 27c).

Pyrroles are accessible via an analogous approach [98]. A range of differentially protected and easily prepared allylic amines underwent efficient CM with monosubstituted enones using **HG2** to provide γ-aminoenone products **102** (Scheme 28a). The CM event is more demanding than for cases involving allylic alcohols, so higher loadings of catalyst were required (10 vs 5 mol%). In these processes it was more efficient to conduct Brønsted acid-catalysed

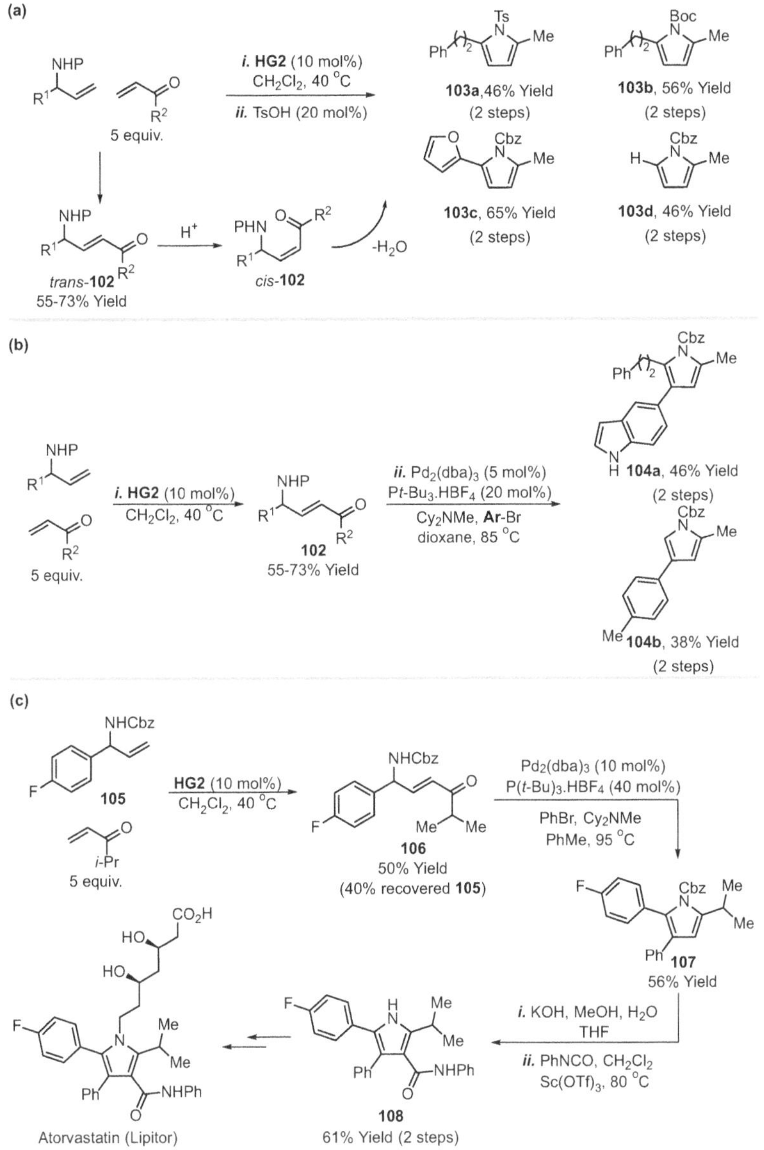

Scheme 28 CM-based approaches to pyrroles and application to atorvastatin. (**a**) CM–isomerisation–condensation approach to pyrroles; (**b**) CM–Heck–condensation approach to pyrroles; (**c**) application to atorvastatin

heteroaromatisation as a separate step and this delivered the target pyrroles (e.g. **103a-d**) in good overall yield. Application to 5-vinyl piperidinones provided access to dihydroindolizinones [98]. In related work, Grela and co-workers developed a tandem catalytic system comprising of **HG2** and $B(OPh)_3$; the latter acts as a

Lewis acid cocatalyst to effect the cycloaromatisation sequence [99]. This approach was effective in delivering directly monosubstituted pyrroles bearing alkyl or aryl groups at C-2. An example involving a 1,1-disubstituted enone was also reported and this provided the target in 70% yield. In a similar vein to that described for furans, Donohoe and co-workers showed that C-3 arylated pyrroles (e.g. **104a**,**b**) can be accessed efficiently by Heck arylation of the γ-aminoenone CM product (Scheme 28b) [98]. The mechanistic rationale for this step is analogous to that shown in Scheme 27b. The CM–Heck approach was employed to a synthesis of the pyrrole core of atorvastatin (Scheme 28c). Here, CM of Cbz-protected allylic amine **105** with *i*-propyl vinyl ketone generated γ-aminoenone **106** in 50% yield. Heck arylation with bromobenzene then delivered pyrrole **107**, which, after *N*-deprotection, underwent smooth Friedel–Crafts reaction with phenylisocyanate to deliver tetrasubstituted pyrrole **108**. In this sequence, full and predictable regiocontrol is achieved, which makes the approach suitable for analogue synthesis.

4.2 *Synthesis of Pyridines by CM*

In the preceding section, pyrroles were accessed by dehydrative cyclisation of γ-aminoenones, which, in turn, were accessed by CM. Donohoe and co-workers have shown that a homologous approach provides pyridines (Scheme 29) [100]. Here, *N*-tosyl homoallylic amines were used and these participated in smooth CM with enones to provide δ-aminoenone intermediates **109**. Acid-catalysed cyclisation, to afford ultimately 2,6-disubstituted pyridines, could not be achieved, but Heck arylation was effective and inverted the alkene geometry to afford adducts

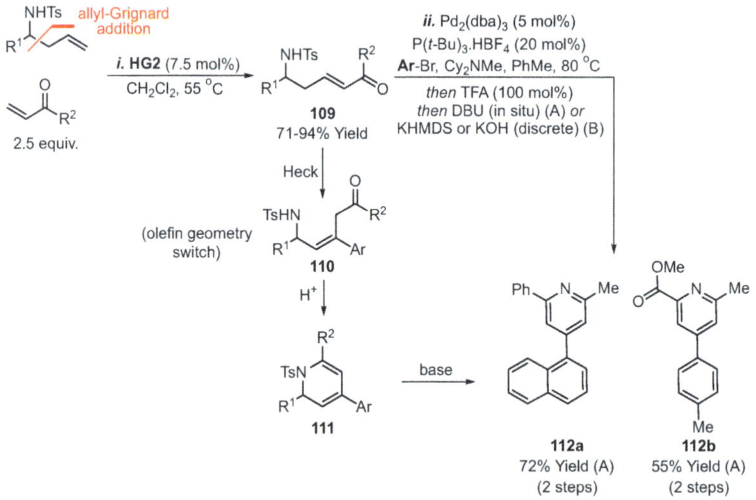

Scheme 29 2,4,6-Trisubstituted pyridines by a CM–Heck–cyclisation–elimination strategy

110. Conversion to the target pyridines involved a sequence of acid-catalysed cyclisation (to **111**) and base-induced sulfinate elimination. In favourable cases, these latter two steps could be telescoped into a one-pot process that also encompassed the Heck arylation. The sulfinate elimination step has E2 character (rather than E1cB) and consequently acidifying groups are not required to facilitate this (cf. Scheme 18a). Using this approach, 2,4,6-trisubstituted pyridines (e.g. **112a**, **b**) were accessed, but other substitution patterns could not be achieved, due to case-specific limitations associated with each step of the sequence.

A more flexible CM-based pyridine methodology, which relies on late-stage introduction of nitrogen, has also been developed by the Donohoe group [101]. Here, a CM-based approach was used to access α,β-unsaturated 1,5-dicarbonyl compounds **113**. These were either modified further or condensed directly with an ammonia source to afford pyridines (Scheme 30). In terms of the strategic considerations highlighted in Scheme 26, the key here is facile olefin isomerisation (from *E* to *Z*) during the condensation event. The most general and controlled approach to intermediates **113** involved CM of homoallylic alcohols with enones (5 equiv.) using the **Zhan-1B** catalyst. The alcohol of the CM product was then oxidised in situ (DMP) to provide dicarbonyl intermediates **113** as an inconsequential mixture of alkene regioisomers. Conversion to the target di- or trisubstituted pyridines was achieved using ammonium acetate under buffered conditions (e.g. **114a**,**b**) (Scheme 30a). More highly substituted pyridines **115a**-**c** were prepared with complete regiocontrol by base-induced alkylation or Pd-catalysed α-arylation [102, 103] of intermediates **113** (Scheme 30b). Aldehydic analogues of **113** are unstable, and so an alternative entry was devised wherein the aldehyde was masked as an acetal (Scheme 30c). The optimal sequence involved CM of β,γ-enones **116** with a large excess of a vinyl acetal partner to deliver intermediates **117** [104]. Exposure of these to ammonium acetate resulted in direct aromatisation to mono- or disubstituted pyridines **118a/b**. It is worth noting that 1,5-dicarbonyls **113** are "classical" intermediates for pyridine synthesis (for a recent example, see [105]) but are rarely exploited because they are relatively difficult to access using conventional methods. A CM-based approach offers good flexibility and control and highlights the synergy between enabling developments in catalysis and classical organic synthesis. To demonstrate applicability, Donohoe and co-workers applied the "CM"–condensation sequence to a short synthesis of (+)-(*R*)-muscopyridine (Scheme 30d) [101]. The most effective approach employed β,γ-enone **119**, which was prepared in enantiopure form over six steps. Macro-RCM, under high dilution conditions, was followed by condensation with ammonium acetate to deliver the natural product in 42% yield for the two-step process. This approach combines the heteroaromatic methodology described here with the established utility of RCM for macrocyclisation [106].

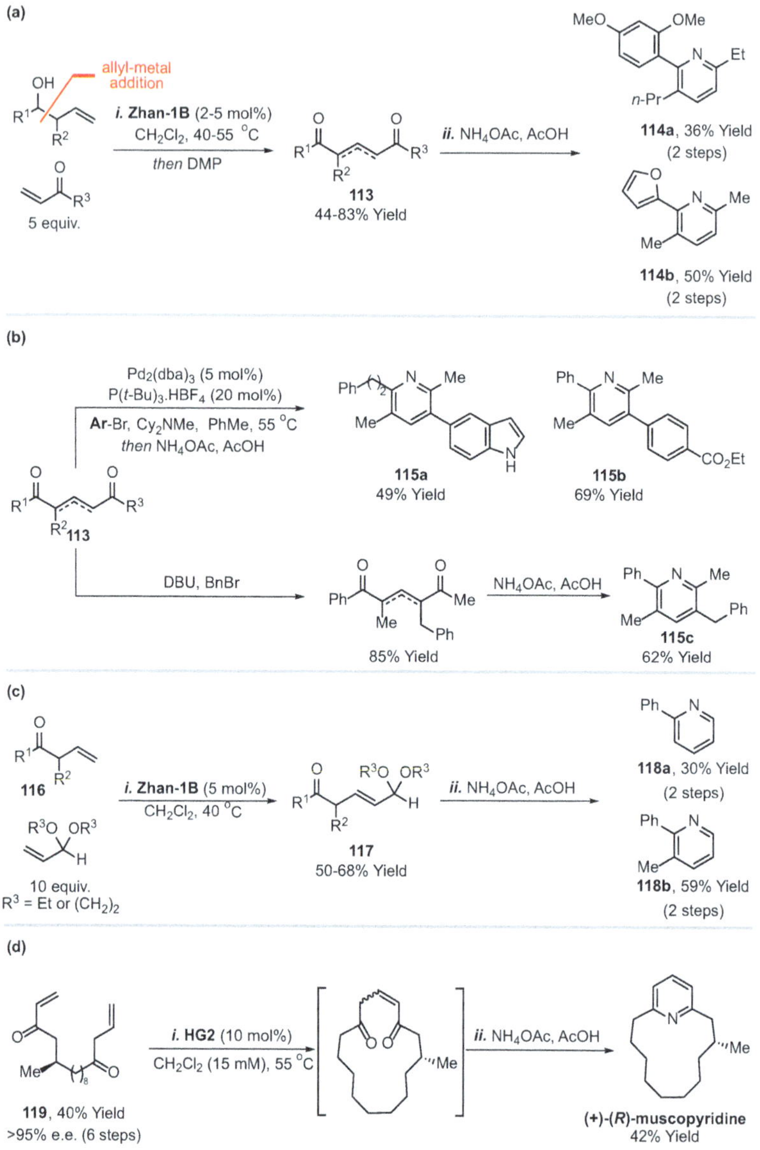

Scheme 30 CM–condensation approaches to pyridines and a related RCM approach to muscopyridine. (**a**) CM–oxidation–condensation approach to pyridines; (**b**) CM–oxidation-funcationalisation–condensation approach to tetrasubstituted pyridines; (**c**) CM–condensation approach to C-6 unsubstituted pyridines; (**d**) macro-RCM–condensation approach (+)-(*R*)-muscopyridine

5 Conclusion

Alkene and enyne metathesis reactions combine simple and readily available precursors to provide powerful entries to functionalised alkene products, which in turn can be advanced to heteroaromatic targets. In approaches that use ring-closing alkene or enyne metathesis, the synthesis of the metathesis precursor is a key consideration. The most general methodologies combine flexible and concise substrate synthesis with reliable heteroaromatisation strategies. In many of these approaches, further modification of the metathesis product can be achieved after cyclisation to provide additional substitution on the eventual heteroaromatic target. Because the metathesis step is intramolecular, high yields are often attainable across a diverse range of alkenes. Methodologies based upon alkene cross-metathesis have been slower to emerge. Key challenges reside in the selection of suitable metathesis partners and in the identification of strategies for the conversion of *E*-alkenes to heteroarenes. This latter aspect is most profitable when alkene isomerisation is accompanied by further functionalisation. However, a key benefit of CM-based strategies is that they combine fragment union with alkene formation and thereby allow shorter and more diversifiable synthetic sequences. Clearly, the two overarching strategies are complementary, with the most suitable likely to be case dependent. For example, CM-based approaches often require an excess of one alkene partner and are therefore most attractive if this component is easily accessible. Conversely, ring-closing metathesis approaches might be preferred for cases where an alkene partner is synthetically valuable or the metathesis event is inherently demanding. Further development in both areas will be dependent upon the availability of increasingly refined catalyst systems. Nevertheless, it is clear that alkene and enyne metathesis reactions already offer a robust platform for de novo heteroarene methodologies, and these, in turn, enable access to targets that might be difficult to prepare by other means.

Acknowledgments N.J.R. thanks the Bristol Chemical Synthesis Doctoral Training Centre, funded by the EPSRC (EP/G036764/1), for the provision of a Ph.D. studentship and the SCI for a postgraduate scholarship. J.F.B. is indebted to the Royal Society for a University Research Fellowship.

References

1. Roth HJ, Kleemann A (eds) (1988) Pharmaceutical chemistry. Drug synthesis, vol 1. Prentice Hall Europe, London
2. Bonnet V, Mongin F, Trécourt F, Breton G, Marsais F, Knochel P, Quéguiner G (2002) Cross-coupling between 3-pyridylmagnesium chlorides and heteroaromatic halides. Synlett 2002:1008–1010
3. Lewis JC, Bergman RG, Ellman JA (2008) Direct functionalization of nitrogen heterocycles via Rh-catalyzed C-H bond activation. Acc Chem Res 41:1013–1025

4. Fairlamb IJS (2007) Regioselective (site-selective) functionalization of unsaturated halogenated nitrogen, oxygen and sulfur heterocycles by Pd-catalyzed cross-couplings and direct arylation processes. Chem Soc Rev 36:1036–1045
5. Donohoe TJ, Orr AJ, Bingham M (2006) Ring-closing metathesis as a basis for the construction of aromatic compounds. Angew Chem Int Ed 45:2664–2670
6. Donohoe TJ, Fishlock LP, Procopiou PA (2008) Ring-closing metathesis: novel routes to aromatic heterocycles. Chem Eur J 14:5716–5726
7. van Otterlo WAL, de Koning CB (2009) Metathesis in the synthesis of aromatic compounds. Chem Rev 109:3743–3782
8. Donohoe TJ, Bower JF, Chan LKM (2012) Olefin cross-metathesis for the synthesis of heteroaromatic compounds. Org Biomol Chem 10:1322–1328
9. Gulevich AV, Dudnik AS, Chernyak N, Gevorgyan V (2013) Transition metal-mediated synthesis of monocyclic aromatic heterocycles. Chem Rev 113:3084–3213
10. Yang Q, Li X-Y, Wu H, Xiao W-J (2006) Microwave-assisted ring-closing metathesis of diallylamines: a rapid synthesis of pyrrole and pyrroline derivatives. Tetrahedron Lett 47:3893–3896
11. Yang C, Murray WV, Wilson LJ (2003) Microwave enabled external carboxymethyl substituents in the ring-closing metathesis. Tetrahedron Lett 44:1783–1786
12. Dieltiens N, Stevens CV, De Vos D, Allaert B, Drozdzak R, Verpoort F (2004) Pyrrole synthesis using a tandem Grubbs' carbene-$RuCl_3$ catalytic system. Tetrahedron Lett 45:8995–8998
13. Dieltiens N, Stevens CV, Allaert B, Verpoort F (2005) A new protocol for pyrrole synthesis by a combination of ring-closing metathesis and in situ oxidative aromatization. ARKIVOC i:92–97
14. Chao W, Weinreb SM (2003) The first examples of ring-closing olefin metathesis of vinyl chlorides. Org Lett 14:2505–2507
15. Sashuk V, Samojłowicz C, Szadkowska A, Grela K (2008) Olefin cross-metathesis with vinyl halides. Chem Commun 2008:2468–2470
16. Gatti M, Drinkel E, Wu L, Pusterla I, Gaggia F, Dorta R (2010) Efficient ring-closing metathesis of alkenyl bromides: the importance of protecting the catalyst during the olefin approach. J Am Chem Soc 132:15179–15181
17. Moonen K, Dieltiens N, Stevens CV (2006) Synthesis of 2-phosphonopyrroles via a one-pot RCM/oxidation sequence. J Org Chem 71:4006–4009
18. Sánchez I, Pujol MD (2006) Synthesis of heterocyclic compounds by ring-closing metathesis (RCM): preparation of oxygenated or nitrogenated compounds. Synthesis 2006:1823–1828
19. Schmidt B, Krehl S, Jablowski E (2012) Assisted tandem catalytic RCM-aromatization in the synthesis of pyrroles and furans. Org Biomol Chem 10:5119–5130
20. Chen W, Wang J (2013) Synthesis of pyrrole derivatives from diallylamines by one-pot tandem ring-closing metathesis and metal-catalyzed oxidative dehydrogenation. Organometallics 32:1958–1963
21. Robertson J, Kuhnert N, Zhao Y (2000) Synthesis of 3-chloro-2-formylpyrrole derivatives. Heterocycles 53:2415–2420
22. Chattopadhyay SK, Sarkar K, Karmakar S (2005) Enantioselective synthesis of dihydrofuranylglycine, furanylglycine, furanylalanine and homo-furanylalanine derivatives. Synlett 2005:2083–2085
23. Schmidt B, Geiβler D (2011) Ru- and Pd-catalysed synthesis of 2-arylfurans by one-flask Heck arylation/oxidation. Eur J Org Chem 2011:4814–4822
24. Taylor JG, Moro AV, Correia CRD (2011) Evolution and synthetic applications of the Heck-Matsuda reaction: the return of arenediazonium salts to prominence. Eur J Org Chem 2011:1403–1428
25. Louie J, Grubbs RH (2002) Metathesis of electron-rich olefins: structure and reactivity of electron-rich carbene complexes. Organometallics 21:2153–2164

26. Bassindale MJ, Hamley P, Leitner A, Harrity JPA (1999) Spirocycle assembly through selective tandem ring closing metathesis reactions. Tetrahedron Lett 40:3247–3250
27. Donohoe TJ, Orr AJ, Gosby K, Bingham M (2005) A metathesis approach to aromatic heterocycles. Eur J Org Chem 2005:1969–1971
28. Donohoe TJ, Fishlock LP, Lacy AR, Procopiou PA (2007) A metathesis-based approach to the synthesis of furans. Org Lett 9:953–956
29. Donohoe TJ, Kershaw NM, Orr AJ, Wheelhouse KMP, Fishlock LP, Lacy AR, Bingham M, Procopiou PA (2008) Flexible metathesis-based approaches to highly functionalized furans and pyrroles. Tetrahedron 64:809–820
30. Rutjes FPJT, Kooistra TM, Hiemstra H, Schoemaker HE (1998) A novel transition metal-catalyzed route to functionalized dihydropyrans and tetrahydrooxepines. Synlett 1998:192–194
31. Donohoe TJ, Ironmonger A, Kershaw NM (2008) Synthesis of (-)-(*Z*)-deoxypukalide. Angew Chem Int Ed 47:7314–7316
32. Kinderman SS, de Gelder R, van Maarseveen JH, Schoemaker HE, Hiemstra H, Rutjes FPJT (2004) Amidopalladation of alkoxyallenes applied in the synthesis of an enantiopure 1-ethylquinolizidine frog alkaloid. J Am Chem Soc 126:4100–4101
33. Desarbre E, Mérour J-Y (1996) Palladium heteroannulation process for synthesis of substituted pyrrolo[2,3-*b*]pyridin-3-ones. Tetrahedron Lett 37:43–46
34. Larock RC, Berrios-Pena NG, Fried CA (1991) Regioselective, palladium-catalyzed hetero- and carboannulation of 1,2-dienes using functionally substituted aryl halides. J Org Chem 56:2615–2617
35. Hirashita T, Kamei T, Horie T, Yamamura H, Kawai M, Araki S (1999) Preparation of γ-heterosubstituted allylindium and diindium reagents and their reactions with carbonyl compounds. J Org Chem 64:172–177
36. Takai K, Kahiuchi T, Kataoka Y, Utimoto K (1994) A novel catalytic effect of lead on the reduction of a zinc carbenoid with zinc metal leading to a geminal dizinc compound. Acceleration of the Wittig-type olefination with the $RCHX_2$-$TiCl_4$-Zn systems by addition of lead. J Org Chem 59:2668–2670
37. Okazoe T, Takai K, Oshima K, Utimoto K (1987) Alkylidenation of ester carbonyl groups by means of a reagent derived from $RCHBr_2$, Zn, $TiCl_4$, and TMEDA. Stereoselective preparation of (*Z*)-alkenyl ethers. J Org Chem 52:4410–4412
38. Declerck V, Ribière P, Martinez J, Lamaty F (2004) Sequential aza-Baylis-Hillman/ring closing metathesis/aromatization as a novel route for the synthesis of substituted pyrroles. J Org Chem 69:8372–8381
39. Benakki H, Colacino E, André C, Guenoun F, Martinez J, Lamaty F (2008) Microwave-assisted multi-step synthesis of novel pyrrolo-[3,2-*c*]quinoline derivatives. Tetrahedron 64:5949–5955
40. Colacino E, André C, Martinez J, Lamaty F (2008) Synthesis of a novel pyrrolo-[3,2-*c*] quinoline *N*-oxide by aza-Baylis-Hillman adduct of *ortho*-nitrobenzaldehyde. Tetrahedron Lett 49:4953–4955
41. Bassetti M, D'Annibale A, Fanfoni A, Minissi F (2005) Synthesis of α,β-unsaturated 4,5-disubstituted γ-lactones via ring-closing metathesis catalyzed by the first-generation Grubbs' catalyst. Org Lett 7:1805–1808
42. Schmidt B, Geißler D (2011) Olefin-metathesis-based synthesis of furans by an RCM/deprotonation/phosphorylation sequence and their Diels–Alder reactions. Eur J Org Chem 2011:7140–7147
43. González-Pérez P, Pérez-Serrano L, Casarrubios L, Domínguez G, Pérez-Castells J (2002) RCM in indoles. A new entry to the mitosene skeleton. Tetrahedron Lett 43:4765–4767
44. Fujimura O, Fu GC, Grubbs RH (1994) The synthesis of cyclic enol ethers via molybdenum alkylidene-catalyzed ring-closing metathesis. J Org Chem 59:4029–4031
45. Krompiec S, Kuźnik N, Bieg T, Adamus B, Majnusz J, Grymel M (2000) Isomerization of allyl-aryl ethers catalyzed by ruthenium complexes. Polish J Chem 74:1197–1200

46. Krompiec S, Pigulla M, Bieg T, Szczepankiewicz W, Kuźnik N, Krompiec M, Kubicki M (2002) Isomerization of *N*-allyl-*N*-arylethanamides catalyzed by ruthenium complexes. J Mol Catal A 189:169–185
47. Krompiec S, Pigulla M, Krompiec M, Baj S, Mrowiec-Białoń J, Kasperczyk J (2004) Highly selective isomerization of *N*-allylamides and *N*-allylamines. Tetrahedron Lett 45:5257–5261
48. Krompiec S, Kuźnik N, Penczek R, Rzepa J, Mrowiec-Białoń J (2004) Isomerization of allyl aryl ethers to their 1-propenyl derivatives catalyzed by ruthenium complexes. J Mol Catal A 219:29–40
49. van Otterlo WAL, Morgans GL, Madeley LG, Kuzvidza S, Moleele SS, Thornton N, de Koning CB (2005) An isomerization-ring-closing metathesis strategy for the synthesis of substituted benzofurans. Tetrahedron 61:7746–7755
50. van Otterlo WAL, Ngidi EL, de Koning CB (2003) Sequential isomerization and ring-closing metathesis: masked styryl and vinyloxyaryl groups for the synthesis of benzo-fused heterocycles. Tetrahedron Lett 44:6483–6486
51. Whitehead A, Moore JD, Hanson PR (2003) A convenient approach to cyclic enol phosphates via ring-closing metathesis. Tetrahedron Lett 44:4275–4277
52. Hekking KFW, Moelands MAH, van Delft FL, Rutjes FPJT (2006) An in-depth study on ring-closing metathesis of carbohydrate-derived α-alkoxyacrylates: efficient syntheses of DAH, KDO, and 2-deoxy-β-KDO. J Org Chem 71:6444–6450
53. Arisawa M, Terada Y, Takahashi K, Nakagawa M, Nishida A (2007) Non-metathesis reactions of ruthenium carbene catalysts and their application to the synthesis of nitrogen-containing heterocycles. Chem Rec 7:238–253
54. Arisawa M, Terada Y, Nakagawa M, Nishida A (2002) Selective isomerization of a terminal olefin catalyzed by a ruthenium complex: the synthesis of indoles through ring-closing metathesis. Angew Chem Int Ed 41:4732–4734
55. Arisawa M, Terada Y, Takahashi K, Nakagawa M, Nishida A (2006) Development of isomerization and cycloisomerization with use of a ruthenium hydride with *N*-heterocyclic carbene and its application to the synthesis of heterocycles. J Org Chem 71:4255–4261
56. Bennasar ML, Roca T, Monerris M, García-Díaz D (2006) Sequential *N*-acylamide methylenation-enamide ring-closing metathesis: construction of benzo-fused nitrogen heterocycles. J Org Chem 71:7028–7034
57. Petasis NA, Bzowej EI (1990) Titanium-mediated carbonyl olefinations. 1. Methylenations of carbonyl compounds with dimethyltitanocene. J Am Chem Soc 112:6392–6394
58. Yoshida K, Hayashi K, Yanagisawa A (2011) Construction of carbocyclic ring of indoles using ruthenium-catalyzed ring-closing olefin metathesis. Org Lett 13:4762–4765
59. Chen Y, Zhang H, Nan F (2004) Construction of a 3-amino-2-pyridone library by ring-closing metathesis of α-amino acrylamide. J Comb Chem 6:684–687
60. Stewart IC, Douglas CJ, Grubbs RH (2008) Increased efficiency in cross-metathesis reactions of sterically hindered olefins. Org Lett 10:441–444
61. Stead D, O'Brien P, Sanderson AJ (2005) Concise synthesis of (±)-cytisine via lithiation of *N*-Boc-bispidine. Org Lett 7:4459–4462
62. Yoshida K, Kawagoe F, Hayashi K, Horiuchi S, Imamoto T, Yanagisawa A (2009) Synthesis of 3-hydroxypyridines using ruthenium-catalyzed ring-closing olefin metathesis. Org Lett 11:515–518
63. Donohoe TJ, Fishlock LP, Procopiou PA (2008) A metathesis-based approach to the synthesis of 2-pyridones and pyridines. Org Lett 10:285–288
64. Donohoe TJ, Fishlock LP, Procopiou PA (2008) Ring-closing metathesis as a key step in the synthesis of 2-pyridones and pyridine triflates. Synthesis 2008:2665–2667
65. Donohoe TJ, Bower JF, Basutto JA, Fishlock LP, Procopiou PA, Callens CKA (2009) Ring-closing metathesis for the synthesis of heteroaromatics: evaluating routes to pyridines and pyridazines. Tetrahedron 65:8969–8980
66. Comins DL, Dehghani A (1992) Pyridine-derived triflating reagents: an improved preparation of vinyl triflates from metallo enolates. Tetrahedron Lett 33:6299–6302

67. Donohoe TJ, Jones CR, Barbosa LCA (2011) Total synthesis of (±)-streptonigrin: de novo construction of a pentasubstituted pyridine using ring-closing metathesis. J Am Chem Soc 133:16418–16421
68. Donohoe TJ, Jones CR, Kornahrens AF, Barbosa LCA, Walport LJ, Tatton MR, O'Hagan M, Rathi AH, Baker DB (2013) Total synthesis of the antitumor antibiotic (±)-streptonigrin: first- and second-generation routes for de novo pyridine formation using ring-closing metathesis. J Org Chem 78:12338–12350
69. Hong SH, Sanders DP, Lee CW, Grubbs RH (2005) Prevention of undesirable isomerization during olefin metathesis. J Am Chem Soc 127:17160–17161
70. Donohoe TJ, Fishlock LP, Basutto JA, Bower JF, Procopiou PA, Thompson AL (2009) Synthesis of substituted pyridines and pyridazines *via* ring closing metathesis. Chem Commun 2009:3008–3010
71. Hoye TR, Jeffrey CS, Tennakoon MA, Wang J, Zhao H (2004) Relay ring-closing metathesis (RRCM): a strategy for directing metal movement throughout olefin metathesis sequences. J Am Chem Soc 126:10210–10211
72. Zhan Z-Y (2007) Preparation of ruthenium complex ligand, ruthenium complexes, supported ruthenium complex catalysts for olefin metathesis. US Patent 20070043180, 22 Feb 2007
73. Kato H, Ishigame T, Oshima N, Hoshiya N, Shimawaki K, Arisawa M, Shuto S (2011) One-pot ring-closing metathesis (RCM)/oxidation by an assisted tandem ruthenium catalysis for the synthesis of 2-quinolones. Adv Synth Catal 353:2676–2680
74. Arisawa M, Theeraladanon C, Nishida A, Nakagawa M (2001) Synthesis of substituted 1,2-dihydroquinolines and quinolines using ene-ene metathesis and ene-enol ether metathesis. Tetrahedron Lett 42:8029–8033
75. Theeraladanon C, Arisawa M, Nishida A, Nakagawa M (2004) A novel synthesis of substituted quinolines using ring-closing metathesis (RCM): its application to the synthesis of key intermediates for anti-malarial agents. Tetrahedron 60:3017–3035
76. Bennasar M-L, Roca T, Monerris M, García-Díaz D (2005) Sequential *N*-acyl amide methylenation-enamide ring-closing metathesis: a synthetic entry to 1,4-dihydroquinolines. Tetrahedron Lett 46:4035–4038
77. Alanine TA, Galloway WRJD, McGuire TM, Spring DR (2014) Concise synthesis of substituted quinolizin-4-ones by ring-closing metathesis. Eur J Org Chem 2014:5767–5776
78. Nuñez A, Abarca B, Cuadro AM, Alvarez-Builla J, Vaquero JJ (2009) Ring-closing metathesis reactions on azinium salts: straightforward access to quinolizinium cations and their dihydro derivatives. J Org Chem 74:4166–4176
79. Nuñez A, Abarca B, Cuadro AM, Alvarez-Builla J, Vaquero JJ (2011) Ring-closing metathesis approach to heteroaromatic cations: synthesis of benzo[*a*]quinolizinium salts. Eur J Org Chem 2011:1280–1290
80. Fürstner A, Szillat H, Gabor B, Mynott R (1998) Platinum- and acid-catalyzed enyne metathesis reactions: mechanistic studies and applications to the syntheses of streptorubin B and metacycloprodigiosin. J Am Chem Soc 120:8305–8314
81. Dieltiens N, Moonen K, Stevens CV (2007) Enyne metathesis-oxidation sequence for the synthesis of 2-phosphono pyrroles: proof of the "yne-then-ene" pathway. Chem Eur J 13:203–214
82. Huang J, Xiong H, Hsung RP, Rameshkumar C, Mulder JA, Grebe TP (2002) The first successful base-promoted isomerization of propargyl amides to chiral ynamides. Applications in ring-closing metathesis of ene-ynamides and tandem RCM of diene-ynamides. Org Lett 4:2417–2420
83. González-Gómez Á, Domínguez G, Pérez-Castells J (2005) Enyne and dienyne metathesis reactions in β-carbolines. Tetrahedron Lett 46:7267–7270
84. Evanno L, Nay B, Bodo B (2005) Unexpected dehydrogenation products in the furan series arising from ruthenium-catalyzed 4-oxo 1,6-enyne metathesis. Synth Commun 35:1559–1565

85. Rosillo M, Domínguez G, Casarrubios L, Amador U, Pérez-Castells J (2004) Tandem enyne metathesis-Diels-Alder reaction for construction of natural product frameworks. J Org Chem 69:2084–2093
86. Mori M, Wakamatsu H, Saito N, Sato Y, Narita R, Sato Y, Fujita R (2006) Synthesis of cyclic dienamide using ruthenium-catalyzed ring-closing metathesis of ene-ynamide. Tetrahedron 62:3872–3881
87. Hayashi K, Yoshida K, Yanagisawa A (2013) Synthesis of 4-vinylindoles using ruthenium-catalyzed ring-closing enyne metathesis. J Org Chem 78:3464–3469
88. Mori M, Sakakibara N, Kinoshita A (1998) Remarkable effect of ethylene gas in the intramolecular enyne metathesis of terminal alkynes. J Org Chem 63:6082–6083
89. Kitamura T, Sato Y, Mori M (2002) Effects of substituents on the multiple bonds on ring-closing metathesis of enynes. Adv Synth Catal 344:678–693
90. Núñez A, Cuadro AM, Alvarez-Builla J, Vaquero JJ (2006) Enyne ring-closing metathesis on heteroaromatic cations. Chem Commun 2006:2690–2692
91. Chatterjee AK, Choi TL, Sanders DP, Grubbs RH (2003) A general model for selectivity in olefin cross metathesis. J Am Chem Soc 125:11360–11370
92. Meek SJ, O'Brien RV, Llaveria J, Schrock RR, Hoveyda AH (2011) Catalytic *Z*-selective olefin cross-metathesis for natural product synthesis. Nature 471:461–466
93. Endo K, Grubbs RH (2011) Chelated ruthenium catalysts for *Z*-selective olefin metathesis. J Am Chem Soc 133:8525–8527
94. Keitz BK, Endo K, Patel PR, Herbert MB, Grubbs RH (2012) Improved ruthenium catalysts for *Z*-selective olefin metathesis. J Am Chem Soc 134:693–699
95. Donohoe TJ, Bower JF (2010) An expedient route to substituted furans via olefin cross-metathesis. Proc Natl Acad Sci USA 107:3373–3376
96. Donohoe TJ, Bower JF, Basutto JA (2010) Olefin cross-metathesis-based approaches to furans: procedures for the preparation of di- and trisubstituted variants. Nat Protoc 5:2005–2010
97. Littke AF, Fu GC (2001) A versatile catalyst for Heck reactions of aryl chlorides and aryl bromides under mild conditions. J Am Chem Soc 123:6989–7000
98. Donohoe TJ, Race NJ, Bower JF, Callens CKA (2010) Substituted pyrroles via olefin cross-metathesis. Org Lett 12:4094–4097
99. Shafi S, Kędziorek M, Grela K (2011) Cross metathesis of *N*-allylamines and α,β-unsaturated carbonyl compounds: a one-pot synthesis of substituted pyrroles. Synlett 2011:124–128
100. Donohoe TJ, Bower JF, Baker DB, Basutto JA, Chan LKM, Gallagher P (2011) Synthesis of 2,4,6-trisubstituted pyridines via an olefin cross-metathesis/Heck-cyclisation-elimination sequence. Chem Commun 47:10611–10613
101. Donohoe TJ, Basutto JA, Bower JF, Rathi A (2011) Heteroaromatic synthesis via olefin cross-metathesis: entry to polysubstituted pyridines. Org Lett 13:1036–1039
102. Johansson CCC, Colacot TJ (2010) Metal-catalyzed α-arylation of carbonyl and related molecules: novel trends in C-C bond formation by C-H bond functionalization. Angew Chem Int Ed 49:676–707
103. Hama T, Hartwig JF (2008) Palladium-catalyzed α-arylation of esters with chloroarenes. Org Lett 10:1545–1552
104. O'Leary DJ, Blackwell HE, Washenfelder RA, Miura K, Grubbs RH (1999) Terminal olefin cross-metathesis with acrolein acetals. Tetrahedron Lett 40:1091–1094
105. Tokuji S, Takahashi Y, Shinmori H, Shinokubo H, Osuka A (2009) Synthesis of a pyridine-fused porphyrinoid: oxopyridochlorin. Chem Commun 2009:1028–1030
106. Gradillas A, Pérez-Castells J (2006) Macrocyclization by ring-closing metathesis in the total synthesis of natural products: reaction conditions and limitations. Angew Chem Int Ed 45:6086–6101

Top Heterocycl Chem (2017) 47: 33–56
DOI: 10.1007/7081_2015_197

Published online: 3 February 2016

The Synthesis of Cyclic Ether-Containing Natural and Non-natural Products by Metathesis Reactions

David M. Lindsay

Abstract This chapter describes recent advances (2005–2015) in the use of metathesis reactions to construct the cyclic ether portions in the total synthesis of natural products. Advances in the synthesis of a range of marine polyether natural products are presented, and various strategic approaches to the bridged pyran-macrocycles neopeltolide and kendomycin are discussed. Finally, the synthesis of a range of biologically active, cyclic ether-based non-natural products is described.

Keywords Bioactive compounds • Cyclisation • Ether • Metathesis reactions • Natural products

Contents

D.M. Lindsay (✉)
Department of Pure and Applied Chemistry, University of Strathclyde, Thomas Graham Building, 295 Cathedral Street, Glasgow G1 1XL, UK
e-mail: david.lindsay@strath.ac.uk

1 Introduction

This chapter describes recent advances in the application of metathesis reactions to the synthesis of cyclic ethers. In this regard, however, the methodological aspects of this topic have been very recently reviewed, within the general context of cyclic ether synthesis, with respect to tetrahydrofurans [1], tetrahydropyrans [2], seven-membered ring ethers [3], eight- to ten-membered ring ethers [4], and spirocyclic ether systems [5]. As a result, this chapter instead focuses on recent uses (2005–2015) of metathesis processes in the total synthesis (for a recent general review of metathesis process in natural product synthesis, not confined to cyclic ether-containing targets, see [6]) of cyclic ether-containing natural and non-natural products.

2 The Application of Metathesis Reactions to the Synthesis of Polyether Natural Products

This section describes recent advances in the synthesis of cyclic ether-containing natural products (for a recent general review on the synthesis of the cyclic ether component of marine natural products, see [7]) that contain the fused ring polyether motif.

2.1 *Gambieric Acids*

The gambieric acids **1** are a family of four related marine natural products, designated A–D, isolated from a marine dinoflagellate (Fig. 1) [8, 9].

Given the complexity of the polyether natural products (for a recent review of polyether natural products, see [10]) and the *pseudo*-symmetry inherent in many of their subdomains, a two-directional approach (for a review of two-directional synthesis, see [11]) to these targets is highly appealing. Combining this strategy with an iterative strategy for cyclic ether formation via ring-closing metathesis leads to a highly elegant approach to these targets, as illustrated by the work of Clark et al. in several different polyether contexts (for model studies establishing this strategy, see [12]). This strategy is highlighted here in the group's first advance [13] towards the gambieric acids.

For the fragment corresponding to the F–J ring system of gambieric acid, Clark et al. targeted intermediate **2** (Fig. 2).

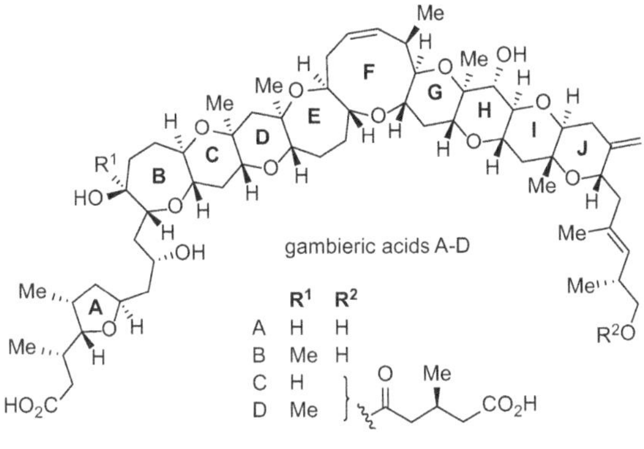

Fig. 1 Gambieric acids A–D

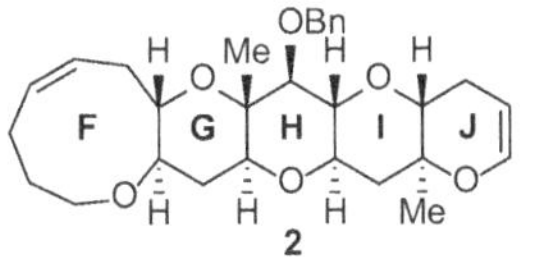

Fig. 2 F–J ring fragment of the gambieric acids

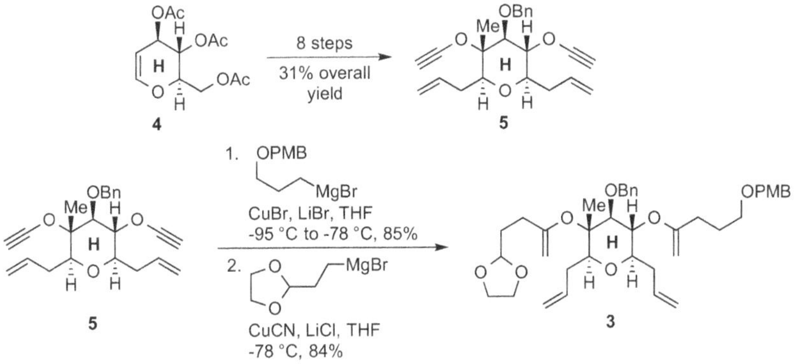

Scheme 1 Preparation of the RCM precursor

Based upon hidden symmetry centred on the H-ring, the strategy for construction of F–J fragment **2** relied on the bidirectional functionalisation of a suitably differentiated H-ring intermediate, **3**. Diyne **5** was prepared in 8 steps from tri-*O*-acetyl D-glucal **4** (Scheme 1). It should be noted that this chosen starting material, **4**, is the antipode of the corresponding stereocentres embedded in the natural material, and therefore all intermediates in this work relate to the absolute stereochemistry of *ent*-gambieric acid. Returning to the synthesis, double RCM precursor **3** was then accessed by two sequential, and remarkably selective, carbocupration reactions,

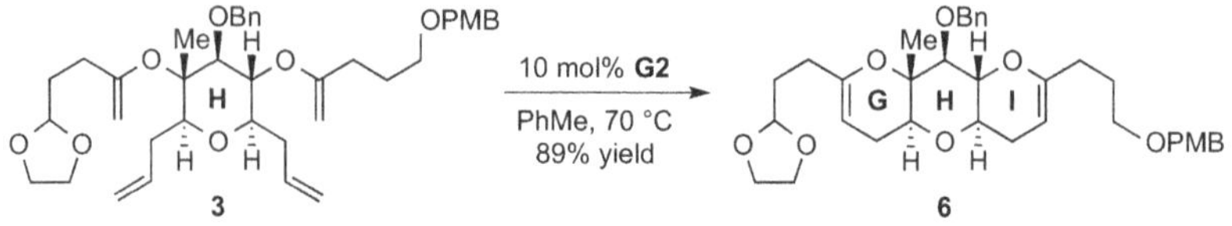

Scheme 2 Double enol ether RCM to construct the gambieric acid GHI fragment

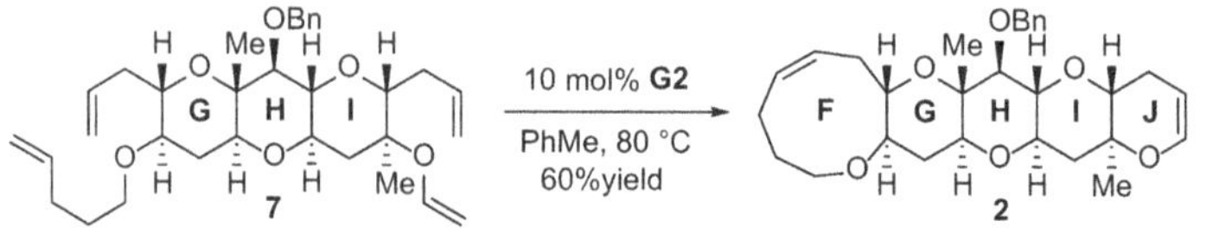

Scheme 3 Diene/enol ether-alkene double RCM to construct the gambieric acid FGHIJ fragment

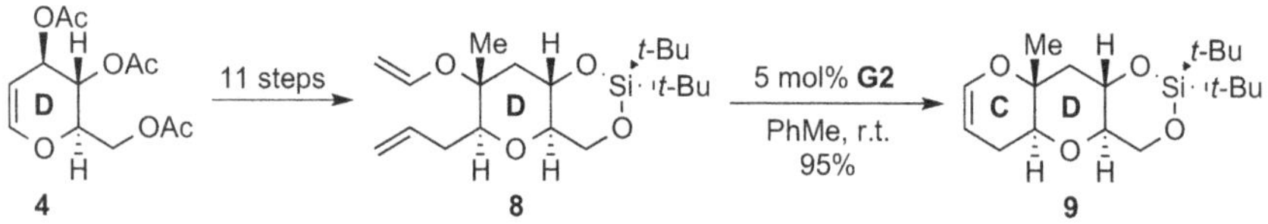

Scheme 4 Construction of the gambieric acid CD ring system by RCM

where the close proximity of the axial methyl group in **5** directed the first carbocupration [14] to the opposite, less hindered alkyne.

With bis(enol ether) **3** in hand, the double RCM proceeded smoothly using 10 mol% of **G2** [15] to deliver GHI fragment **6** in an excellent 89% yield (Scheme 2).

A further 14 chemical steps transformed RCM product **6** into **7**, the precursor for the second double RCM of the synthesis. Again, using 10 mol% of **G2**, a good 60% yield of the FGHIJ fragment **2** was obtained (Scheme 3).

Although F–J ring fragment **2** does not contain a functional handle for further elaboration of the F ring, the very late-stage introduction of the pent-4-enyl unit, subsequently used to close the F ring, means that incorporation of functionality to enable coupling of the A–E portion would be relatively straightforward. In addition, the methyl group missing from the F ring could also be readily incorporated in a total synthesis. Moreover, the enol ether in ring J provides the functionality required to install both the J ring's side chain and exocyclic methylene unit.

More recently, Clark et al. have utilised the RCM reaction in the synthesis of the BCD ring system of the gambieric acids [16]. With the D ring based on tri-*O*-acetyl D-glucal **4**, this starting material was elaborated to diene **8** in 11 steps. The enol ether RCM then proceeded in excellent yield, under mild conditions, to establish the C ring and deliver cyclic enol ether **9** in 95% yield (Scheme 4).

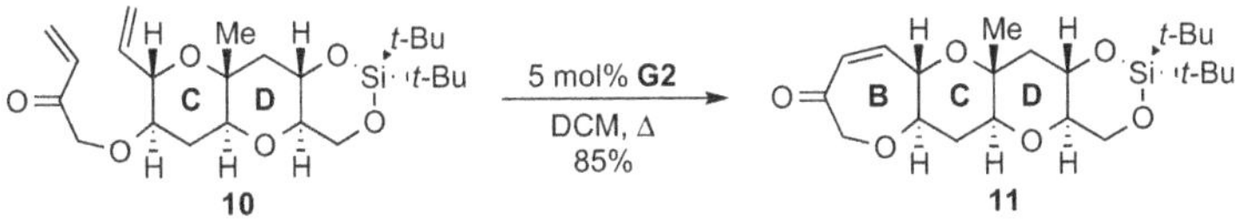

Scheme 5 Construction of the gambieric acid BCD ring system by RCM

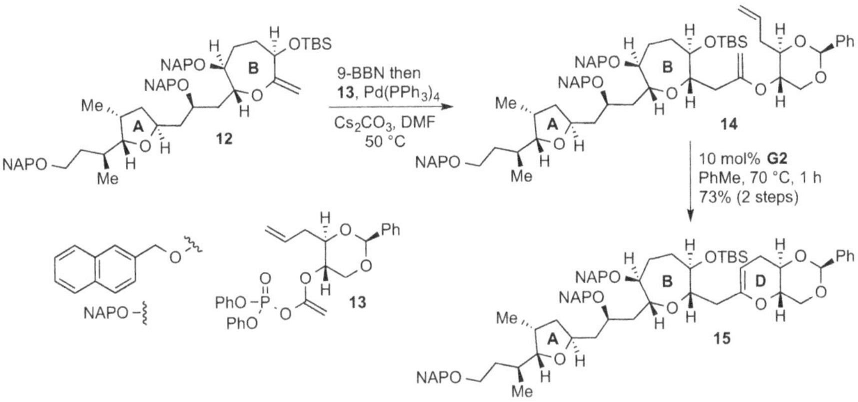

Scheme 6 Formation of the gambieric acid D ring by enol ether RCM

In addition to the RCM process, it is worth noting that this sequence of chemical steps begins and ends with an enol ether, and the sequence of epoxidation, allylation, enol ether formation and RCM represents a powerful iterative process to extend the polyether chain by one ring. A similar sequence of chemical steps then converted enol ether **9** to enone **10**, primed for the formation of the B ring (Scheme 5). RCM proceeded smoothly to deliver BCD fragment **11** in a very high yield of 85% using the **G2** catalyst.

Clark et al. then described the attachment of the gambieric acid A ring via an elegant modification of the Tsuji-Trost allylation [17] pioneered by Stoltz [18–20]. In addition, the D ring portion of this fragment contains a protected diol functionality designed to enable formation of the E ring via coupling with the FGHIJ fragment.

The first total synthesis of gambieric acid A was reported by Sasaki et al. in 2012 [21], allowing the absolute and relative stereochemistry of the natural product to be assigned, as depicted in Fig. 1. This landmark total synthesis relied on the RCM reaction at two stages of the synthetic route. In the first instance, Sasaki's group developed a two-step enol phosphate cross-coupling/RCM process [22] which was employed to construct the D ring. Specifically, cross-coupling of the alkylborane derived from **12** with enol phosphate **13**, laying the foundations of the C ring, was followed by RCM of the resulting enol ether **14**, using 10 mol% **G2** catalyst, to

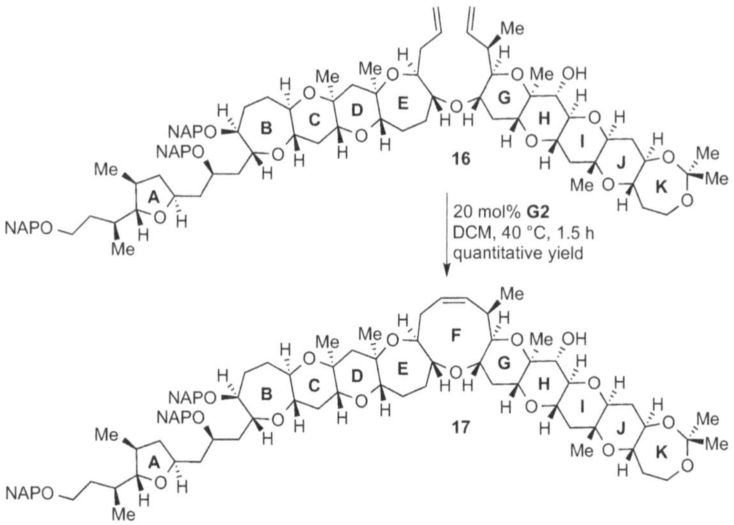

Scheme 7 Construction of the F ring of gambieric acid A by RCM

close the D ring and deliver ABD fragment **15** (Scheme 6) in 73% yield over the two steps.

The second instance of the RCM reaction in Sasaki's synthesis was a spectacular, late-stage forging of the final ring of the polyether framework, the challenging nine-membered F ring, from precursor **16** (Scheme 7). A 20 mol% loading of the **G2** catalyst was required for this RCM process, and the full A–J framework **17** was obtained in quantitative yield after only 1.5 h at reflux in DCM.

Following this outstanding result, a further 13 steps were required to install the J ring side chain and carry out the functional group manipulations to complete the total synthesis of gambieric acid A.

2.2 Hemibrevetoxin B

Hemibrevetoxin B (**18**, Fig. 3) is a tetracyclic polyether marine toxin, isolated from a red tide organism [23]. Like many other members of the polyether family, RCM has been widely deployed in synthetic approaches towards this natural product.

For example, Clark et al. have employed multiple RCM reactions to access the tetracyclic core of hemibrevetoxin B, culminating in a double RCM reaction [24]. Starting from a commercially available glucose derivative, the precursor for the first RCM reaction, enone **19**, was accessed in four steps and in high yield. Using only 3 mol% of **G2**, the 7-membered enone **20** corresponding to the B ring was formed in an excellent 94% yield following refluxing in dichloromethane (Scheme 8).

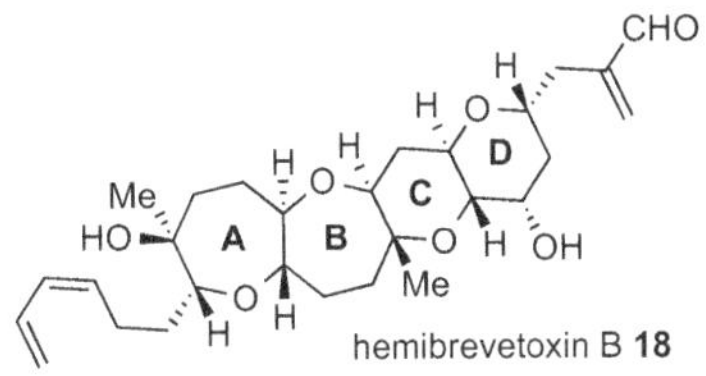

Fig. 3 Polyether natural product hemibrevetoxin B

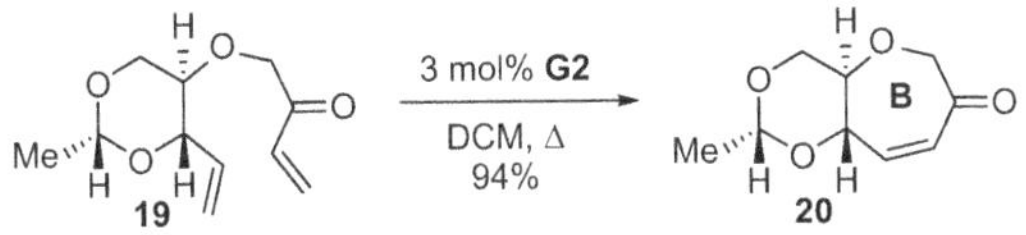

Scheme 8 Formation of the B ring of hemibrevetoxin B by RCM

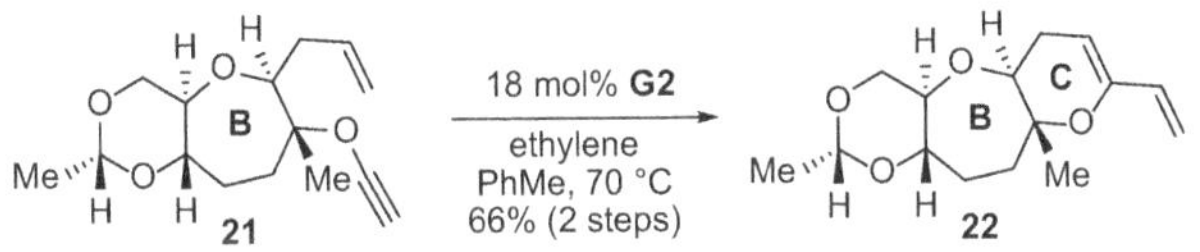

Scheme 9 Ynol ether enyne metathesis towards hemibrevetoxin B

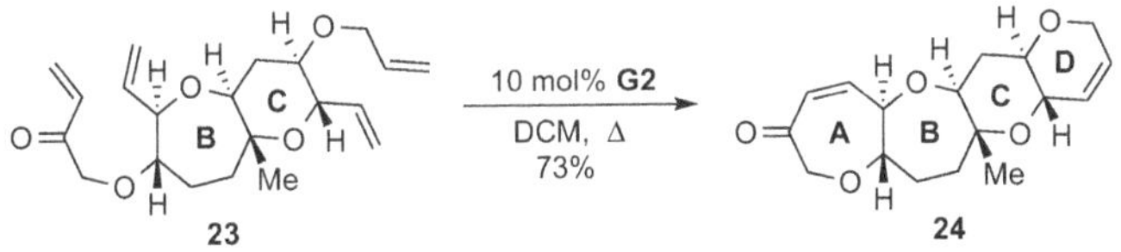

Scheme 10 Double RCM to complete the tetracyclic core of hemibrevetoxin B

Further along the synthetic route, an interesting ring-closing enyne metathesis was carried out on ynol ether **21** (Scheme 9) to form the C ring. A relatively high loading of 18 mol% **G2**, preactivated with ethylene, provided vinyl enol ether **22** in a reported 66% yield over two steps (the metathesis reaction and the preceding ynol ether formation from the corresponding alcohol). However, this process was occasionally complicated by side reactions depending on the scale and reaction conditions and an alternative route was thus established.

Further elaboration of vinyl enol ether **22**, over 12 chemical steps, delivered the final, double RCM substrate **23**. Using 10 mol% of the **G2** catalyst, a very good 73% yield of the tetracyclic core of hemibrevetoxin B **24** was obtained (Scheme 10), with the seven-membered A ring enone formation on one end of the molecule reminiscent of the previous RCM in this synthesis (**19–20**, Scheme 8), while the other concurrent metathesis process to form the D ring was a more conventional six-membered ether formation.

2.3 Brevenal

Following the report of the isolation of marine polyether brevenal (**25**, Fig. 4) in 2005 [25], there has been considerable interest in this molecule, not only due to its structural complexity but also due to its noteworthy biological properties.

In 2010, Crimmins et al. reported the synthesis of an advanced AB–E ring intermediate of brevenal **25** [26]. The RCM reaction was employed in the closure of two of these three rings, namely, the seven-membered B and E rings. As shown in Scheme 11, the formation of AB intermediate **27** occurred smoothly from functionalised A ring compound **26**, in an excellent 94% yield with only 5 mol% of **G2** catalyst, with no issues resulting from the nearby nitrile functionality.

In contrast to many other RCM-based approaches to polyethers, where a cyclic enol ether is formed, in this case, the cyclic allylic ether motif was targeted. As a result of this astute design, dihydroxylation of the resulting cyclic alkene can be used to simultaneously install both the B ring secondary hydroxyl group and the ether oxygen for construction of the C ring.

The seven-membered E ring was constructed by another high yielding RCM on diene **28** (Scheme 12). Although slightly more forcing conditions were required compared to the formation of AB ring intermediate **27** (Scheme 11), the relatively low 5 mol% loading of **G2** catalyst was maintained and delivered E ring intermediate **29** in 91% yield, in spite of the quite significant steric congestion adjacent to one of the olefin coupling partners.

Following the first two total syntheses of brevenal **25** [27–29], Rainier et al. reported a total synthesis in 2011, which relied extensively on ring-forming

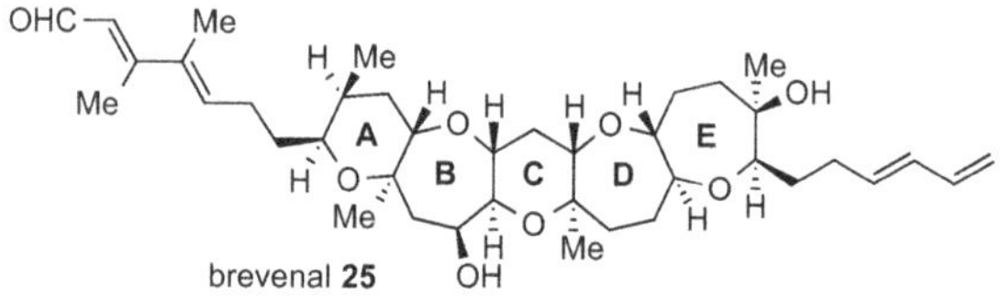

Fig. 4 Polyether natural product brevenal

Scheme 11 Synthesis of the AB ring system of brevenal by RCM

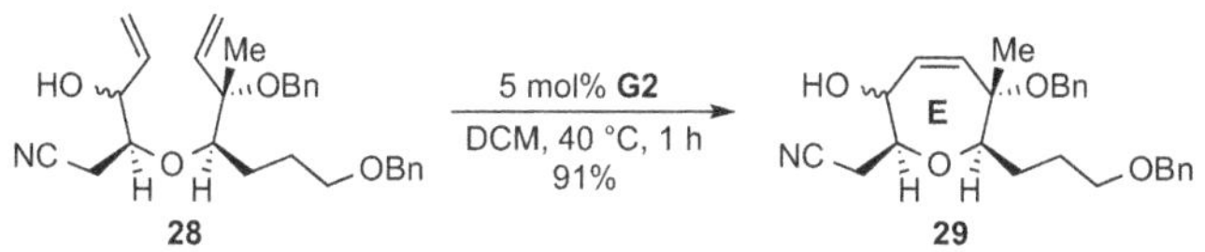

Scheme 12 Formation of the brevenal E ring by RCM

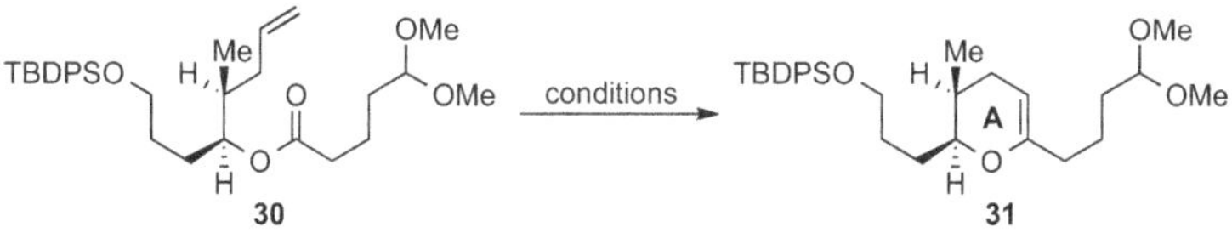

A: 1. $TiCl_4$, Zn, $PbCl_2$, CH_2Br_2, THF, TMEDA, DCM, 65 °C; 2. **G2** (75%, 2 steps)
B: $TiCl_4$, Zn, $PbCl_2$, CH_3CHBr_2, THF, TMEDA, DCM, 65 °C, (88%)

Scheme 13 OLEC process to construct brevenal ring A

metathesis processes [30]. Specifically, Rainier et al. employed the olefin-ester cyclisation (that they abbreviate as OLEC), which they have developed to great effect in several other total syntheses (see, for example, [31]). The OLEC transformation involves the Takai-Utimoto-type olefination of an ester, which then undergoes RCM with the olefinic partner. Given the facile ester disconnection to alcohol and acid, this represents a powerful method for the convergent synthesis of functionalised cyclic systems, and, unsurprisingly, this method has been extensively employed in total synthesis since the pioneering work of Nicolaou to develop the initial concept in polyether synthesis [32].

The A ring of brevenal was constructed from olefinic ester **30**, as depicted in Scheme 13. The two-step procedure, where Takai-Utimoto olefination is followed by RCM using catalyst **G2** (Scheme 13, conditions A), was compared to the direct OLEC process (Scheme 13, conditions B), with A ring intermediate **31** being formed in 75% yield over the two-step process, compared to an increased 88% yield for the one-step OLEC method.

Following this success, the seven-membered E ring was constructed by a similar OLEC process on olefinic ester **32** (Scheme 14). E ring intermediate **33** was obtained in a moderate 66% yield, along with 22% of the corresponding acyclic enol ether **34** resulting from olefination of the ester.

Following several synthetic transformations, union of the AB and E ring intermediates delivered ABE system **35**, primed for a further OLEC reaction to forge the C ring. Indeed, under standard conditions, this methodology delivered advanced ABCE intermediate **36** in an outstanding 83% yield (Scheme 15).

Unfortunately, subsequent attempts to close the D ring from intermediate **36** were unsuccessful. A new strategy was thus adopted, based on constructing the D ring first from a modified ABE ring intermediate, followed by formation of the C ring.

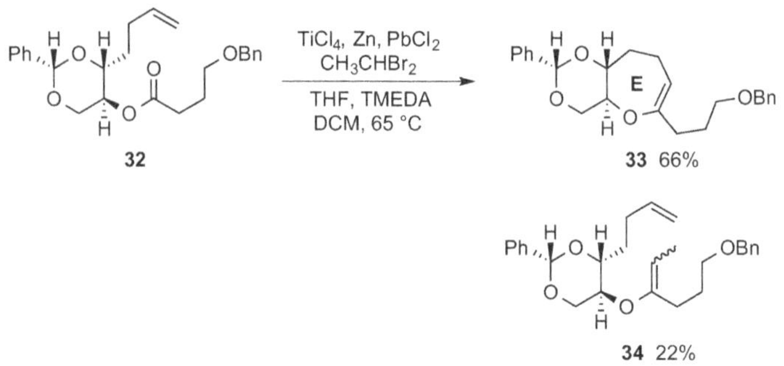

Scheme 14 OLEC process to construct the brevenal E ring

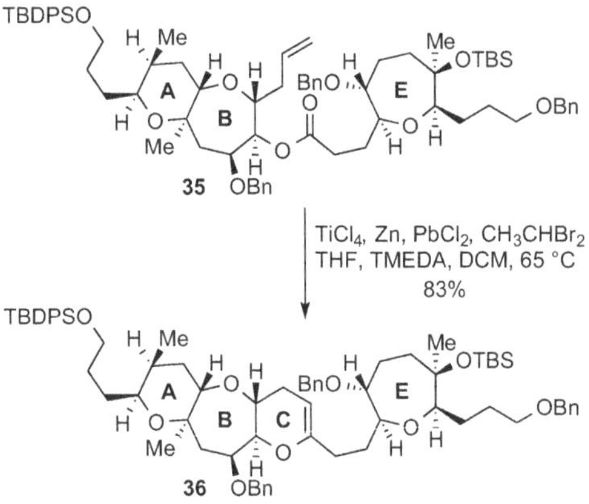

Scheme 15 Assembly of the ABCE ring system of brevenal using OLEC

To this end, ABE ring precursor **37** was reacted under modified OLEC conditions (no DCM in the solvent mixture) to give ABDE ring intermediate **38** in a disappointing 30% yield (Scheme 16). The balance of material (70%) was again the acyclic enol ether resulting from olefination of the ester. This by-product could be converted to the desired product **38** in 35% yield using catalyst **G2**, resulting in a 65% overall yield of the ABDE ring fragment **38**.

With ABDE fragment **38** in hand, a further 15 chemical steps were then required to forge the C ring and construct the two side chains to complete the total synthesis of brevenal.

It is testament to the flexibility of both the OLEC process, and the olefinic ester starting materials, that such a setback at an advanced stage of the synthesis could be overcome by a relatively straightforward alteration of existing AB and E ring intermediates to give a new OLEC precursor.

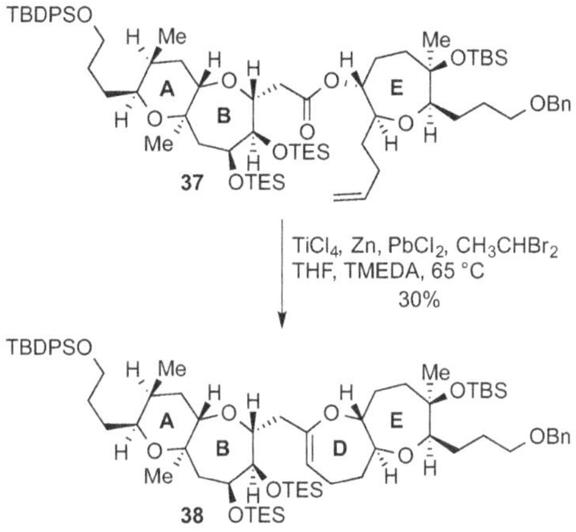

Scheme 16 Formation of the brevenal ABDE ring system by OLEC

2.4 Ciguatoxin CTX3C

Ciguatoxin CTX3C **39** (Fig. 5) is a member of the large class of ciguatoxin marine polyethers [33]. Given its incredible complexity, it is unsurprising that it has been the focus of much synthetic effort, including a total synthesis by Hirama et al., first reported in 2001 [34–36]. Although employing RCM reactions, detailed discussion of these synthetic efforts is outwith the scope of this chapter, and instead, more recent developments in the synthesis of fragments towards **39** will be discussed.

A bidirectional strategy was used by Clark et al. in their synthesis of the A–E rings of ciguatoxin CTX3C **39**. In this instance, the double RCM process was deployed early in the synthesis to establish the A and C rings from a heavily functionalised B ring framework [37]. Double RCM substrate **40** was accessed in eight steps from an advanced starting material. The double RCM reaction then proceeded in 58% yield using 10 mol% of **G2** at an elevated temperature in toluene (Scheme 17). The seven-membered A ring ether was established through a classical RCM process, whereas the six-membered C ring enol ether in **41** was formed by enyne metathesis onto an ynol ether. As with the related transformation in their synthesis of the hemibrevetoxin B core [24], this latter cyclisation required activation of the catalyst with ethylene.

Given the relatively moderate yield of this double RCM process, an iterative approach to the formation of the ABC ring system was also investigated. Thus, RCM of intermediate **42** delivered the AB ring system **43** in an outstanding 98% yield, using only 3 mol% of the simple **G1** catalyst (Scheme 18), reflecting the relatively unchallenging nature of this metathesis reaction.

Further modification of **43** through seven chemical steps delivered the ynol ether substrate **44**, required for the second metathesis ring closure (Scheme 19). Upon application of similar conditions to the corresponding double RCM attempt (10 mol

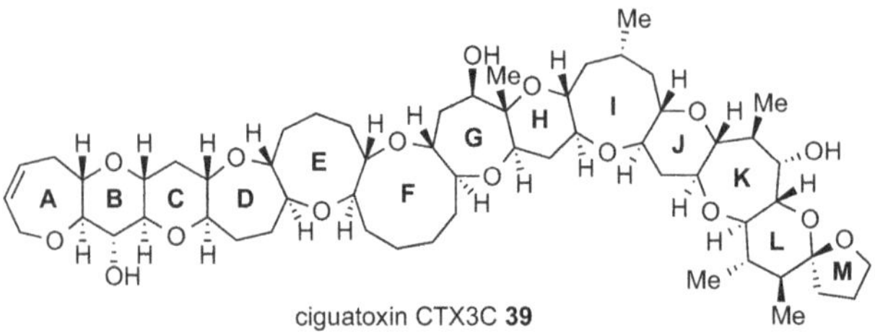

Fig. 5 Marine polyether ciguatoxin CTX3C

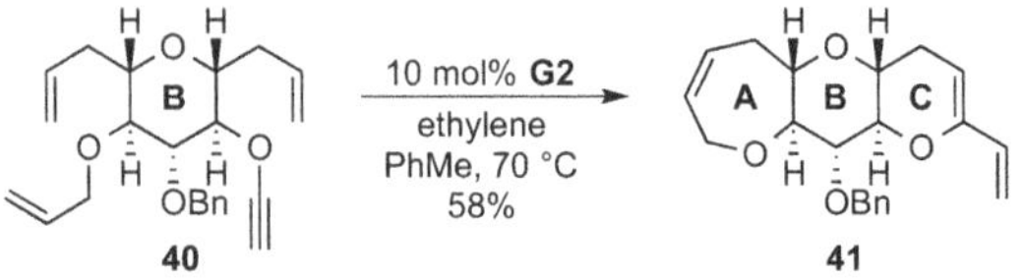

Scheme 17 Diene/ynol ether-alkene double RCM towards ciguatoxin CTX3C

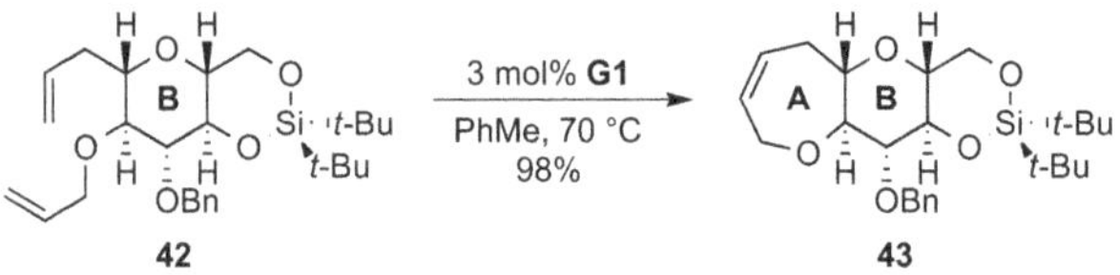

Scheme 18 Iterative approach to the ABC ring system of ciguatoxin CTX3C – first step

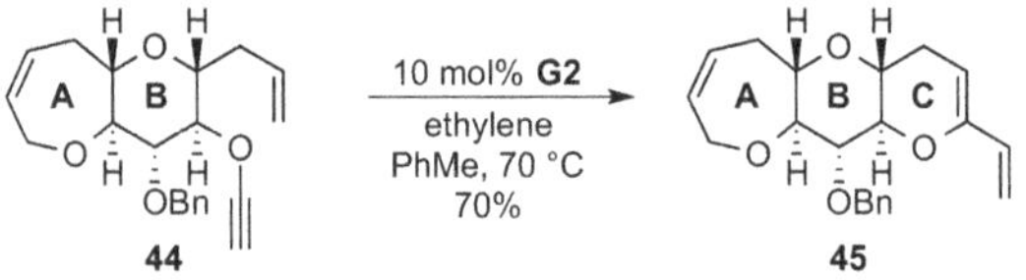

Scheme 19 Iterative approach to the ABC ring system of ciguatoxin CTX3C – second step

% **G2**, ethylene activation, toluene, 70°C), ABC ring system **45** was obtained in 70% yield. This represents a moderate improvement over the 58% yield of the double RCM process (Scheme 17), taking into account the near quantitative yield of the first RCM to form AB system **43** (Scheme 18).

Metathesis processes were also employed to subsequently forge both the D and E rings. In this regard, a four-step elaboration of ABC fragment **45** gave diene **46**. The seven-membered enone-forming RCM process, employing 5 mol% of **G2**, delivered the tetracyclic ABCD system **47** in a good 70% yield (Scheme 20).

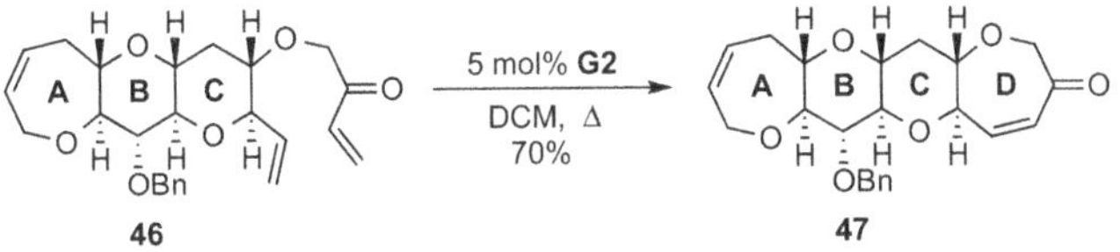

Scheme 20 Formation of the ABCD ring system of ciguatoxin CTX3C

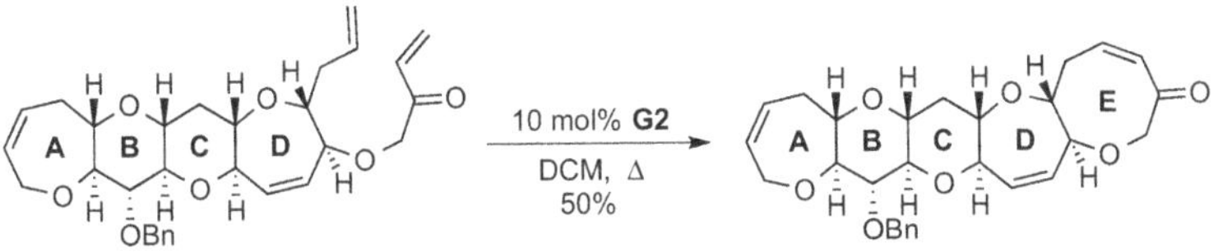

Scheme 21 Completion of the A–E ring system of ciguatoxin CTX3C

Following a further six manipulations, diene **48** then underwent RCM to give the ciguatoxin CTX3C A–E ring system **49** using 10 mol% of **G2** (Scheme 21). The more challenging nature of this eight-membered enone system was reflected in the moderate 50% yield for this process.

2.5 *Maitotoxin*

The structural and stereochemical complexity of the marine polyether natural product family culminates in the stunning 32-ring assembly of maitotoxin, featuring six domains of fused polyether units (see [38, 39] and references therein). Nicolaou et al. have been engaged in efforts towards the total synthesis of this remarkably challenging target molecule, and their report on the preparation of the WXYZA′ fragment **50** of maitotoxin (Fig. 6) employs an RCM process as part of an ester-olefin cyclisation [40].

WZA′ ring intermediate **51** was treated with the standard Ti/Zn/Pb Takai-Utimoto olefination mixture to form the seven-membered Y ring and deliver WYZA′ intermediate **52** (Scheme 22). Initial attempts, with bulky TBS protection on the W ring secondary alcohol, resulted only in the recovery of starting material, and so an optimization of the process was undertaken, including a screen of reaction time, temperature, solvent and the nature of protecting group for this W ring secondary alcohol, as well as the free alcohol itself. Under the optimised conditions, an impressive 67% yield of the WYZA′ ring intermediate **52** was obtained, with only 17% of the uncyclised enol ether, resulting from olefination of the ester, obtained as a by-product, as is often unavoidably the case in these processes. Remarkably, this optimum yield was obtained on the free W ring secondary alcohol, with the corresponding TMS ether delivering a considerably lower yield

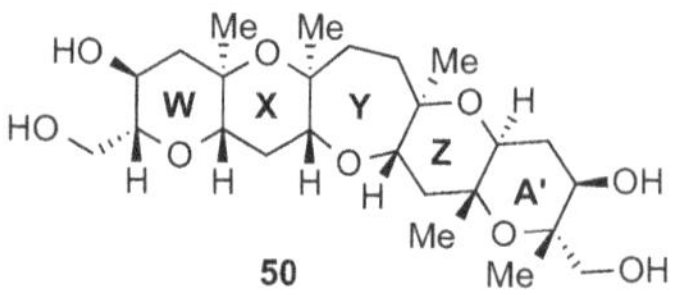

Fig. 6 The WXYZA′ domain of maitotoxin

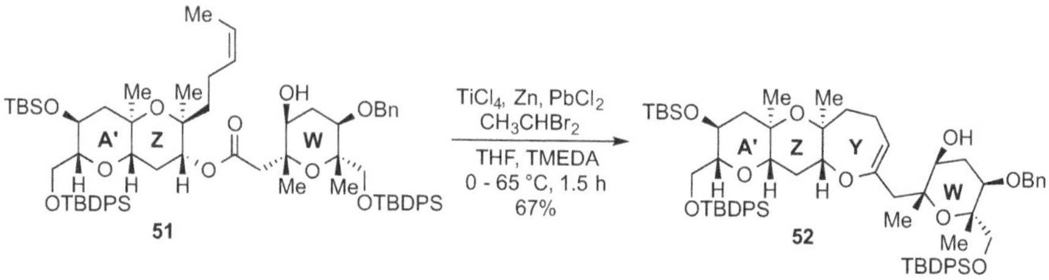

Scheme 22 Formation of the Y ring of maitotoxin

of WYZA′ intermediate. As noted also by Rainier et al. in their brevenal work [30], omitting DCM from the solvent mixture also led in this case to the optimum yield of product.

Intermediate **52** was then converted in a further 8 steps to the fully deprotected WXYZA′ ring fragment **50**. Notably, in a subsequent study on the coupling of the WXYZA′ and QRSTUV fragments, an improved 78% yield is reported for this closure of the Y ring to form intermediate **52** [41].

2.6 *Access to* Cis-*Fused Polyethers by RCM*

In the overwhelming majority of the ladder polyether natural products, the ring junctions are *trans*-fused. In contrast, a small number of polyether natural products feature *cis*-fused ring junctions, for example, the C32–33 and C35–36 ring junctions of halichondrin B [42], and the LM and NO domains of maitotoxin [43]. A recent report by Singh et al. discloses a method for the formation of such *cis*-fused polyethers using the RCM reaction. A Lewis acid-mediated oxocarbenium ion cyclisation was employed to establish 6,6-*cis*:*anti*:*cis* system **53**, and RCM was used to form both 6,6,7- and 6,6,8-systems **54** and **55** (Scheme 23), using the simple **G1** catalyst, in 85% and 63% yields, respectively.

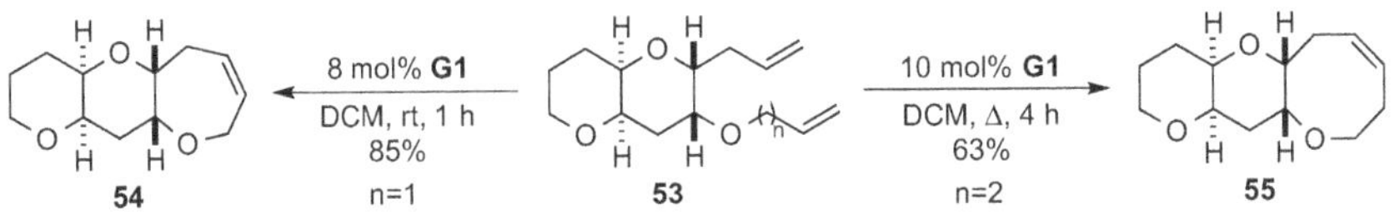

Scheme 23 Synthesis of *cis*-fused polyethers by RCM

3 The Application of Metathesis Reactions to the Synthesis of Other Natural Products

This section describes recent advances in the synthesis of cyclic ether-containing natural products which do not contain the fused ring polyether motif.

3.1 Neopeltolide

Following total syntheses by Panek et al. [44] and Scheidt et al. [45], the structure of marine macrolide (+)-neopeltolide was revised from the proposed structure to **56**, shown in Fig. 7. Of specific interest here is the pyran ring (for a general review on the synthesis of tetrahydropyran rings in natural products, see [46]), across the ether portion of which is bridged a 12-membered macrolactone. This natural product has inspired several different metathesis-based approaches, which are discussed below.

In the total synthesis of neopeltolide reported by Fuwa and Sasaki et al. [47, 48], the cyclic ether portion of the pyran-macrolide bridging system was assembled by the group's developed enol phosphate cross-coupling/RCM process (Scheme 24) [14]. Enol phosphate **57**, readily prepared by *O*-phosphorylation of the corresponding acetate ester enolate, was cross-coupled with the *B*-alkyl-9-BBN reagent derived from iodide **58**, following iodine-lithium exchange and transmetallation with *B*-OMe-9-BBN. The resulting substituted enol ether **59** underwent smooth RCM using 10 mol% of catalyst **G2** to deliver dihydropyran **60** in 78% yield over the two steps. The carbon skeleton of the macrolide portion was designed into fragments **57** and **58**, and, following their union to give **60**, only a few simple protecting group removal and oxidation steps were then required to deliver the macrolactonisation precursor.

Fuwa et al. then developed a second-generation total synthesis of neopeltolide, cutting the longest linear sequence from 24 to 13 steps [49]. In this new approach, the enol phosphate cross-coupling/RCM strategy for construction of the pyran ring was abandoned in favour of an oxa-Michael reaction, and the RCM process was instead employed in the construction of the macrocycle.

An ambitious strategy was envisioned, whereby the macrocyclic 12-membered ether/lactone ring would be closed by RCM. Additionally, the diene substrate was chosen such that the ring closure would occur across C8–C9 (Scheme 25). With the

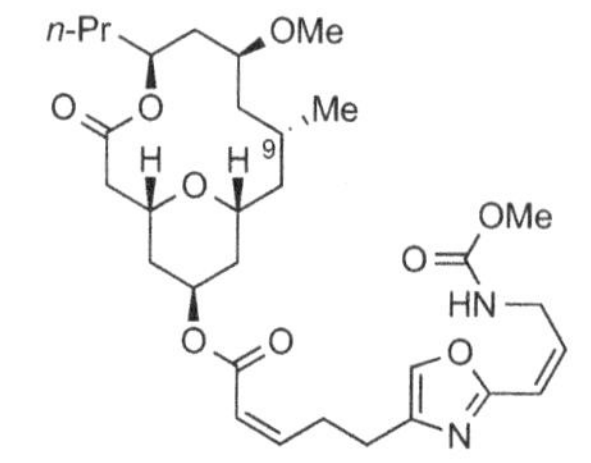

56 neopeltolide

Fig. 7 Revised structure of neopeltolide

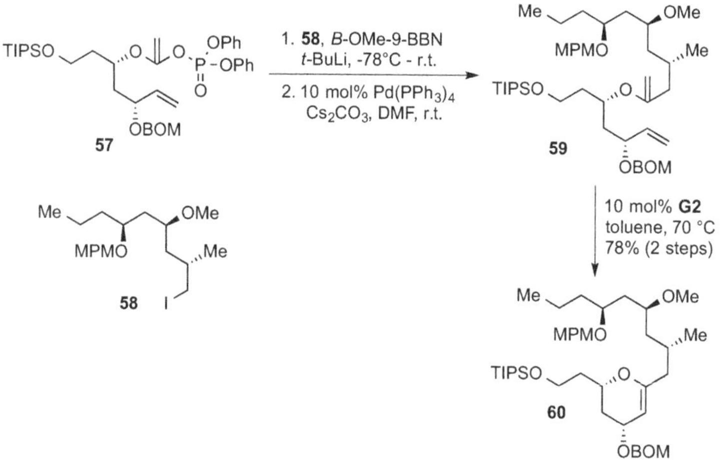

Scheme 24 Tandem enol phosphate cross-coupling/RCM in the total synthesis of neopeltolide

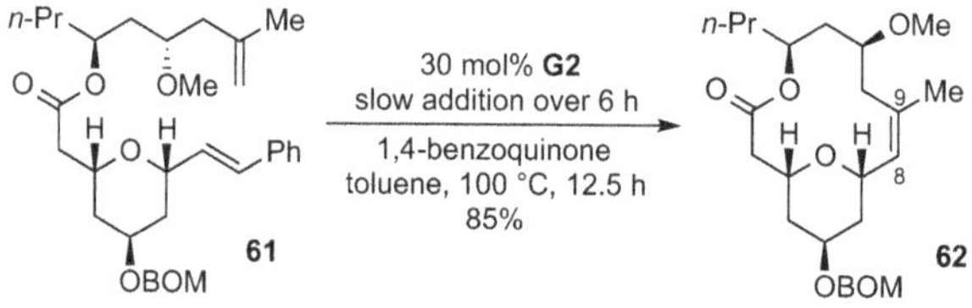

Scheme 25 The Fuwa et al. macrocyclic RCM approach towards neopeltolide

C9 methyl group already installed, this would translate into a challenging RCM macrocyclisation to form a trisubstituted alkene.

Diene precursor **61** underwent RCM to form macrocyclic product **62** in 85% yield, following application of some relatively specialised reaction conditions to this highly challenging substrate (Scheme 25). In terms of the substrate, rather than employ the more usual terminal alkene, the C8 alkene was substituted with a phenyl

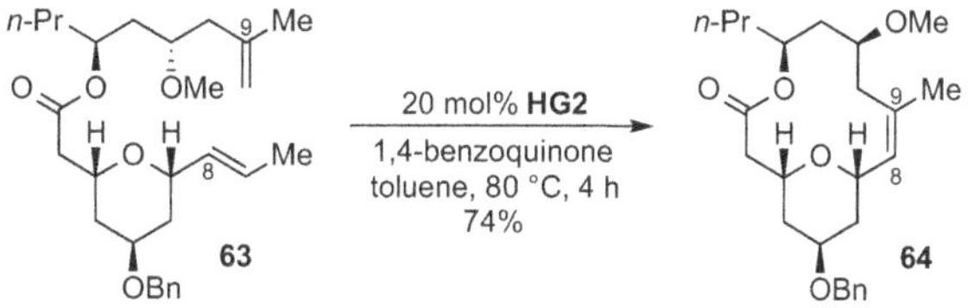

Scheme 26 The She et al. macrocyclic RCM approach towards neopeltolide

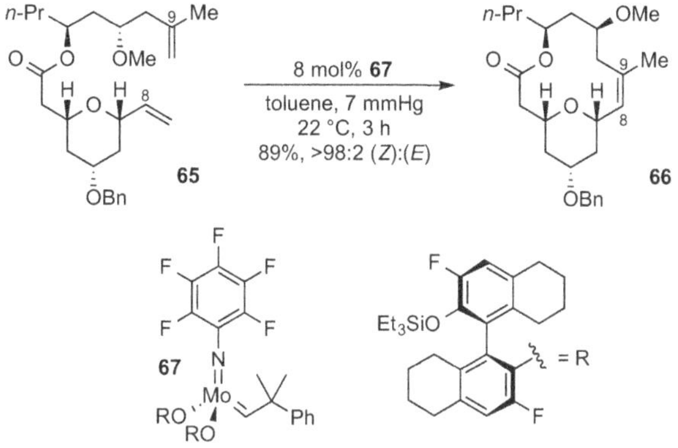

Scheme 27 The Hoveyda et al. macrocyclic RCM approach towards neopeltolide using molybdenum catalyst **67**

group on the carbon which is cleaved during the RCM reaction. In addition, the well-established 1,4-benzoquinone additive [50] was employed. More importantly, the 30 mol% of catalyst **G2** that was used was added slowly over a period of 6 h.

Hydrogenation of trisubstituted alkene **62** then delivered the required stereochemistry at C9, according to precedent established in previous synthetic studies by Floreancig et al. [51], who arrived at a related neopeltolide intermediate using a more conventional macrolactonisation approach. It is worth noting that this overall strategy of macrocyclic RCM, followed by a stereoselective hydrogenation of the resulting macrocyclic trisubstituted alkene, is reminiscent of the approach to the total synthesis of fluvirucin B_1 aglycon by Hoveyda et al. [52], one of the earliest examples of the use of RCM in natural product synthesis.

The challenge of this demanding RCM has also been confronted by two other formal syntheses of neopeltolide. In the first of these reports, and following shortly after the work of Fuwa et al. [39], She et al. studied very closely related substrate **63**, which differed from **61** only in the nature of the protecting group for the pyran hydroxyl, and in the second substituent of the putative C8 alkene, which in this case was a methyl group (Scheme 26) [53].

In this system, despite only small differences in comparison to the Fuwa substrate, the **G2** catalyst failed to deliver any cyclised product. Instead, the **HG2** catalyst [54] was required for successful RCM, and, using 20 mol% of this catalyst, a 74% yield of macrocycle **64** was obtained.

A very similar RCM precursor was employed in an elegant and concise synthesis of neopeltolide by Hoveyda et al. [55]. In this case, the C8 alkene in RCM substrate **65** (Scheme 27) was derived from a terminal alkene, with the epimeric C5 protected alcohol being the only other difference compared to the She et al. substrate **63**.

However, in a departure from the previous studies, the macrocycle-forming RCM was in this case effected not by a ruthenium-based metathesis catalyst but by molybdenum bis(aryloxide) complex **67** [56]. At room temperature in toluene, and under a slight vacuum to drive the removal of ethylene, macrocycle **66** was formed in an outstanding 89% yield, with >98% selectivity for the desired (*Z*)-alkene isomer. More conventional Mo-based metathesis catalysts resulted in significantly lower yields of product, and, in comparison, use of the **G2** and **HG2** catalysts at elevated temperature resulted in only a 49% and 54% yield of macrocycle **66**, respectively.

Thus, the natural product neopeltolide serves as a useful case study for the use of the RCM reaction in total synthesis. From the enol ether RCM to form the pyran ring to more ambitious and challenging macrocyclisations, the flexibility of the RCM reaction allows for a variety of strategic approaches. In terms of the tactics used to accomplish the overall strategy, the various approaches to the macrocyclisation in Schemes 25–27 showcase the broad range of highly active metathesis catalysts now available to the synthetic chemist.

3.2 *Kendomycin*

Kendomycin **68** (Fig. 8) is an intriguing natural product, featuring both a pyran ring and a quinone methide/lactol unit fused to its central, 16-membered macrocyclic core (See, for example: [57]).

A recent study by Fürstner et al. sought to construct the macrocycle in **68** via a ring-closing alkyne metathesis (RCAM) [58]. In the first instance (Scheme 28), diyne substrate **69** was selected as the RCAM candidate. This substrate represented an ambitious choice, with substitution both adjacent to the alkyl-substituted alkyne and *ortho*- to the aryl alkyne. Indeed, of several catalysts screened, only molybdenum alkylidyne complex **71** [59], in large quantities, offered any meaningful conversion to macrocyclic alkyne **70**, with a maximum of 62% yield. Moreover, this RCAM also proved to be capricious, and so a second-generation synthesis, based on an alternative diyne substrate, was devised.

Fig. 8 Kendomycin

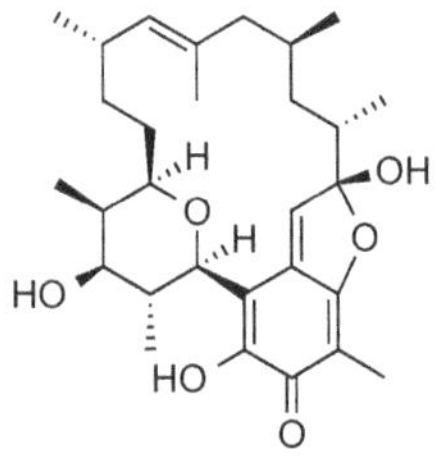

kendomycin **68**

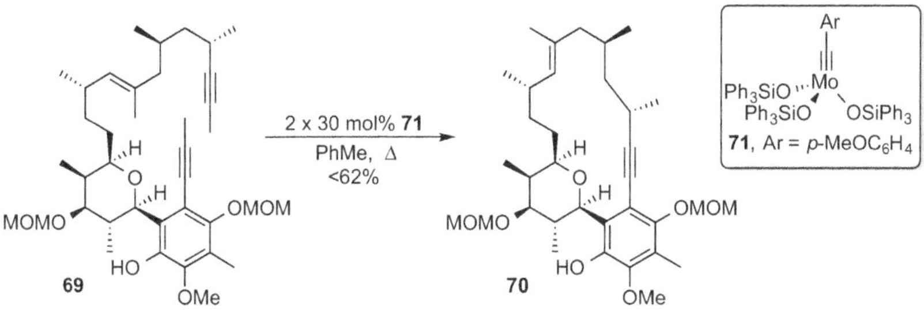

Scheme 28 Initial attempt at RCAM towards kendomycin

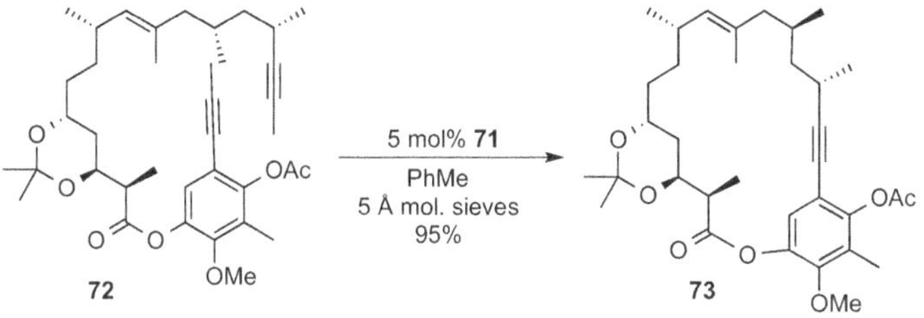

Scheme 29 Second-generation RCAM towards kendomycin

The modified substrate was designed to relieve strain in the product by modifying the macrocycle to incorporate the aromatic moiety via a *meta*-, rather than an *ortho*-, substitution pattern. Following this design modification, diyne **72** underwent smooth RCAM using only 5 mol% of catalyst **71** to give the macrocyclic alkyne **73** in an excellent 95% yield (Scheme 29).

Following ester hydrolysis of alkyne **73** and subsequent gold-catalysed furan formation, the synthesis converged with a prior total synthesis by Mulzer [60], which employed an elegant photo-Fries rearrangement to re-establish the correct *ortho*-connectivity of the macrocycle across the aromatic ring.

Although outwith the scope of this review and thus not the subject of detailed discussion here, Fürstner et al. have also recently employed catalyst **71** in a RCAM-based approach towards the proposed structure of macrolactone natural product mandelalide A [61].

4 The Application of Metathesis Reactions to the Synthesis of Non-natural Products

During the review period, metathesis processes have also been applied to the synthesis of a range of unnatural products with potential biological activity, and selected examples are discussed below.

4.1 Carbohydrate-Linked Polyethers

Mandal et al. have prepared a range of polyether-linked macrocycles based on carbohydrates [62]. Starting from an intermediate based on D-glucose, dimerisation was effected via cross metathesis, and a double *O*-allylation delivered diene substrate **74**. RCM was mediated by an unspecified amount of **G1** catalyst in refluxing benzene, and, following hydrogenation of the resulting alkene, 12-membered macrocycle **75** was obtained in 56% yield over the two steps (Scheme 30).

Following the success of this approach, even larger macrocyclic systems were targeted. These included a 13-membered dimeric species and a 17-membered trimeric analogue. More spectacularly, as shown in Scheme 31, RCM of diene **76** gave 21-membered tetrameric analogue **77** in 55% yield, using an unspecified amount of catalyst **G2**.

Following RCM and reduction, the carbohydrate portions of the macrocycles were then coupled with nucleosides, and their interaction with a range of small molecules was studied.

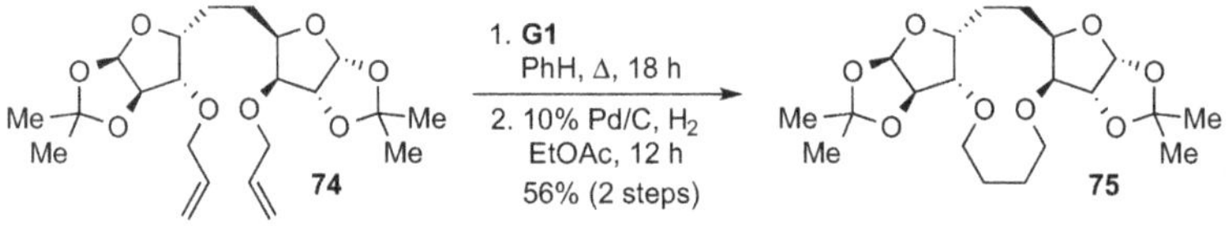

Scheme 30 Formation of a 12-membered macrocyclic ether by RCM

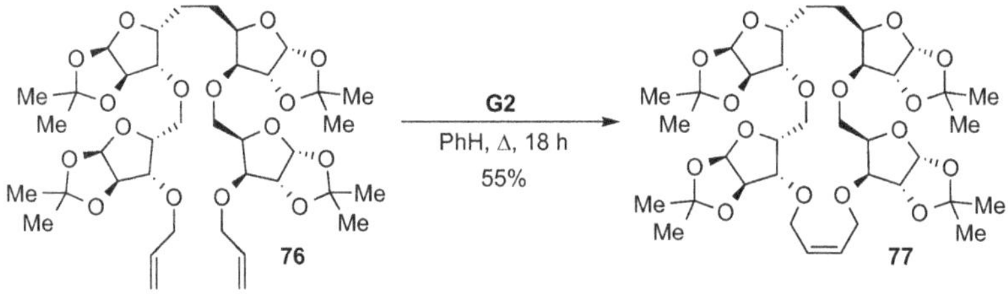

Scheme 31 Formation of a 21-membered macrocyclic ether by RCM

Scheme 32 Ring-closing enyne metathesis in the preparation of oxepanes

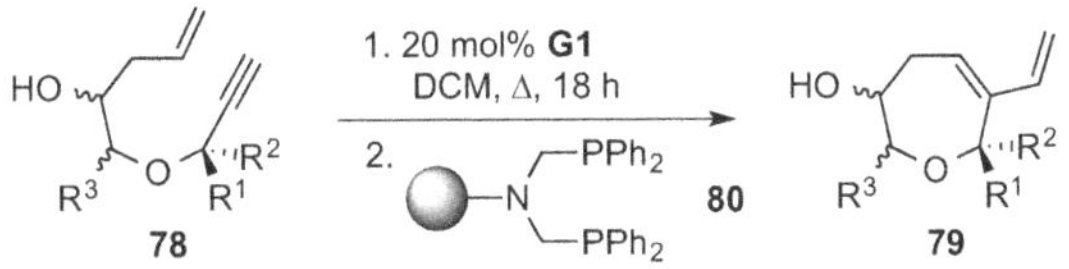

4.2 Synthesis of a Natural Product-Like Oxepane Library

Waldmann et al. have used the ring-closing enyne metathesis (RCEYM) of a range of propargyl bis(homoallyl) ethers to generate a 90-member library of oxepanes, inspired by structurally similar natural products [63]. As shown in Scheme 32, RCEYM of substrates **78**, using 20 mol% of the simple **G1** catalyst, gave oxepanes **79**.

Notably, this library of compounds was prepared in solution, with polymeric scavenging agents being used to remove the reagents at each step. For the RCEYM, polymeric phosphine **80** was employed to scavenge the **G1** catalyst.

The diene functionality resulting from the RCEYM was then harnessed to generate a range of diverse structures, with this functional group being used for a range of Diels-Alder as well as cross-metathesis reactions.

5 Conclusions

Interest in the use of metathesis processes for the construction of natural products, and unnatural bioactive products, continues to be high. A wide variety of metathesis catalysts are finding use, including those based on molybdenum, as well as the more commonplace ruthenium-based catalysts. A variety of metathesis process are also being employed in the synthesis of these target molecules, with ring-closing enyne metathesis (RCEYM) being employed alongside the more routine ring-closing diene metathesis (RCM) and ring-closing alkyne metathesis (RCAM) also finding continued use in total synthesis. Undoubtedly, the synthesis of natural products and other bioactive targets will continue to provide fertile ground for the application of developed metathesis processes, as well as for the innovation of new, metathesis-based methods.

References

1. Rainier JD (2014) Synthesis of substituted tetrahydrofurans. In: Cossy J (ed) Synthesis of saturated oxygenated heterocycles I, vol 35, Topics in heterocyclic chemistry. Springer, Heidelberg, pp 1–41

2. Perry MA, Rychnovsky SD, Sizemore N (2014) Synthesis of saturated tetrahydropyrans. In: Cossy J (ed) Synthesis of saturated oxygenated heterocycles I, vol 35, Topics in heterocyclic chemistry. Springer, Heidelberg, pp 43–95
3. Piva O (2104) Synthesis of seven-membered ring ethers and lactones. In: Cossy J (ed) Synthesis of saturated oxygenated heterocycles I, vol 35, Topics in heterocyclic chemistry. Springer, Heidelberg, pp 283–320
4. Soriano E, Marco-Contelles J (2014) Synthesis of eight- to ten-membered ring ethers. In: Cossy J (ed) Synthesis of saturated oxygenated heterocycles II, vol 36, Topics in heterocyclic chemistry. Springer, Heidelberg, pp 321–368
5. Brimble M, Stubbing L (2014) Synthesis of 5,6- and 6,6-spirocyclic compounds. In: Cossy J (ed) Synthesis of saturated oxygenated heterocycles I, vol 35, Topics in heterocyclic chemistry. Springer, Heidelberg, pp 189–267
6. Prunet J (2011) Eur J Org Chem 2011:3634–3647
7. Martin T, Padrón JI, Martín VS (2014) Synlett 25:12–32
8. Nagai H, Torigoe K, Satake M, Murata M, Yasumoto T, Hirota H (1992) J Am Chem Soc 114:1102–1103
9. Nagai H, Murata M, Torigoe K, Satake M, Yasumoto T (1992) J Org Chem 57:5448–5453
10. Nicolaou KC, Frederick MO, Aversa RJ (2008) Angew Chem Int Ed 47:7182–7225
11. O'Connell KMG, Díaz-Gavilán M, Galloway WRJD, Spring DR (2012) Beilstein J Org Chem 8:850–860
12. Clark JS, Hamelin O (2000) Angew Chem Int Ed 39:372–374
13. Clark JS, Kimber MC, Robertson J, McErlean CSP, Wilson C (2005) Angew Chem Int Ed 44:6157–6162
14. Denmark SE, Guagnano V, Dixon JA, Stolle A (1997) J Org Chem 62:4610–4628
15. Scholl M, Ding S, Lee CW, Grubbs RH (1999) Org Lett 1:953–956
16. Clark JS, Romiti F, Sieng B, Paterson LC, Stewart A, Chaudhury S, Thomas LH (2015) Org Lett 17:4694–4697
17. Trost BM, Crawley ML (2003) Chem Rev 103:2921–2943
18. Behenna DC, Stoltz BM (2004) J Am Chem Soc 126:15044–15045
19. Behenna DC, Mohr JT, Sherden NH, Marinescu SC, Harned AH, Tani K, Seto M, Ma S, Novák Z, Krout MR, McFadden RM, Roizen JL, Enquist JA Jr, White DE, Levine SR, Petrova KV, Iwashita A, Virgil SC, Stoltz BM (2011) Chem Eur J 17:14199–14223
20. Hong AY, Stoltz BM (2013) Eur J Org Chem 2013:2745–2759
21. Fuwa H, Ishigai K, Hashizume K, Sasaki M (2012) J Am Chem Soc 134:11984–11987
22. Fuwa H, Sasaki M (2008) Org Lett 10:2549–2552
23. Prasad AVK, Shimizu Y (1989) J Am Chem Soc 111:6476–6477
24. Clark JS, Grainger DM, Ehkirch AA-C, Blake AJ, Wilson C (2007) Org Lett 9:1033–1036
25. Bourdelais AJ, Jacocks HM, Wright JLC, Bigwarfe PM Jr, Baden DG (2005) J Nat Prod 68:2–6
26. Crimmins MT, Shamszad M, Mattson AE (2010) Org Lett 12:2614–2617
27. Fuwa H, Ebine M, Bourdelais AJ, Baden DG, Sasaki M (2006) J Am Chem Soc 128:16989–16999
28. Ebine M, Fuwa H, Sasaki M (2008) Org Lett 10:2275–2278
29. Takamura H, Kukuchi S, Nakamura Y, Yamagami Y, Kishi T, Kadota L, Yamamoto Y (2009) Org Lett 11:2531–2534
30. Zhang Y, Rohanna J, Zhou J, Iyer K, Rainier JD (2011) J Am Chem Soc 133:3208–3216
31. Johnson HWB, Majumder U, Rainier JD (2005) J Am Chem Soc 127:848–849
32. Nicolaou KC, Postema MHD, Claiborne CF (1996) J Am Chem Soc 118:1565–1566
33. Swift AEB, Swift TR (1993) J Toxicol Clin Toxicol 31:1–29
34. Hirama M, Oishi T, Uehara H, Inoue M, Maruyama M, Oguri H, Satake M (2001) Science 294:1904–1907
35. Inoue M, Miyazaki K, Uehara H, Maruyama M, Hirama M (2004) Proc Natl Acad Sci 101:12013–12018

36. Inoue M, Hirama M (2004) Acc Chem Res 37:961–968
37. Clark JS, Conroy J, Blake AJ (2007) Org Lett 9:2091–2094
38. Murata M, Yasumoto T (2000) Nat Prod Rep 17:293–314
39. Nonomura T, Sasaki M, Matsumori N, Murata M, Tachibana K, Yasumoto T (1996) Angew Chem Int Ed Engl 35:1675–1678
40. Nicolaou KC, Baker TM, Nakamura T (2011) J Am Chem Soc 133:220–226
41. Nicolaou KC, Heretsch P, Nakamura T, Rudo A, Murata M, Konoki K (2014) J Am Chem Soc 136:16444–16451
42. Hirata Y, Uemura D (1986) Pure Appl Chem 58:701–710
43. Murata M, Naoki H, Iwashita T, Matsunaga S, Sasaki M, Yokoyama A, Yasumoto T (1993) J Am Chem Soc 115:2060–2062
44. Youngsaye W, Lowe JT, Pohlki F, Falifo P, Panek JS (2007) Angew Chem Int Ed 46:9211–9214
45. Custer DW, Zabawa TP, Scheidt KA (2008) J Am Chem Soc 130:804–805
46. Nasir NM, Ermanis K, Clarke PA (2014) Org Biomol Chem 12:3323–3335
47. Fuwa H, Naito S, Goto T, Sasaki M (2008) Angew Chem Int Ed 47:4737–4739
48. Fuwa H, Saito A, Naito S, Konoki K, Yotsu-Yamashita M, Sasaki M (2009) Chem Eur J 15:12807–12818
49. Fuwa H, Saito A, Sasaki M (2010) Angew Chem Int Ed 49:3041–3044
50. Hong SC, Sanders HP, Lee CW, Grubbs RH (2005) J Am Chem Soc 127:17160–17161
51. Tu W, Floreancig PE (2009) Angew Chem Int Ed 48:4567–4571
52. Houri AF, Xu Z, Cogan DA, Hoveyda AH (1995) J Am Chem Soc 117:2943–2944
53. Yang Z, Zhang B, Zhao G, Yang J, Xie X, She X (2011) Org Lett 13:5916–5919
54. Garber SB, Kingsbury JS, Gray BL, Hoveyda AH (2000) J Am Chem Soc 122:8168–8179
55. Yu M, Schrock RR, Hoveyda AH (2015) Angew Chem Int Ed 127:217–222
56. Wang C, Haeffner F, Schrock RR, Hoveyda AH (2013) Angew Chem Int Ed 52:1939–1943
57. Tanak K, Watanabe M, Ishibashi K, Matsuyama H, Saikawa Y, Nakata M (2010) Org Lett 12:1700-1703 and references therein
58. Hoffmeister L, Persich P, Fürstner A (2014) Chem Eur J 20:4396–4402
59. Heppekausen J, Stade R, Kondoh A, Seidel G, Goddard R, Fürstner A (2012) Chem Eur J 18:10281–10299
60. Magauer T, Martin HJ, Mulzer J (2009) Angew Chem Int Ed 48:6032–6036
61. Willwacher J, Fürstner A (2014) Angew Chem Int Ed 53:4217–4221
62. Das SN, Rana R, Chatterjee S, Kumar GS, Mandal SB (2014) J Org Chem 79:9958–9969
63. Basu S, Ellinger B, Rizzo S, Deraeve C, Schürmann M, Preut H, Arndt H-D, Waldmann H (2011) Proc Natl Acad Sci 108:6805–6810

Top Heterocycl Chem (2017) 47: 57–110
DOI: 10.1007/7081_2015_143

Published online: 16 June 2015

Metathetic Synthesis of Common and Medium-Sized Lactones: The State of the Art

Mauro Bassetti and Andrea D'Annibale

Abstract This review describes the use of intramolecular olefin metathesis procedures for the synthesis of five- to eleven-membered lactones, with emphasis on the literature from the last five years. The application of diene ring-closing metathesis to the synthesis of natural lactones is discussed, including highlights of the reaction coupled to new synthetic tools.

Keywords Acrylates · Allylic alcohols · Butenolides · Homoallylic alcohols · Lactones · Pentenolides · Ring-closing metathesis · Ruthenium carbene catalysts

Contents

M. Bassetti (✉)
Istituto di Metodologie Chimiche del CNR, Sezione Meccanismi di Reazione, Dipartimento di Chimica, P.le A. Moro 5, 00185 Roma, Italy
e-mail: mauro.bassetti@uniroma1.it

A. D'Annibale (✉)
Dipartimento di Chimica, Università degli Studi di Roma "La Sapienza", P.le A. Moro 5, 00185 Roma, Italy
e-mail: andrea.dannibale@uniroma1.it

1 Introduction

Among the natural organic compounds, lactones of different ring size represent one of the most important classes. This is mainly due to the wide occurrence of lactones in nature, as well as to the relevance of their biological and pharmaceutical properties. Even though it is a hard challenge to give a thorough survey of their presence as natural products, it still is useful to establish a relationship between the lactone ring size and the occurrence in nature [1].

With this regard, a particular attention must be given to five- and six-membered lactones, since they are by far the most abundant, showing at the same time a variety of different biological activities. One of the main activities of chiral γ- and δ-lactones is related to their strong odorous properties. Indeed, they are present as odorants in many food categories, like meat products, milk, butter and dairy derivatives, vegetables and fruits [2–4], as well as in wines and spirits [5–8] (Fig. 1a–d). The same olfactory properties of these compounds account for their occurrence as pheromone molecules designated to the sexual communication between insects (Fig. 1e, f) [9].

As a matter of fact, five and six-membered lactones are an important part of the defensive system in plants, as witnessed by sesquiterpene lactones, mainly found in plants of *Compositae* family, but still detectable in a lot of different plant families

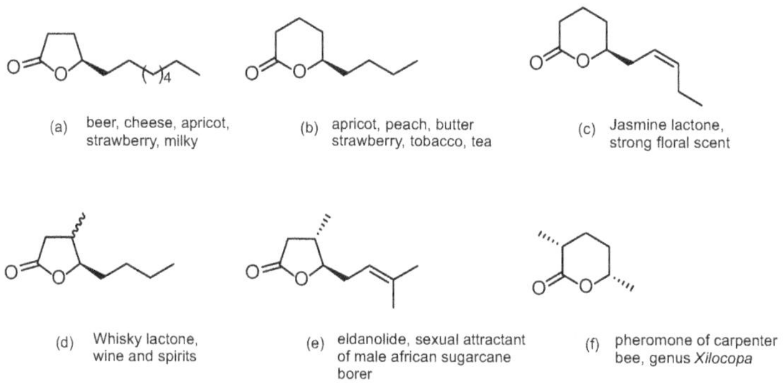

Fig. 1 Selected examples of five- and six-membered lactones active as fragrances and pheromones

[1]. These sesquiterpenoid lactones include more than 3,000 different structures, grouped into several main classes (Fig. 2), the majority of them showing important activities [10], such as cytotoxicity [11, 12], antifeedant [13], and allelopathic activity [14].

A structural feature shared by sesquiterpene lactones is an exocyclic alkylidene moiety alpha to carbonyl that proves to be fundamental to ensure their bioactivity.

The analogous but endocyclic α,β-unsaturated carbonyl moiety confers a wide range of properties also to conjugated butenolides [15, 16] and pentenolides [17]. Indeed, such lactone compounds are known to possess important antibacterial and cytotoxic properties as well as insecticidal, fungicidal, antibiotic, anticancer, antiinflammatory, antiallergic and antipsoriasis ones. They also behave as inhibitors toward cyclooxygenase and phospholipase A2. Selected examples of butenolides and pentenolides are shown in Fig. 3.

As outlined above, the explanation to such a wide range of bioactivities lies in the conjugated double bond functionality able to act as a Michael acceptor, thus making lactones prone to bind specifically to biological molecules such as enzymes and proteins, mainly through their thiol groups.

As regards seven-membered lactones, their natural presence is negligible with respect to the other smaller sized lactones, and just few examples of triterpenoid

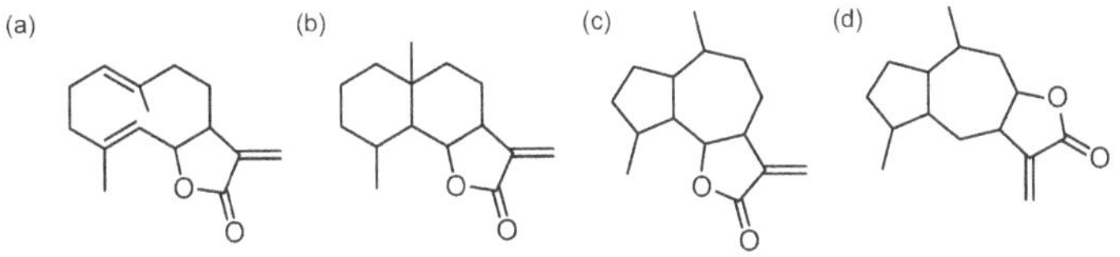

Fig. 2 General structures of sesquiterpene lactones: (**a**) germacranolides, (**b**) eudesmanolides, (**c**) and (**d**) guaianolides

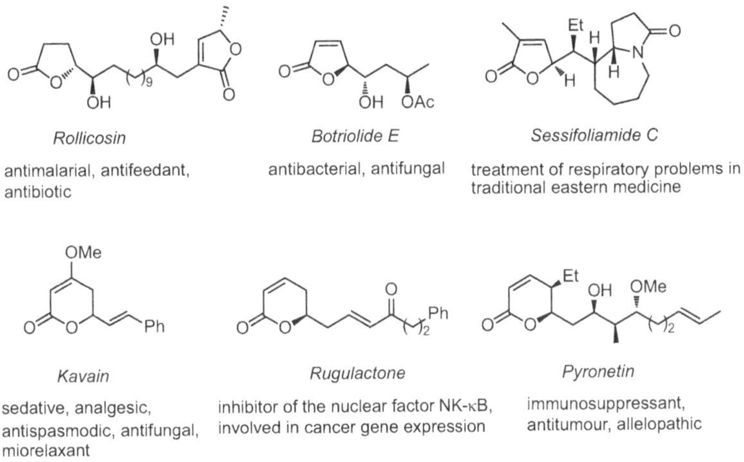

Fig. 3 Examples of conjugated natural butenolides and pentenolides

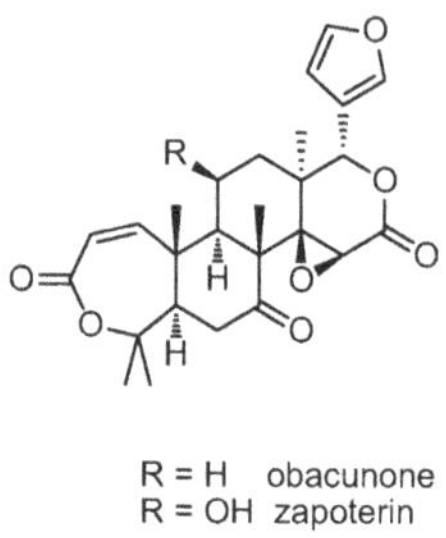

Fig. 4 Examples of natural seven membered lactones

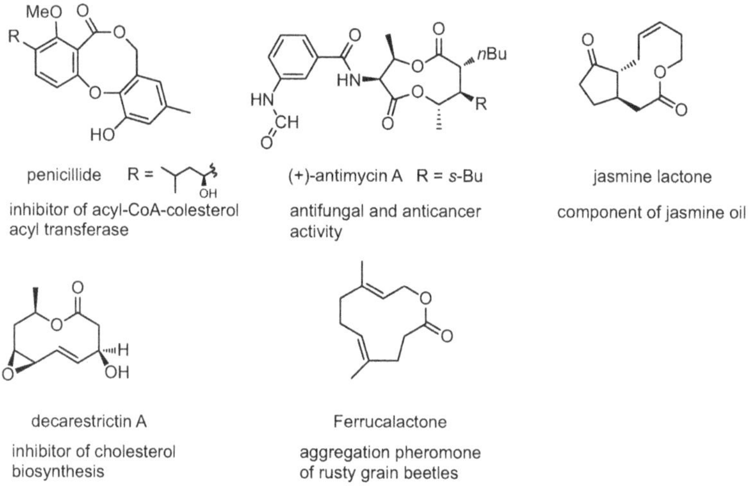

Fig. 5 Bioactive medium-size lactones

lactones such as obacunone and zapoterin, extracted from Citrus and other plants of Rutaceae family, can be mentioned, even though their biological anticancer properties [18] are quite interesting (Fig. 4).

Switching to medium-sized lactones, that is, the ones from eight- to eleven-membered rings, the number of products isolated from natural source is not large, and literature about both presence in nature and bioactivities is excellently reviewed [19–21].

Eight-membered lactones isolated from natural source are no more than a few tens, and the nine- and eleven-membered lactone families only count a few cases. Biological activities found among the members of these classes are either low or have been scarcely investigated. On the other hand, ten-membered lactones are far more numerous and present a wide range of biological properties [22]. Selected examples of bioactive members of these classes are shown in Fig. 5.

The importance of lactones is not only confined to the natural occurrence as bioactive compounds. Lactones are precious synthetic intermediates as well, and various methodologies were developed using lactones as starting materials [23]. If needed, lactones can be obtained easily in high enantiomeric purity and then used as chiral building blocks [24–28], en route to more valuable targets.

For all these reasons, in the last two decades the old-fashioned methodologies for the construction of the lactone ring have been revisited [29–36] and new synthetic tools have been developed [37–40]. Most often, the ring-closure reaction represents the crucial step in synthetic pathways to important natural products. In this respect, methods based on organometallic reagents or catalysts, such as carbonylation, carboxylation, and hydroacylation processes, have become increasingly important due to the possibility to control and direct the efficiency and selectivity of the lactone ring formation by appropriate tuning of the metal coordination sphere [41, 42].

When the development of well-defined ruthenium and molybdenum carbene complexes in the early nineties had disclosed the great potential of olefin metathesis (OM) in the synthesis of cyclic molecules [43–46], the ring-closing metathesis (RCM) approach to unsaturated lactones has witnessed increasing success. Various accounts have included details on the synthesis of lactones by RCM [47–52].

This review focuses on the recent developments in the synthesis of common (5–7) and medium (8–11) ring-sized unsaturated lactones by RCM including recent selected applications to the construction of complex molecular backbones and natural products. The nature and the role of the catalyst are highlighted, especially when the choice is critical for the reactions of substrates with electron deficient and/or sterically hindered double bonds or deactivating functional groups. To this aim, the structures of the ruthenium catalysts mostly used in alkene metathesis are shown at the beginning of this book. Furthermore, the cases of ameliorating and innovative synthetic tools used in conjunction with the metathesis reaction will be discussed.

The formation of lactones by RCM relies on the use of an open chain ester with terminal double bonds. This class of substrates is easily accessible by coupling of carboxylic acid derivatives with unsaturated alcohols. Currently, various protocols developed for lactonization reactions are used efficiently in the bimolecular version for the synthesis of diene esters [53]. This approach bypasses the synthetic difficulties implied in the end-group differentiation of α,ω-bifunctional molecules and facilitates the introduction of functional groups and stereocenters in the molecular skeleton. Furthermore, due to the nature of the RCM reaction, the position of the double bond in the ring with respect to the ester function, a structural element which can affect the bioactivities of the target molecule, can been regulated a priori by choice of the coupling partners. Therefore, the synthetic strategy based on RCM can be regarded as extremely versatile and amenable to the introduction of the desired functionalities within the carbon chain. As in the example in Scheme 1, the esterification reaction of acid and alcohol partners has been adopted in the synthetic route toward the ten-membered ring natural product cytospolide P [54]. The diene ester thus obtained features five stereogenic centers which are not affected in the subsequent RCM step.

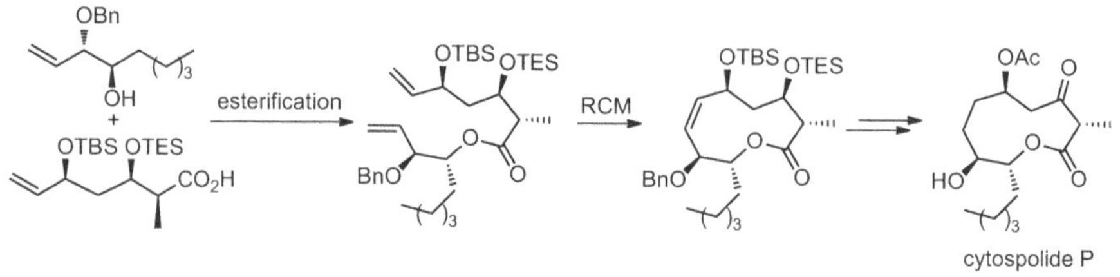

Scheme 1 Synthesis of a diene ester featuring five stereogenic centers and its RCM reaction en route to cytospolide P

2 Synthesis of Five-Membered Lactones

2.1 The "Acrylate Approach": The Catalysts

The preparation of conjugated butenolides by RCM involves invariably the ring closure of acrylate dienic esters. The involvement in the metathetic process of the acrylate conjugated double bond, in principle rather unreactive in the metathesis due to its low electron density, brings the reader directly to a very significant topic, that is, the strategies and tools needed for a successful outcome of the intramolecular alkene metathesis, when carried out on deactivated substrates. Another important important factor to be considered is the further reduction of the reactivity of the acrylate ester provided by an increasing alkyl substitution on either the conjugated or the other double bond. Since the synthesis of many lactones, no matter what their ring size is, can be approached starting from an acrylate ester, the findings described in literature about this issue are particularly significant. The experimental conditions, the specific catalyst, or the stratagems used for the metathetic transformation of acrylates have been recently reviewed [52]. The general concepts will be summarized here, by taking first into account the effects played by the carbene ruthenium complex. The first-generation Grubbs catalyst **G1** appeared to promote RCM of acrylates only with difficulty, thus requiring $Ti(O\textit{i}Pr)_4$ as a cocatalyst in order to reduce the inhibitory effect of carbonyl coordination on ruthenium [55]. Even in the latter case, a conformational predisposition in the acrylate diene was necessary for a successful ring closure. The second generation carbene complex **G2** gave by far better results. A comparison between the two catalysts was reported by Carda [56] in the RCM of acrylate **1** to lactone **2**, a precursor of (−)-muricatacin (Scheme 2a), and subsequently by Fujii [57] in the chemoenzymatic stereoselective synthesis of 5-alkyl substituted butenolides from acrylates **3** and crotonates **4** (Scheme 2b).

According to these results, the catalyst **G1** seemed to be unfit to promote the acrylate cyclization, in the absence of a Lewis acid. However, by performing the reaction under diluted conditions (0.01 M in ester), it was found that the cyclization of methallyl acrylates could be performed with satisfactory results [58], most likely due to suppression of concentration-dependent catalyst deactivation processes [59]. The expected lactones were obtained in good yields (Scheme 3).

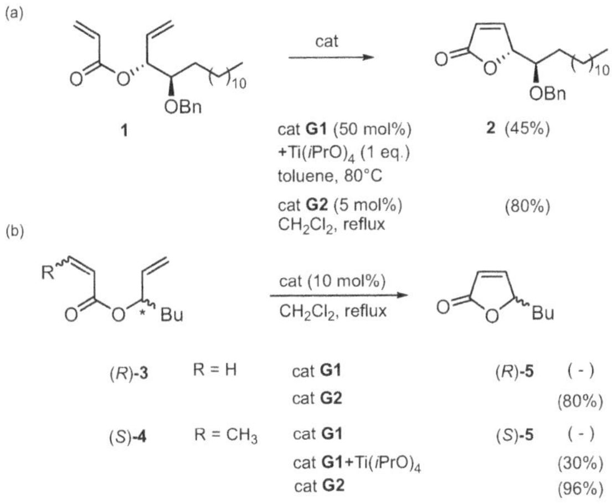

Scheme 2 RCM of allylic acrylates and crotonates by first- or second-generation Grubbs catalysts

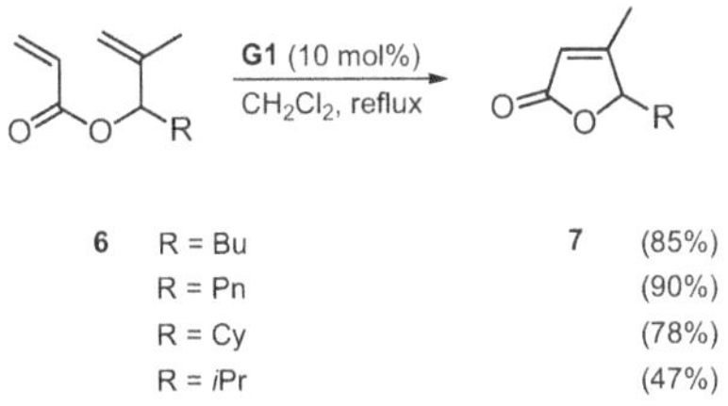

Scheme 3 RCM of methallyl acrylates in diluted conditions

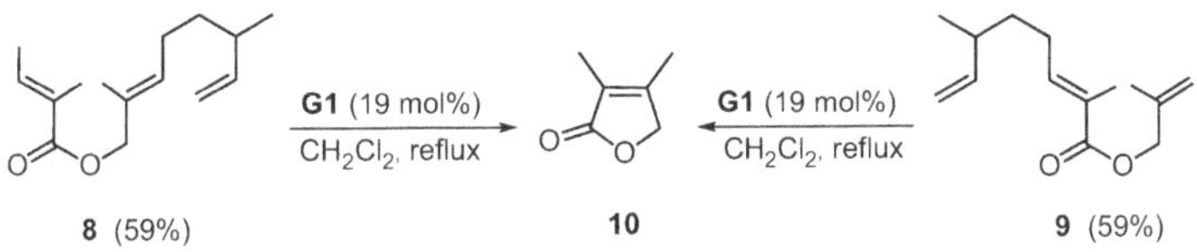

Scheme 4 Formation of 3,4-dimethyl-2-furanone **10** by RRCM

To overcome the reluctance of some double bonds to undergo intramolecular metathesis when the first-generation catalyst **G1** was used, Hoye and coworkers developed the so-called "relay RCM" concept (RRCM). In this case, the substrate is carefully tailored in such a way that the metal is guided on the double bonds through a series of cascade metathesis events [60]. This approach was used in the cyclization of two related starting materials, **8** and **9**, bearing two spatially close trisubstituted double bonds, to form the butenolide **10** in similar yields (Scheme 4).

Noteworthy, the synthesis of **10** is at the best of our knowledge the only case of a butenolide with a tetrasubstituted double bond ever obtained by RCM promoted by a first-generation Grubbs catalyst.

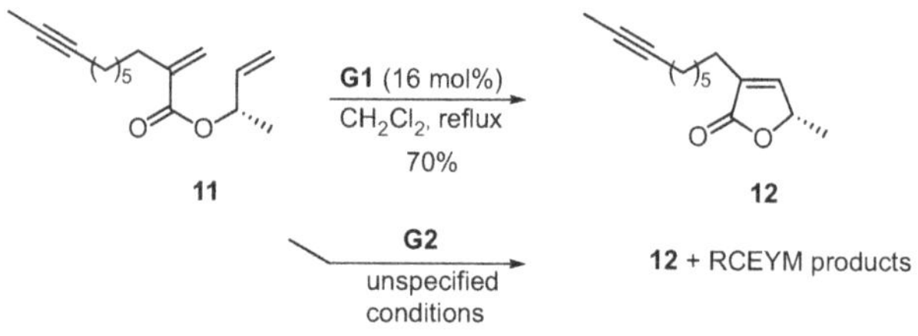

Scheme 5 Chemoselectivity in acrylate RCM

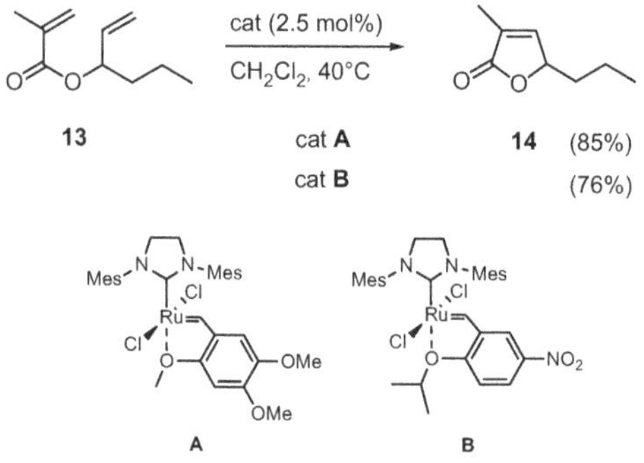

Scheme 6 Grela catalysts in the cyclization of methacrylate **13**

The increased robustness, activity, and functional group tolerance, general peculiarities of second-generation catalysts, resulted in major performances in olefin metatheses [61]. An important feature of these catalysts was the ability to promote the formation of tri- or even tetra-substituted cyclic olefins containing heteroatom functionalities [62–64]. In general, the carbene complex **G2** was the catalyst of choice for the cyclization of allyl acrylates, except when its activity triggered competitive or subsequent metathetic events. By way of example, the RCM of ester **11** proceeded with good chemoselectivity with catalyst **G1**, affording solely lactone **12**, while catalyst **G2** led in parallel to ring-closing enyne metathesis (RCEYM) byproducts (Scheme 5) [65]. It is worth mentioning that in the presence of a terminal alkyne, the same selectivity in favor of the diene metathesis can be achieved by protecting the alkyne moiety with $Co_2(CO)_8$ [66, 67].

By taking into account the common features involved in the RCM of allyl acrylates (high catalyst loadings, elevated temperatures and diluted conditions), a systematic investigation was carried out by Schmidt and Geißler in order to find the optimal conditions for the use of catalyst **G2** [68]. This work pointed out the occurrence of a critical starting concentration of the acrylate diene system and the irrelevance of excessively high catalyst loadings.

The Hoveyda–Grubbs catalyst **HG2** showed the same compatibility as **G2** with alkyl substitution on the vinylic carbons of acrylates, being particularly effective in the cyclization of methacrylates. The aromatic core present in **HG2** offered a further possibility of tailoring the catalyst structure and changing its reactivity by varying the electronic features of the isopropoxy-benzylidene ligand. This was the

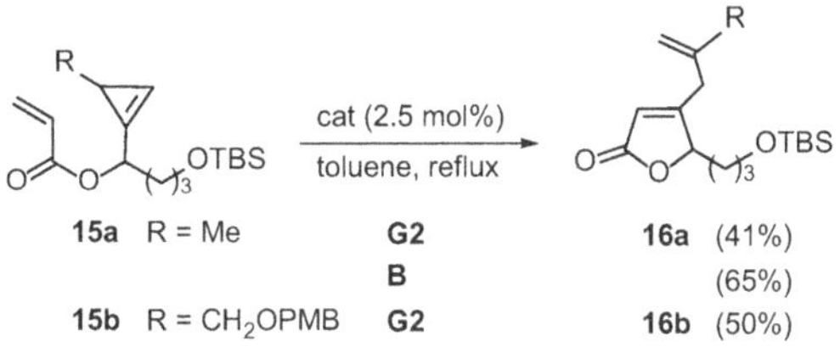

Scheme 7 ROM/RCM (RRM) of acrylates **15**

case of catalysts **A** and **B**, developed by Grela [69, 70], which successfully promoted RCM of methacrylate **13** to butenolide **14** (Scheme 6).

Grela's catalyst **B** proved to be effective also in the process of ring-rearrangement metathesis (RRM) of acrylates **15** bearing a cyclopropene ring, with formation of the conjugated lactones **16**. In scheme 7 a comparison of the effectiveness between the catalyst **B** and the most popular **G2** on substrate **15a** is reported [71].

2.2 The "Acrylate Approach": Synthesis of Natural Products

In this section we will focus our attention on the metathesis-based synthesis of valuable butenolides reported in the period from 2008.

The most important family of naturally occurring bioactive butenolides are the acetogenins, extracted from the Annonaceae plants typical of tropical and subtropical climates [72]. Acetogenins have cytotoxic activity, being strong inhibitors of mitochondrial respiratory chain complex I (NADH–ubiquinone oxidoreductase). They selectively cause tumor cell death, and they are also good antifeedant, antimalarial, antiparasitic, antimicrobial, and pesticide agents [73]. A really original RCM-based approach to the acetogenin skeleton was the recent convergent synthesis of pyranicin **17** [74], one of the few acetogenins bearing a tetrahydropyran moiety in its backbone, active against human pancreatic adenocarcinomal cell lines (PACA-2). The formation of the butenolide subunit of **17** was obtained by a relay RCM (RRCM) process of the chiral acrylate **18**. The reason for the choice of a relay strategy was in the ease of stereoselective preparation, isolation, and purification of the enantiopure allylic alcohol **19**, used as precursor of **18**. The RCM step involves the loss of dihydrofuran, and a subsequent CM step promoted by **HG2** finishes to build the pyranicin structure (Scheme 8).

As it can be seen in the structure of pyranicin, the acetogenins bear a lactone ring as the terminal group, a central hydrophilic core, in which hydroxyl groups and/or cyclic ether moieties can be found, and a long alkyl chain as the other end of the molecule. This is the case of muricatacin **20**, a potent antitumor agent isolated from the seeds of *Annona muricata*. In its structure only one hydroxyl group can be found in the central core, so the main challenge in the RCM-based synthesis was to obtain the metathesis acrylate precursor **22** with the right stereocenter configurations. This

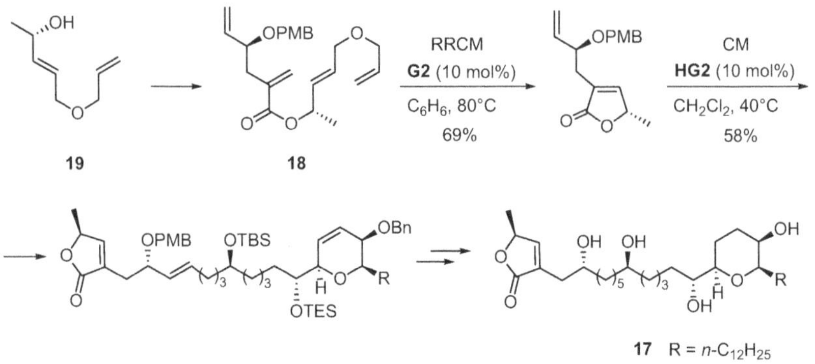

Scheme 8 Synthesis of pyranicin **17**

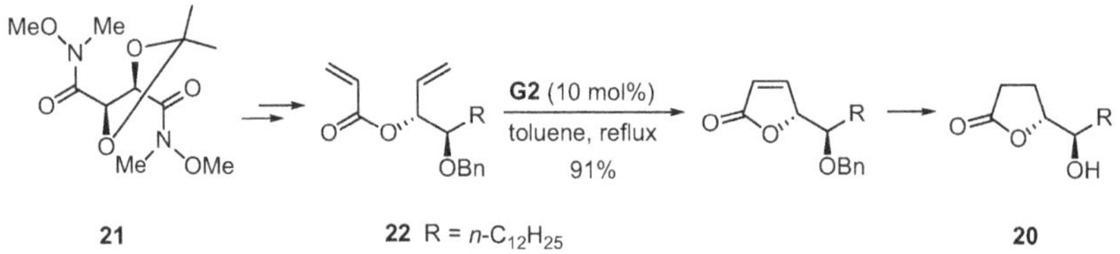

Scheme 9 Synthesis of muricatacin **20**

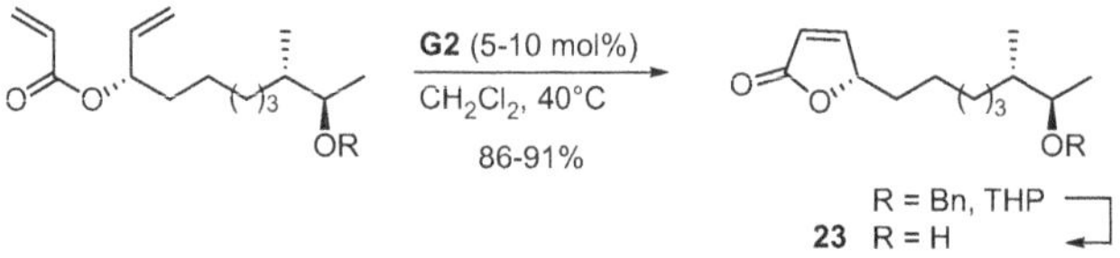

Scheme 10 Synthesis of marine butenolides **23**

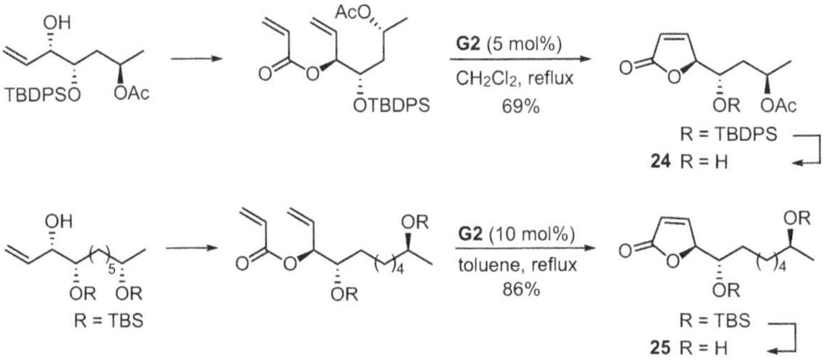

Scheme 11 Synthesis of botryolide E **24** and isocladospolide B **25**

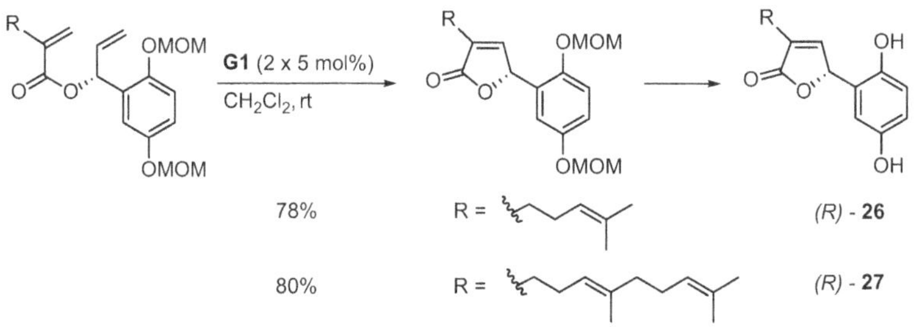

Scheme 12 Synthesis of fornicin A **26** and ganomycin I **27**

was done using the chiral pool approach, that is, starting from the Weinreb *bis*-amide **21** of L-(+) tartaric acid (Scheme 9) [75].

A similar structure, even if the single hydroxy group was at a bigger distance in the alkyl chain, was found in the marine butenolide diastereomers **23**, isolated from *Streptomyces* genus strains of marine sediments, all showing cytotoxic activity against humane leukemia K562 and murine lymphoma P388 cell lines. All the diastereomers were synthesized by metathetic processes (Scheme 10) [76–78].

The same hydroxylated side-chain motif can be found in two other similar products prepared by RCM, botryolide E **24** [79, 80], and isocladospolide B **25** [81] (Scheme 11). The first was extracted from cultures of a fungicolous isolate of *Botryotrichum* sp. (NRRL 38180) and is active against *Staphylococcus aureus* and *Escherichia coli*, while the second is a marine product isolated from Red Sea sponge *Cladosporium* sp.

In the last years, an increasing attention was given to the hydroquinones that were found as secondary metabolites from a wide number of microorganisms, due to their interesting bioactivities. In particular, prenylated hydroquinones fornicin A **26** and ganomycin I **27**, containing a butenolide in the prenylated side chain, were isolated from basidiomycete *Ganoderma* genus sp. The former compound is a cytotoxic agent against human larynx carcinoma (Hep-2) cells, while the second is an HIV protease inhibitor. Both of them were prepared by a RCM reaction, and it is worth mentioning the surprising ineffectiveness of the second-generation catalysts **G2** and **HG2** in the ring closure, while **G1** showed a nice performance, despite the *gem*-disubstitution on the conjugated acrylate double bond (Scheme 12) [82].

It is possible to find saturated and unsaturated γ-lactones also in alkaloid structures, just like the ones from Stemonaceae family plants. The RCM reaction allowed to build the butenolide system of sessifoliamide C **28** (Scheme 13a) [83], found in *Stemona sessilifolia*, a perennial herb used to treat respiratory diseases. An interesting synthesis of stemaphylline **29** and stemaphylline *N*-oxide **30**, insecticidal compounds from the root of *Stemona aphylla*, involved the RRCM of the ester **31**, which allowed to obtain in a tandem process both the butenolide system (to be further reduced) and the typical pyrrolo-[1,2-a]azepine core, specific of Stemona alkaloids, in the presence of the previously mentioned Grela catalyst **B** and two equivalents of camphorsulfonic acid (CSA) (Scheme 13b) [84].

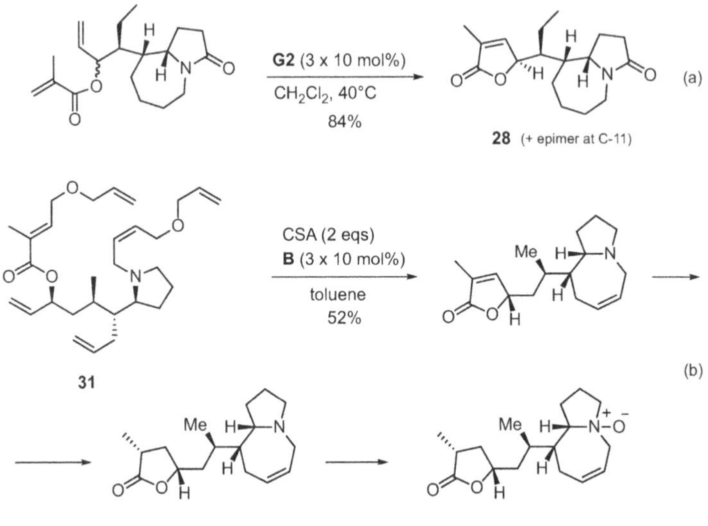

Scheme 13 Synthesis of sessifoliamide C **28**, stemaphylline **29**, and stemaphylline *N*-oxide **30**

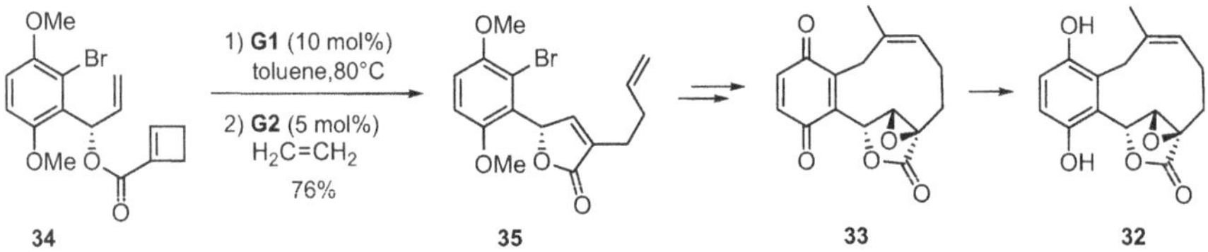

Scheme 14 Synthesis of clavilactones A **32** and B **33**

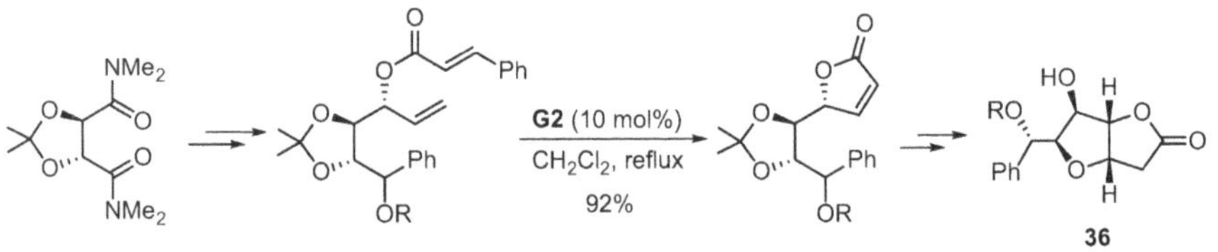

Scheme 15 Synthesis of crassalactone C **36**

In the next two syntheses of five-membered lactones, the final products involved will be targets in which the lactone ring is a part of a fused polycyclic system. The first example is the case of clavilactones A **32** and B **33**, isolated from the fungus *Clitocybe clavipes*, having antifungal and antibacterial properties. The key step in their metathetic preparation is a RRM (ROM/RCM sequence) on the cyclobutene carboxylate **34** in high-diluted solution in order to prevent dimerization of **35** (Scheme 14) [85].

The last case is the preparation of (−)-crassalactone C **36**, a styryl lactone extracted from *Polyalthia crassa*, starting from the *bis-N,N*-dimethylamide of tartaric acid [86] (Scheme 15).

3 Synthesis of Six-Membered Lactones

3.1 Acrylate-Type Ring Closure: The Catalysts

The cyclization of homoallyl acrylates to pentenolides catalyzed by first-generation catalyst **G1** seemed to proceed efficiently. Ghosh et al. reported several years ago a comparison of reactivity of allyl and homoallyl acrylates in RCM promoted by **G1** with or without $Ti(OiPr)_4$ as additive (see selected data in Table 1) [87].

The formation of six-membered rings appeared markedly favored with respect to that of five-membered analogs. The dienes featured unsubstituted terminal double bonds and the reaction involved a 10 mol% catalyst loading.

These data were confirmed by the work of three independent groups, and by way of example, we report in Table 2 the results obtained in RCM of the chiral homoallyl acrylates or crotonates affording 6-undecyl-5,6-dihydro-2-pyranone **37** [88–90].

Despite the positive results described so far, the **G1** catalyst seemed to be ineffective when the substrates were methacrylates, as reported by different authors ([91–93], and references therein). On the other hand, the same catalyst was found to be compatible in RCM involving homomethallyl acrylates **38** bearing a geminally substituted double bond. As shown by the yields of products **39**, the reaction was favored by alkylic substitution at the 1-homoallylic carbon, probably due to a suitable conformational arrangement in the starting diene (Scheme 16) [94].

In the same work, it was demonstrated by ^{1}H and ^{31}P-NMR that ruthenium complex **G1** interacts with acrylate **40** by selective attack at the homoallylic double bond with formation of a detectable alkylidene intermediate, which can evolve in

Table 1 Synthesis of α,β-unsaturated five- and six-membered lactones by RCM in the presence of catalyst **G1**

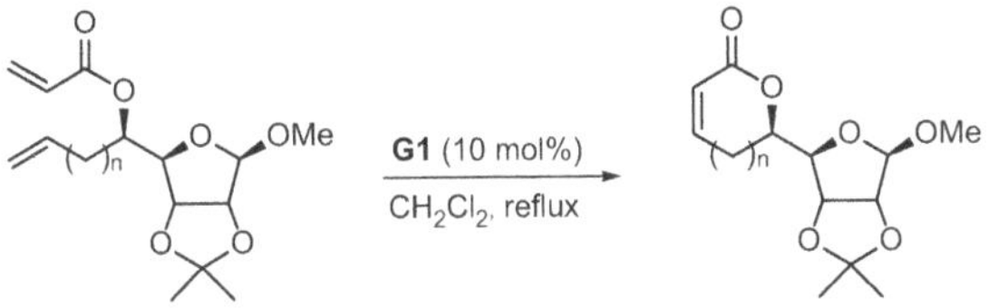

Entry	*n*	$Ti(OiPr)_4$ (eqs)	Time (h)	Yield (%)
1	0	–	15	40
2	1	–	10	88
3	0	3	15	72
4	1	0.3	2	94

Table 2 RCM of homoallylic acrylates or crotonate to afford 6-undecyl-5,6-dihydro-2-pyranone **37**

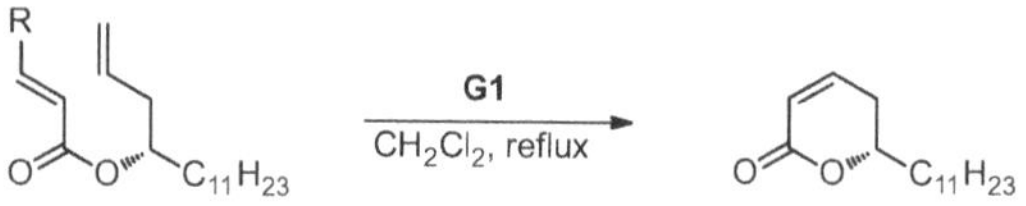

37

R	**G1** (mol%)	Ti(O*i*Pr)$_4$	Time (h)	Yield (%)	References
H	10	–	6	86	[88]
Me	15	0.3 eqs	24	70	[89]
H	10	0.3 eqs	15	93	[90]
H	10	–	15	50	[90]

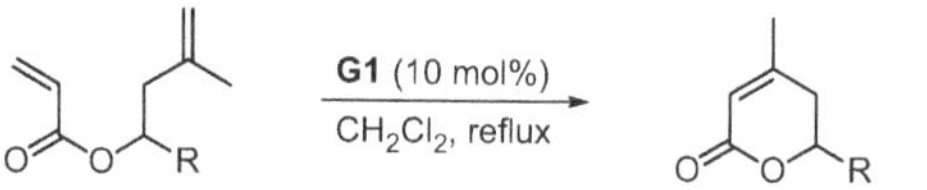

38 R = H
R = C_4H_9
R = C_9H_{19}

39 (20%)
(55%)
(74%)

Scheme 16 RCM of homomethallylic acrylates promoted by catalyst **G1**

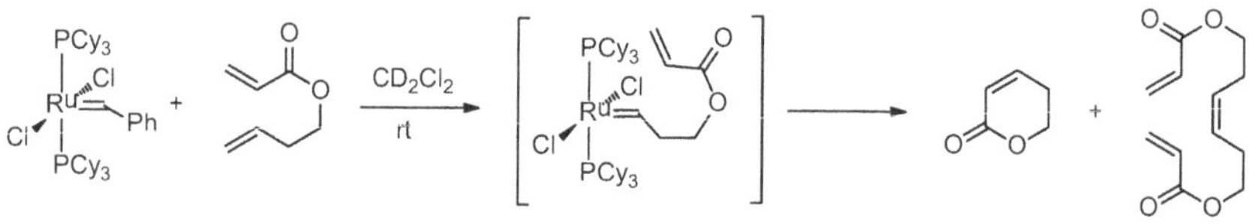

G1 **40** **41** **42** **43**

Scheme 17 Activation stage in the RCM of acrylate **40** promoted by complex **G1**

the product **42** or in dimer **43** (Scheme 17). The formation of homodimers accounted for the lactone yields being limited in the range 60–70%.

Even though the first-generation catalyst **G1** worked well in the RCM of homoallylic acrylates, especially when double bonds in the precursors were unsubstituted, the benefits arising from the use of the better performing *N*-heterocyclic carbene ruthenium complexes made them the catalysts of choice for the formation of six-membered lactones. Lower catalyst loadings (1–5 mol%), shorter reaction times (1–12 h), and most often absence of the cocatalyst Ti(O*i*Pr)$_4$ effectively counterbalanced both the higher commercial cost and synthetic effort in the preparation of the *N*-heterocyclic carbene ruthenium complexes.

A nice comparison of catalyst efficiency, **G1** vs **G2**, can be found in the literature about the key step in the synthesis of (+)-boronolide **48**, which is the RCM of structurally similar acrylates derived from differently protected polyols [95–100]. The results reported as an example in Scheme 18 for acrylates **44** [91]

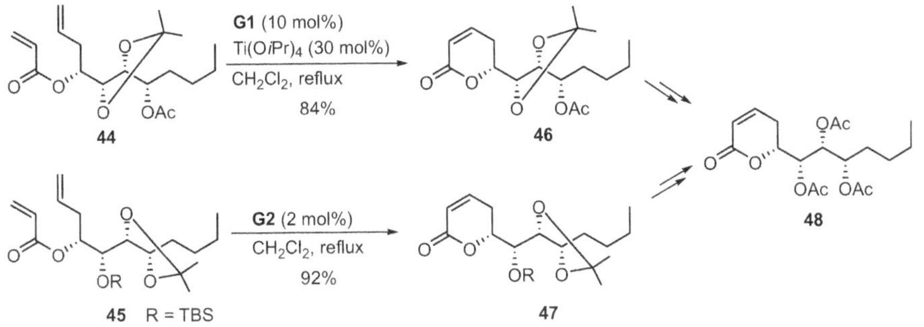

Scheme 18 The role of catalyst in the synthesis of (+)-boronolide precursors **46** and **47**

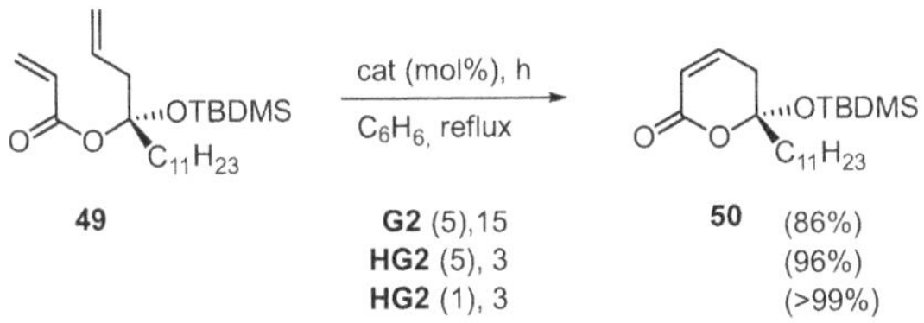

Scheme 19 Effectiveness of second-generation catalysts **G2** and **HG2** on RCM of model **49**

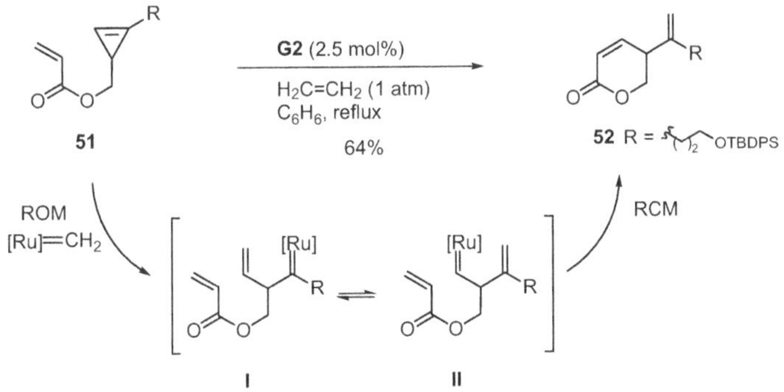

Scheme 20 RRM of cyclopropene-based acrylates **51** to lactones **52**

and **45** [92] clearly show that the use of **G2** involved a lower catalyst loading for the preparation of **47**.

The second-generation Hoveyda–Grubbs catalyst **HG2** has also been extensively used in the cyclization of homoallyl acrylates, in some cases exhibiting higher activity than **G2** even at very low loadings, as observed in the RCM of **49** to **50** (Scheme 19) [101].

A limit of the RCM of homoallyl acrylates is the creation of conjugated pentenolides with a low degree of functionalization. An alternative approach is represented by the ring-rearrangement metathesis (RRM) of cyclopropene-containing acrylates **51**, developed by Cossy et al. and depicted in Scheme 20

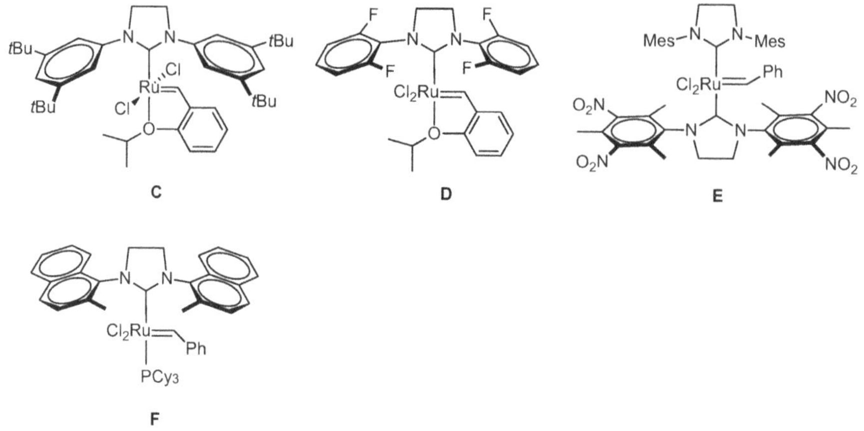

Fig. 6 Modified (NHC)-Ru carbenes as catalysts for the formation of sterically hindered olefins

[69]. This process leads regioselectively to lactones **52** with an exocyclic unsaturation available for further synthetic manipulations. The selective formation of **52** was explained by the authors as the result of an initial ROM process on the substrate, leading to equilibrating ruthenium carbene intermediates **I** and **II**, being the latter of which more stable and undergoing RCM to lactone **52**.

A very important issue is the reactivity of two *gem*-disubstituted alkenes toward formation of tetrasubstituted cyclic olefins. Thus, the structure of (NHC)-carbene ruthenium catalysts was modified in order to specifically address this matter [102, 103]. In Fig. 6 are reported just some selected new catalysts **C-F** developed in the last years [104–107]. The catalysts shown in Fig. 6 were tested in the RCM of the model compound **53** that has been adopted throughout the years for the evaluation of metathesis catalyst performance. The results obtained using NHC-ruthenium catalysts are reported in Table 3.

The use of complex **F** [102] in hexafluorobenzene or of the phenylindenylidene complex **M2** in combination with microwave irradiation (MW) in octafluorotoluene [108] allowed the conversion of the homomethallylic methacrylic ester **53** into the tetrasubstituted lactone **54** in 80% isolated yield. The results indicate a synergic effect of catalyst modified structure with the use of fluorinated solvents.

3.2 Acrylate-Type Ring Closure: Synthesis of Natural Products

The great relevance of RCM of homoallyl acrylates resides in the widespread occurrence of tetrahydrohydropyran-2-one and 5,6-dihydro-2H-pyran-2-one subunits in a huge number of naturally occurring bioactive compounds. The number of literature articles concerning the RCM-based syntheses of natural pentenolides

Table 3 Performance of second-generation catalysts in the formation of 3,4-dimethyl-5,6-dihydropyran-2-one **54**

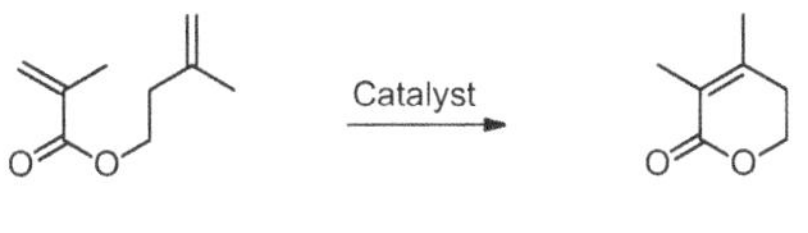

Catalyst (mol%)	Solvent	T (°C)	Time (h)	Yield	References
GN (5)	C_6H_5Me	80	40	42	[63]
G2 (10 + 10)	CH_2Cl_2	40	48	40	[92]
HG2 (5)	C_6D_6	60	24	nr[a]	[102]
C (5)	C_6D_6	60	24	nr[a]	[102]
D (5)	C_6D_6	60	24	nr[a]	[102]
E (0.5)	C_6H_5Me	80	3	26	[105]
F (5)	C_6F_6	80	12	80	[106]
M2 (2 + 2) MW	$C_6F_5CF_3$	120	0.5	80	[108]

[a]No reaction

starting from homoallyl acrylates are numerous in comparison to the analogous metatheses of allyl butenoates that afford nonconjugated pentenolides. It is worth pointing out that many of the six-membered natural lactones prepared by RCM of acrylates share some structural motifs, as well as biological activities. We will group them according to natural source or structural criteria, for a more concise description.

The group of pentenolides most spread across plant families, such as Lauraceae, Piperaceae, and Ranunculaceae, is represented by styryl lactones, bearing a styrene function at the end of a side chain linked to the C-6 position of the pentenolide core. Compounds of such group are strongly cytotoxic especially against cancer cell lines. Its most simple but still strongly bioactive representative is goniothalamin **55**, first extracted from *Cryptocarya caloneura*, which is a powerful antitumor agent without significant side effects on benign cells. Recently, this compound was made by an asymmetric synthesis of its acrylate precursor **56**, which undergoes a subsequent RCM catalyzed by **G1** (Scheme 21) [109].

Another important member of the styryl lactone family is kavain **57**, extracted from Kava plant (*Piper methysticum*), having analgesic, anesthetic, antifungal, antithrombotic, anticonvulsive, and muscle-relaxing bioactivities. The synthesis of **57** by RCM is peculiar since the metathesis precursor is the acrylate **58** bearing an enol ether function. In this case, the delicate metathetic matching between two double bonds with opposite electron density in compound **58** called for repetitive additions of 5 mol% portions of **G2** until a total 55 mol% loading, to afford the intermediate **59** (Scheme 22) [110].The amazingly high loading of catalyst **G2** is most probably due to the low reactivity of vinyl ethers in alkene metathesis processes [62].

Styryl lactones, just like cryptopyranmoscatone A1 **60**, cryptomoscatone D2 **61**, and cryptofolione **62**, are present also in another wide family of pentenolides, that

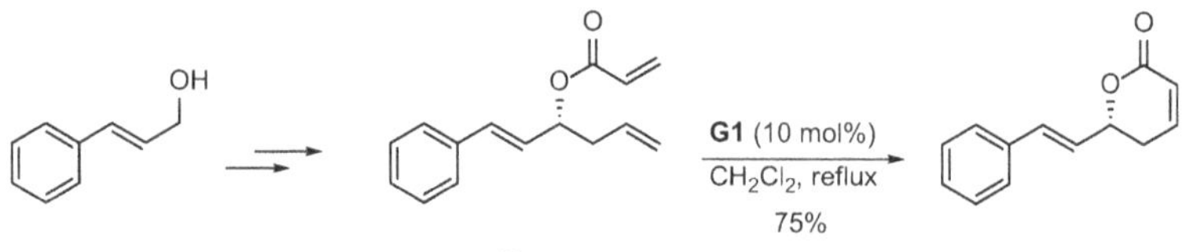

56

55

Scheme 21 Synthesis of goniothalamin **54**

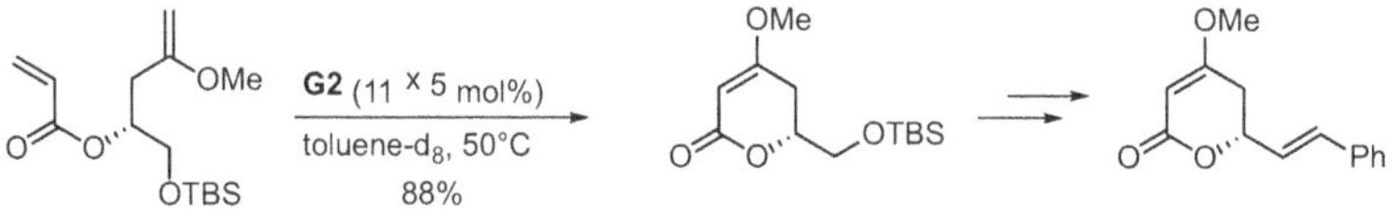

58

59

57

Scheme 22 Synthesis of kavain **57**

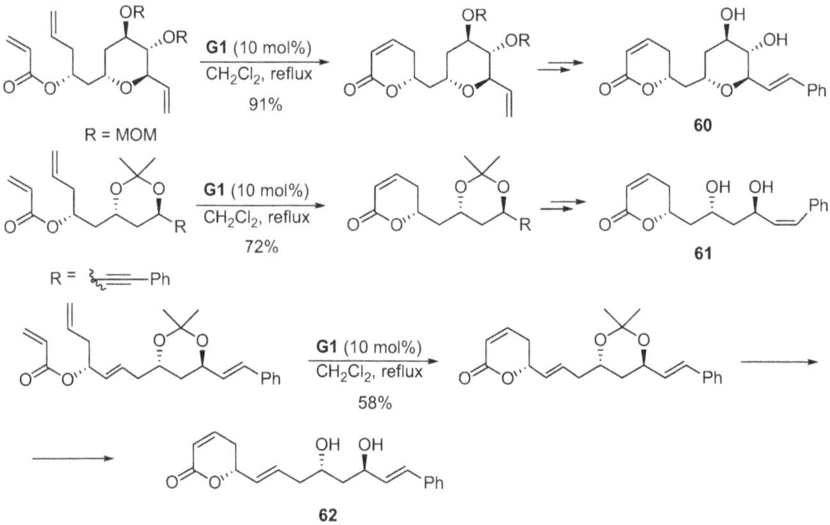

60

61

62

Scheme 23 Synthesis of compounds **60**, **61**, and **62**

is, the ones extracted from plants of genus *Cryptocarya* (Lauraceae family), spread mostly in tropical and subtropical regions. Their biological activities are important, compound **60** being an efficient inhibitor of cyclooxygenase-1 and cyclooxygenase-2 and **61** and **62** inhibitors of cell G2 DNA damage check point. Interestingly, in all the cases the RCM key step of their synthesis [111–113] was carried out with **G1** catalyst (Scheme 23), confirming once again the narrowness of the usage borders between first- and second-generation catalysts.

First-generation Grubbs catalyst **G1** was the catalyst of choice also in the synthesis of two styryl lactones substituted at ring C-5, bitungolides E **63** [114]

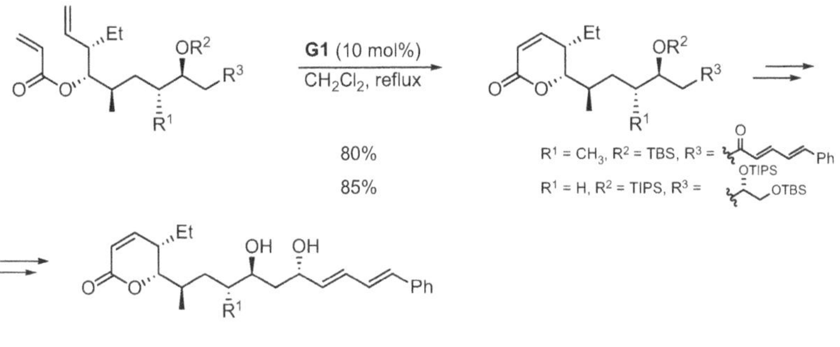

Scheme 24 Synthesis of bitungolides E **63** and F **64**

and F **64** [115, 116], two cytotoxic agents extracted from marine sponge *Theonella* cf. *swinhoei* (Scheme 24).

Phenyl group at the end of the chain seems to be a crucial element in determining bioactivity in another group of compounds, whose six-membered lactone ring was prepared by RCM, that is, the anticancer and insect repellant goniodiol **65** [117, 118] and rugulactone **66** [119], inhibiting activation pathway of nuclear factor (NK-κB) involved in gene expression in cancer and diabetes. Another member of this group is the powerful antifungal lactone **67** from *Ravensara crassifolia* that was synthesized by RCM of the same exact acrylate by three independent groups [120–122]. The metathetic pathways for preparation of **65–67** are shown in Scheme 25. The comparison of catalysts, loadings, and yields of RCM in the case of compound **67** gives a clear evidence that in many synthetic pathways, reaction conditions for RCM are not really systematically explored.

Two other natural pentenolides with an aryl-terminated side chain that received much attention from a metathetic standpoint are the vasorelaxant agent dodoneine **68** [123–126] from *Tapinanthus dodoneifolius* and antifungal strictifolione **69** [127–129] from *Cryptocarya strictifolia*. In Scheme 26 selected metathetic pathways for the two compounds are reported, among which the chemoselective RCM on diacrylate **70** by first-generation catalyst **G1** en route to strictifolione.

In general, common feature of all lactones taken into consideration inside this section is the alkyl side chain placed at C-6 position of the lactone ring. This side chain can contain more than a single unsaturation, just like in coibacins A **71** and B **72**, isolated from marine cyanobacteria and showing antileishmanial and cytotoxic activity. The RCM-based synthetic sequences are summarized in Scheme 27 [130].

Otherwise the side chain can contain oxygenated functions, mainly hydroxyl groups, and this is the case of a quite large family of polyhydroxylated δ-lactones present in the plants of *Cryptocarya* genus. These compounds have important biological activities, such as antibacterial, antifungal properties, as well as activity against cancer and pulmonary diseases. Important members of this group prepared by metathetic pathways are cryptocarya diacetate **73** [131], cryptocarya triacetate

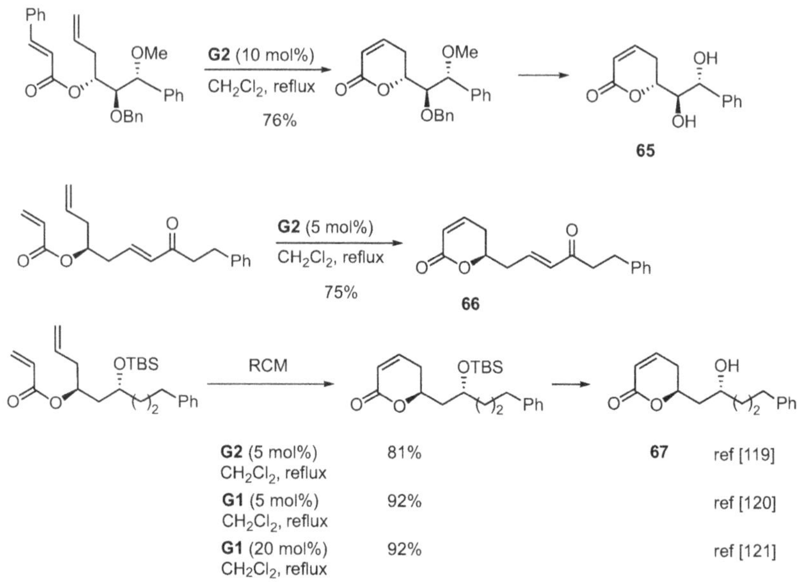

Scheme 25 Synthesis of goniodiol **65**, rugulactone **66**, and compound **67**

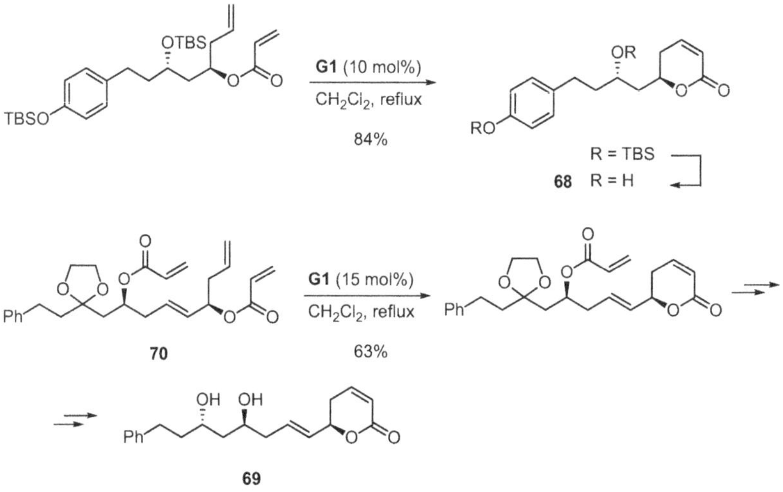

Scheme 26 Synthesis of dodoneine **68** and strictifolione **69**

74 [132, 133], cryptocaryolone **75** [128, 129], cryptocaryolone diacetate **76** [128, 129], and cryptocaryols A **77** [134, 135] and B **78** [131]. The latter two compounds gained synthetic importance due to their effect as stabilizers of programmed cell death PDCP4, that is, apoptosis. In Scheme 28 there are some selected RCM-based pathways to compounds **73–78**.

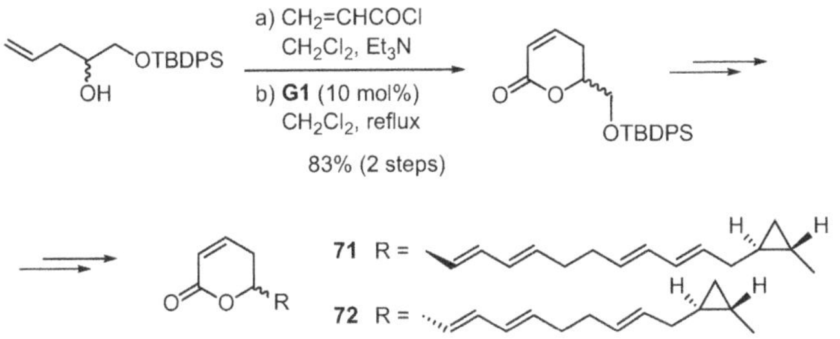

Scheme 27 Synthesis of coibacins A **71** and B **72**

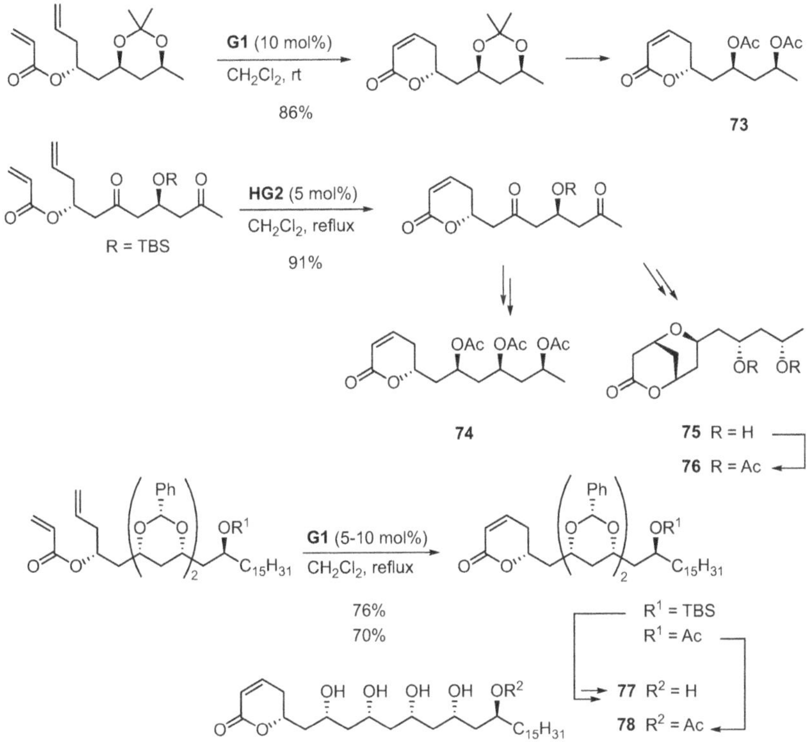

Scheme 28 Synthesis of lactones from Cryptocarya family **73–78**

Another lactone of the Cryptocarya group is (+)-obolactone **79**, from *Cryptocarya obovata*, showing activity against *Trypanosoma brucei brucei*. Its structure exhibits a styryl function instead of the typical oxygenated side chain. Its synthesis is reported in Scheme 29 [136].

A polyhydroxylated side chain, typical of the above mentioned cryptocaryols, is shown also by passifloricin A **80**, from *Passiflora foetida*, having antiprotozoal and antifungal properties. Some RCM-based syntheses for this compound using both

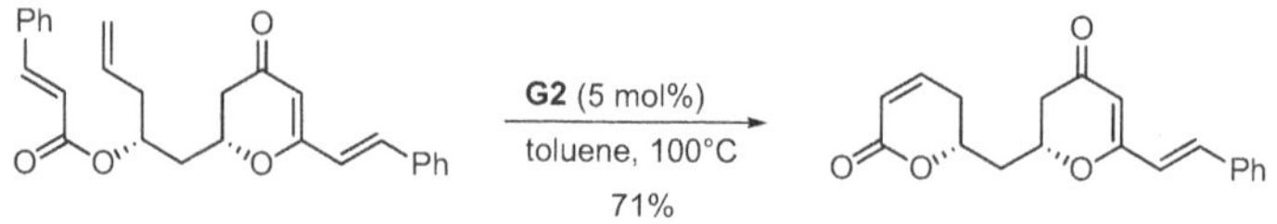

79

Scheme 29 Synthesis of obolactone **79**

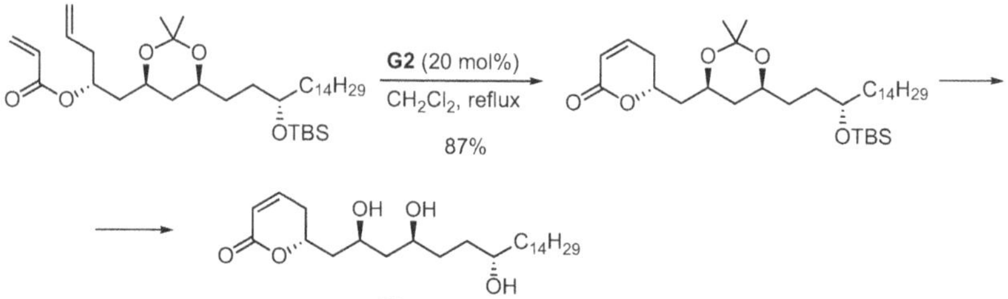

80

Scheme 30 Synthesis of passifloricin A **80**

G1 and **G2** were recently described [137–139], one of which is shown in Scheme 30.

Another crowded group of six-membered lactones being preferential targets for RCM are the ones extracted from plants of the genus *Syncolostemon* and *Hyptis*, in the Lamiaceae family. Such lactones have numerous important bioactivities, especially antimicrobial, antifungal, phytotoxic, and cytotoxic towards tumor cells. Pentenolides belonging to such class share a side chain with a double bond and several acetate groups in various positions. (+)-Hyptolide **81** [140, 141], (−)-spicigerolide **82** [142], synargentolides A **83** [143–146] and B **84** (present as a diastereomeric mixture) [147, 148], synparvolide C **85** [149], and (−)-synrotolide **86** [150] are selected δ-lactones belonging to this group and recently prepared by a RCM reaction as the key step. Metathetic syntheses of lactones **81–86** are shown in Scheme 31.

Very often, naturally occurring conjugated pentenolides prepared by RCM of acrylates show in their structure not only a functionalized side chain at C-6, as we have seen so far, but also a substituent at ring C-5. It could be a hydroxyl group, mainly masked as an ester, and this is the case for (−)-asperlin **87** [151], cleistenolide **88** [152–154], and pectinolides A **89** [155] (Scheme 32).

An alkyl substitution at C-5 is shown also by some members of the so-called "phoslactomycin family," that is, a group of antifungal, antitumour, and antibiotics lactones, isolated from the soil bacteria *Streptomyces* sp. The most important compounds in this group are fostriecin **90** and cytostatin **91**, strong inhibitors of phosphatase 2A, an enzyme implicated in the regulation of many crucial biological events including cell division, and this inhibition effect is supposed to be the key to their potent antitumor activity. Worth mentioning are also phoslactomycin B **92** and leustroducsin B **93**, produced by *Streptomyces platensis*.

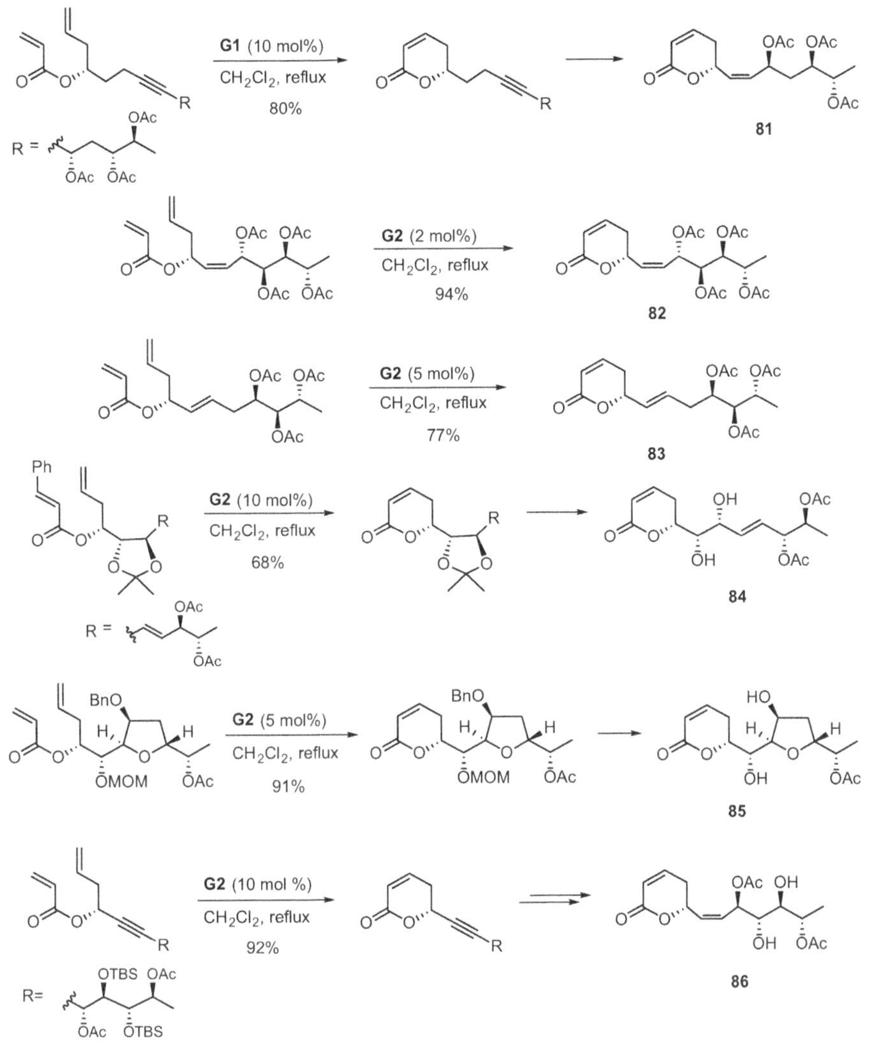

Scheme 31 Synthesis of lactones from Lamiaceae family **81–86**

RCM of acrylates was the key step to build the pentenolide subunit in many formal and total syntheses of fostriecin **90** [156–160], as well as in the preparation of cytostatin **91** [161], phoslactomycin B **92** [162–164], and leustroducsin B **93** [67] (Scheme 33).

The ethyl substituent at the pentenolide C-5, generally shown by phoslactomycin family members, seems to be a common structural feature for all the compounds produced by *Streptomyces* sp, as in the case of pironetin **94**. This compound is very important since it is a potent immunosuppressant with an activity comparable to cyclosporins, and it has a plant growth regulation activity. Moreover, it binds to the

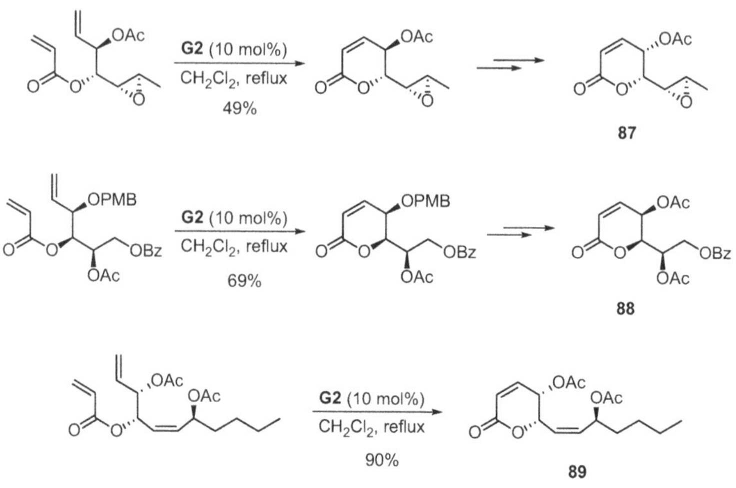

Scheme 32 Synthesis of compounds **87–89**

α-subunit of tubulin, thus presenting itself as a strong antitumor agent, and for this reason, it is an important target, also for RCM-based syntheses (Scheme 34) [165].

The last two applications of intramolecular metathesis to the synthesis of bioactive compounds that we report here involve the RCM under unusual reaction conditions for the preparation of compounds in which the pentenolide moiety is linked to other nonaromatic cyclic substructures. The first case is (+)-dumetorin **95** [166], isolated from tubers of *Dioscorea dumetorum Pax*, and used for a long time in African folk medicine, and obtained by RCM of an acrylate under flow chemistry conditions with PEG-supported second-generation Hoveyda–Grubbs catalyst. The second compound is EBC 23 **96**, isolated from *Cinnamon laubatii*, which is an inhibitor of the growth of androgen-independent prostate tumor cells. The preparation of **96** involves a RCM step aided by microwave irradiation [167, 168] (Scheme 35).

3.3 "Non-acrylate"-Type Ring Closure

Another possible approach to unsaturated six-membered lactones involves the RCM of allyl 3-butenoate, leading to nonconjugated 3,6-dihydro-2H-pyran-2-ones. The first RCM of allyl butenoates **97** either by catalysts **G1** or **G2** was reported in 2002 by Wang et al. as the key step of a synthesis of 2,6-dideoxysugars **99** (Scheme 36) [169].

The results obtained with the two catalysts were similar, except for the predictable case of the formation of a tetrasubstituted double bond (R^1=R^4=CH_3). Since then, the RCM of butenoates has been used in methodology and synthetic applications [170–180]. It is quite amazing that despite this kind of ring closure appears to

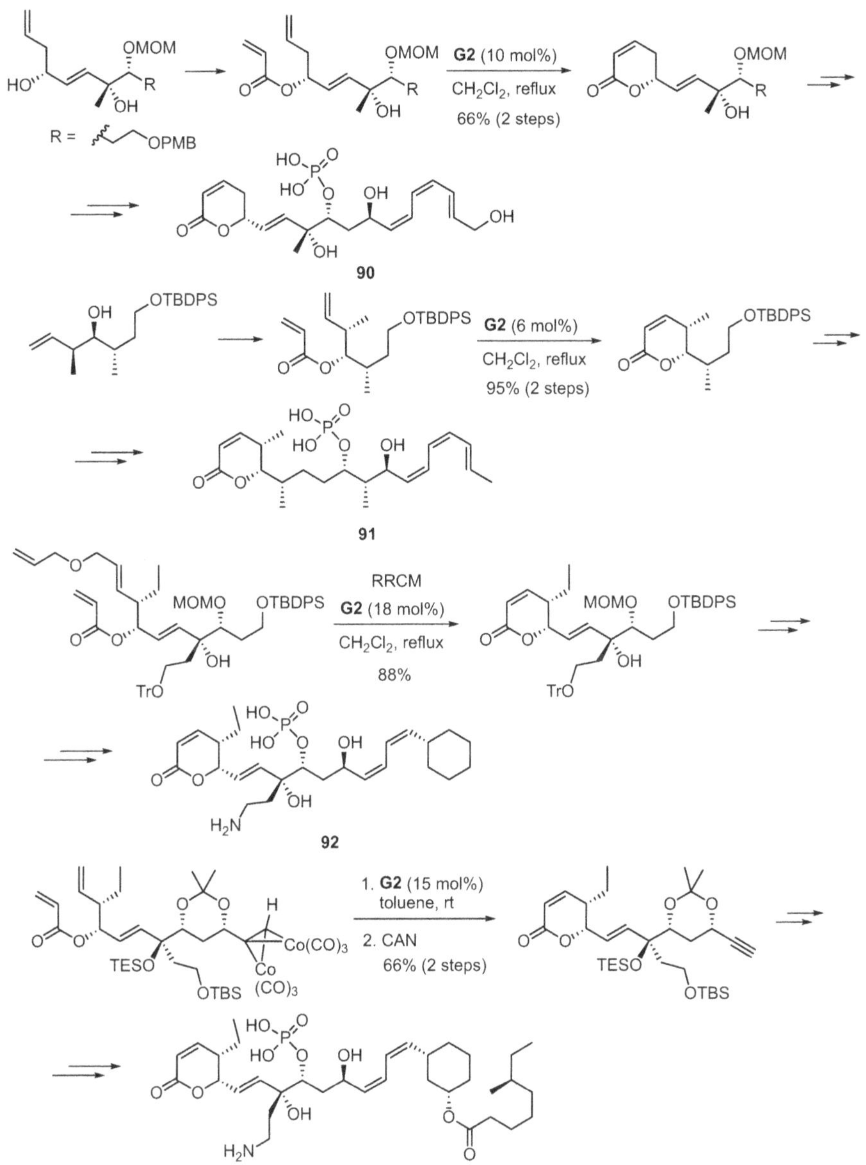

Scheme 33 Synthesis of selected lactones of the phoslactomycin group **90–93**

be not very demanding, nevertheless the literature items about this topic are by far fewer than the ones involving RCM of acrylate esters.

Very recently, the performance of thirteen different commercially available ruthenium catalysts was evaluated and compared in the RCM of the simple allyl butenoate **100**. Such screening was due to the importance of its RCM product, the

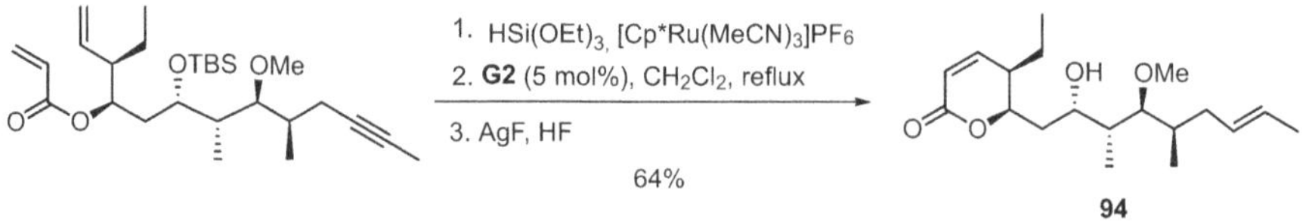

Scheme 34 Synthesis of pironetin **94**

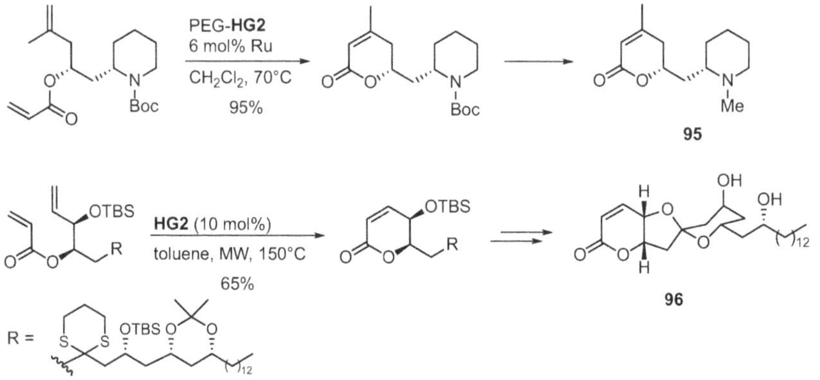

Scheme 35 Synthesis of dumetorine **95** and EBC 23 **96**

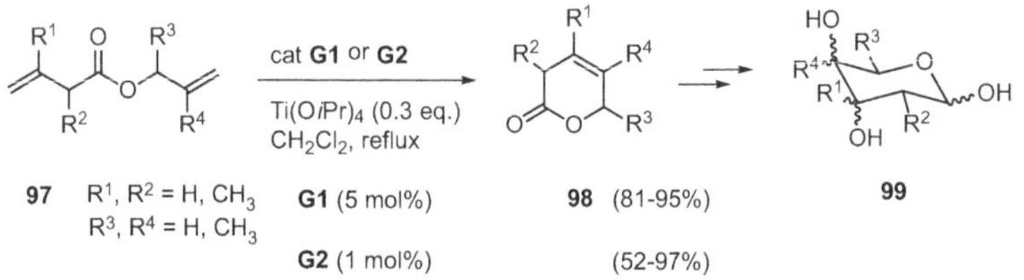

Scheme 36 RCM of allyl butenoates **97**

pentenolide **101**, that is the direct precursor of δ-decalactone, largely used in food and beverage industry. This study tried to clarify which was the best and fastest catalyst for this transformation in the light of the development of an industrial RCM-based synthesis of δ-decalactone. Some of the best performing catalysts used in this experiment are shown in Fig. 7.

The experimental parameters of this RCM reaction were chosen in view of convenient large scale processes. So, the catalysts were used in really low loadings (0.1 mol%) on one millimole of ester and for a short reaction time (2 h). Some selected data are reported in Table 4 [181].

A strong push toward the RCM of allyl butenoates came from the further possibility to isomerize the endocyclic double bond of the pyrone system formed by RCM in the presence of a suitable base (in most cases DBU and KO*t*Bu) or even in the presence of trifluoroacetic acid (Scheme 37) [182–184]. Such an

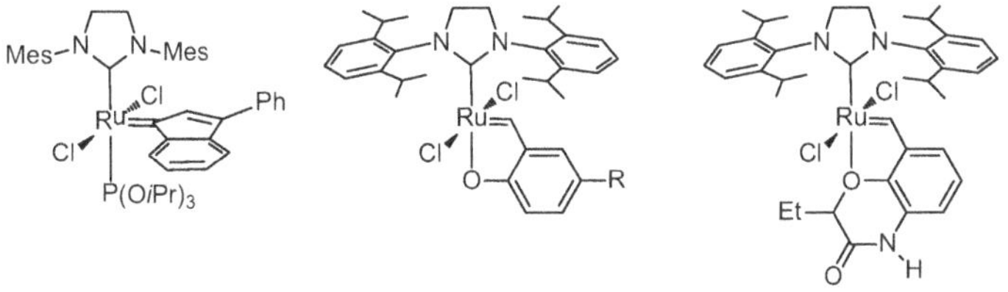

H **I** R = NHCOCF **J**

Fig. 7 Some catalysts screened in the RCM of ester **100**

Table 4 RCM of butenoate **100** with various catalysts

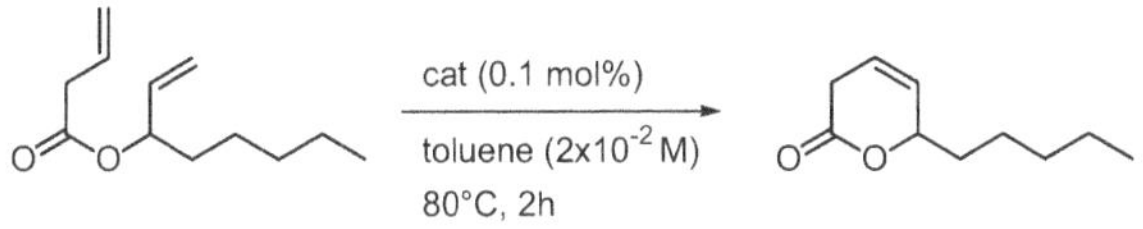

100 **101**

Catalyst	Lactone/ester ratio	Yield (%)
G2	47/53	nd[a]
HG2	79/21	56
H	>99:1	77
I	>99:1	87
J	88/12	77

[a]Not determined

isomerization allows to effectively turn γ,δ-unsaturated into conjugated valerolactones, thus overcoming all the problems related to the RCM of acrylates.

Indeed, in that way it was possible to afford the same conjugated lactones from allyl butenoates as well as from homoallyl acrylates, but using lower loadings in catalysts and higher concentrations in substrates. For these reasons, the sequence RCM/isomerization has been widely exploited in the synthesis of natural products or useful intermediates [185–188], for example, in the synthesis of polyrachitides A **101** and B **102**, secondary metabolites from the Chinese ant species *Polyrhachis lamellidens* (Scheme 38) [189].

The same isomerization was also observed in tandem RCM/CM of allyl butenoate **103** catalyzed by **G2** when prolonged reaction times were used, affording a mixture of lactones **104** and **105**, but interestingly in this case isomerization took place without the need to add any base or acid (Scheme 39). The same isomerization was not observed when **HG2** was the catalyst [190]. These findings were explained by supposing that metathesis catalyst **G2** could be turned by the prolonged warming into a ruthenium isomerization catalyst [191], probably through the conversion into a ruthenium hydride species.

In a recent paper, Schmidt et al. investigated the effect of various "chemical triggers," that is, species which are able to turn the Ru carbene into a Ru hydride [192]. The best trigger proved to be triethylsilane, affording conjugated pentenolide **107** from ester **106** in high yields (Scheme 40).

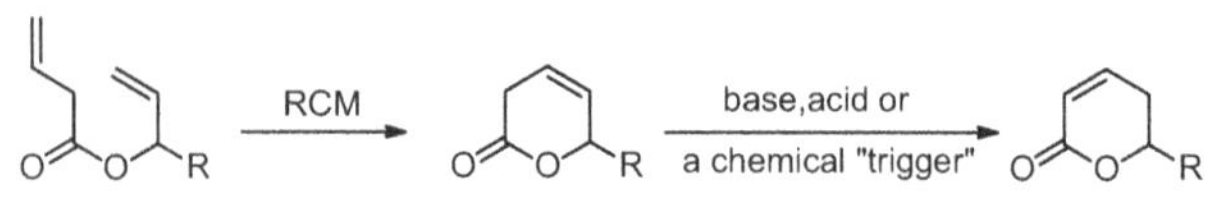

Scheme 37 RCM/isomerization sequence of allyl butenoates

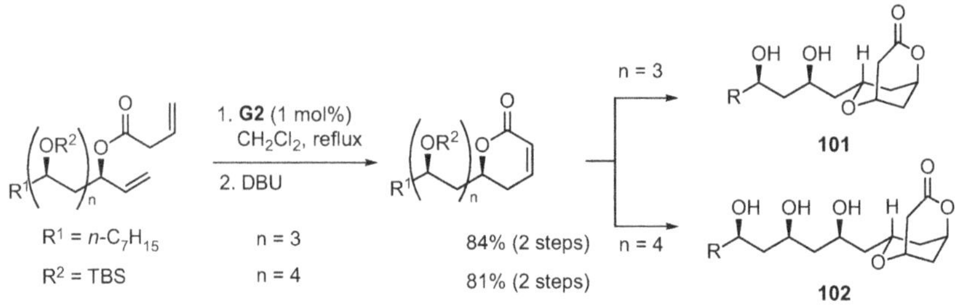

Scheme 38 Synthesis of polyrachitides A **101** and B **102** by RCM/isomerization

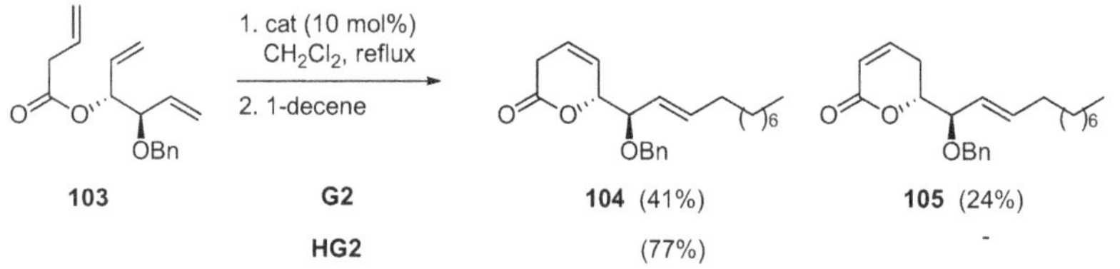

Scheme 39 RCM/CM isomerization of ester **103** in the presence of **G2** catalyst

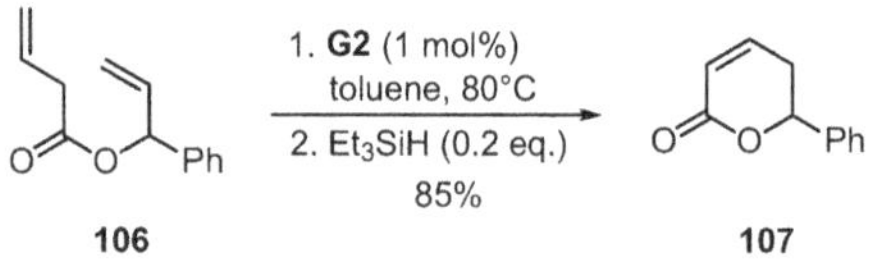

Scheme 40 Effect of chemical trigger on RCM/isomerization of ester **106**

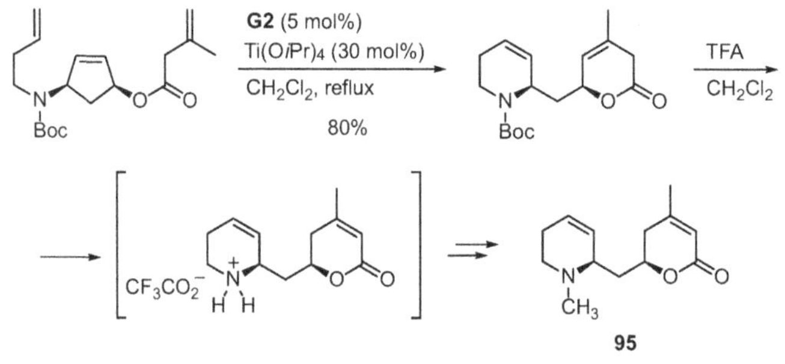

Scheme 41 Synthesis of dumetorine **95** by RRM/isomerization sequence

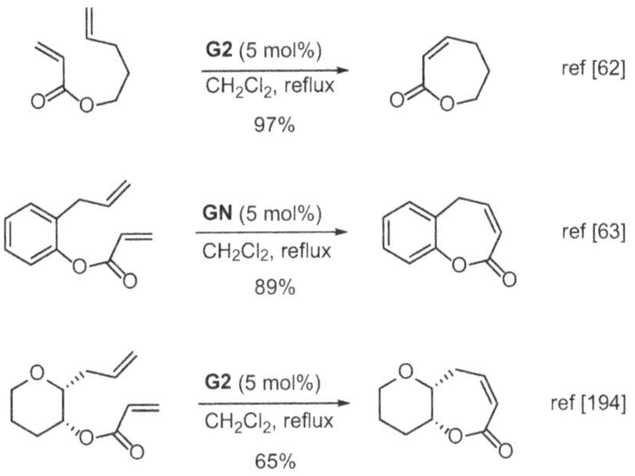

Scheme 42 RCM of 3-butenyl acrylates

The synthesis of dumetorin **95** by ring-rearrangement metathesis followed by double-bond isomerization promoted by an acid is shown in Scheme 41 [193].

4 Synthesis of Seven-Membered Lactones

4.1 Acrylate Ring Closure

The preparation of seven-membered lactones by RCM of acrylates was described only in the methodological studies of RCM by Grubbs et al. [62, 194] and Fürstner et al. [63]. Despite the positive results shown in Scheme 42, at the best of our knowledge, only a few synthetic applications have followed in the last years.

4.2 But-3-enoate Ring Closure

The RCM of 3-butenyl butenoate has been rarely used in the last years for the synthesis of bioactive compounds, and few papers have dealt with this topic. In the first one, the reaction of ester **108** with catalyst **G2** is described to afford the ε-lactone **109** [195], used as an intermediate in the formal synthesis of *trans*-kumausine **110**, present in nature in the red algae of the genus *Laurencia*. The molecule shows several important activities, such as antitumor, antimicrobial, immunosuppressant, antifeedant, and pesticidal properties (Scheme 43). The same lactone **109** was used in the same work in the synthesis of *cis*-Hagen gland lactones **111**, fragrances from some parasitic wasps which can be useful in biocontrol of fruit fly population.

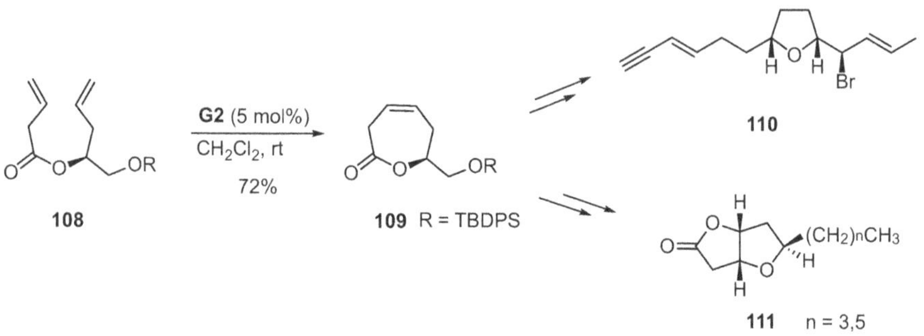

Scheme 43 Formal synthesis of compounds **110–111** by RCM of ester **108**

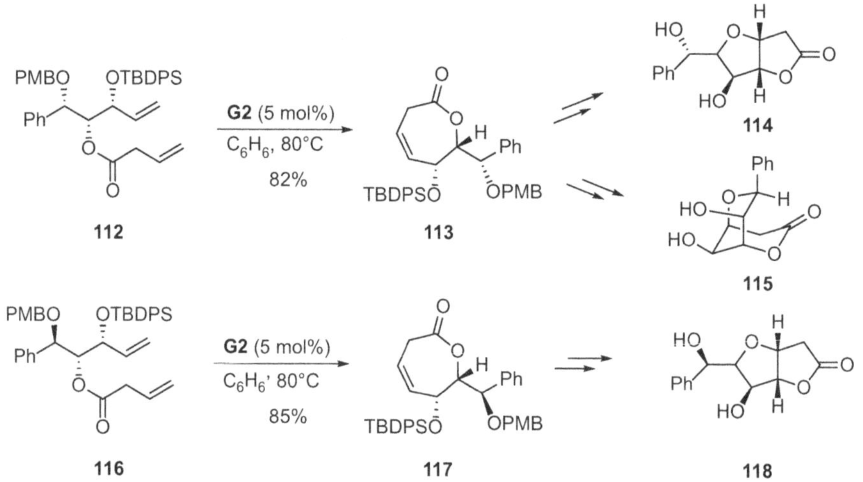

Scheme 44 Synthesis of compounds **114**, **115**, and **118** by RCM of **112** and **116**

The RCM carried out on the ester **112** and its diastereomer **116** afforded the seven-membered lactones **113** and **117**, respectively, used in the total synthesis of *epi*-goniofurone **114**, (+)-goniofurone **118,** and (+)-goniopyrone **115** [196] (Scheme 44). Compounds **118** and **115** are powerful antitumor agents.

The difficulties for a wider application of RCM to the synthesis of ε-lactones arise from the side reactions which can be involved in the process, such as the intermolecular CM and the acyclic diene metathesis (ADMET), yielding dimers or oligomers, which are favored by high substrate concentrations or specific conformational arrangements of the two alkene moieties. In 2008, Nguyen et al. reported on the possibility to prevent such intermolecular side reactions by "encapsulating" the RCM substrate within a bulky cone-like Lewis acid, aluminum tris (2,6-diphenyl)phenoxide (ATPH) [197]. As shown in Table 5, the effect of ATPH on the model substrate **119** was to favor the formation of the RCM product **120** over

Table 5 Effect of Lewis acid on the RCM/ADMET product ratio

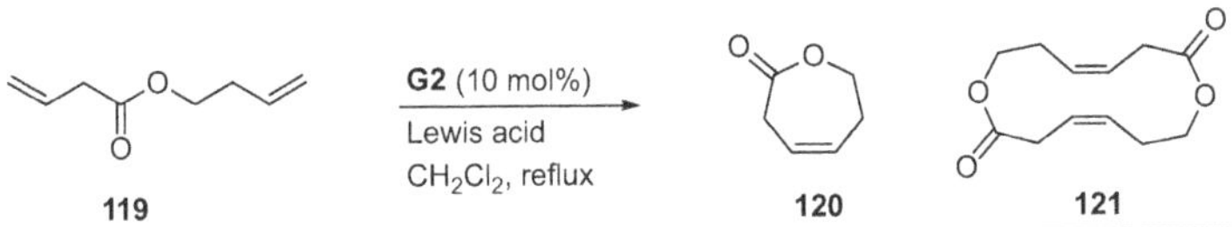

Substrate concentration (mM)	Lewis acid (eq.)	**120/121** molar ratio
20	Ti(O*i*Pr)$_4$ (1.05)	0/100
20	ATPH (0.1)	31/79
20	ATPH (1.05)	87/13
0.1	ATPH (1.05)	90/10

Table 6 Effect of ATPH in the minimization of ADMET product

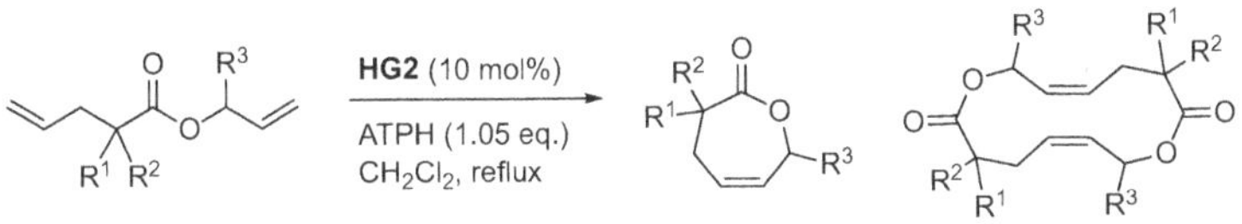

R^1	R^2	R^3	Yield of **123** (%)[a]	**123/124** molar ratio
H	H	H	60	97/3
H	CH_3	H	80	99/1
CH_3	CH_3	H	48	88/12
H	H	CH_3	53	93/7

[a]Isolated yield

the cyclic dimer **121** (Table 5). In the same work this approach was successfully extended to the preparation of γ,δ-unsaturated seven-membered lactones **123** from allyl pentenoates **122**, minimizing ADMET product **124** in the presence of the **HG2** catalyst (Table 6).

4.3 *Pent-4-enoate Ester Ring Closure*

As regards the cyclization of allyl pentenoates affording γ,δ-unsaturated ε-lactones, there are not many literature reports on the topic so far. In 2008, the synthesis of α- and β-glucopyranosyl serine **126** by RCM of ester **125** was reported by Nolen et al. (Scheme 45) [198].

The RCM process was then used to create derivatives of cyclic β-amino acids **128** bearing a carbohydrate side chain from the precursor **127** (Scheme 46) [199].

After optimization, the best conditions for RCM were found to be using catalyst **G2** in toluene at 80°C, for 20 h. Under analogous conditions, nine-membered aza-lactones were also obtained, although in lower yields.

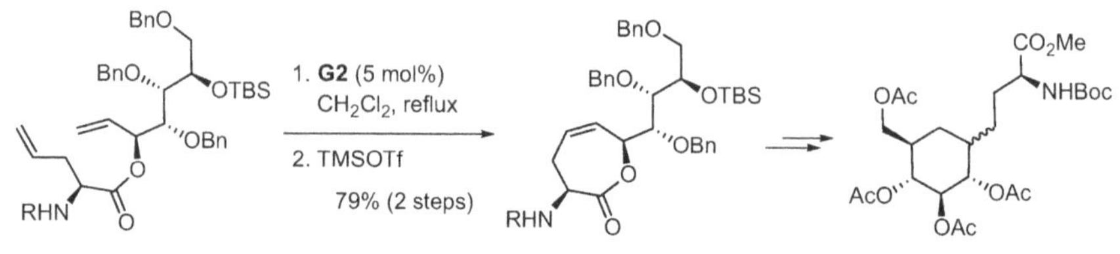

Scheme 45 Synthesis of diastereomers α and β of the glucopyranosyl serine **126**

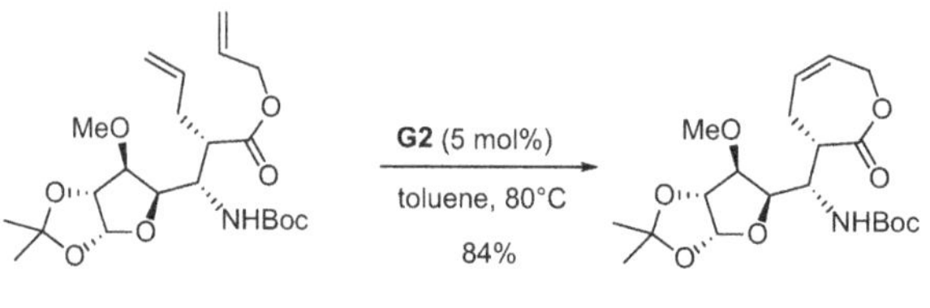

Scheme 46 Synthesis of β-amino acid cyclic derivative **128**

5 Synthesis of Medium (Eight- and Nine-membered) Ring-Sized Lactones

Medium-sized lactones are structural motifs occurring in a variety of biologically active natural products [19], although much less abundant with respect to both common and macrocylic lactones. The developments in synthetic approaches toward eight- and nine-membered lactones, based essentially on lactonization methods from carboxylic acids or their activated derivatives and including various cases of total synthesis of naturally occurring compounds, have been described in the reviews by Shina [53] and by Longo Jr and coworkers [21].

It should be emphasized that the formation of medium rings, with respect to either smaller or larger size rings, is the most difficult to effect via the ring-closure reaction of α,ω-bifunctionalized molecules. With the aim to understand the dependence of ring formation on structural and energetic factors, different classes of reactions, including lactonization of ω-bromoalkane-carboxylate ions, were analyzed by the use of a kinetic approach in which the rate constants of ring- closure (k_{intra}) or the values of effective molarity ($\mathrm{EM} = k_{\mathrm{intra}}/k_{\mathrm{inter}}$) (The ratio between the first-order rate constant of an intramolecular reaction (k_{intra}, s^{-1}) and the second-order rate constant of the model intermolecular reaction (k_{inter}, $\mathrm{s}^{-1}\ \mathrm{m}^{-1}\ \mathrm{L}$) is called effective molarity (EM) and is expressed with the unit of concentration ($\mathrm{m\ L}^{-1}$): [200]) were compared against the ring size for rings with $n = 3$–23 [201]. The EM parameter represents a quantitative measure of the ease of ring closure and corresponds to the concentration of the bifunctional substrate below which the ring formation predominates over oligomerization. It was found that the activation energy involved in the formation of lactones is highest for eight- and nine-membered rings [202] and derives essentially from ring strain due to imperfect

Table 7 Formation of medium-ring lactones as a function of ring size (*n*), of related values of effective molarity (EM), and of substrate concentration

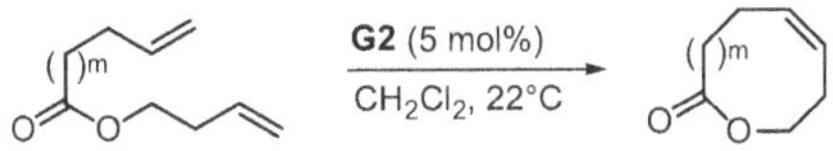

Substrate	m	[diene] (mM)	EM	Product	Lactone ring size *n*	% RCM[a]
129	0	0.50	55	**130**	7	78
131	1	0.05	0.57	**132**	8	0
133	3	0.50	1.9	**134**	10	95
133	3	5	1.9	**134**	10	41

[a]Determined at equilibrium

staggering, deformation of ring bond angles, and adverse transannular interactions. In such unfavorable cases, formation of oligomers is highly competitive. In the specific case of RCM of unsubstituted diene esters **129**, **131** and **133**, values of EM were compared with the molar percentages of oligomers and lactone for rings with $n = 7$, 8, 10 in the presence of catalyst **G2** [203]. Indeed, the formation of the expected γ,δ-unsaturated eight-membered lactone 3,4,7,8-tetrahydro-*H*-oxocin-2-one **132** from 3-buten-1-yl-4-pentenoate **131** failed even at substrate concentration as low as 0.05 mM, whereas lactones **130** ($n = 7$) and **134** ($n = 10$) formed in good yields (Table 7).

It is worth mentioning that performing reactions at low substrate concentrations represents a classical procedure which was developed in early preparative studies in order to favor intramolecular cyclization with respect to intermolecular processes and is based on the high dilution principle [204]. Even more significantly, the work of Fogg et al. showed that the oligomers generated in the early stages of the reaction are not adverse to RCM but rather represent key intermediates for efficient synthesis of medium or large rings starting from different classes of dienes. This is the result of a concentration dependent "backbiting" reaction of the initial ADMET products (Scheme 47).

The relative contributions of the direct RCM reaction and the indirect pathway (ADMET + CDP) toward the cyclic products were then evaluated in a series of *R*-but-3-enoate esters by a quantum mechanical approach in conjunction with thermochemical data, thus yielding the free energy changes associated with each route, ΔG_{RCM} and ΔG_{ADMET} [205]. It was found that formation of lactones is endothermic in all cases for *n* in the range 5–20 and that the eight-membered lactone and systems with rings greater than $n = 13$ imply the largest enthalpy cost.

5.1 8-Membered Lactones

In agreement with these considerations, the synthesis of eight-membered-ring lactones had been regarded as extremely challenging and naturally occurring medium-sized lactones as rare species of organic molecules [206]. However, with

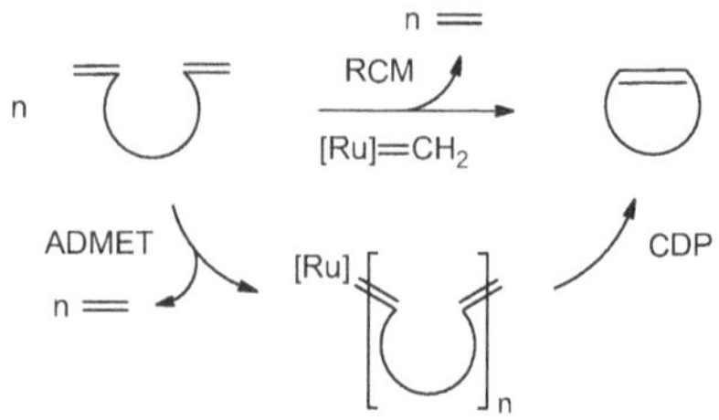

Scheme 47 Concurrent direct pathway by RCM and indirect pathway by ADMET and cyclodepolymerization (CDP) to cyclic alkenes promoted by $[Ru]{=}CH_2$ catalytic species

the advent of the second-generation Grubbs catalysts, the RCM reaction showed a good potential for the synthesis of the eight-membered lactone ring, at least comparable to lactonization procedures. In fact, the unsubstituted α,β-unsaturated lactone 5,6,7,8-tetrahydro-(3*Z*)-*H*-oxocin-2-one (**136**, 2-heptenolide) was obtained from the corresponding acrylate ester **135** by using a second-generation Grubbs catalyst (5 mol%) and isolated in high yield (92%) without apparent limitations with respect to the smaller α,β-unsaturated seven- **138** and six-membered **140** lactones (Table 8) [62]. Although this result may appear in contrast with the already mentioned lack of formation of the γ,δ-unsaturated analog **132** [203], it is very likely that conjugation between the carbonyl and the double bond in 2-heptenolide acts as a stabilizing factor during the transition state of ring closure. To the best of our knowledge, the unsubstituted β,γ- [200], δ,ε-, or ε,ζ-[207] unsaturated eight-membered lactones have not been prepared by RCM reactions.

The olefin metathesis approach first appeared on the stage of natural products chemistry in 2002 for the synthesis of the eight-membered lactone core of octalactins, which are potent antitumor agents (Fig. 8). (−)-Octalactins A **141** and B **142** have solicited considerable interests among organic chemists due to their structural complexity and interesting biological properties [53].

(−)-Octalactin A was obtained by Buszek and his group via a synthetic route involving the formation of an oxocene ring by RCM of acyclic diene esters. The RCM reactions were performed using the first-generation catalyst **G1**, in the absence of Lewis acid cocatalysts or of additional cyclic or conformational constraints in the diene **143** (Scheme 48) [208]. The yields of the eight-membered δ,ε-unsaturated lactone **144** and its various epimers (20–90%) are remarkable in consideration of the inherent difficulty of formation of such rings and are comparable to those based on a mixed-anhydride lactonization procedure affording the corresponding saturated ring systems [209].

The authors suggested that restriction of the bond angles due to the stereocenters caused a favorable disposition for ring closure. As a matter of fact, different epimers gave different yields of the corresponding unsaturated ring. This work represented the first example of RCM applied to a monocyclic eight-membered δ,ε-unsaturated lactone **144**, which was further elaborated to obtain (−)-octalactin A **141**.

Other examples of metathetic synthesis of natural octalactins were described in the following years. By using sequential Evans–Tishchenko and RCM reactions, eight-membered rings were obtained in excellent yields using **G1** (5 mol%) for the

Table 8 Formation of α,β-unsaturated six- to eight-membered lactones via RCM of acrylic esters

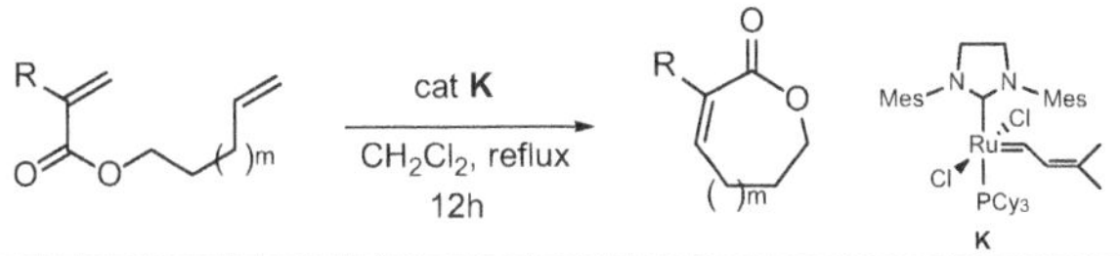

Substrate	m	R	[diene] (mM)	Cat **K** (mol %)	Product	Lactone ring size *n*	Yield (%)
135	2	H	20	4.8	**136**	8	92
137	1	H	20	5.6	**138**	7	97
139	0	Me	50	4.5	**140**	6	86

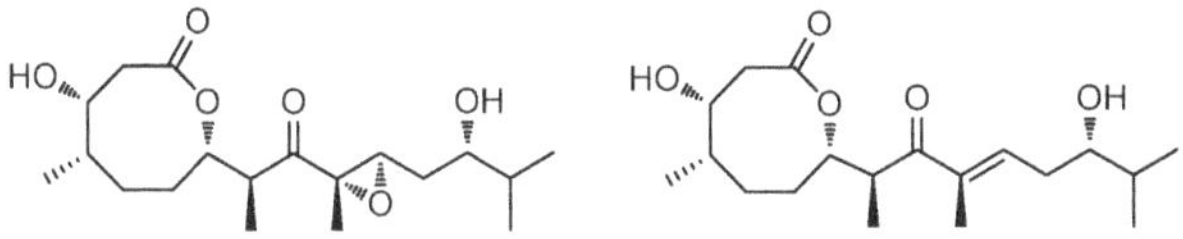

(−)-octalactin A **141** (−)-octalactin B **142**

Fig. 8 Structure of octalactins A and B

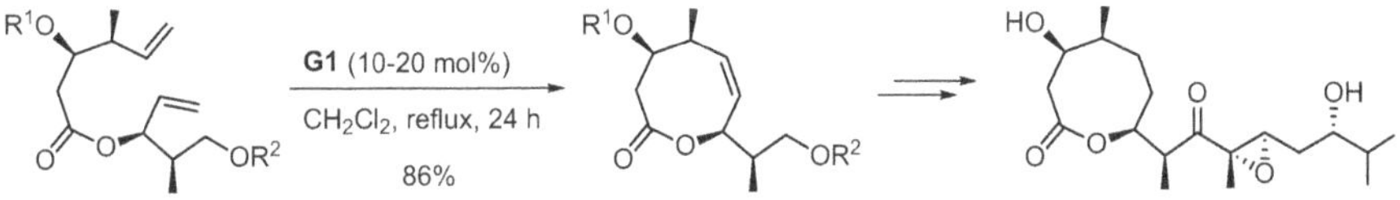

143 R^1 = MPM
R^2 = TBDPS
144
141

Scheme 48 Synthesis of the oxocene ring of octalactin A via RCM

RCM of a TBS protected diene **145** affording a γ,δ-unsaturated lactone **146** (99%) or **G2** (20 mol%) for the reaction of more demanding unprotected substrates **147** and **149** affording δ,ε-unsaturated lactone **148** and **150** (Scheme 49). The fully functional octalactin core was obtained after an eight-step linear sequence with 70% yield in the final RCM reaction using catalyst **G2** (20 mol%) in the presence of $Ti(O\mathit{i}Pr)_4$ (10 mol%) [210]. This Evans–Tishchenko–RCM approach provided a highly convergent strategy to the synthesis of medium rings.

Chattopadhyay and coworkers proposed a protocol starting from (*R*)-cyclohexylideneglyceraldehyde and involving its diastereoselective crotylation, thus furnishing building blocks with the desired stereochemistry suitable for further conversion [211]. The target δ,ε-unsaturated octalactin lactone **152** was obtained by RCM of **151** in 84% yield using catalyst **G2** (Scheme 50).

It is worth mentioning that the CM reaction has been used by Cossy et al. to construct the C5-C6 bond in the formal synthesis of octalactins A and B. In the initial steps of the synthetic route, the hydroxyl ester **153**, formed by sequential ozonolysis and enantioselective crotyl titanation of methyl-3-butenoate, was

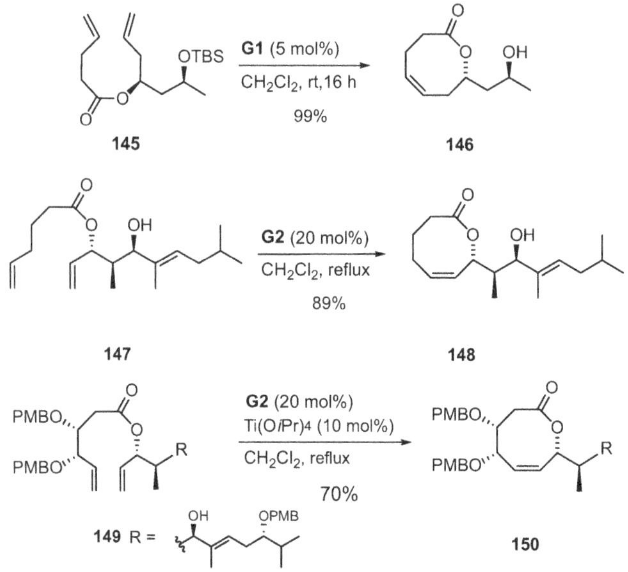

Scheme 49 Formation of eight-membered lactones and of the fully functionalized core of octalactin A by RCM

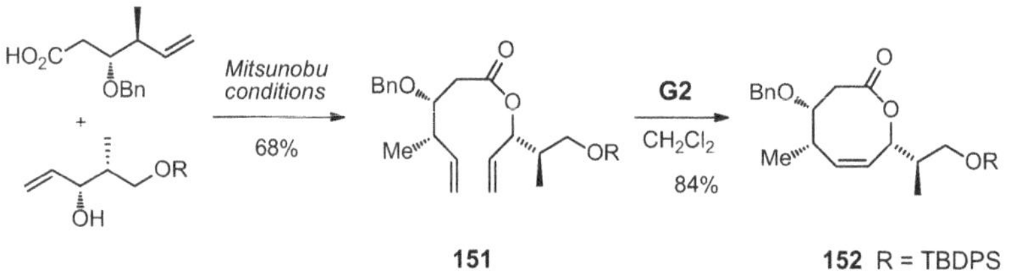

Scheme 50 Sequence of esterification (Ph_3P/DEAD/Et_2O:tol 2:1) and RCM reactions toward an advanced precursor of octalactins; R=TBDPS

subjected to CM with acrolein to afford the desired unsaturated aldehyde **154** (86%), in the presence of the Hoveyda–Grubbs complex **HG2** (Scheme 51) [212]. By applying this protocol in a subsequent work, (−)-octalactin B was obtained in 14.5% overall yield and 12 steps, as one of the shortest syntheses of the octalactin core [213].

5.2 *Nine-Membered Lactones*

A description of the early syntheses of nine- and ten-membered lactones by RCM can be found in the account by Deiters and Martin published in 2004 [47]. As in the case of the eight-membered ring, the formation of unsubstituted nine-membered

Scheme 51 Use of CM for the synthesis of an aldehyde precursor of (−)-octalactin B

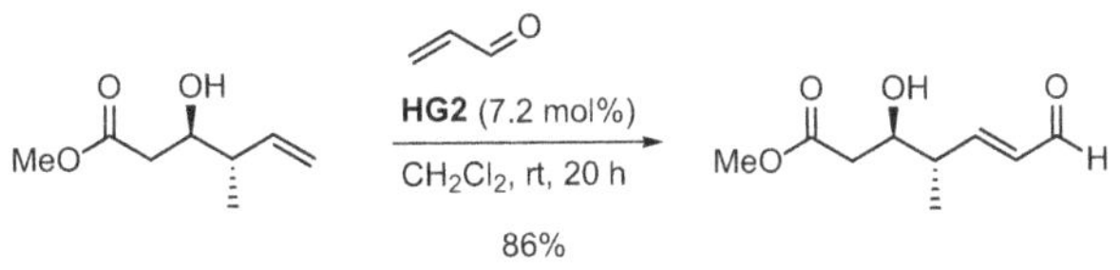

153 **154**

lactones by RCM appeared rather difficult. For example, treatment of the 1-ω-diene ester $CH_2{=}CH(CH_2)_2CO_2(CH_2)_3CH{=}CH_2$, pent-4-enoic acid pent-4-enyl ester, under typical metathetic conditions (**G1**, CH_2Cl_2, Δ) afforded only an inseparable isomeric mixture of unsaturated eighteen-membered diolides (30%), without traces of the nine-membered lactone [214].

It is worth recalling that the total syntheses of the naturally occurring halicolactone **156**, reported by Takemoto et al. [215, 216] and by Kitahara et al. [217], remained for several years as the only example of formation of a nine-membered lactone built by RCM (Scheme 52). The metathesis reaction was performed using a large loading of **G1** under high dilution conditions (0.1 mM), since higher concentrations of the diene **155** (e.g. 1.0 mM) afforded similar amounts of the desired lactone and of dimer (20–30%). The reaction was highly stereoselective since the corresponding *E*-isomer could not be detected under any circumstances. The role played by the protecting groups in the diene was not minimal. In fact, an analogous substrate bearing SEM and TBS groups in place of acetates gave a mixture of starting material (45%), nine-membered lactone (19%), and dimer (8%).

Since then, few cases of nine-membered lactones obtained by RCM have been reported in the literature. Functionalized nonanolactones were obtained by sequential esterification and RCM which resulted in stereoselective olefin formation, as the result of the interaction of homoallylic or allylic hydroxyl groups with the ruthenium metal center of the **G2** catalyst [218]. Furthermore, the precursor of the A/B ring analog of the antitumor compounds eleutherobin **157** and sarcodictyns **158a**,**b** (Fig. 9) was obtained by a diastereoselective Diels–Alder reaction of a dienophile containing a furanosugar moiety with cyclopentadiene followed by sequential ROM of the adduct **159** and RCM of the acrylate **160** affording lactone **161** (Scheme 53) [219].

(*Z*)-4,5,8,9-Tetrahydro-3*H*-oxonin-2-one is the core of naturally occurring topsentolides, which are cytotoxic oxylipins isolated from a marine sponge of the genus *Topsentia*. Topsentolide analogs were obtained by the use of an enantiodivergent approach involving sequential lipase-catalyzed kinetic resolution and RCM (Scheme 54) [220]. The formation of the nine-membered ring of (*S*)-topsentolide **163** required the addition in portions of both the allylic ester **162** and the **G1** catalyst and three consecutive cycles of 24 h each. Similar results were obtained starting from either the enantiomeric esters (65–70%) or from the racemate (98%). Surprisingly, the use of the **G2** catalyst with the racemic ester as starting material, afforded only the dimeric eighteen-membered macrolide **164**.

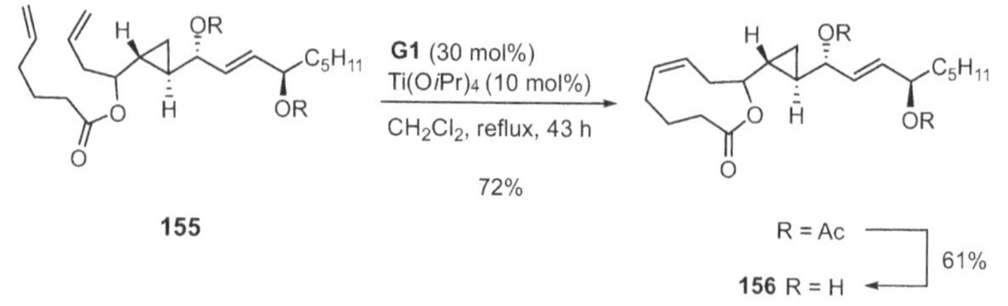

Scheme 52 RCM step in the synthesis of halicolactone

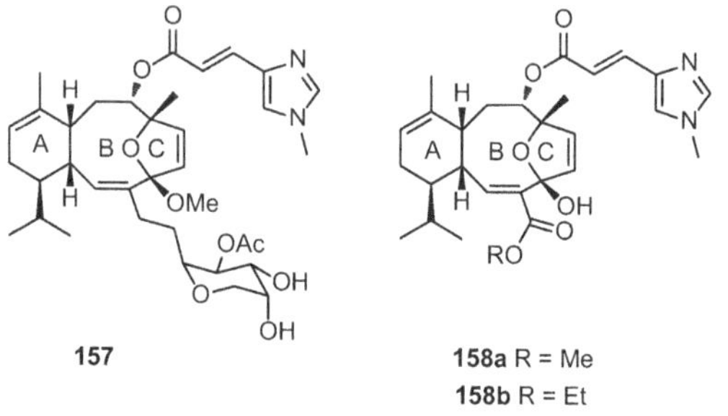

Fig. 9 Structures of eleutherobin **157** and sarcodictyns **158**

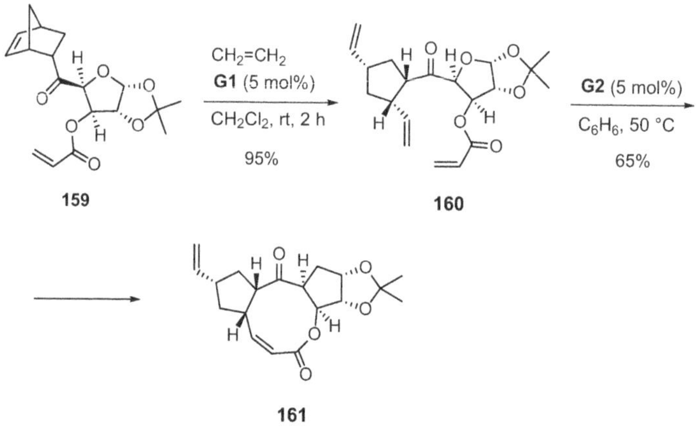

Scheme 53 Synthesis of the core structure of eleutherobin and sarcodictyns by ROM–RCM reactions

This result is due to increased activity of **G2** in CM with respect to **G1** [221]. The lactones exhibited poor activity against HL-60 cells.

By contrast, **G2** catalysts gave better results in the formation of functionalized nonanolactones [213] and in the first total synthesis of topsentolide B3 **166**. In this case, the formation of the nine-membered ring was accomplished in 78% yield in the presence of **G2** (10 mol%) (Scheme 55) [222]. It is worth pointing out that the internal double bonds in the ester substrate **165** remained inert under metathetic

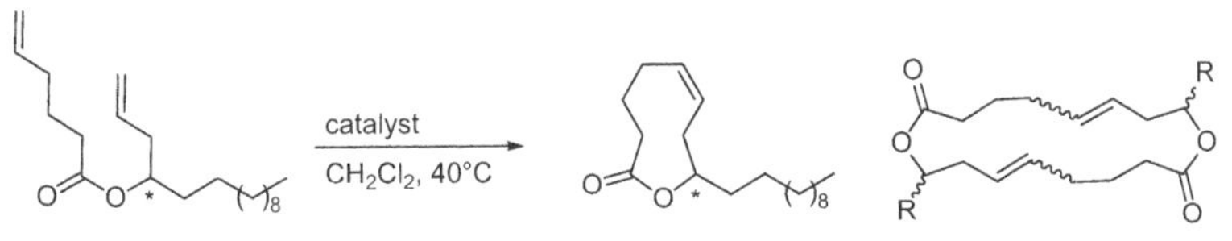

(*S*)-**162**	**G1** in portions (overall 52 mol%)	(*S*)-**163** (70%)	-
(*R*)-**162**	"	(*R*)-**163** (68%)	-
rac-**162**	**G2** (5 mol%)	-	**164** (20%) R = *n*-$C_{12}H_{25}$

Scheme 54 Synthesis of the stereoisomers of topsentolide analog **163** by RCM promoted by **G1**

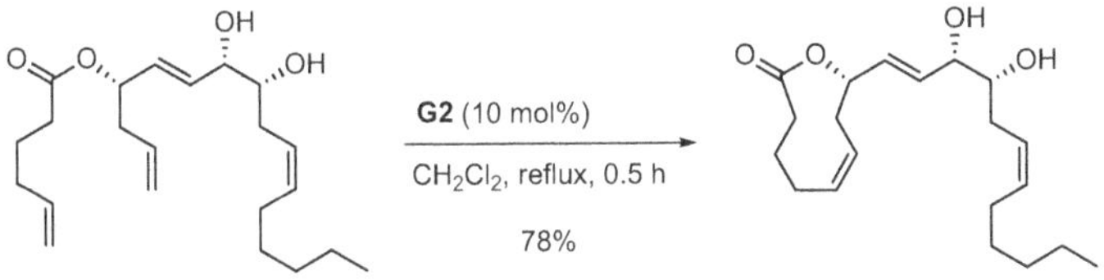

165 **166**

Scheme 55 Synthesis of topsentolide B3 **166** by RCM promoted by **G2**

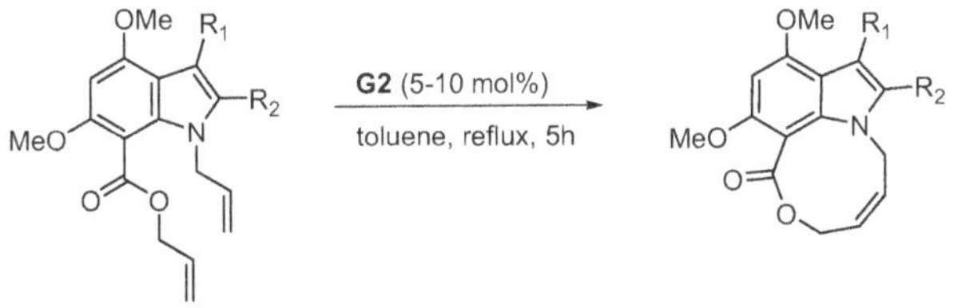

167a	38%	**168a** $R^1 = R^2 = Ph$
167b	78%	**168b** R^1 = 4-BrC_6H_4, R^2 = Ph
167c	15%	**168c** R^1 = 4-ClC_6H_4, R^2 = Ph

Scheme 56 RCM strategy toward functionalized 1,7-annulated 4,6-dimethoxyindoles **168**

conditions and that the hydroxyl functions may have acted as anchoring groups to the metal center thus assisting the ring-closure process with respect to CM.

Indoles annulated with medium rings are interesting structures, some of which are also found in natural products. In the development of RCM strategies towards 1,7-annulated-4,6-dimethoxyindoles, nine-membered lactones **168** have been prepared from bis-allyl-*N*,*O* precursors **167** in the presence of catalyst **G2** (5–10 mol %), as outlined in Scheme 56.

The structure of one of these indoles, which exhibited a fluxional behavior in the NMR time scale due to the flexibility of the nine-membered ring, was confirmed by X-ray structural determination [223]. It is worth mentioning that the corresponding lactam could not be obtained via a similar approach since the *bis*-allyl amine precursor ring-closed to form the preferred five-membered ring.

6 Synthesis of Medium (Ten- and Eleven-Membered) Ring-Sized Lactones

6.1 Ten-Membered Lactones

The destabilization interactions which affect the energy of medium rings, in particular eight- and nine-membered, are progressively released with increasing ring size, thus accounting for the relative easiness of macrolide formation [202, 204]. For instance, the *Z*-isomer of the ten-membered lactone 3,4,5,6,9,10-hexahydro-2*H*-oxecin-2-one **134**, with ε,ζ-unsaturation, is formed in 95% relative yield by RCM reaction of the diene ester precursor 3-butenyl-6-heptenoate **133**, in the presence of **G2** (5 mol%, equilibration in CH_2Cl_2 at reflux, 45 min) (Table 7) [203].

This situation corresponds to abundance of naturally occurring nonanolides, a large family of secondary metabolites featuring a ten-membered macrolide subunit [224]. Additional structural characteristics are a C-9 alkyl group, either methyl or longer chains and stereochemically pure hydroxyl and epoxy functionalities. These medium-sized macrolides are classified in monocyclic polyketides, monocyclic oxylipins, or aliphatic bicyclic and aromatic bicyclic lactones. Following from the structural characterization of the jasmine keto lactone, isolated in 1964 from *Jasminum grandiflorum*, many other nonanolides have been discovered in the following years. Among these, ten-membered-ring macrolides displaying biological activities are aspinolide B, pinolidoxin, decarestrictines A–D, herbarumins I–III, stagonolides A–I, diplodialides, phoracanthonolides, pyrenolides, and microcarpolides. The well-defined geometry of the double bond and the presence of stereochemically pure hydroxyl functions have represented significant synthetic challenges. A review concerning natural occurrence, biological properties, and total synthesis of ten-membered lactones has been published recently [225], and synthetic routes involving RCM reactions have been described by Ferraz et al. [226]. Only selected contributions within the last five years are discussed in this chapter.

A series of aspinolides (A, B, and C) were first isolated in 1997 from the cultures of *Aspergillus ochraceus*. The first total synthesis of aspinolide A **169** has been achieved using RCM as a key step, with generation of stereogenic centers by means of Jacobsen hydrolytic kinetic resolution of racemic epoxides [227]. The synthetic route adopted by Kumar et al. is depicted in Scheme 57.

The final step was performed using the first-generation Grubbs catalyst **G1** under high-dilution conditions. The spectroscopic features of the synthetic compound are identical to those of the natural product.

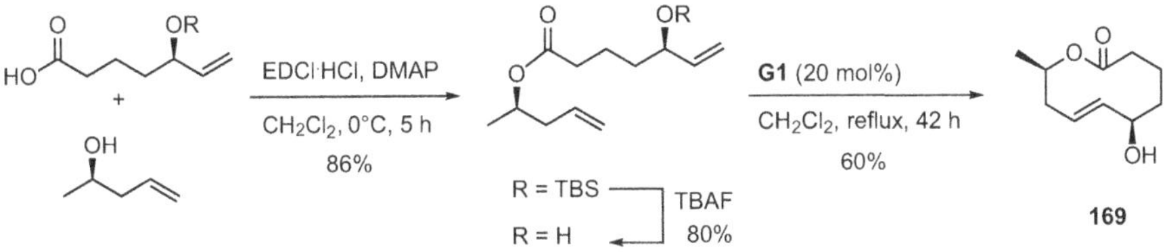

169

Scheme 57 RCM approach to aspinolide A **169**

Decarestrictines are ten-membered lactones produced by different strains of *Penicillium*, featuring a *Z*-configured double bond located at either C-4 or C-5, and have shown inhibitory effect on cholesterol biosynthesis in cell line tests with HEP-G2 liver cells [228]. The first stereoselective total syntheses of the polyketide decarestrictine I **170** and of the biosynthetically related analog botryolide B **171** were reported by Krishna and Rao, using a convergent synthetic strategy based on a common epoxy alcohol intermediate (Scheme 58), obtained in turn by kinetic resolution of an epoxide. The key steps involved Yamaguchi esterification of the alcohol with either enantiomer of the PMB-protected 3-hydroxy-4-pentenoic acid, RCM of the respective dienoic ester, and final deprotective reactions. In the course of the last step, decarestrictine I formed by intramolecular epoxide rearrangement and botryolide B by simple hydrolysis of the PMB group [229].

A convergent total synthesis of decarestrictine I has been also proposed by Yadav and coworkers adopting the synthetic route outlined in Scheme 58. In this latter case, both the alcohol and acid precursors used in step **a** were prepared from L-malic acid as a common starting material. The chiral epoxy alcohol was obtained via a nine-reaction sequence and overall 14.9% yield, involving Still–Gennari olefination for chain elongation and Sharpless asymmetric epoxidation, among the key steps. The chiral pentenoic acid was obtained via a five-reaction sequence, including *cis*-selective Wittig epoxidation, in 39.8% overall yield [230]. The target ten-membered natural lactone was obtained in 5.4% total yield.

The first total syntheses of natural products (3*R*,5*S*)- and (3*R*,5*R*)-sonnerlactone **172** have been accomplished in eleven steps and 22% total yield starting from enantiomerically pure (*R*)-propylene oxide. Hydrolytic kinetic resolution, Sharpless epoxidation, reductive elimination of iodo epoxide, and RCM are the key steps of this synthetic route [231]. The reaction conditions for conclusive RCM and olefin reductions are shown in Scheme 59.

Nonenolides are secondary metabolites of terrestrial and marine organisms, such as bacteria, fungi, and plants. Some new nonenolides have been isolated recently from the liquid culture filtrate of *Stagonospora cirsii*. One of these, the biological active stagonolide-C **174** was then synthesized via a reaction sequence involving RCM of the ester **173** in the final step, as depicted in Scheme 60 [232].

The decanolides seimatopolide A and B were isolated from the fungus *Seimatosporium discosioides*. The enantioselective total synthesis of both enantiomers of seimatopolide A **175**, proposed as potential therapeutic agent against type-2 diabetes, has been recently reported by Schmidt et al. and is based on a synthetic

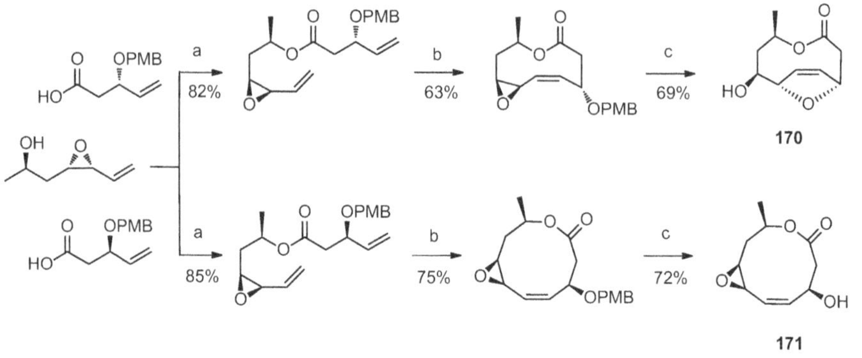

Scheme 58 RCM approach to decarestrictine I **170** and botryolide B **171**. Reagents and conditions: (**a**) 2,4,6-trichlorobenzoyl chloride, Et_3N, THF, 0°C–rt, 4 h, then DMAP, enantiopure acid, toluene, 0°C–rt, 12 h; (**b**) **G2** (10 mol%), CH_2Cl_2, reflux, 12 h; (**c**) DDQ, CH_2Cl_2–H_2O, 0°C–rt, 1 h

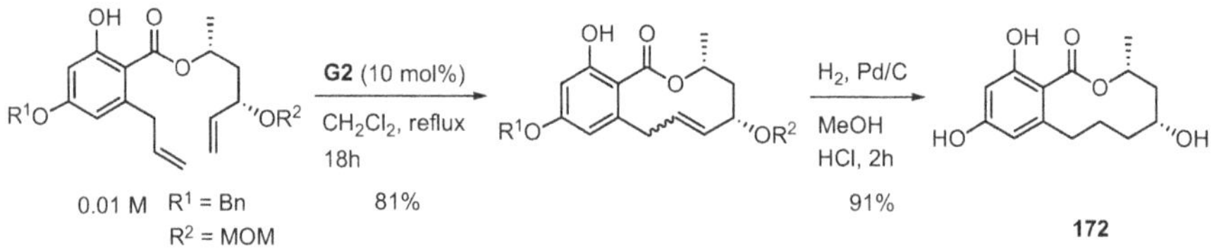

Scheme 59 Final steps in the stereoselective total synthesis of (3*R*,5*R*)-sonnerlactone **172**

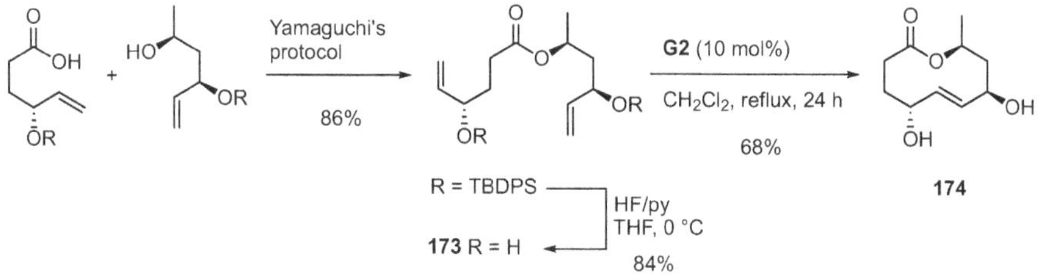

Scheme 60 RCM approach for the stereoselective total synthesis of stagonolide-C **174**

route involving both a CM and an *E*-selective RCM [233]. The starting materials, C_2-symmetric (*R*,*R*)- or (*S*,*S*)-hexa-1,5-diene-3,4-diol, were derived from the chiral pool compounds D-mannitol or L-tartrate, thus allowing for a reliable assignment of the absolute configuration of the natural products. Esterification, RCM, and global deprotection to form (−)-(3*S*,6*S*,7*S*,9*R*)-seimatopolide A **175** are shown in Scheme 61.

The ten-membered-ring lactones stagonolide E **176** and curvulide A **177** were synthesized starting from the ex-chiral pool building block (*R*,*R*)-hexa-1,5-diene-3,4-diol and adopting a bidirectional OM functionalization of the terminal double bonds. The essential steps are site-selective cross metathesis of the diene and

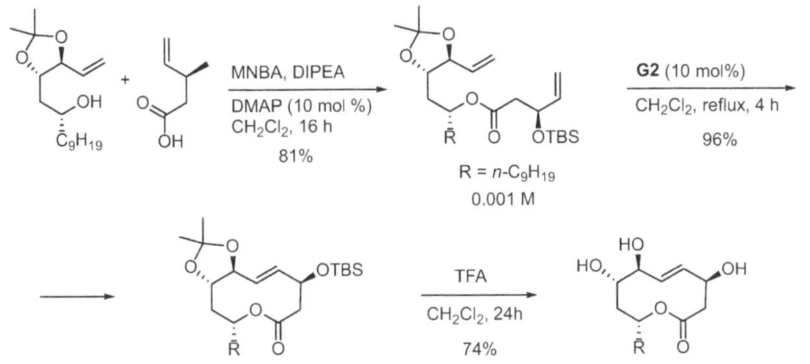

175

Scheme 61 Synthesis of (−)-(3*S*,6*S*,7*S*,9*R*)-seimatopolide A **175**

one-pot RCM/base-induced ring-opening sequence. Ring closure was then accomplished by Yamaguchi lactonization of the MOM protected precursor to furnish stagonolide E, which was further elaborated to curvulide A through Sharpless epoxidation [234]. The one pot RCM/ring-opening sequence and subsequent steps are shown in Scheme 62. The desired configuration at C-9 was obtained by Ru–lipase-catalyzed dynamic kinetic resolution.

Five new nonanolides, cytospolides A–E, were isolated in 2011 from the endophytic fungus *Cytospora* sp., found in the evergreen shrub *Ilex canariensis*. Their carbon skeleton contains 15 carbon atoms including a unique C-2 methyl substitution. These compounds have shown antitumor activities. The first total synthesis of the *Z*-isomer of cytospolide E **178**, based on Evans aldol reaction, Sharpless kinetic resolution, and RCM, was reported in 2012 by Yadav and coworkers (Scheme 63) [235].

The RCM reaction occurred under standard metathetic conditions of catalyst **G2**. The structure and geometric isomerism of the decanolide were determined by NMR spectroscopy. The asymmetric total syntheses of (*Z*)-cytospolides D, E and six of their stereoisomers have then been achieved by Rey and Nanda [236]. In these cases, the alcohol fragment was obtained from alternative chemoenzymatic processes while the acid partner by enantioselective enzymatic desymmetrization of prochiral 2-methyl-1,3-propanediol and stereoselective carbonyl reduction promoted by the CBS catalyst. The yields of the RCM reaction were near 70% for the different stereoisomeric lactones, indicating no influence of configuration on the cyclization step. A short and convergent total synthesis of cytospolide P **181** has then been reported by Goswami and coworkers. This nonanolide is one of the most active in the family of cytospolides and exhibited inhibitory activity against human carcinoma cell lines. The salient features of the synthetic strategy included modified Crimmins aldol reactions, Yamaguchi esterification, and RCM reaction [54]. Ring closure of the diene **179** was affected by the presence of the protecting group and also depended markedly on the catalyst (Scheme 64).

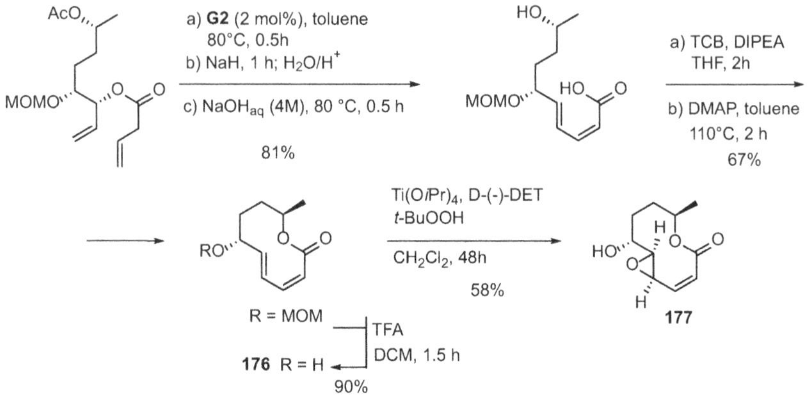

Scheme 62 Synthesis of stagonolide E **176**, highlighting the one-pot RCM/ring-opening sequence, and of curvulide A **177**

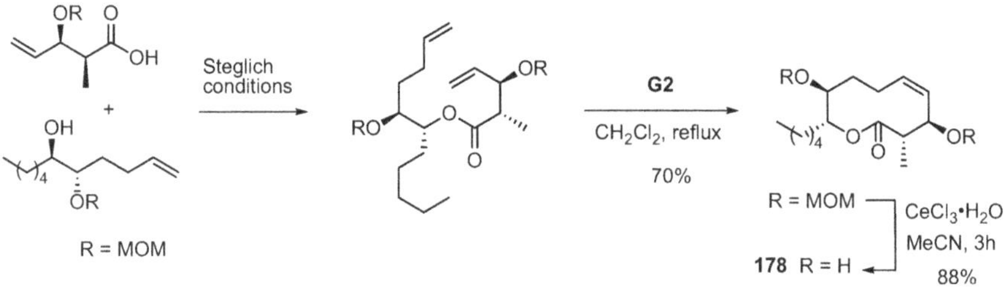

Scheme 63 Stereoselective synthesis of the *Z*-isomer of cytospolide E **178**

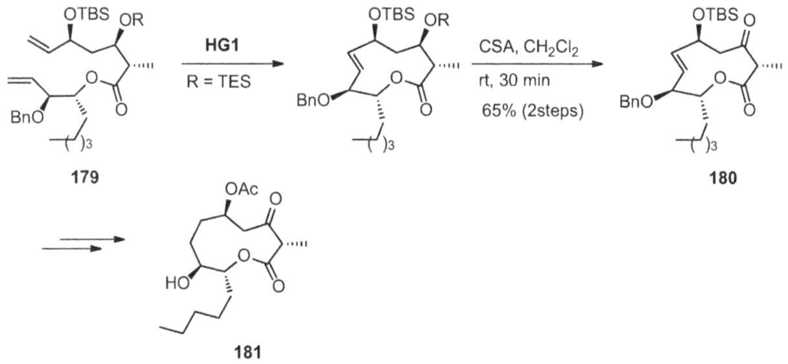

Scheme 64 Stereoselective synthesis of cytospolide P **181**

In the case of the diene featuring a free hydroxyl group at the 3-carbon, **G1**, **G2**, or **HG1** catalyst failed to give the desired lactone. The corresponding TES protected diene underwent cyclization in refluxing toluene (4d) only in the presence

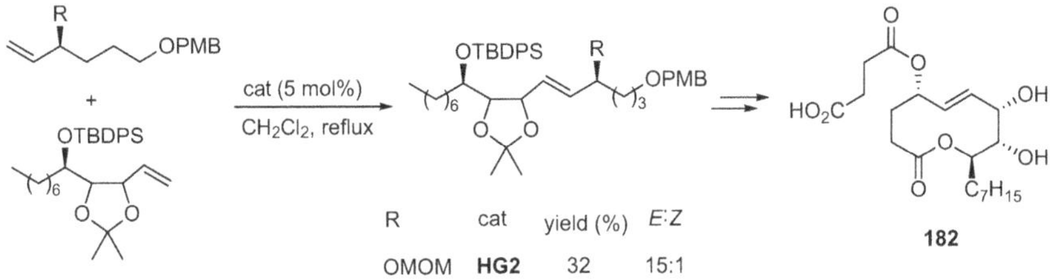

R	cat	yield (%)	E:Z
OMOM	**HG2**	32	15:1
OH	**HG2**	65	15:1
OH	**G2**	30	1:1
OH	**HG1**	nr	

182

Scheme 65 *E*-selective CM reaction, along the asymmetric total synthesis of xyolide **182**

of catalyst **HG1**. The crude product was elaborated further by cleavage of the protecting group and the desired β-keto ester **180** obtained in overall 65% yield.

Regarding the use of CM, the asymmetric total synthesis of xyolide **182** was achieved through an *E* selective CM reaction between two appropriate alkene fragments followed by lactonization under Shiina conditions. After optimization, the CM reaction of the two olefins, one bearing a free hydroxyl group, was performed using catalyst **HG2**, which gave best yields and selectivity in comparison with **G2** and **HG1** (Scheme 65). The approach described by Nanda has been presented as an efficient and alternative strategy for the construction of macrolide natural products with respect to the more conventional esterification of alcohol and acid fragments followed by a late stage RCM reaction.

6.2 Eleven-Membered Lactones (Oxacycloundecane-2-ones)

Few classes of eleven-membered lactones are found in nature, the most known being pyrrolizidine alkaloids, ferrulactones, suspensolide, and aspercyclides [19]. It is worth mentioning that applications of RCM procedures to the synthesis of eleven-membered carbon and heterocycles are relatively scarce and often rather low yielding. An additional effect involved in RCM of increasingly larger rings is reduced selectivity in the geometry of the double bond due to the possibility to accommodate both isomers within the ring. Under analogous RCM conditions in which formation of the nine-membered lactone failed, the eleven-membered derivative, oxacycloundec-6-en-2-one, was obtained in 58% yield from hex-5-enoic acid hex-5-enyl ester [214], in agreement with the decrease of ring strain involved in the formation of the larger ring, although with *E*/*Z* ratio of 1:1.

The aspercyclides are some fungal metabolites that were extracted from the cultural broth of an *Aspergillus* sp. derived from a soil sample in Tanzania. They consist of an eleven-membered unsaturated lactone moiety doubly fused with oxygen-substituted substituted aryl rings (Fig. 10). These natural products have

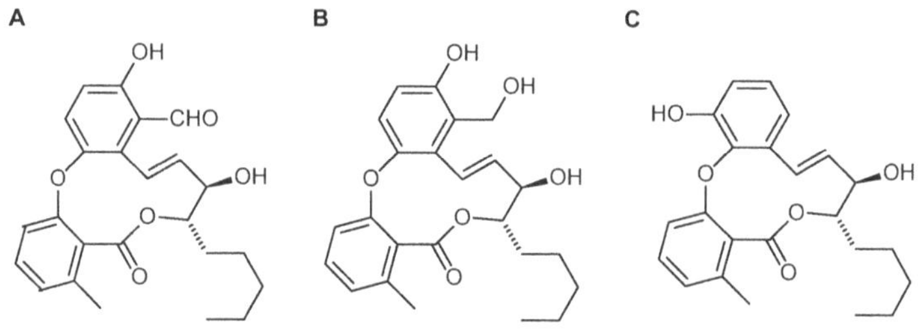

183

Fig. 10 Structure of aspercyclides

shown moderate activity in disrupting the interaction between large proteins with implications in anti-IgE therapy (IgE = immunoglobulin E).

The first total synthesis of aspercyclide C **183** was reported by Fürstner and coworkers using a kinetically controlled RCM reaction to form the eleven-membered unsaturated lactone **183** [237]. At first, the RCM reaction was performed using catalyst **G1** on the precursor **184** with the aim to form selectivity only one geometric isomer, due to relative inertness of this catalyst toward internal alkenes and hence toward equilibration processes of the kinetic product. Indeed, the *E* isomer of lactone **185** formed selectively although with incomplete conversion (<50%) and high catalyst loadings (50 mol%). More conveniently, a mixture of the two isomers **185a,b** (69%) was obtained using **G2** (20 mol%) in refluxing toluene (Scheme 66).

A sample of the *E*-isomer did not equilibrate further in the presence of fresh catalyst, thus indicating high kinetic barrier toward isomerization. Such an approach based on "thermodynamic *vs* kinetic control," or vice versa, can represent the key strategy to impose the desired selectivity of the double bond. The same group presented in a subsequent work alternative routes to aspercyclides, involving either a RCM reaction or a highly diastereoselective Cr-mediated Nozaki–Hiyama–Kishi reaction. Although the RCM approach allowed the synthesis of aspercyclide C, it was not appropriate for the preparation of seemingly related molecules [238]. In addition, due to the high loading of the ruthenium catalyst, the isolation of pure products proved to be difficult.

A family of sesquiterpenes including botcinic and botcineric acids, the ten-membered lactones botcinin E and botrilactone, and the related eleven-membered 3,4-dihydroxy-2,4,6,8-tetramethyldec-8-enolide **186** (Fig. 11) are among the major phytotoxic metabolites which have been obtained from cultures of *Botrytis cinerea*.

The regioselective RCM of a triene precursor has allowed for the asymmetric synthesis of compound **187**, advanced analog of the eleven-membered lactone [239], obtained in a short sequence and 26% overall yield (Scheme 67). The last step implied the use of **G2** under high-dilution conditions in anhydrous CH_2Cl_2 and the isolation of the lactone **187** in *E*:*Z* stereoisomeric ratio of 9:1.

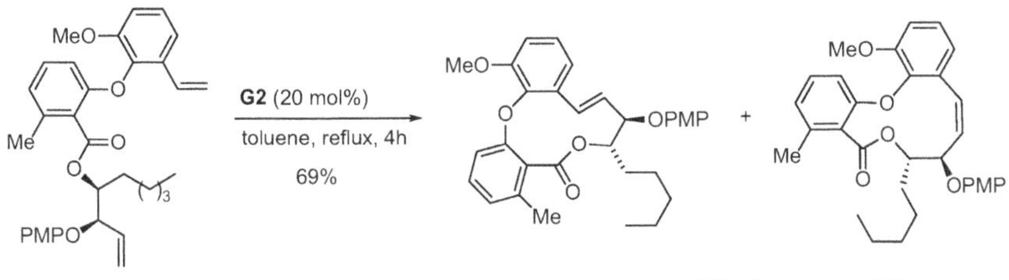

184 (2 mM) **185a** *E*:*Z* = 5:1 **185b**

Scheme 66 RCM synthesis of aspercyclide C precursors

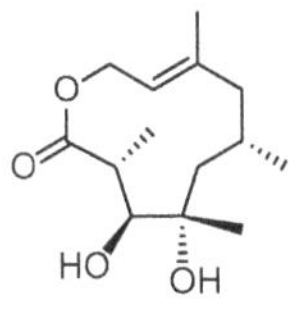

186

Fig. 11 Eleven-membered lactone **186** from culture of *Botrytis cinerea*

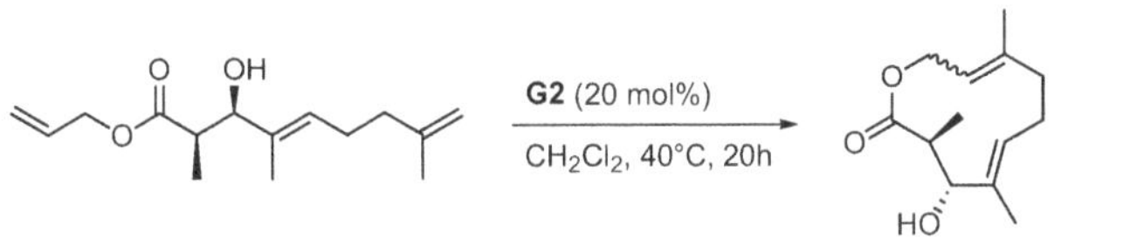

187 *E*:*Z* = 9:1

Scheme 67 Stereoselective synthesis of (*E*)-(2*R*,3*R*,4*E*,8*E*)-3-hydroxy-2,4,8-trimethyldeca-4,8-dienolide **187**

The compound was biotransformed by two mutants strains of *Botrytis cinerea* into four new highly oxygenated eleven-membered lactones, thus increasing significantly the structural diversity of these family of polyketides.

7 Summary, Conclusions, and Outlook

The RCM approach based on ruthenium carbene promoters represents nowadays a powerful tool for the synthesis of unsaturated lactones and related natural products. In general, second-generation catalysts are by far the most effective for lactone formation in comparison with first-generation analogs, in terms of catalyst loading, reaction times, and product yields, except when catalyst endurance and/or high activity induce competitive CM or undesired subsequent transformations due to activation of internal double bonds. The concept related to match or mismatch between catalyst activity and efficiency in RCM reactions has been recently discussed in the article by Caijo et al. [181]. In terms of OM reaction strategies which have addressed current issues related to atom and pot economy, the RRCM approach [60] and the RCM/ring-opening [233] sequence can be regarded as leading cases in the development of sustainable methodologies for the synthesis of lactones. The issue of double bond selectivity, which gains relevance at increasing ring size, was addressed on the basis of "thermodynamic *vs* kinetic control," by taking advantage of the activity profile of different OM catalysts [237].

One of the advances of the RCM approach in the synthesis of natural lactones is a facilitated access to the rare eight- and nine-membered macrolides, whose formation is hindered by intrinsic structural and energetic constrains. Restriction of the diene bond angles due to stereocenters, nature and position of free hydroxyl and/or protecting groups, and α,β-unsaturation in the forming ring have proved to be elements assisting the successful formation of such unfavorable rings by RCM. To the best of our knowledge, comprehension based on systematic or computational studies of these effects is still missing.

In spite that the experimental procedures involved in the RCM of diene esters have reached a significant degree of maturity, some issues deserve further attention and evolution. For instance, although high dilution of the diene ester substrates can be regarded as a convenient method to favor cyclization versus oligomerization, the use of large amount of solvent represents nowadays a serious concern. In addition, the catalyst loading is most often near 5 mol%, which can be regarded as relatively high when compared to various OM processes running nowadays at significantly lower loadings [181]. Some works have presented systematic approaches aimed at minimizing the dose of the ruthenium catalysts in the synthesis of lactones and it is now recognized that high loadings can often be detrimental to efficiency or selectivity. In light of the biological properties and applications of lactones obtained from synthetic routes involving OM reactions, it should also be mentioned that the removal of catalyst impurities from the organic product represents a challenging task. The initial and latest experimental strategies adopted for the separation of the ruthenium impurities have been described in recent review articles [240, 241], whose contents can be particularly useful in the synthesis of natural products.

References

1. Albrecht L, Albrecht A, Yanecki T (2014) α-Alkylidene-γ- and δ-lactones and lactams. In: Yanecki T (ed) Natural lactones and lactams, Chap. 4. Wiley-VCH, Weinheim
2. Maga JA (1976) Lactones in foods. CRC Crit Rev Food Sci Nutr 8:1
3. Guth H, Buhr K, Fritzler R (2001) ACS Symposium Series, 794 (Aroma Active Compounds in Foods): 93
4. Velisek J (2014) The chemistry of food. Blackwell, Wiley
5. Waterhouse AL, Towey JP (1994) J Agr Food Chem 42:1971
6. Guth H (1996) Helv Chim Acta 79:1556
7. Ferreira V, Lopez R, Escudero A, Cacho JF (1998) J Sci Food Agr 77:259
8. Abbott N, Puech J-L, Bayonove C, Baumes R (1995) Am J Enol Viticult 46:292

9. Pirkle WH, Adams PE (1979) J Org Chem 44:2169
10. Chadwick M, Trewin H, Gawthrop F, Wagstaff C (2013) Int J Mol Sci 14:12780
11. Merfort I (2011) Curr Drug Targets 12:1560
12. Hehner SP, Heinrich M, Bork PM, Vogt M, Ratter F, Lehmann V, Schulze-Osthoff K, Dröge W, Schmitz ML (1998) J Biol Chem 273:1288
13. Passreiter CM, Isman MB (1997) Biochem Syst Ecol 25:371
14. Macias FA, Fernandez A, Varela RM, Molinillo JMG, Torres A, Alves PLCA (2006) J Nat Prod 69:795
15. Hanson JR (2000) Nat Prod Rep 17:423
16. Ugurchieva TM, Veselovsky VV (2009) Russ Chem Rev 78:337
17. Marco JA, Carda M (2014) Recent advances in the field of naturally occurring 5,6-dihydropyran-2-ones. In: Janecki T (ed) Natural lactones and lactams, Chap. 2. Wiley-VCH, Weinheim
18. Ozaki Y, Fong C, Herman Z, Maeda H, Miyake M, Ifuku Y, Hasegawa S (1991) Agr Biol Chem 55:137
19. Rousseau G (1995) Tetrahedron 51:2777
20. Shiina I, Nakata K (2014) Medium size lactones. In: Janecki T (ed) Natural lactones and lactams, Chap. 5. Wiley-VCH, Weinheim
21. Ferraz HMC, Bombonato FI, Sano MK, Longo LS Jr (2008) Quimica Nova 31:885
22. Riatto VB, Pilli RA, Victor MM (2008) Tetrahedron 64:2279
23. Boyd GV (1979) The chemistry of lactones and lactams. In: Patai S (ed) Acid derivatives, vol 1. Wiley, Chichester
24. Bazan-Tejeda B, Georgy M, Campagne, J-M (2004) Synlett 720
25. Boeckman RK Jr, Pero JE, Boehmler DJ (2006) J Am Chem Soc 128:11032
26. Brasholtz M, Macdonald JM, Saubern S, Ryan JH, Holmes AB (2010) Chem Eur J 16:115471
27. Watanabe T, Odagi M, Furukori K, Nagasawa K (2014) Chem Eur J 20:591
28. Fischer T, Pietruszka J (2012) Adv Synth Cat 354:2521
29. Ogliaruso MA, Wolfe JF (1993) Synthesis of lactones and lactams. Wiley, Chichester
30. Rao YS (1976) Chem Rev 76:625
31. Ranganathan S, Muraleedharan KM, Vaish NK, Jayaraman N (2004) Tetrahedron 60:5273
32. Nolsoe J M, Hansen TV (2014) Eur J Org Chem 3051
33. Bolm C, Luong TKK, Beckmann O (2001) Oxidation of carbonyl compounds: asymmetric Baeyer-Villiger oxidation. In: Katsuki T (ed) Asymmetric oxidation reactions, Oxford University Press, Oxford
34. Mihovilovic M (2006) Curr Org Chem 10:1265
35. Maier ME (2006) Product class 6: lactones. Sci Synthesis 20b:1421
36. Leisch H, Morley K, Lau PCK (2013) Chem Rev 111:4165
37. Collins I (1999) J Chem Soc Perkin Trans 1(11):1377
38. Paryzek Z, Skiera I (2007) Org Prep Proc Int 39:203
39. Palanichamy K, Kaliappan KP (2014) Synthesis of saturated six-membered ring lactones. In: Cossy J (ed) Synthesis of saturated oxygenated heterocycles II, Topics in heterocyclic chemistry vol 35. Springer-Verlag, Berlin
40. Piva O (2014) Synthesis of seven-membered ring ethers and lactones In: Cossy J (ed) Synthesis of saturated oxygenated heterocycles II, Topics in heterocyclic chemistry, vol 35. Springer-Verlag, Berlin
41. Yeung CS, Dornan PK, Dong VM (2011) Transition metal catalyzed approaches to lactones involving C-O bond formation. In: Yudin AK (ed) Catalyzed carbon-heteroatom bond formation. Wiley-VCH, Weinheim
42. Coulter MM, Dong VM (2012) Enantioselective synthesis of lactones via Rh-catalyzed ketone hydroacylation. In: Christmann M, Brase S (eds) Asymmetric synthesis II. Wiley-VCH, Weinheim
43. Fürstner A (2000) Angew Chem Int Ed 39:3012
44. Grubbs RH (2004) Tetrahedron 60:7117

45. Armstrong SK (1998) J Chem Soc Perkin Trans 1:371
46. Prunet J (2011) Eur J Org Chem 3634
47. Deiters A, Martin SF (2004) Chem Rev 104:2199
48. Edlin CD, Faulkner J, Fengas D, Helliwell M, Knight CK, House D, Parker J, Preece I, Quayle P, Raftery J, Richards SN (2006) J Organomet Chem 691:5375
49. Schall A, Reiser O (2008) Eur J Org Chem 2353
50. Majumdar KC, Muhuri S, Islam RU, Chattopadhyay B (2009) Heterocycles 78:1109
51. Marco JA, Carda M (2011) Nat Prod Comm 6:505
52. Bassetti M, D'Annibale A (2013) Curr Org Chem 17:2654
53. Shiina I (2007) Chem Rev 107:239
54. Chatterjee S, Guchhait S, Goswami RK (2014) J Org Chem 79:7689
55. Fürstner A, Langemann K (1997) J Am Chem Soc 119:9130
56. Carda M, Rodríguez S, González F, Castillo E, Villanueva A, Marco JA (2002) Eur J Org Chem 2649
57. Fujii M, Fukumura M, Hori Y, Hirai Y, Akita H, Nakamura K, Toriizuka K, Ida Y (2006) Tetrahedron Asymmetry 17:2292
58. Bassetti M, D'Annibale A, Fanfoni A, Minissi F (2005) Org Lett 7:1805
59. Ulman M, Grubbs RH (1999) J Org Chem 64:7202
60. Hoye TR, Jeffrey CS, Tennakoon MA, Wang J, Zhao H (2004) J Am Chem Soc 126:10210
61. Samojłowicz C, Bieniek M, Grela K (2009) Chem Rev 109:3708
62. Chatterjee AK, Morgan JP, Scholl M, Grubbs RH (2000) J Am Chem Soc 122:3783
63. Fürstner A, Thiel OR, Ackermann L, Schanz H-J, Nolan SP (2000) J Org Chem 65:2204
64. Garber SB, Kingsbury JS, Gray BL, Hoveyda AH (2000) J Am Chem Soc 122:8168
65. Fürstner A, Dierkes T (2000) Org Lett 2:2463
66. Ono K, Nagata T, Nakagawa M, Nishida A (2003) Synlett 1207
67. Greszler SN, Malinowsky JT, Johnson JS (2011) Org Lett 13:3206
68. Schmidt B, Geißler D (2010) ChemCatChem 2:423
69. Grela K, Kim M (2003) Eur J Org Chem 963
70. Michrowska A, Bujok R, Harutyunyan S, Sashuk V, Dolgonos G, Grela K (2004) J Am Chem Soc 126:9318
71. Miege F, Meyer C, Cossy J (2010) Org Lett 12:248
72. Bermejo A, Figadère B, Zafra-Polo MC, Barrachina I, Estornell E, Cortes D (2005) Nat Prod Rep 22:269
73. McLaughlin JL (2008) J Nat Prod 71:1311
74. Crimmins MT, Jacobs DL (2009) Org Lett 11:2695
75. Prasad KR, Gandi V (2008) Tetrahedron Asymmetry 19:2616
76. Karlsson S, Andersson F, Breistein P, Hedenström E (2007) Tetrahedron Lett 48:7878
77. Dai W-M, Shi L, Li Y (2008) Tetrahedron Asymmetry 19:1549
78. Wang Y, Dai W-M (2010) Tetrahedron 66:187
79. Veeranjaneyulu B, Srilatha M, Reddy GC, Das B (2012) Helv Chim Acta 95:1152
80. Rao DC, Reddy DK, Shekhar V, Venkateswarlu Y (2013) Tetrahedron Lett 54:828
81. Prasad KR, Gandi VR (2011) Tetrahedron Asymmetry 22:499
82. Yajima A, Urao S, Katsuta R, Nukada T (2014) Eur J Org Chem 731
83. Hoye AT, Wipf P (2011) Org Lett 13:2634
84. Schulte ML, Turlington ML, Phatak SS, Harp JM, Stauffer SR, Lindsley CW (2013) Chem Eur J 19:11847
85. Takao K-I, Nanamiya R, Fukushima Y, Namba A, Yoshida K, Tadano K-I (2013) Org Lett 15:5582
86. Prasad KR, Kumar SM (2013) Synthesis 45:785
87. Ghosh AK, Cappiello J, Shin D (1998) Tetrahedron Lett 39:4651
88. Ramachandran PV, Reddy MVR, Brown HC (2000) Tetrahedron Lett 41:583
89. Dirat O, Kouklovsky C, Langlois Y, Lesot P, Courtieu J (1999) Tetrahedron Asymmetry 10:3197

90. Ghosh AK, Liu C (1999) Chem Commun 1743
91. Cossy J, Bauer D, Bellosta V (1999) Tetrahedron Lett 40:4187
92. Du Y, Wiemer DF (2001) Tetrahedron Lett 42:6069
93. Carda M, Rodríguez S, Castillo E, Bellido A, Díaz-Oltra S, Marco JA (2003) Tetrahedron 59:857
94. D'Annibale A, Ciaralli L, Bassetti M, Pasquini C (2007) J Org Chem 72:6067
95. Ghosh AK, Bilcer G (2000) Tetrahedron Lett 41:1003
96. Trost BM, Yeh VSC (2002) Org Lett 4:3513
97. Naidu SV, Gupta P, Kumar P (2005) Tetrahedron Lett 46:2129
98. Raghavan S, Krishnaiah V (2006) Tetrahedron Lett 47:7611
99. Boruwa J, Barua NC (2006) Tetrahedron 62:1193
100. Kumar P, Naidu SV (2006) J Org Chem 71:3935
101. Mizutani H, Watanabe M, Honda T (2002) Tetrahedron 58:8929
102. Deshmukh PH, Blechert S (2007) Dalton Trans (24):2479
103. Vougioukalakis GC, Grubbs RH (2010) Chem Rev 110:1746
104. Berlin JM, Campbell K, Ritter T, Funk TW, Chlenov A, Grubbs RH (2007) Org Lett 9:1339
105. Stewart IC, Ung T, Pletnev AA, Berlin JM, Grubbs RH, Schrodi Y (2007) Org Lett 9:1589
106. Rost D, Porta M, Gessler S, Blechert S (2008) Tetrahedron Lett 49:5968
107. Vorfalt T, Leuthäuer S, Plenio H (2009) Angew Chem Int Ed 48:5191
108. Samojłowicz C, Borré E, Mauduit M, Grela K (2011) Adv Synth Cat 353:1993
109. Harsh P, O' Doherty G (2009) Tetrahedron 65:5051
110. Pospíšil J, Markó IE (2008) Tetrahedron Lett 49:1523
111. Sabitha G, Reddy SSS, Yadav JS (2011) Tetrahedron Lett 52:2407
112. Reddy GC, Balasubramanyam P, Salvanna N, Reddy TS, Das B (2012) Bioorg Med Chem Lett 22:2415
113. Kumar RN, Meshram HM (2011) Tetrahedron Lett 52:1003
114. Shashidhar J, Reddy KM, Ghosh S (2011) Tetrahedron Lett 52:3106
115. Ghosh S, Kumar SU, Shashidhar J (2008) J Org Chem 73:1582
116. ElMarrouni A, Joolakanti SR, Colon A, Heras M, Arseniyadis S, Cossy J (2010) Org Lett 12:4074
117. Kiran INC, Reddy RS, Suryavanshi G, Sudalai A (2011) Tetrahedron Lett 52:438
118. Paioti PHS, Coelho F (2011) Tetrahedron Lett 52:6180
119. Das B, Srinivas Y, Sudhakar C, Reddy PR (2011) Helv Chim Acta 94:1290
120. Das B, Laxminarayana K, Krishnaiah M, Kumar DN (2009) Bioorg Med Chem Lett 19:6396
121. Narasimhulu M, Krishna AS, Rao JV, Venkateswarlu Y (2009) Tetrahedron 65:2989
122. Dwivedi N, Tripathi D, Kumar P (2011) Tetrahedron Asymmetry 22:1749
123. Álvarez-Bercedo P, Falomir E, Murga J, Carda M, Marco JA (2008) Eur J Org Chem 4015
124. Das B, Suneel K, Satyalakshmi G, Kumar DN (2009) Tetrahedron Asymmetry 20:1536
125. Rauniyar V, Hall DG (2009) J Org Chem 74:4236
126. Chinnababu B, Reddy SP, Rao CB, Rajesh K, Venkateswarlu Y (2010) Helv Chim Acta 93:1960
127. Tang S, Xie X, Wang X, He L, Xu K, She X (2010) J Org Chem 75:8234
128. Kumar P, Pandey M, Gupta P, Naidu SV, Dhavale DD (2010) Eur J Org Chem 6993
129. Sabitha G, Vangala B, Reddy SSS, Yadav JS (2010) Helv Chim Acta 93:329
130. Carneiro VMT, Avila CM, Balunas MJ, Gerwick WH, Pilli RA (2014) J Org Chem 79:630
131. Sabita G, Reddy NM, Prasad MN, Yadav JS (2009) Helv Chim Acta 92:967
132. Wang X, Wang W, Zheng H, Su Y, Jiang T, He Y, She X (2009) Org Lett 11:3136
133. Melillo B, Smith AB III (2013) Org Lett 15:2282
134. Reddy DS, Mohapatra DK (2013) Eur J Org Chem 1051
135. Wang Y, O'Doherty GA (2013) J Am Chem Soc 135:9334
136. Walleser P, Brückner R (2013) Org Lett 15:1294
137. Chandrasekhar S, Rambabu C, Reddy AS (2008) Tetrahedron Lett 49:4476
138. Kumar P, Pandey M, Guptaa P, Dhavale DP (2012) Org Biomol Chem 10:1820

139. Reddy CR, Veeranjaneyulu B, Nagendra S, Das B (2013) Helv Chim Acta 95:505
140. Chakraborty TK, Purkait S (2008) Tetrahedron Lett 49:5502
141. Sabitha G, Raju A, Reddy CN, Yadav JS (2014) RSC Adv 4:1496
142. Georges Y, Ariza X, Garcia J (2009) J Org Chem 74:2008
143. Kamal A, Balakrishna M, Reddy PV, Faazil S (2010) Tetrahedron Asymmetry 21:2517
144. Prasad KR, Penchalaiah K (2010) Tetrahedron Asymmetry 21:2853
145. Yadav JS, Thirupathaiah B, Singh VK, Ravishashidhar V (2012) Tetrahedron Asymmetry 23:931
146. Enders D, Reichenbach LF (2013) Synthesis 45:959
147. Sabitha G, Shankaraiah K, Yadav JS (2013) Eur J Org Chem 4870
148. Prasad KR, Gutala P (2013) J Org Chem 78:3313
149. Sabitha G, Sandeep A, Rao AS, Yadav JS (2013) Eur J Org Chem 6702
150. Sabitha G, Rao AS, Sandeep A, Yadav JS (2014) Eur J Org Chem 455
151. Akaike H, Horie H, Kato K, Akita H (2008) Tetrahedron Asymmetry 19:1100
152. Schmidt B, Kunz O, Biernat A (2010) J Org Chem 75:2389
153. Ramesh P, Meshram HM (2011) Tetrahedron Lett 52:2443
154. Babu DC, Selavam JJP, Reddy DK, Shekhar V, Venkateswarlu Y (2011) Tetrahedron 67:3815
155. Yadav JS, Mandal SS (2011) Tetrahedron Lett 52:5747
156. Hayashi Y, Yamaguchi H, Toyoshima M, Okado K, Toyo T, Shoji M (2008) Org Lett 10:1405
157. Sarkar SM, Wanzala EN, Shibahara S, Takahashi K, Ishihara J, Hatakeyama S (2009) Chem Commun 5907
158. Shibahara S, Fujino M, Tashiro Y, Okamoto N, Esumi T, Takahashi K, Ishihara J, Hakateyama S (2009) Synthesis 2935
159. Gao D, O'Doherty GA (2010) Org Lett 12:3752
160. Hayashi Y, Yamaguchi H, Toyoshima M, Okado K, Toyo T, Shoji M (2010) Chem Eur J 16:10150
161. Salit A-F, Meyer C, Cossy J, Delouvrié B, Hennequin L (2008) Tetrahedron 64:6684
162. Shibahara S, Fujino M, Tashiro Y, Takahashi K, Ishihara J, Hakateyama S (2008) Org Lett 10:2139
163. Druais V, Hall MJ, Corsi C, Wendeborn SV, Meyer C, Cossy J (2009) Org Lett 11:935
164. Druais V, Hall MJ, Corsi C, Wendeborn SV, Meyer C, Cossy J (2010) Tetrahedron 66:6358
165. Bressy C, Vors J-P, Hillebrand S, Arseniyadis S, Cossy J (2008) Angew Chem Int Ed 47:10137
166. Riva E, Rencurosi A, Gagliardi S, Passarella D, Martinelli M (2011) Chem Eur J 17:6221
167. Dong L, Gordon VA, Grange RL, Johns J, Parsons PG, Porzelle A, Reddell P, Schill H, Williams CM (2008) J Am Chem Soc 130:15262
168. Dong L, Schill H, Grange RL, Porzelle A, Johns JP, Parsons PG, Gordon VA, Reddell PW, Williams CM (2009) Chem Eur J 15:11307
169. Andreana PR, McLellan JS, Chen Y, Wang PG (2002) Org Lett 4:3875
170. Roulland E, Ermolenko MS (2005) Org Lett 7:2225
171. Zhu L, Kedenburg JP, Xian M, Wang PG (2005) Tetrahedron Lett 46:811
172. Ko1nig CM, Harms K, Koert U (2007) Org Lett 9:4777
173. Ichige T, Okano K, Kanoh N, Nokata M (2007) J Am Chem Soc 129:9862
174. Hoye TR, Jeon J, Kopel LC, Ryba TD, Tennakoon MA, Wang Y (2010) Angew Chem Int Ed 49:6151
175. Ichige T, Okano Y, Kanoh N, Nakata M (2009) J Org Chem 74:230
176. Cros F, Ruiz J, Pelotier B, Piva O (2010) Synlett (17):2621
177. Pisani L, Superchi S, D'Elia A, Scafato P, Rosini C (2012) Tetrahedron 68:5779
178. Scott RW, Mazzetti C, Simpson TJ, Willis CL (2012) Chem Commun 48:2639
179. Takii K, Kanbayashi N, Onitsuka K (2012) Chem Commun 48:3872
180. Cooksey JP (2013) Org Biomol Chem 11:5117

181. Caijo F, Tripoteau F, Bellec A, Crevisy C, Basle O, Mauduit M, Briel O (2013) Cat Sci Tec 3:429
182. Virolleaud A, Piva O (2004) Synlett (12):2087
183. Baktharaman S, Selvakumar S, Singh VK (2005) Tetrahedron Lett 46:7527
184. Binder JT, Kirsch SF (2007) Chem Commun 4164
185. Cros F, Pelotier B, Piva O (2010) Eur J Org Chem 5063
186. Kirsch SF, Klahn P, Menz H (2011) Synthesis (22):3592
187. Schmidt B, Kunz O (2012) Eur J Org Chem 1008
188. Nodwell MB, Menz H, Kirsch SF, Sieber SA (2012) ChemBioChem 13:1439
189. Menz H, Kirsch SF (2009) Org Lett 11:5634
190. Quinn KJ, Curto JM, McGrath KP, Biddick NA (2009) Tetrahedron Lett 50:7121
191. Alcaide B, Almendros P, Luna A (2009) Chem Rev 109:3817
192. Schmidt B, Kunz O (2012) Synlett (23):851
193. Rückert A, Deshmukh PH, Blechert S (2006) Tetrahedron Lett 47:7977
194. Lee CW, Grubbs RH (2001) J Org Chem 66:7155
195. Agrawal D, Sriramurthy V, Yadav VK (2006) Tetrahedron Lett 47:7615
196. Yadav VK, Agrawal D (2007) Chem Commun 5232
197. Pentzer EB, Gadzikwa T, Nguyen ST (2008) Org Lett 10:5613
198. Nolen EG, Donahue LA, Greaves R, Daly TA, Calabrese DA (2008) Org Lett 10:4911
199. Sundararaju B, Sridhar T, Achard M, Sharma GVM, Bruneau C (2010) Eur J Org Chem 6092
200. Illuminati G, Mandolini L, Masci B (1977) J Am Chem Soc 99:6308
201. Illuminati G, Mandolini L (1981) Acc Chem Res 14:95
202. Galli C, Mandolini L (2000) Eur J Org Chem 3117
203. Conrad JC, Eelman MD, Duarte Silva JA, Monfette S, Parnas HH, Snelgrove JL, Fogg DE (2007) J Am Chem Soc 129:1024
204. Ziegler K (1955) In: Müller E (ed) Methoden der Organischen Chemie (Houben-Weyl), vol. 4/2. Georg Thieme Verlag, Stuttgard
205. Burt MB, Crane AK, Su N, Rice N, Poirier RA (2010) Can J Chem 88:1094
206. Kao CM, McPherson M, McDaniel RN, Fu H, Cane DE, Khosla C (1997) J Am Chem Soc 119:11339
207. Reignier T, de Berardinis V, Petit J-L, Mariage A, Duquesne K, Alphand V (2014) Chem Commun 50:7793
208. Buszek KR, Sato N, Jeong Y (2002) Tetrahedron Lett 43:181
209. Shiina I, Hashizume M, Yamai Y-S, Oshiumi H, Shimazaki T, Takasuna Y-J, Ibuka R (2005) Chem Eur J 11:6601
210. Aird JI, Hulme AN, White JW (2007) Org Lett 9:631
211. Sharma A, Gamre S, Roy S, Goswami D, Chattopadhyay A, Chattopadhyay S (2008) Tetrahedron Lett 49:3902
212. Dinh M-T, BouzBouz S, Peglion J-L, Cossy J (2005) Synlett 2851
213. Dinh M-T, Bouzbouz S, Peglion J-L, Cossy J (2008) Tetrahedron 64:5703
214. Hon T-S, Wong Y-C, Chang C-P, Hsieh C-A (2007) Tetrahedron 63:11325
215. Takemoto Y, Baba Y, Saha G, Nakao S, Iwata C, Tanaka T, Ibuka T (2000) Tetrahedron Lett 41:3653
216. Baba Y, Saha G, Nakao S, Iwata C, Tanaka T, Ibuka T, Ohishi H, Takemoto Y (2001) J Org Chem 66:81
217. Takahashi T, Watanabe H, Kitahara T (2002) Heterocycles 58:99
218. Ramírez-Fernández J, Collado IG, Hernández-Galán R (2008) Synlett 339
219. Malik CK, Hossain MF, Ghosh S (2009) Tetrahedron Lett 50:3063
220. Wetzel I, Krauss J, Bracher F (2012) Lett Org Chem 9:169
221. Chatterjee AK, Choi T-L, Sanders DP, Grubbs RH (2003) J Am Chem Soc 125:11360
222. Sreedhar E, Venkanna A, Chandramouli N, Babu KS, Rao JM (2011) Eur J Org Chem 2011:1078
223. Wood K, Black DSC, Kumar N (2011) Tetrahedron 67:4093

224. Dräger G, Kirschning A, Thieriche R, Zerlin M (1996) Nat Prod Rep 13:365
225. Sun P, Lu S, Ree TV, Krohn K, Li L, Zhang W (2012) Curr Med Chem 19:3417
226. Ferraz HMC, Bombonato FI, Longo LS Jr (2007) Synthesis 3261–3285
227. Chowdhury PS, Gupta P, Kumar P (2009) Tetrahedron Lett 50:7018–7020
228. Mayer M, Thiericke R, Hoechst AG (1993) J Antibiot 46:1372–1380
229. Krishna PR, Rao TJ (2010) Org Biomol Chem 8:3130
230. Yadav JS, Venkatesh M, Kumar AS, Reddy PAN, Reddy BVS, Prasad AR (2014) Helv Chim Acta 97:830
231. Thirupathi B, Gundapaneni RR, Mohapatra DK (2011) Synlett 2667
232. Venkatesham A, Nagaiah K (2012) Tetrahedron Asymmetry 23:1186
233. Schmidt B, Kunz O, Petersen MH (2012) J Org Chem 77:10897
234. Schmidt B, Kunz O (2013) Beilstein J Org Chem 9:2544
235. Yadav JS, Pandurangam T, Kumar AS, Reddy PAN, Prasad AR, Reddy BVS, Rayendraprasad K, Kunwar AC (2012) Tetrahedron Lett 53:6048
236. Rej RK, Nanda S (2014) Eur J Org Chem 860
237. Fürstner A, Müller C (2005) Chem Comm 5583
238. Pospîsil J, Müller C, Fürstner A (2009) Chem Eur J 15:5956
239. Botubol-Ares JM, Durán-Peña MJ, Macías-Sánchez AJ, Hanson JR, Collado IG, Hernández-Galán R (2014) Org Biomol Chem 12:5304
240. Vougioukalakis GC (2012) Chem Eur J 18:8868
241. Kotha S, Dipak MK (2012) Tetrahedron 68:397

Top Heterocycl Chem (2017) 47: 111–154
DOI: 10.1007/7081_2014_139

Published online: 20 January 2015

Synthesis of Amine-Containing Heterocycles by Metathesis Reactions: Recent Advances and Opportunities

Philippe Compain and Damien Hazelard

Abstract Metathesis is one of the most powerful methods to access amine-containing heterocycles. However, the ability of amine to coordinate to the metal centre of almost all metathesis catalysts may limit the efficiency of this strategy. To date, the main approach to overcome these unwanted coordination events has been based on deactivation of the nitrogen atom by an electron-withdrawing protecting group. Basic amines are nevertheless not incompatible with metathesis reactions as shown by numerous recent examples described in the literature. The purpose of this review is to provide an overview of successful metathesis reactions performed with amine-containing substrates to access azacycles. The focus will be made on the different parameters that may favor the metathesis process including steric effects, amine basicity, and the nature of the catalyst.

Keywords Amine • Azacycles • Bioactive compounds • Metathesis reactions • Natural products

Contents

P. Compain (✉) and D. Hazelard
Laboratoire de Synthèse Organique et Molécules Bioactives (SYBIO), Université de Strasbourg/CNRS (UMR 7509), Ecole Européenne de Chimie, Polymères et Matériaux, 25 rue Becquerel, 67087 Strasbourg, France
e-mail: philippe.compain@unistra.fr

1 Introduction

Saturated cyclic amines play a pivotal role in (bio)organic and medicinal chemistry due to their importance as synthetic targets and building blocks and their potential as drug candidates. These heterocycles are indeed found in many alkaloid natural products [1, 2] and glycomimetics (iminosugars) [3]. Not surprisingly, saturated cyclic amines have been used as privileged scaffolds for library design and drug discovery leading to thousands of leads evaluated in clinical and preclinical studies over the past decades [4]. The structural features of azacycles make them also valuable tools in asymmetric synthesis as ligands or chiral auxiliaries. Olefin metathesis and related reactions are currently among the best synthetic options for constructing saturated *N*-heterocycles [5–11]. In addition to several advantages including high levels of chemo-, regio- and stereoselectivity, these catalytic carbon–carbon bond-forming reactions lead to alkene products. These versatile building blocks in organic synthesis offer many synthetic opportunities for further structural elaboration including cycloadditions or oxidation reactions. However, the efficiency of olefin metathesis based strategies for the synthesis of cyclic amines may be limited by the presence of basic amino groups [12]. A major drawback associated with the use of amines concerns indeed their ability to coordinate to metal-alkylidene complexes and to interfere unproductively with catalytic activity. The main strategy to overcome these unwanted coordination events is based on deactivation of the nitrogen atom by an electron-withdrawing protecting group. This chemical answer, probably the first one coming into chemist's mind, is however not a panacea; the carbonyl oxygen of the protected amines may also form unproductive metallacycle intermediates, leading to poor conversion in the desired products [13]. More importantly, the efficiency of the whole synthetic sequence is reduced by the addition of protection group manipulations. More straightforward approaches have been developed from deactivated amine substrates obtained directly by protonation of the amino group. Metathesis reactions are then carried out on the corresponding ammonium salts or with the free amines in the presence of a Brønsted acid. The reaction could be also performed in the presence of Lewis acids as shown for the first time by Xiao and Yu and others [14, 15]. These direct deactivation strategies give nevertheless contrasted results and are not applicable to acid-sensitive substrates. All things considered, is there a more straightforward alternative to efficiently synthesize cyclic amines by way of olefin metathesis reactions? Sometimes, the solution may be simpler than expected if one goes beyond conditioned response patterns of thinking. To paraphrase

T. W. Greene and P. G. M. Wuts, no protective group may indeed be the best protective group [16]. Based on literature precedents and on the abovementioned efforts to prevent coordination to the catalyst, the consensus view that has prevailed in the synthetic chemist community is that amines are incompatible with metathesis reactions. However, as highlighted in 2007 in a recent review [13], increasing number of examples of amine-containing compounds that are good substrates for metathesis has been reported in the literature. For example, when the reaction is performed with tertiary amines, the nature of the catalyst may play an important role. Although molybdenum catalysts such as **S** are more sensitive to moisture and atmospheric oxygen than classical ruthenium catalysts including **G1**, **G2**, **HG1**, **HG2,** or **GN**, they are relatively tolerant to amine-containing compounds [5, 13]. In general, successful metathesis amine substrates have structural and/or electronic features that prevent coordination of the amino group to the catalyst's metal center. Typical examples are sterically hindered amines and/or amines α-substituted with an electron-withdrawing group which decreases the electronic density on the nitrogen atom [13]. This review, which covers the literature from January 2007 to January 2014, provides an overview of successful metathesis reactions performed with amine-containing substrates to access saturated cyclic amines. (The synthesis of *N*-heterocycle-containing polymers is not presented in this review. For example, see [17–19]). Our intent is not to be exhaustive but rather to focus on the formation of the heterocycle structures and on the different parameters that may favor the metathesis reaction including steric effects, amine basicity, or the nature of the catalyst. The examples described in this review are discussed in five sections. The first ones are arranged according to the size of the *N*-heterocycles. The last section is devoted to tandem processes with the metathesis reaction as the key step.

2 Five-Membered Cyclic Amines

Del Valle and Ranatunga reported the synthesis of highly constrained fully substituted carbocyclic α-amino acid derivatives via the successful ring-closing metathesis reaction (RCM) of amine **1** [20]. After some optimization, pyrrolidine **2** is obtained in 78% yield by using 7 mol% of **G2** (added in two portions) in toluene at 60°C. The reaction is favored by the steric hindrance around the nitrogen atom which is exacerbated by the neopentyl-like character of the amine (Scheme 1).

Another example of such an effect has been reported in the total synthesis of (±)-Lundurine [21] (Scheme 2). RCM of diene **3** in the presence of **G2** catalyst provided in an excellent yield 3-pyrroline **4** which contains the entire carbon framework of the final target. It is important to note, however, that a large amount of ruthenium carbene complex is used in this process (20 mol%).

In contrast, the RCM of sterically congested amine **5** was performed in the presence of 30 mol% of $Ti(O\textit{i}Pr)_4$ [15] (Scheme 3). The authors made no mention of any attempts without such additives, and it is therefore difficult to evaluate the pertinence of using a Lewis acid in this case.

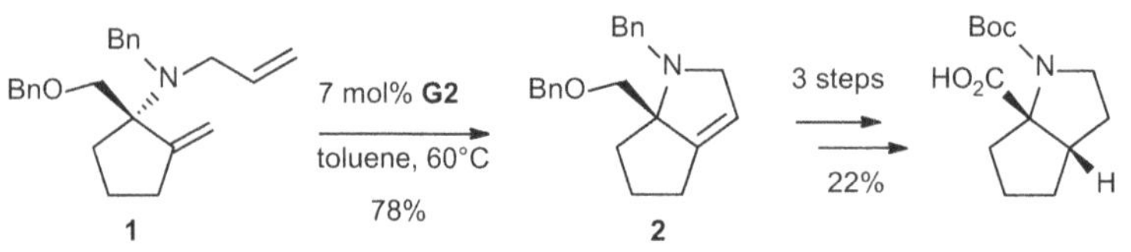

Scheme 1 Synthesis of a constrained α-amino acid precursor by RCM

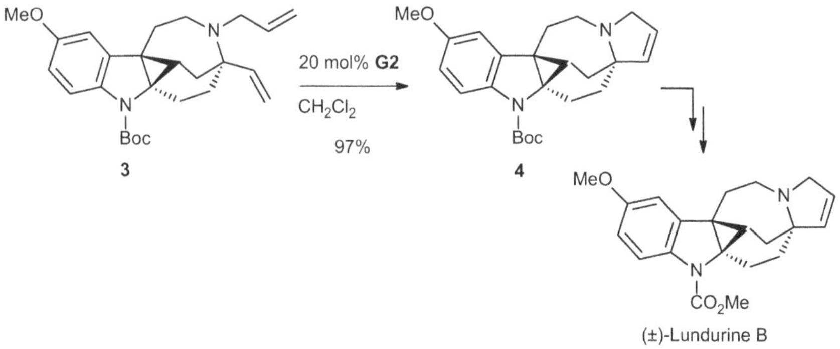

Scheme 2 Synthesis of the pyrroline ring of (±)-Lundurine

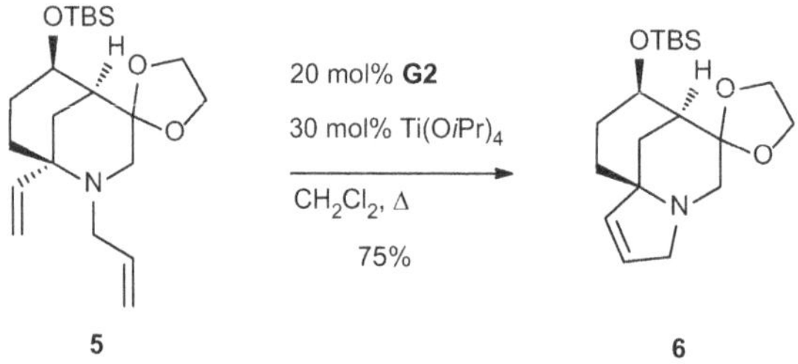

Scheme 3 RCM synthesis of polycyclic amine **6** using Ti(O*i*Pr)$_4$

Endocyclic amines of 2,5-disubstituted pyrrolidines and 2,6-disubstituted piperidines are also known to be particularly crowded. These compounds are therefore promising substrates for metathesis reactions as shown with amines **7** and **10a** (Schemes 4 and 5). However, in the case of amino diene **7**, first attempts using **G1** in DCM at 40°C afforded the desired product **9** in less than 10% yield. The major product was keto pyrrole **8** obtained in 62% yield [22]. Indolizidine **9** could be obtained as the major product in 69% yield by reducing the temperature and the reaction time (Scheme 4). According to the authors, pyrrole **8** could be obtained from **9** by way of a retro-Mannich reaction. The Grubbs catalyst is believed to promote the process by associating with the piperidone oxygen. After further synthetic manipulation including a Barton-McCombie radical deoxygenation, indolizidine **9** afforded the poison frog alkaloid (-)-221T.

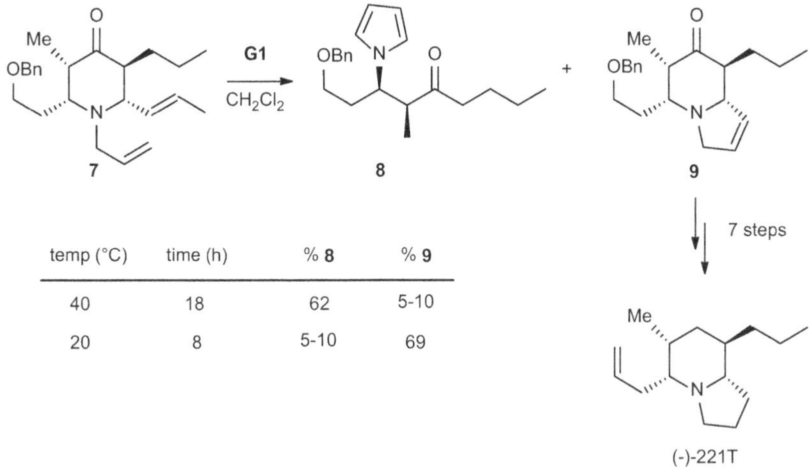

temp (°C)	time (h)	% **8**	% **9**
40	18	62	5-10
20	8	5-10	69

Scheme 4 Synthesis of poison frog alkaloid (-)-221T by way of RCM

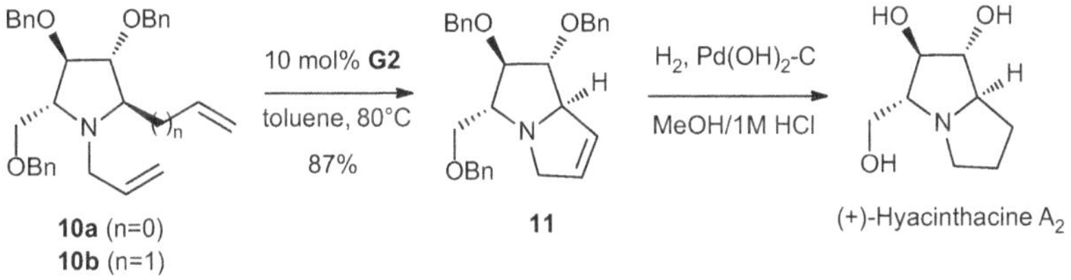

Scheme 5 Synthesis of (+)-Hyacinthacine A_2 via RCM of 2,5-disubstituted pyrrolidine **10a**

Hyacinthacine A_2 was synthesized by way of RCM of 2,5-disubstituted pyrrolidine **10a** [23]. The metathesis reaction was performed using 10 mol% of **G2** in toluene at 80°C to give the expected pyrrolizidine **11** in 87%. Polyhydroxylated indolizidine derivatives were also obtained following a similar strategy from **10b** [23].

The access to pyrrolizidine skeleton was performed from monosubstituted pyrrolidine **12a** en route to the total synthesis of (-)-Lentiginosine (Scheme 6) [24]. However, the yield reported for the RCM reaction is lower than the one observed for the corresponding more hindered 2,5-disubstituted analogue **10a** (Scheme 5). It is noteworthy that all attempts to access pyrroloazepine derivatives by way of RCM from **12b** failed under a variety of conditions (**G1**, **G2,** or **HG2** catalysts) [24].

The group of Rao also reported the synthesis of bicyclic iminosugars by way of RCM reactions from acyclic amines [25, 26] (Scheme 7). Excellent yields were obtained for the formation of the expected cyclized products **15** by treatment of **14** with 10 mol% of **G1** catalyst in DCM.

Jarvis and Charette recently reported a novel methodology for the asymmetric synthesis of 3-substituted piperidines based on a novel irreversible dihydropyrrole–tetrahydropyridine ring expansion [27]. The key pyrrolinol intermediates **17** and **19**

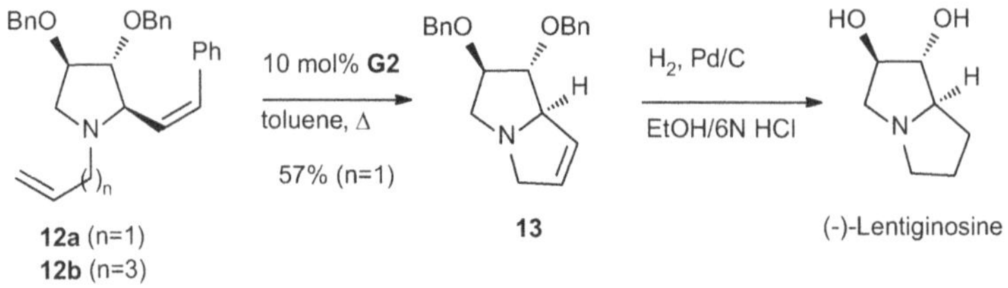

Scheme 6 Synthesis of (-)-Lentiginosine via RCM of monosubstituted pyrrolidine **12a**

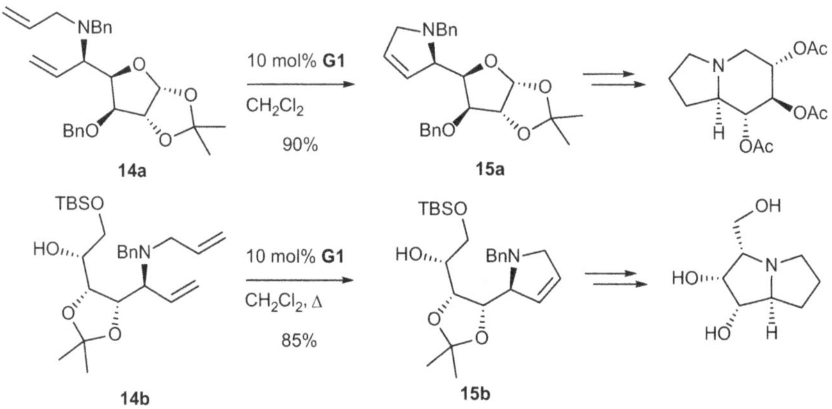

Scheme 7 Synthesis of bicyclic iminosugars **15** by way of RCM reaction

were prepared following a Petasis–Mannich condensation/RCM sequence (Scheme 8) [27]. The RCM of the Petasis–Mannich products **16** and **18** proceeded well using **G2** catalyst in refluxing DCM. However, RCM reaction with secondary amine **18** required protonation of the substrate amino group with 4-toluenesulfonic acid.

The synthesis of densely substituted bicyclic nitrogen heterocycles was reported recently by the group of Vicario. The synthetic strategy made use of efficient RCM reactions for building up the pyrrolizidine ring from pyrrolidine substrate **20** (Scheme 9) [28]. Here, two favorable effects are combined, that is, steric hindrance around the amine and the presence of electron-withdrawing groups which decrease the electronic density on the nitrogen atom. Remarkably, whereas the RCM of diallylamine **20** proceeded in reasonable yield, no cyclized product could be obtained from the corresponding amide derivative **22** by treatment with **G1** or **G2** catalysts. This result may be explained by the formation of a favorable six-membered chelate ring that suppresses catalytic turnover (Scheme 9). This study confirms that systematic deactivation of an amine by an electron-withdrawing group such as an amide may not be the first solution to be envisaged in a retrosynthetic analysis!

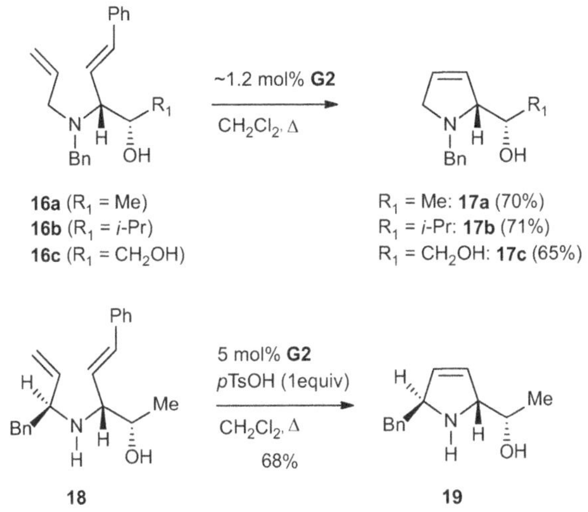

Scheme 8 Synthesis of enantiopure substituted pyrrolinols by RCM reactions

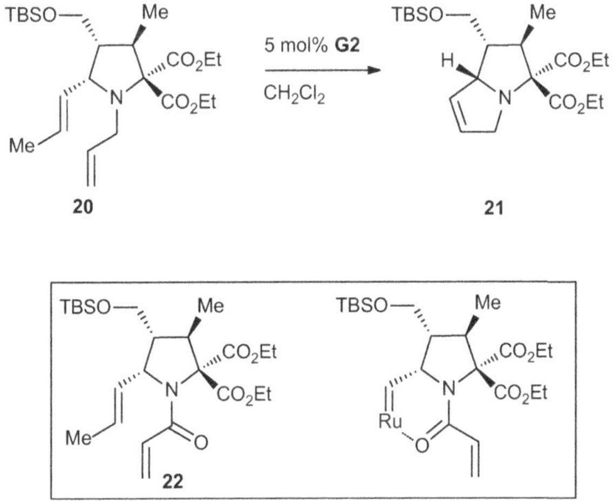

Scheme 9 Synthesis of pyrrolizidine **21** by RCM reactions

The decrease of the electron density on the nitrogen may be also achieved with arylamines in which the nitrogen lone pair is partially conjugated into the benzene ring. For example, fast RCM of diallylamine derivatives of coumarin, quinolone, pyridine, and substituted benzene leading to the corresponding five-membered azacycles in very good yields has been recently reported (Scheme 10) [29] (for related recent examples see [30–33]). Preparation of pyrroline **25** required to block the nitrogen functionality of the pyridine moiety by the formation of the pyridinium salt with 4-toluenesulfonic acid.

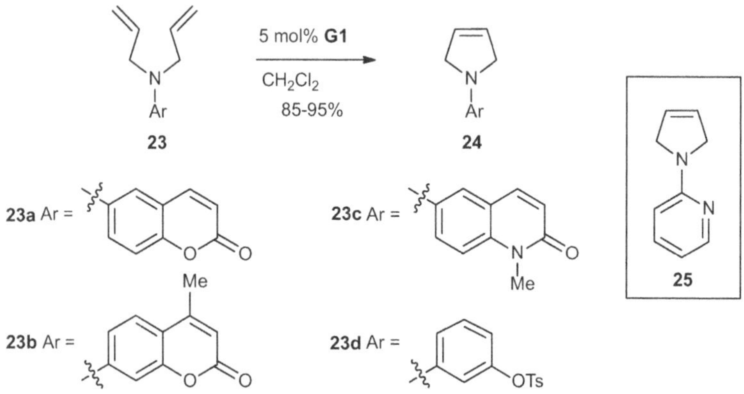

Scheme 10 Synthesis of cyclic arylamines by RCM reactions

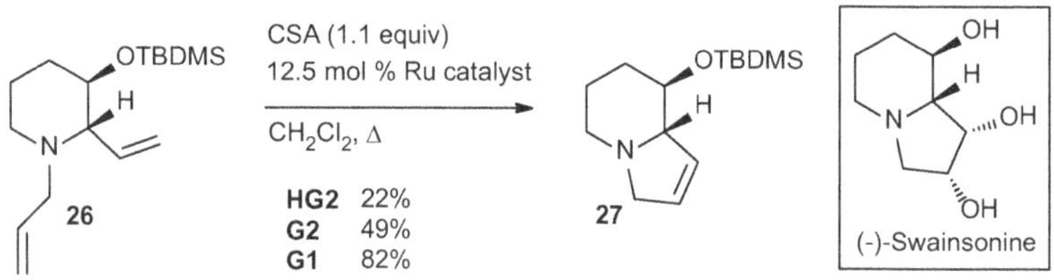

Scheme 11 Synthesis of a (-)-Swainsonine precursor by RCM reactions

Several synthesis of 8-hydroxy-indolizidine derivatives have been performed by RCM of the ammonium salt of the corresponding piperidines obtained by treatment with a Brønsted acid (camphorsulfonic acid or 4-toluenesulfonic acid) [34–37]. Gomez Pardo and Cossy et al. reported, for example, the synthesis of the (-)-Swainsonine precursor **27** from protected 3-hydroxypiperidine **26** in 82% yield. The best conditions found made use of 1.1 equivalent of camphorsulfonic acid and 12.5 mol% of **G1** catalyst added in three portions (Scheme 11) [34]. Much lower yields were obtained using **G2** or **HG2** catalysts.

The importance of the choice of the Brønsted acid was nicely highlighted by comparing the RCM reactivity of the corresponding HBr and HCl salts of secondary diallylamine **28** (Scheme 12) [38] (for examples highlighting the influence of the counterion in ruthenium–alkylidene-catalyzed cross metathesis of ammonium salts, see [39]). Chiral pyrroline **29** could be obtained in 79% yield by treatment of the HBr salt of **28** with 4 mol% of **HG2** catalyst whereas only degradation was observed from the corresponding HCl salt.

Synthesis of five- or six-membered azacycles were also performed by RCM in aqueous media from the corresponding chloride ammonium salt using different ruthenium catalysts [40–44] (for a review on olefin metathesis in aqueous media, see [45]). Ionic liquids have also been used in metathesis reaction as potentially recyclable, green solvents (for a review on olefin metathesis in ionic liquids,

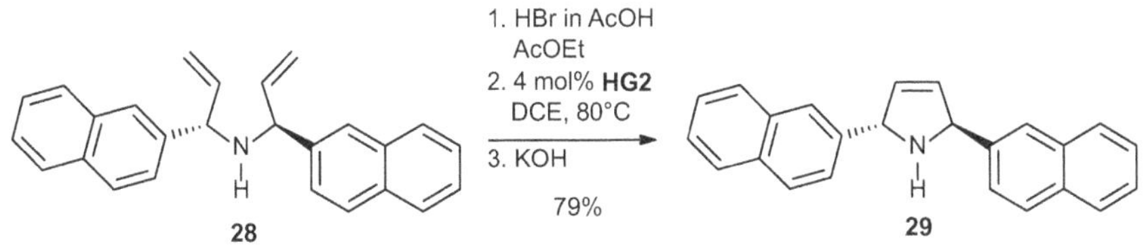

Scheme 12 Synthesis of secondary cyclic amine **29** by RCM

see [46]). Following this approach, *N*-benzyl 3-pyrroline has been obtained from the corresponding diallylamine by treatment with 5 mol% of **S** catalyst in ionic liquids [47].

3 Six-Membered Cyclic Amines

3.1 RCM of Hindered Amines

Ring-closing metathesis was used by several groups as a key step to construct the final E ring of Meloscine [48–53]. Mukai et al. reported an interesting highly stereoselective approach in which only one of the two diastereotopic vinyl groups was engaged to provide the pentacycle of (±)-Meloscine [51] (Scheme 13).

Another example of a good RCM substrate based on a neopentylic endocyclic amine has been reported by Prasad and Nidhury en route to (+)-Eburnamonine (Scheme 14) [54]. Treatment of diene **30** with 5 mol% of **G2** catalyst in DCM at rt afforded the corresponding indolo[2,3-a]quinolizine derivative **31** in a very good yield. A related RCM reaction from close analogues of **30** with less sterically hindered amino groups was performed by the same group to provide in lower yields azepane **33**, an advanced intermediate in the synthesis of (-)-Quebrachamine (Scheme 14).

Schrock and Hoveyda et al. reported the enantioselective total synthesis of the same alkaloid using a new class of chiral Mo-based complexes [55] (Scheme 15). The tetracyclic core of this natural adrenergic blocking agent was synthesized from neopentylic endocyclic amine **34** by highly enantioselective RCM catalyzed by chiral molybdenum catalyst **36**.

The synthesis of the six-membered D ring of (+)-Isolysergol was described using a similar approach [56]. It is noteworthy that reaction of triene **37** with **G1**, **G2,** or **S** catalysts either at rt or 100°C failed to give the corresponding tetracycles **38–39**. Heating **37** using microwave irradiation (300 W) in the presence of molybdenum catalyst **S** however induced a diastereoselective RCM to give **38** and **39** in 44% yield (dr 4.5/1 in favor of **38**). Better yields were finally obtained with chiral Schrock–Hoveyda catalyst **40** (75% yield, dr ~ 3/1 in favor of **38**) (Scheme 16).

Formation of polycyclic systems has been also performed by RCM reactions from pyrrolidines or piperidines with crowded endocyclic amines di- or trisubstituted in α

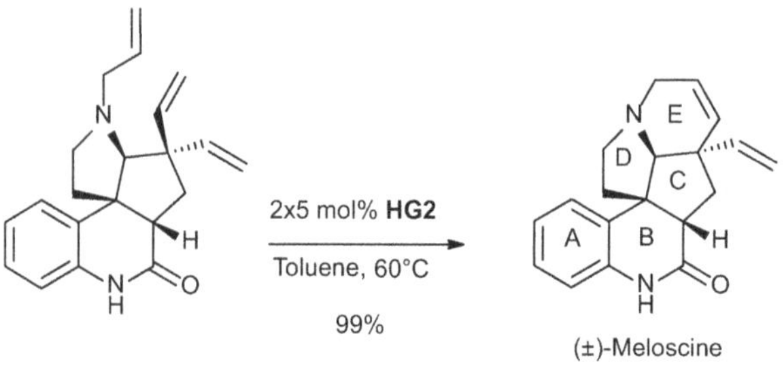

Scheme 13 Diastereoselective formation of the E ring of (±)-Meloscine by RCM

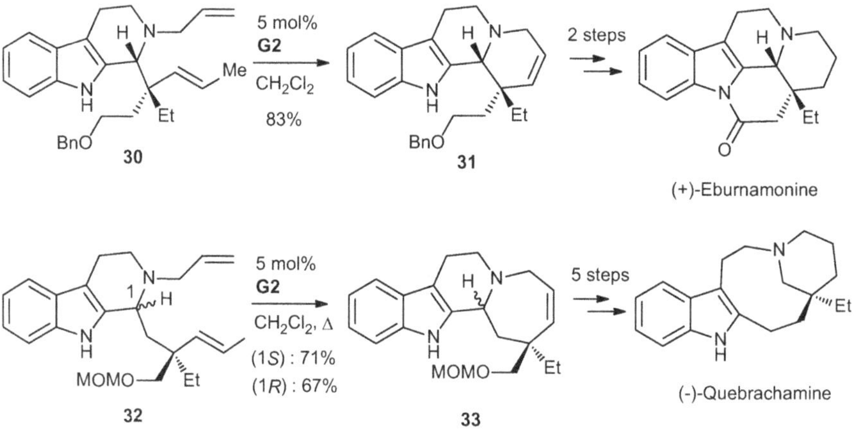

Scheme 14 Synthesis of (+)-Eburnamonine and (-)-Quebrachamine by way of RCM

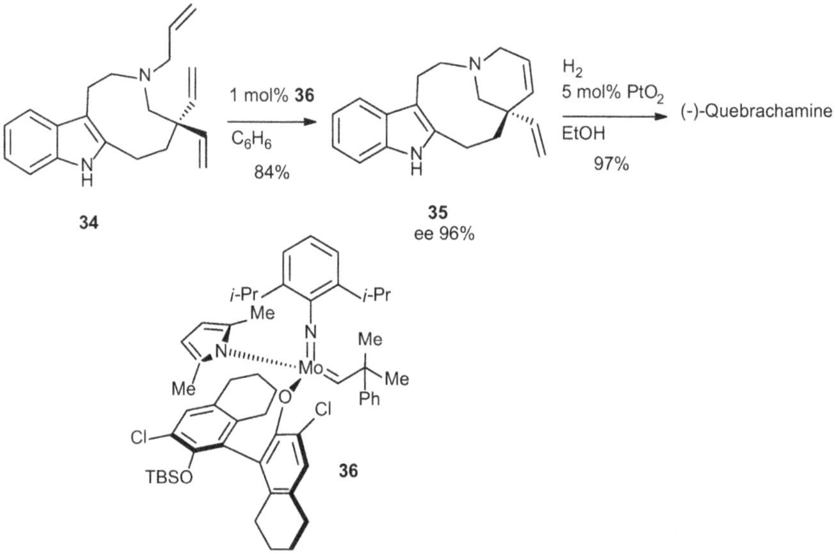

Scheme 15 Synthesis of (-)-Quebrachamine by way of enantioselective RCM

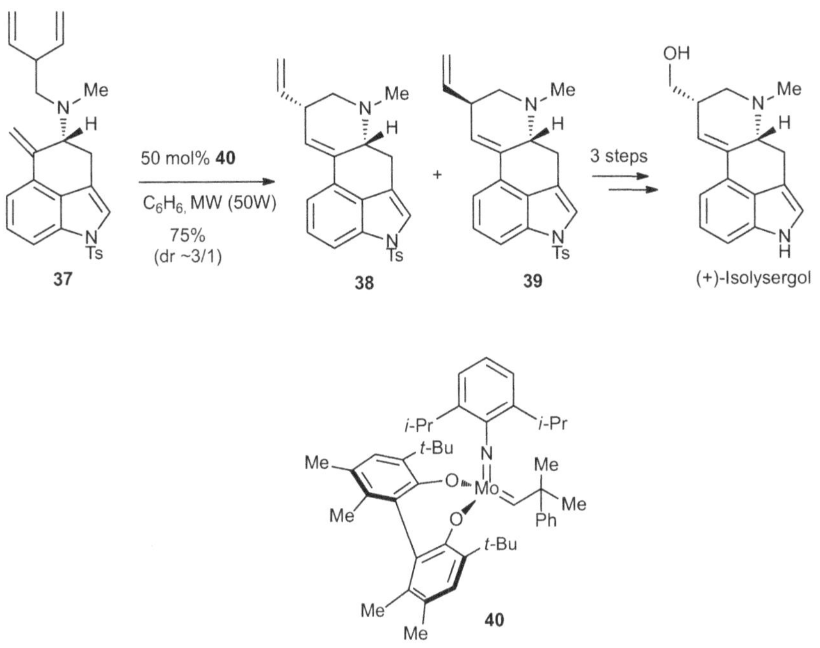

Scheme 16 Synthesis of (+)-Isolysergol by way of diastereoselective RCM

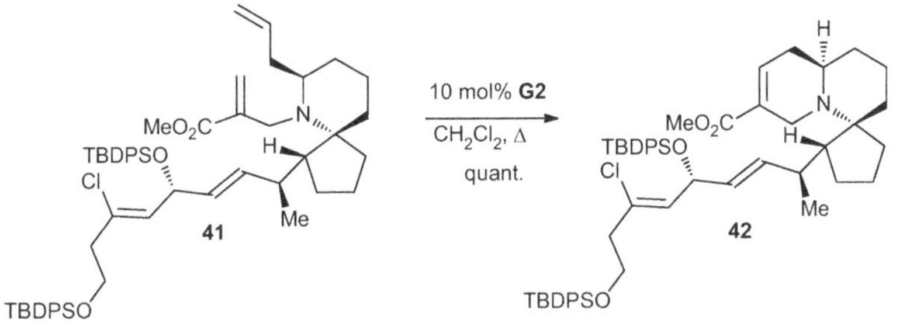

Scheme 17 Synthesis of a tricyclic precursor of Halichlorine

and α′ positions [57–59]. For example, Arimoto et al. reported the construction of tricyclic compound **41**, an advanced intermediate in the synthesis of Halichlorine, from the azaspiranic derivative **42** in quantitative yield (Scheme 17) [57].

The tricyclic core of frog alkaloid (-)-205B was synthesized by RCM of **43** conducted with 5 mol% of **G2** catalyst at 55°C in *tert*-butyl methyl ether (Scheme 18) [58].

RCM reactions of less hindered cyclic amine substrates have been also reported [60–63]. However, these compounds required in general the use of more drastic conditions as shown with *N*-allyl pyrrolidine **45** [60, 61]. For example, upon

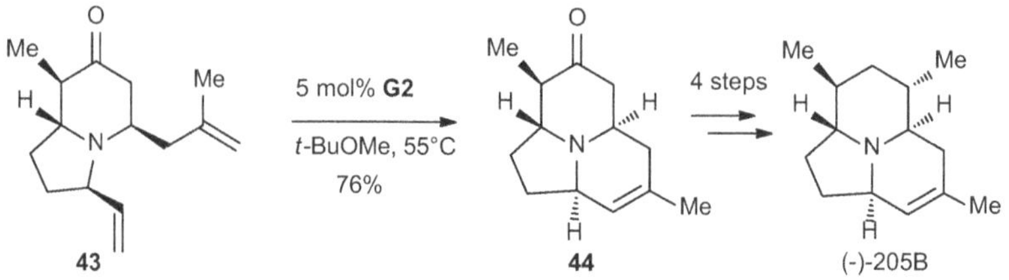

Scheme 18 Synthesis of a tricyclic precursor of frog alkaloid (-)-205B

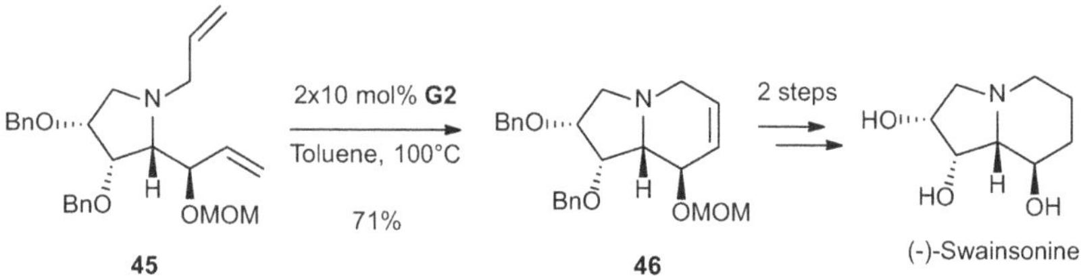

Scheme 19 Synthesis of the indolizidine core of (-)-Swainsonine

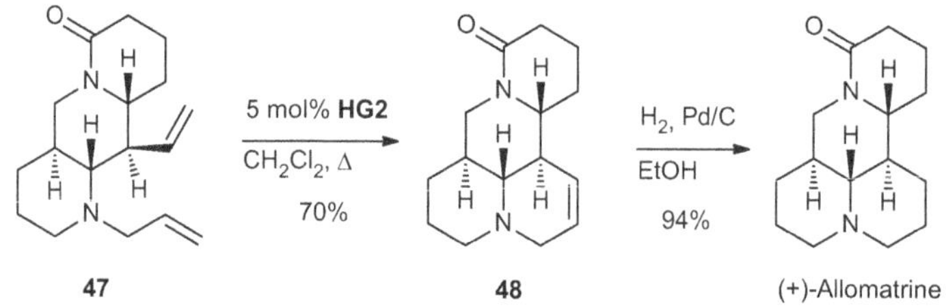

Scheme 20 Synthesis of the B ring of (+)-Allomatrine by RCM

treatment with 2 × 10 mol% of **G2** catalyst in toluene at 100°C, **45** afforded the indolizidine skeleton of (-)-Swainsonine in 71% yield (Scheme 19) [60].

For more conformationally constrained cyclic amine substrate, such as tricyclic compound **47**, the RCM reaction proceeded in good yield under classical conditions using 5 mol% of **HG2** catalyst in DCM (Scheme 20) [63].

The challenging construction of highly strained amine-containing polycyclic structures may be achieved by RCM as demonstrated by the group of Tokuyama (Scheme 21) [62]. In addition to the presence of an endocyclic basic amino group, substrate **49** is prone to oxidation and olefin migration. The oxidation of the tertiary amine to the corresponding enamine with Grubbs' catalysts was found to be accelerated by the strained character of the piperidine ring in **49**. After extensive optimization study, compound **50**, the potential precursor of strictamine, was obtained in 52% yield (66% based on recovered starting material) by RCM; treatment of **49** with 10 mol% **HG2** catalyst and 20 mol% of 1,4-benzoquinone (1,4-BQ) in toluene at 70°C under microwave irradiation was found to strongly

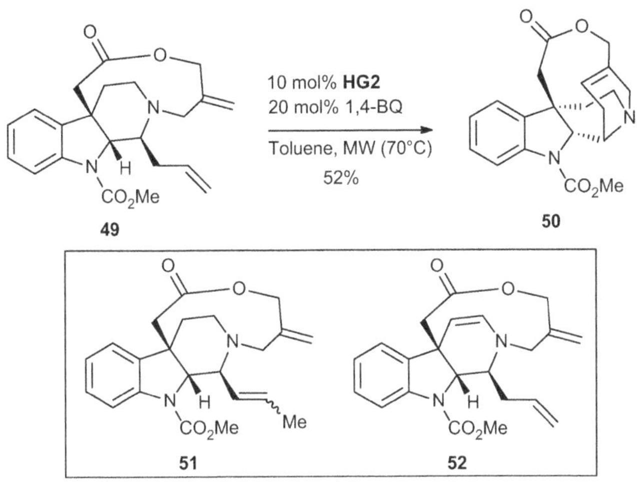

Scheme 21 Optimized conditions for the synthesis of highly strained amine-containing polycycle **50** by RCM

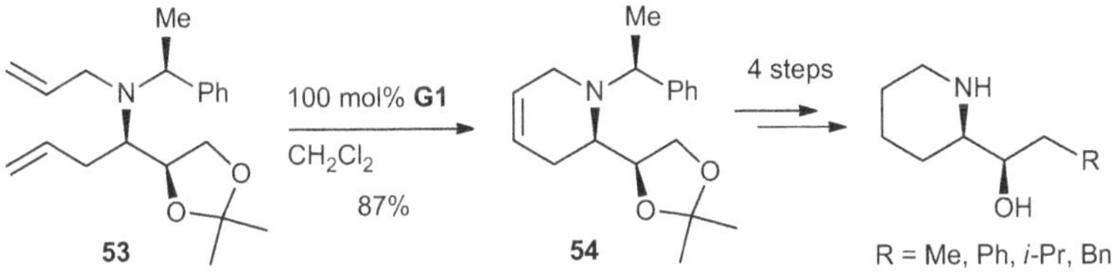

Scheme 22 Synthesis of β-(+)-Conhydrine (R=Me) and its analogues by RCM

reduce isomerization (compound **51**) and the formation of the oxidation product (compound **52**).

Introducing steric hindrance around the amino group by using *N*-α-methylbenzyl protecting group may represent an interesting alternative to deactivation of the nitrogen atom by electron-withdrawing groups [13]. This approach avoids the potential formation of unproductive metallacycles due to the presence of a chelating carbonyl group. Following this strategy, Gálvez et al. have reported the synthesis of β-(+)-Conhydrine and several analogues by way of RCM of acyclic amine **53** (Scheme 22) [64]. A large amount of **G1** catalyst (one equivalent) was used in the conversion of **53** to piperidine **54** which is quite surprising by comparison with related RCM reactions using similar substrates and catalytic amount of ruthenium catalysts [13].

Hindered amines that require the use of a Brønsted or Lewis acid to provide six-membered azacycles by RCM have been reported in the literature [65, 66]. For example, RCM of pyrrolidinol derivatives **56** by using either **G1** or **G2** catalysts in toluene or DCM (rt to reflux) failed to give the expected cyclized products **57** [66]. However, the mixture of stereoisomers **56** underwent RCM with 6 mol% of **G2** catalyst in the presence of two equivalents of 4-toluenesulfonic acid in refluxing

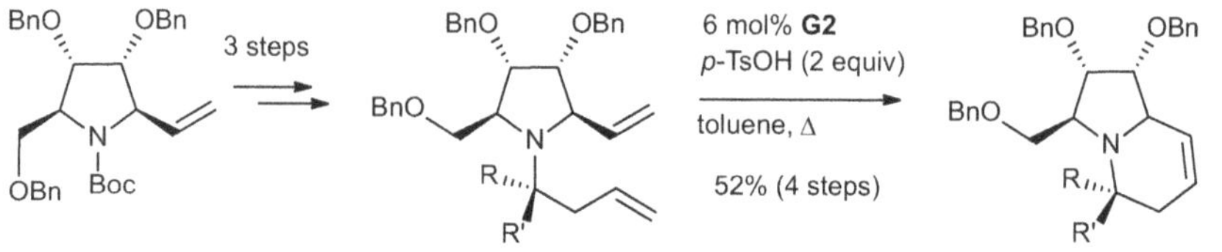

Scheme 23 Formation of the six-membered cycle of indolizidine derivatives by RCM

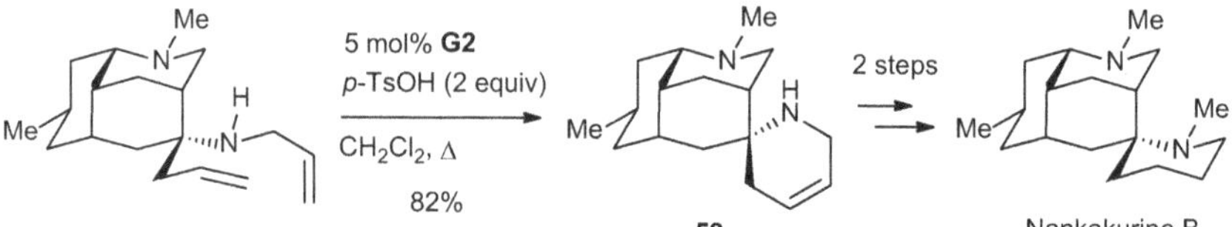

Scheme 24 Formation of the six-membered cycle of Nankakurine B by RCM

toluene to give indolizidine derivatives **57** in 52% yield over four steps from **55** (Scheme 23). It is noteworthy that related analogues of **56**, such as **10**, **12**, or **45** (Schemes 5, 6, and 19), underwent RCM in acceptable to very good yields without the use of Brønsted acid.

In situ formation of ammonium salt by addition of a Brønsted acid was found to be a powerful strategy with **58**, a challenging substrate bearing two basic amino groups including a secondary amine [67] (for a related example using Ti(*i*OPr)$_4$ additive, see [68]) (Scheme 24). Other examples of RCM reactions with hindered amines using a Brønsted acid or Ti(*i*OPr)$_4$ have been published in the literature (see, e.g., [69, 70]). Unfortunately, the authors did not report if the corresponding reactions from the free amines have been performed and consequently if the use of such additives was mandatory.

3.2 Ruthenium-Catalyzed RCM of Amines α-Substituted with an Electron-Withdrawing Group

Acyclic amines α-substituted with one or two electron-withdrawing groups have been found to be good substrates for RCM reactions [71–75]. For example, in the context of a new strategy to access 3-hydroxypyridines, Imamoto and Yanagisawa et al. have reported the synthesis of 1,6-dihydro-2*H*-pyridin-3-ones **61a–d** by RCM of α-amino ketones **60a–d** (Scheme 25) [71]. As expected, increasing the steric hindrance of the substituent around the nitrogen atom was found to improve the yield of cyclized products **61a–d** (Scheme 25). The best results were nevertheless

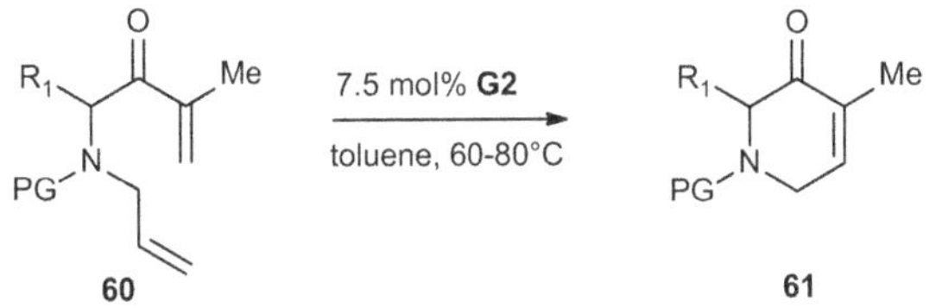

60 **61**

Substrate	temp (°C)	R_1	PG	Product	Yield
60a	60	CH_3	Bn	**61a**	49%
60b	60	CH_2OMOM	Bn	**61b**	71%
60c	60	CH_2Ph	Bn	**61c**	83%
60d	80	CH_2Ar	Bn	**61d**	92%
60e	60	CH_2Ph	Ts	**61e**	97%

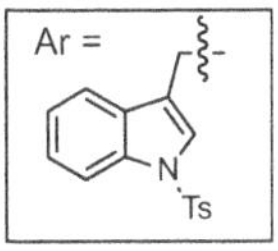

Scheme 25 Synthesis of 1,6-dihydro-2*H*-pyridin-3-ones by RCM

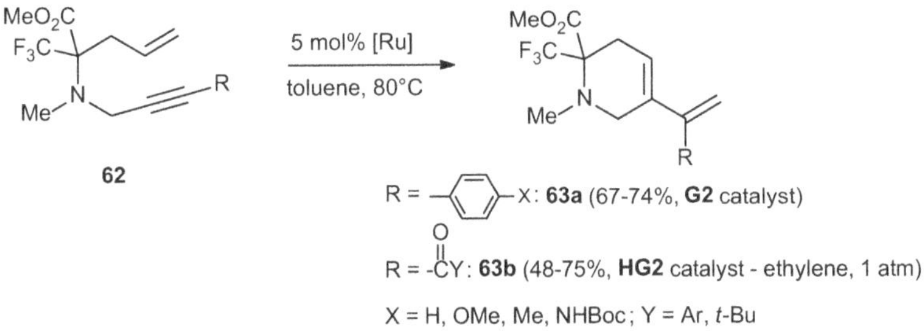

Scheme 26 Synthesis of α-CF_3-substituted pipecolic acid derivatives by ring-closing enyne metathesis

obtained for sulfonamide derivatives as exemplified by the yields obtained for the preparation of **61c** and its corresponding *N*-tosyl analogue **61e** [71].

Ring-closing metathesis has been also reported with acyclic amines α-substituted with two electron-withdrawing groups [72, 73]. Osipov et al. have, for example, reported the ring-closing metathesis of aryl- or acyl-substituted 1,7-enynes **62** (Scheme 26) [72]. This process allows the access to α-CF_3-substituted pipecolic acid derivatives **63** containing the synthetically useful 1,3 diene moiety. 1,7-Enynes with electron-withdrawing acyl groups displayed low reactivity, acceptable yields being obtained only with **HG2** catalyst under an ethylene atmosphere.

Not surprisingly, and despite the presence of two ester groups in α-position to the nitrogen atom, secondary amine **64** was converted to its corresponding chloride ammonium salt before treatment with **G2** catalyst (Scheme 27) [73]. After a basic work-up, the mono-unsaturated piperidine **65** was obtained in high yield.

Conversion of cyclic α-amino ketones was also reported as shown with the synthesis of isoquinolone **67** obtained in high yield from the corresponding *N*-allyl

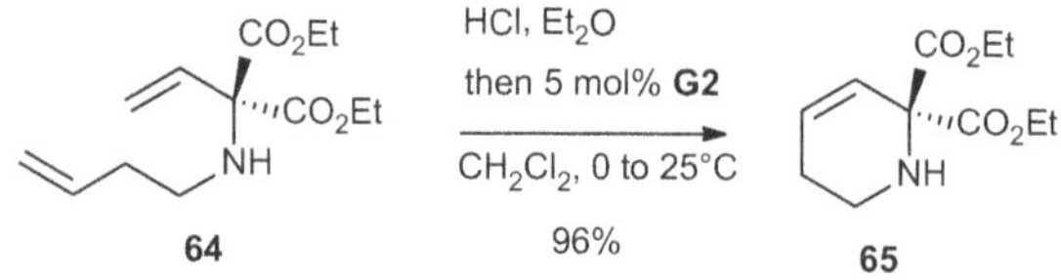

Scheme 27 Synthesis of piperidine **65** by RCM

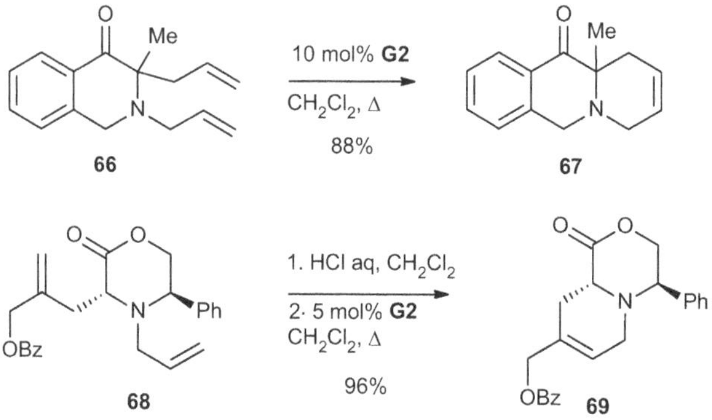

Scheme 28 Synthesis of the piperidine ring of polycyclic systems by RCM

isoquinolone **66** (Scheme 28) [74]. This substrate indeed combines two favorable effects (decrease electronic density around the amine and steric hindrance). Despite the same advantage, the RCM of morpholine **68** was performed on the corresponding chloride ammonium salt to provide an advanced intermediate in the asymmetric synthesis of substituted pipecolic esters (Scheme 28) [75].

3.3 RCM of Phenylamines and Related Analogues

As shown with the synthesis of pyrroline (Scheme 10), phenylamine and related analogues are in general good substrates for ruthenium-catalyzed RCM [13, 29–33]. The construction of the piperidine ring of di- to tricyclic systems has been performed following this process in high yields [76–78]. For example, Yoshida et al. have reported the efficient conversion of diene **70** to produce the corresponding strained tricycle, azabicyclo[3.3.1]nonene **71**, in 90% yield (Scheme 29) [76].

RCM of arylamines α-substituted with an electron-withdrawing group was also reported with good to very high yields [79–83]. Ganguly et al. have recently published a novel synthesis of *ortho–ortho* disubstituted biphenyls containing eight-membered lactam rings of therapeutic interest [79–81]. The key RCM step afforded the expected tetracycles **73** in 67–76% yield (Scheme 30).

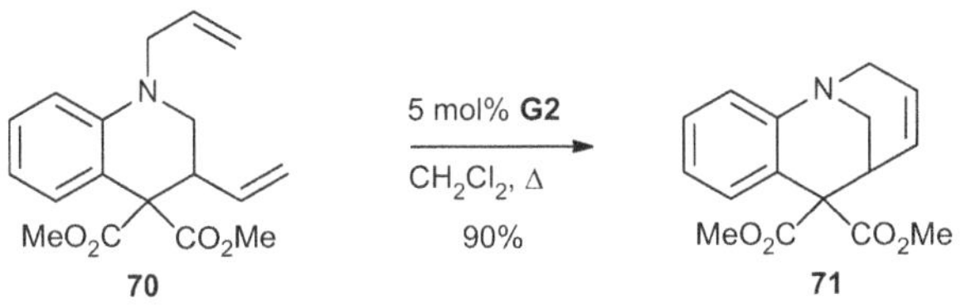

Scheme 29 Synthesis of the piperidine ring of azabicyclo[3.3.1]nonene **71** by RCM

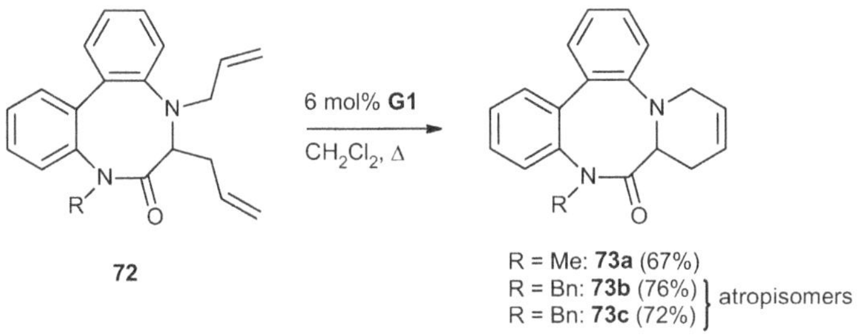

Scheme 30 Synthesis of the piperidine ring of tetracycles **73** by RCM

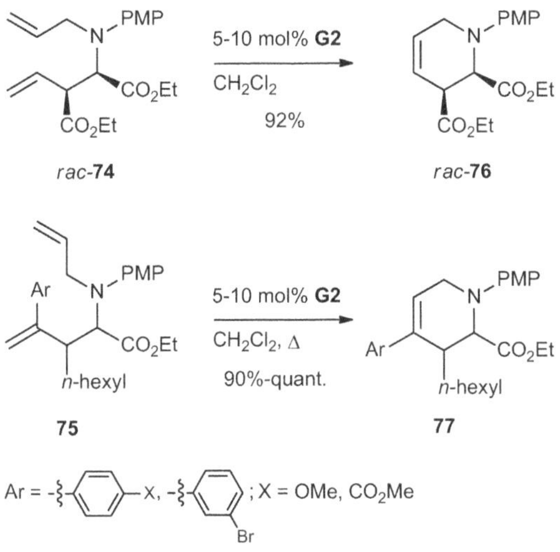

Scheme 31 Synthesis of polysubstituted pipecolic analogues **76–77** by RCM

Synthesis of polyfunctionalized pipecolinic acid analogues **76–77** was also reported by way of RCM of tertiary amines **74–75** protected with a *p*-$MeOC_6H_4$ (PMP) group (Scheme 31) [82, 83] .

Enantioselective RCM of phenylamine-containing dienes has been described with chiral metathesis catalysts [84, 85]. To date, molybdenum complexes exhibit

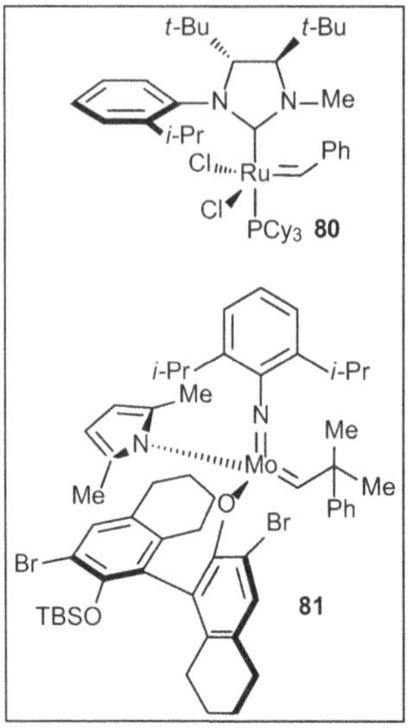

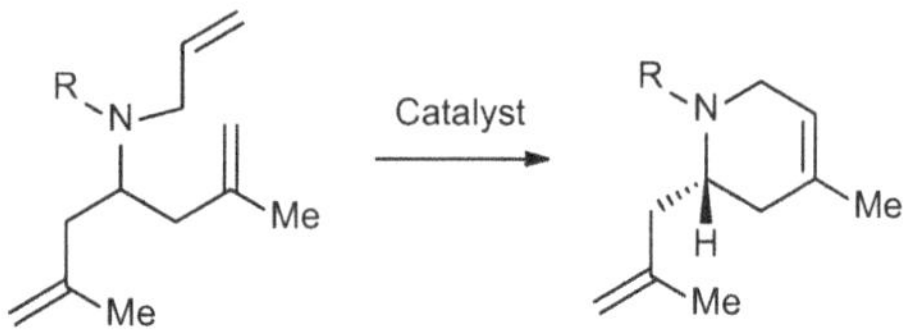

78a (R = Ph)
78b (R = H)

79

Catalyst	Solvent	Temp	R	e. e.	Yield (Product)
80 (2.5 mol%)	CH_2Cl_2	40°C	Ph	9%	39% (**79a**)
81 (1 mol%)	benzene	22°C	Ph	93%	91% (**79a**)
81 (2.5 mol%)	benzene	22°C	H	67%	89% (**79b**)

Scheme 32 Synthesis of piperidine **79** by enantioselective RCM

the best results in terms of yield and enantioselectivity as exemplified by tertiary amine **78a** (Scheme 32). Remarkably, catalyst **81** is able to promote RCM of secondary allylamine **78b** although the ee drops to 67% [84]. This result constitutes a rare example of secondary amines that are good substrates for metathesis reactions without the help of Brønsted or Lewis acid (for another example see Scheme 35) [13].

3.4 Exceptions to the Rule?

Based on literature precedents, it was established as a rule that direct RCM synthesis of cyclic amines could be successfully performed by preventing coordination to ruthenium–alkylidene complexes following two main strategies. The first one exploits the steric hindrance around the substrate amino function. The second strategy is aimed at decreasing the electronic density around the nitrogen atom by using adjacent electron-withdrawing groups or by exploiting the lower basicity of phenylamine derivatives [13]. However, exceptions to this rule do exist as shown with the smooth conversion of unhindered acyclic amines **82a–83** into the corresponding piperidine derivatives **84a–85** (Scheme 33) [86, 87] (for examples of RCM synthesis of 2-substituted piperidine derivatives, see [88, 89]; for example of RCM synthesis of quinolizidine derivatives from unhindered substrates, see [90]). Interestingly, all attempts to convert **82b**, a close analogue of **82a**, into the corresponding piperidine failed under a variety of RCM reaction conditions including the use of Lewis acid to prevent nonproductive chelate formation [14, 15]. This result indicates that simple modification of the amine-containing starting material

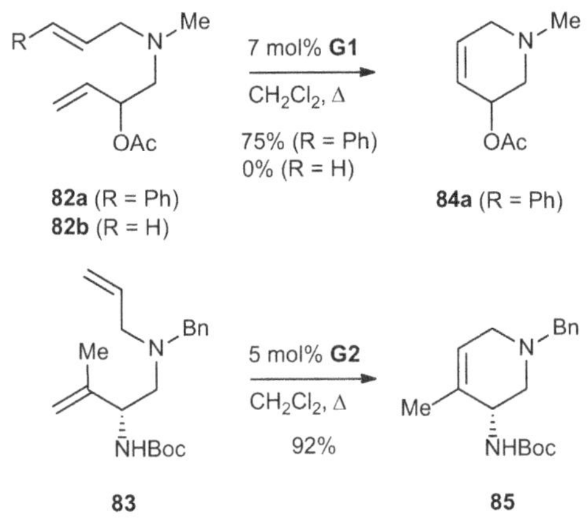

Scheme 33 Synthesis of piperidine derivatives from unhindered substrates

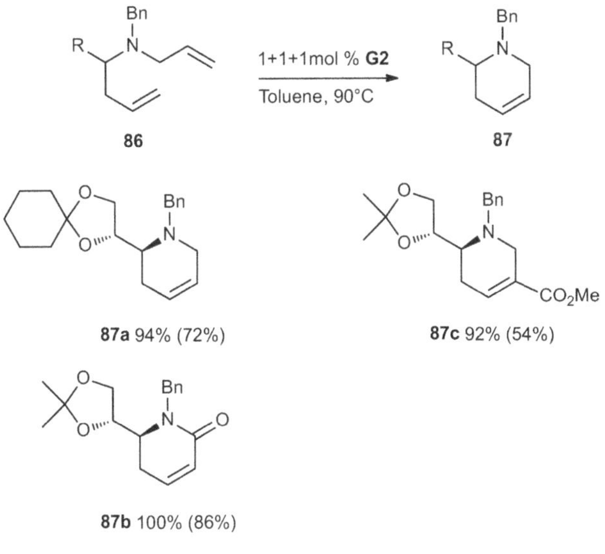

Scheme 34 Improved RCM synthesis of piperidine derivatives using portion-wise addition of 3 mol% of **G2** catalyst (yields in brackets correspond to a single addition of 3 mol%)

by incorporation of a phenyl olefin substituent may represent a useful option to enable successful RCM reactions. The choice of the olefin substituent is indeed important because it directly affects the stability and the reactivity of the resultant catalytic alkylidenes [91].

In the context of the synthesis of nitrogen-containing glycomimetics, Vankar et al. developed an improved method for performing RCM in the presence of basic amines [92]. By adding portion-wise 3 mol% of **G2** catalyst to a solution of dienes **86** in toluene at 90°C (3 portions at equal time intervals) instead of a single addition, a significant increase in yield (up to 38%) was observed (Scheme 34). Slight

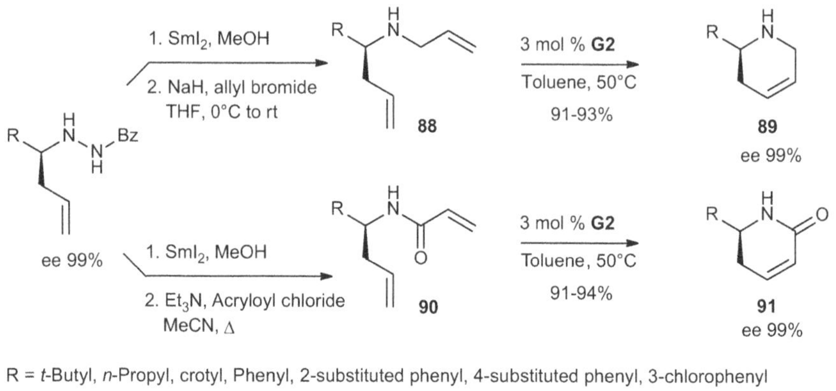

Scheme 35 Synthesis of 2-substituted piperidine derivatives from secondary allylamines

increase in yield was also observed for the synthesis of **87b**, a close analogue of **87a**, with a deactivated nitrogen lone pair.

Quite surprisingly, Jang et al. reported the synthesis of 2-substituted piperidines **89** by way of RCM of secondary allylamines **88** in more than 91% yield (Scheme 35) [93]. Similarly to the procedure reported by Vankar et al. (Scheme 34), 3 mol% of **G2** catalyst was added to a solution of the reaction mixture in toluene, but in a single addition and at a lower temperature (50°C). It is worth mentioning that RCM of **90**, the amide analogues of **88**, provided the corresponding cyclized products **91** in also more than 91% yield.

4 Seven to Nine-Membered Cyclic Amines

As described in Sects. 2 and 3, arylamines are one of the substrates of choice for the synthesis of medium-size cyclic amines by the way of ring-closing metathesis [94–98]. An interesting example has been reported by Benedetti and Penoni et al. for the synthesis of benzazepine derivatives (Scheme 36) [98]. Depending on the condition reactions, desired products **93** and **94** or dimeric products **95** and **96** were synthesized. Expected products **93** and **94** were obtained only when the ring-closing ene–yne metathesis (RCEYM) was run in the presence of **HG2** catalyst under an ethylene atmosphere. Classical inert atmosphere favored dimeric products **95** and **96** by RCEYM/homo cross coupling reaction. It is worth mentioning that no dimeric products were obtained when the amine substrate was protected as an amide or a tosylate.

Stevens et al. reported the synthesis of 1-phosphonylated benzazocines using RCM as the key step [99]. Compound **97** in the presence of **G1** gave desired products **98** in good yield (75%) whereas RCM with **G2** catalyst afforded unexpected results. In addition to the desired compounds **98**, isomerized products **99** were indeed also obtained (Scheme 37). In order to have full conversion toward

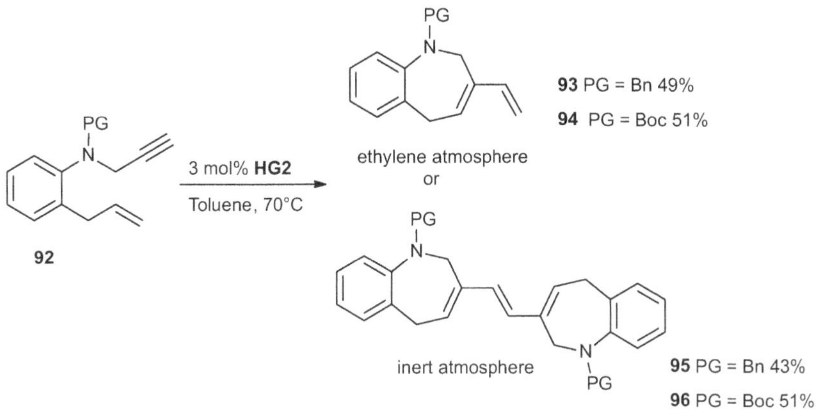

Scheme 36 Influence of the reaction conditions for the synthesis of benzazepine derivatives

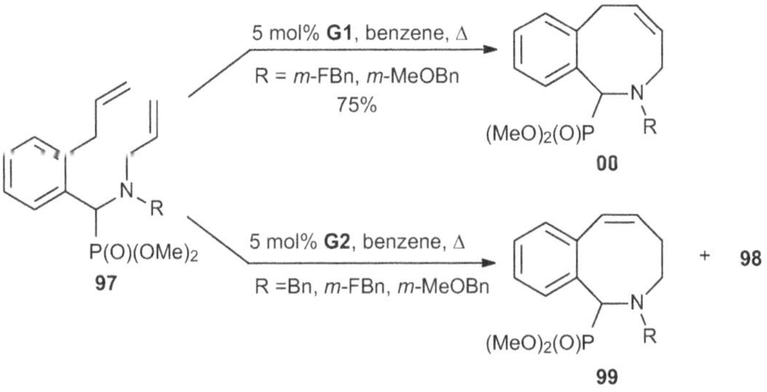

Scheme 37 Influence of the metathesis catalyst in the synthesis of phosphonylated benzazocines

compounds **99**, an isomerization catalyst ($RuClH(CO)(PPh_3)_3$) was added to the metathesis mixture. In these conditions, benzazocines **99** were obtained in 70–74% yields. The authors explained the isomerization step by the generation of Ru–H complex from **G2**. It is indeed known that **G2** can be decomposed toward complex **100** and that this complex can isomerize double bonds (Fig. 1) [100].

This example highlights the importance of selecting the right catalyst for any given reaction. Furthermore, isomerization is one of the most frequent side reactions during RCM (see also Scheme 21). This side reaction can also result from the purification or isolation step. Residual ruthenium species may indeed cause isomerization, decomposition, or other undesired side reactions. Consequently, strategies to remove residual ruthenium from RCM mixture have been developed [101]. For example, Davies et al. removed the spent catalyst by using $P(CH_2OH)_3$ for the preparation of eight-membered ring **101**, an intermediate in the synthesis of polyhydroxylated pyrrolizidines (Scheme 38) [102, 103]. This example further

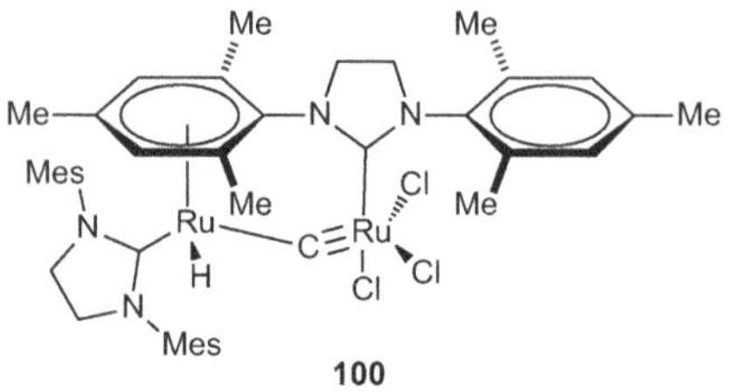

100

Fig. 1 Structure of complex **100** formed by the thermal decomposition of Grubbs 2 catalyst

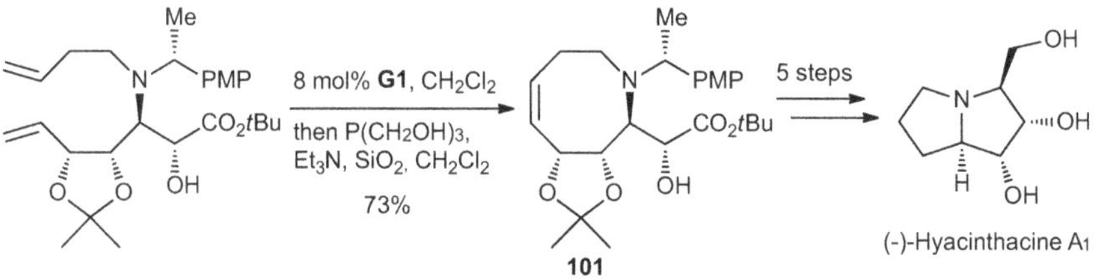

101

Scheme 38 Synthesis of (-)-Hyacinthacine A_1 using $P(CH_2OH)_3$ for the removal of Grubbs 1 catalyst

highlights the use of *N*-α-methylbenzyl protecting group in RCM (see also Scheme 22).

Synthesis of cyclic β-amino alcohols [104], azabicyclic derivatives [105], and polyhydroxylated alkaloid analogues [106] has been performed by RCM of the corresponding ammonium salt. In 2011, Dixon et al. reported an elegant synthesis of natural product Nakadomarin A based on RCAM (ring-closing alkyne metathesis) followed by *syn* reduction of the alkyne for the construction of the larger ring. The synthesis of the medium ring is performed by means of an olefin–olefin RCM [107]. Thus, alkyne ring-closing metathesis of **102** catalyzed by Schrock tungsten neopentylidyne catalyst followed by selective alkyne hydrogenation afforded **103**. RCM on this intermediate proved problematic and amide functions were reduced to give *bis*-amine **104**. Treatment of **104** with **G1** in the presence of (+)-CSA afforded the expected natural product in good yield (Scheme 39).

Synthesis of medium-size amines by way of RCM follows the same rules establishing in Sects. 2 and 3: compounds with steric hindrance around the amino function (Scheme 38) (for the synthesis of azetidine-fused eight-membered cyclic amine, see [108]; for another examples highlighting the use of *N*-α-methylbenzyl protecting group for the synthesis of medium cyclic amines by the way of RCM, see [109, 110]) or with a decreasing of the electron density on the nitrogen atom (Schemes 36 and 37) are generally good substrates for RCM. However surprising results have been described. For example, the groups of Tambar and Titanyuk have independently reported that RCM of amines α-substituted with electron-withdrawing group did not proceed without the presence of Brønsted or Lewis acid. [111, 112]. Tambar et al. reported the synthesis of seven-membered ring

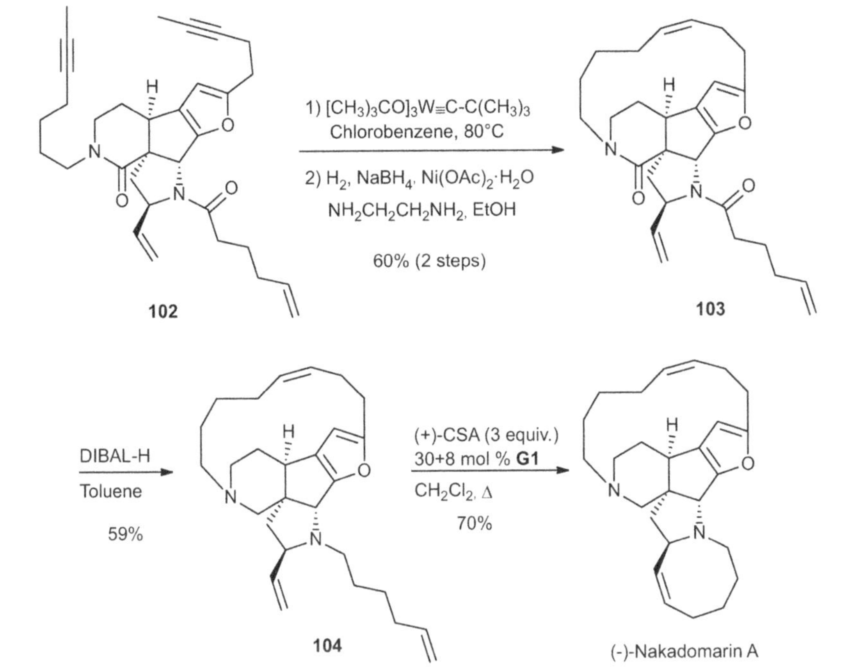

Scheme 39 Synthesis of (-)-Nakadomarin A

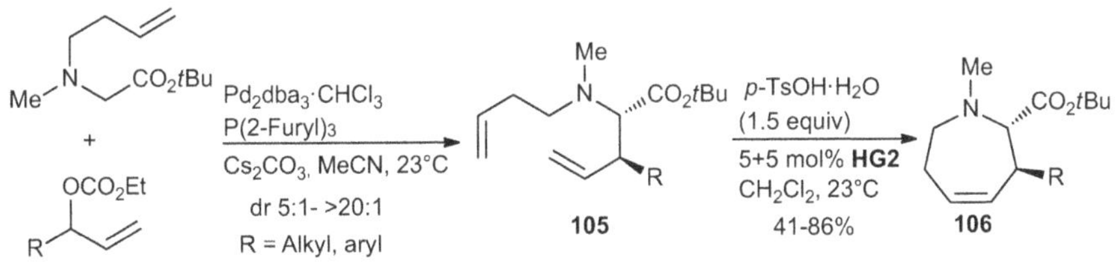

Scheme 40 Synthesis of cyclic amino acid derivatives **106** by RCM

cyclic amino acid derivatives **106** by a two-step strategy combining tandem allylic amination/Stevens rearrangement to give **105** followed by a ring-closing metathesis (Scheme 40) [111]. The tandem allylic amination/Stevens rearrangement occurred with good to excellent diastereoselectivity. For RCM, additive is necessary as the reaction did not proceed in the presence of the basic nitrogen atom. The authors examined Ti(O-*i*Pr)$_4$ and *p*-TsOH. The cyclization occurred only with the Brønsted acid. It is also interesting to notice that Grubbs 2 was ineffective as catalyst for the ring-closing metathesis of **105**. One example of eight-membered ring was also reported with the same efficiency [111].

Surprisingly, Titanyuk et al. reported that Grubbs 2 in the presence of Ti(O-*i*Pr)$_4$ can promote the formation of amino acid derivatives **108** by way of RCM, although substrates **105** and **107** are quite similar. Six- and seven-membered rings were

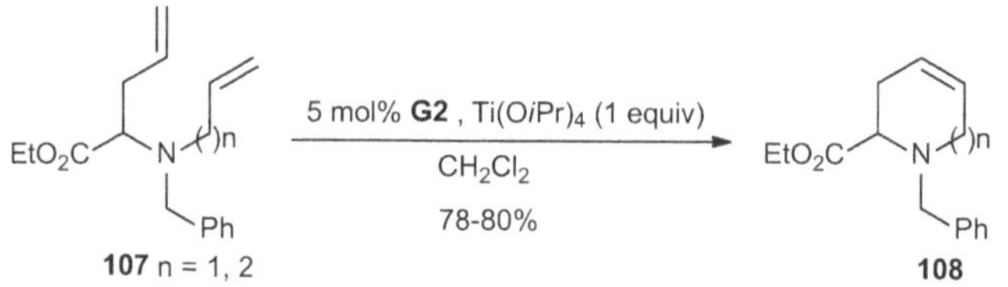

Scheme 41 Synthesis of cyclic amino acid derivatives **108** by way of RCM

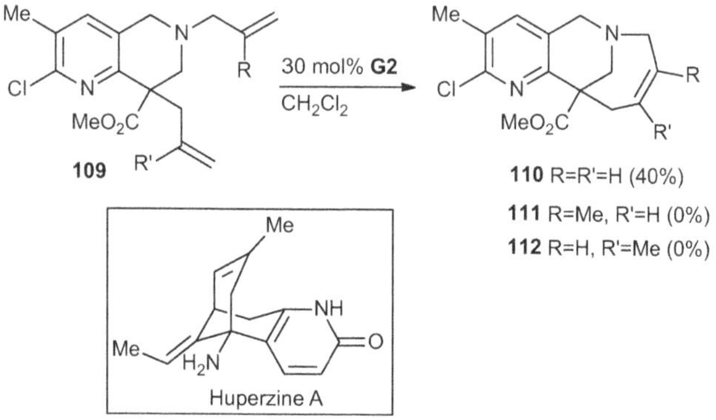

Scheme 42 Synthesis of an analogue of Huperzine A

obtained in good yields (Scheme 41), but this reaction did not proceed cleanly for the formation of eight membered rings [112].

Compernole et al. reported the synthesis of **110**, an analogue of Huperzine A, a natural product which is a potent, reversible, and selective inhibitor of AChE (acetylcholinesterase) (Scheme 42). This enzyme is one of the most popular targets for Alzheimer's disease treatment. The key step for the synthesis of **110** is a late stage RCM. Thus, in the presence of 30 mol% of **G2**, precursor **109** afforded the target molecule in a moderate yield of 40%. Under the same conditions, more substituted analogues **111** and **112** were not obtained showing that steric hindrance has detrimental effects. Furthermore, the authors explained the moderate yield of **110** and the need of large loading of catalyst by the possible formation of a stable complex (six-membered ring) involving the ester group and the proximate allyl function [113].

Van der Eycken et al. reported the synthesis of nine-membered ring analogues of Buflavine, a natural product exhibiting interesting α-adrenolytic and anti-serotonin activities. The key step was RCM of biaryl amines **113**. The best conditions found made use of 3 mol% of **G2** under microwave irradiation (Scheme 43). Other catalysts or conventional heating conditions afforded no conversion or low yields in the desired products **114** [114].

RCM of unhindered amines **115** has been reported with good yields using large loading of catalyst **G1** (Scheme 44) [115].

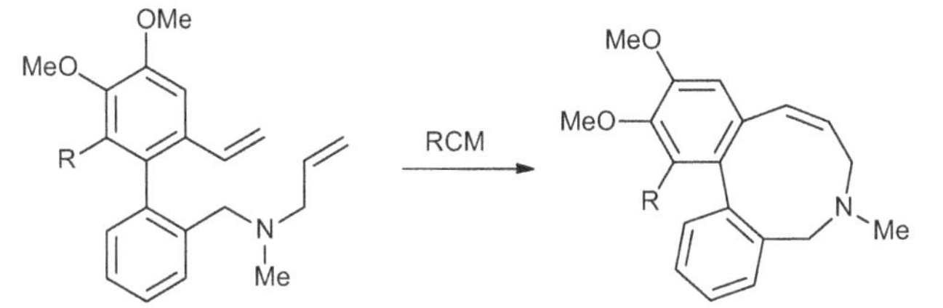

113 114

Substrate	R	Reaction conditions	Product	Yield
113a	H	Toluene, reflux, 3 mol% **G1** or **HG1**	**114a**	0%
113a	H	Toluene, 125°C, μW, 3 mol% **G1** or **HG1**	**114a**	0%
113a	H	Toluene, reflux, 3 mol% **G2**	**114a**	15%
113a	H	Toluene, 125°C, μW, 3 mol% **G2**	**114a**	55%
113b	OMe	Toluene, 125°C, μW, 3 mol% **G2**	**114b**	54%

Scheme 43 Synthesis of nine-membered ring analogues of Buflavine

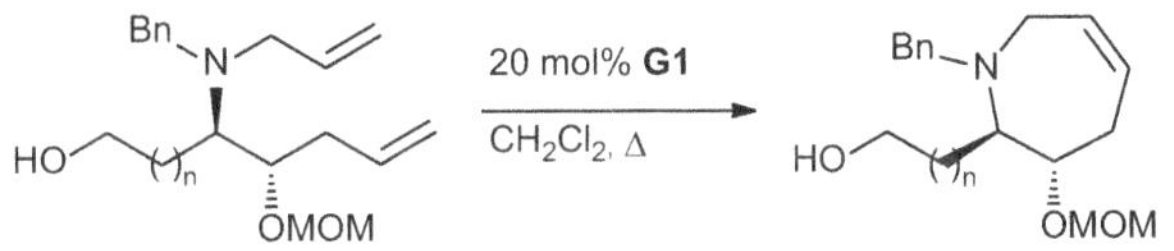

115a (n=1)
115b (n=2)

n = 1 : **116a** (73%)
n = 2 : **116b** (51%)

Scheme 44 Synthesis of azepane derivatives by RCM

During their synthesis of natural product (+)-Manzamine A, Fukuyama et al. built the eight-membered ring by way of RCM (Scheme 45) [116]. Ring-closing metathesis of intermediate **117** is quite challenging due to the high density of reactive functionalities. Reproducible yields were only obtained when a stoichiometric amount of catalyst **119** was used in the presence of *p*-methoxyphenol.

Enantioselective synthesis of azepane derivatives has been described with similar catalysts than the ones used for asymmetric synthesis of six-membered cyclic amines (Schemes 15 and 32). The strategy reported by Schrock et al. was based on an enantioselective RCEYM of diallyl **120** (Scheme 46) [117]. Generally, *endo* products are preferentially obtained by enyne metathesis catalyzed by Mo or W complexes whereas *exo* products are favored with Ru catalysts. Desymmetrization of **120** using chiral catalyst **123** afforded azepane **121** in good yield, enantioselectivity, and diastereoselectivity when benzene was used as solvent. Ether was the optimal solvent in terms of enantioselectivity (*ee* = 91%) although the yield dropped to 35%.

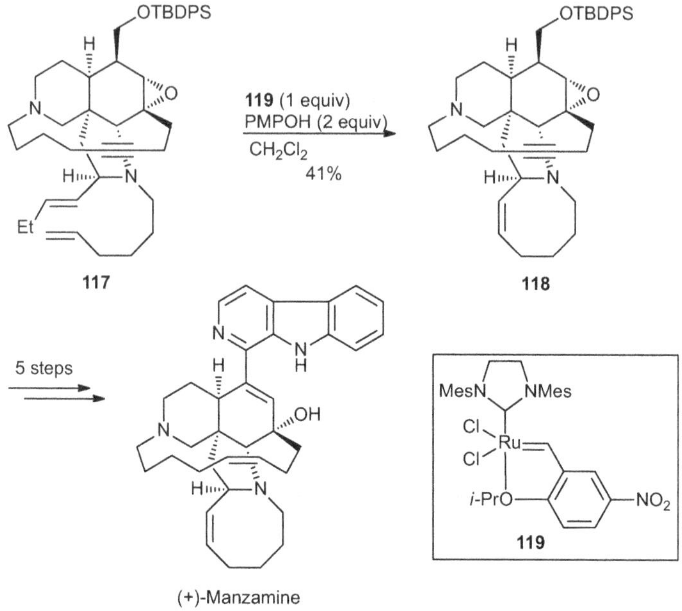

Scheme 45 Synthesis of the natural product (+)-Manzanine A

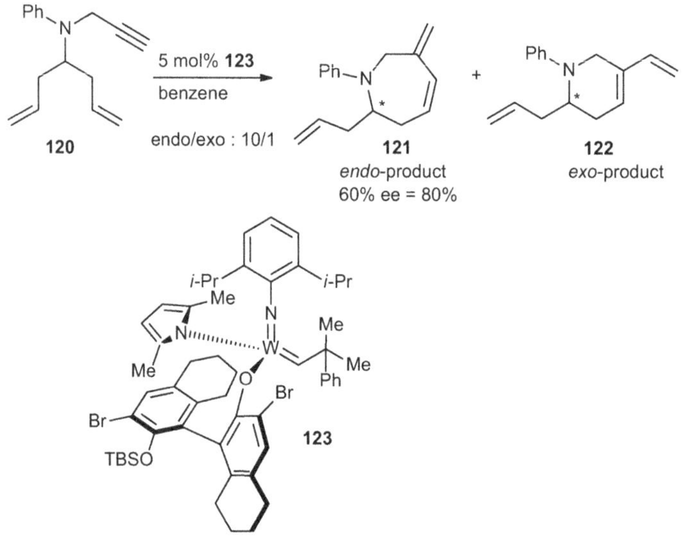

Scheme 46 Asymmetric synthesis of azepane **121** by enantioselective enyne metathesis

5 Macrocyclic Amines

Tokuyama et al. reported the synthesis of both enantiomers of Petrosin, an alkaloid-possessing activity against HIV [118]. Formation of the 16-membered ring was made by RCM of the diamine **124** (Scheme 47). This intermediate contains two amine functions and the metathesis afforded key molecule **125** in high yield. It is worth mentioning that this reaction is very fast since complete conversion of **124** was observed after only 15 min. Furthermore, when the reaction was performed without 1,4 benzoquinone (1,4 BQ), **125** was obtained in 45% yield due to side reactions involving isomerization of the terminal double bonds.

Synthesis of a macrocyclic analogue of Cinchonine has been performed by RCM of amine **126**, obtained in one step from alkaloid cinchonine. The metathesis of conformationally constrained amine **126** proceeded in a modest yield of 30% to afford macrocycle **127** (Scheme 48) [119].

As with smaller azacyles, in situ formation of ammonium salt is a good strategy for the synthesis of macrocyclic amine by RCM [120–125]. For example, a 57-membered cyclic amine has been obtained following this strategy en route to catenane derivatives [123].

Synthesis of Taxol analogues possessing a macrocylic amine core has been reported by Kingston et al. using RCM as the key step [124]. For example, treatment of **128** and **129** with 10 mol% of **G2** catalyst in the presence of three equivalents of camphorsulfonic acid afforded, after protection of the secondary alcohol, intermediates **130** and **131** in, respectively, 70% and 19% yield. These results show that the increase of the chain length is detrimental for the efficacy of the RCM process (Scheme 49). Interestingly, the RCM of close analogues of **128** and **129**, compounds **132** and **133**, did not proceed using several catalysts and conditions. The authors suggested that this lack of reactivity is due to the *O*-benzoyl group which interposes between the two olefins during the metathesis.

Sulikowski et al. reported the racemic synthesis of tetrahydrohaliclonacyclamine A [125], the common reduction product of natural product Haliclonacyclamine

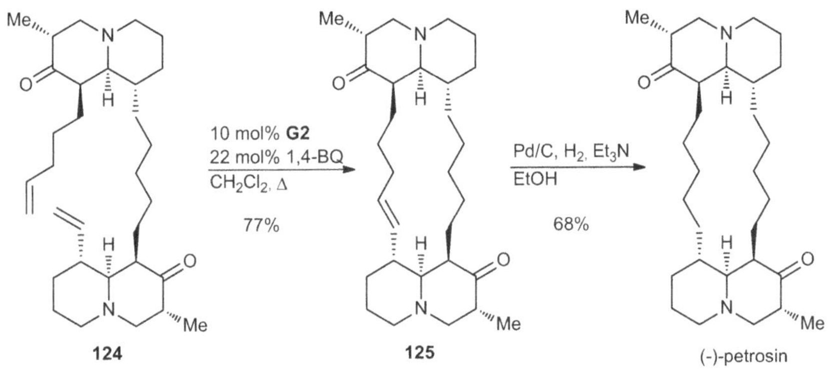

Scheme 47 Synthesis of (-)-Petrosin

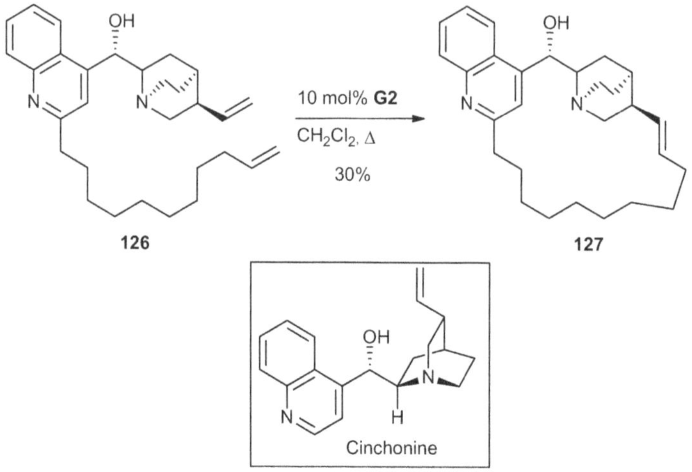

Scheme 48 Synthesis of macrocyclic analogue of Cinchonine

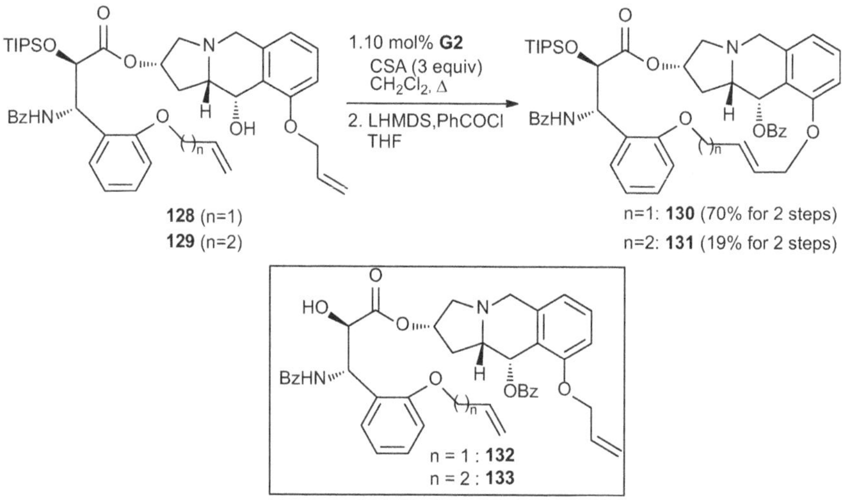

Scheme 49 Synthesis of Taxol analogues

A and its congeners B, C, and D. Haliclonacyclamine A and B display antibiotic, antifungal, and cytotoxic properties. The synthesis was based on two RCM reactions in order to build the two macrocycles of the target molecule (Scheme 50). While RCM of free amine **134** failed, treatment of hydrochloride salt of **134** with Fürstner ruthenium–indenylidene catalyst **M1** afforded **135** in 80% yield in high stereoselectivity in favor of the *trans* product. The authors indicated that RCM reaction proceeded with comparable efficiency when **G1** was used as the catalyst. Hydrogenations with concomitant debenzylation followed with oxidation and Wittig reactions gave **136** and its diastereomer in a modest ratio of 1.3/1 in favor of the

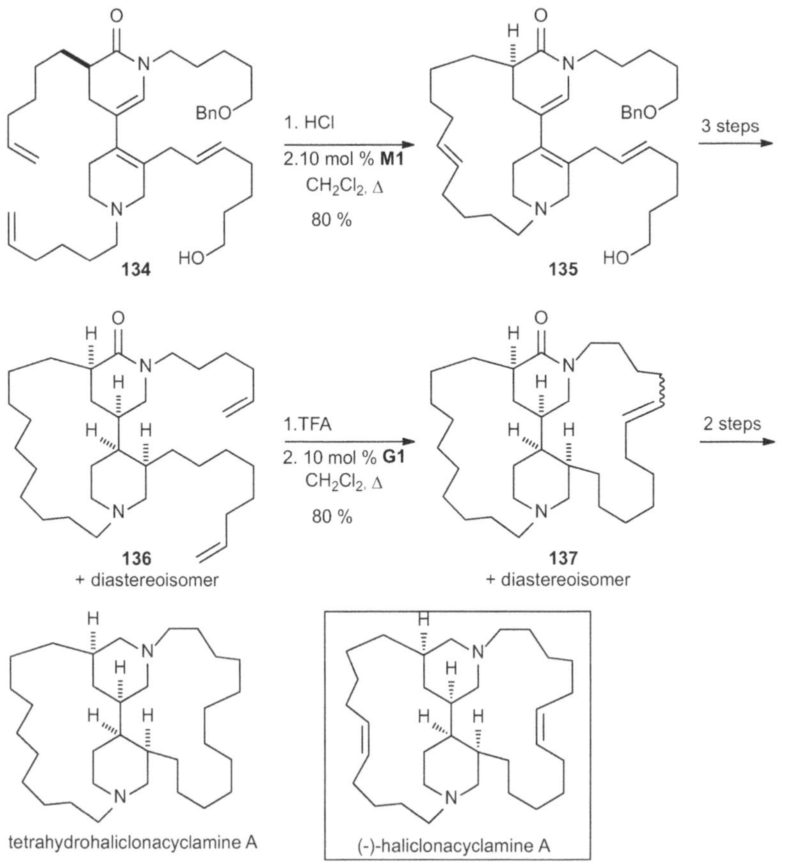

Scheme 50 Synthesis of tetrahydrohaliclonacyclamine A by way of two RCM reactions

desired product **136**. Ring-closing metathesis on a mixture of TFA salts derived from **136** and its diastereomer led to the desired product **137** and its diastereomer in good yield. At this stage, the two isomers can be separated and **137** (6:1 *E*/*Z*) was isolated in 50% yield. Tetrahydrohaliclonacyclamine A was obtained after reduction of alkene and carbonyl functions.

Synthesis of the larger ring of (-)-Nakadomarin A by way of RCM has been reported by several groups [126–130] (for the preparation of the eight-membered ring, see Scheme 39). Different results in terms of *Z*/*E* selectivity have been obtained depending on reaction conditions (Scheme 51). Thus, treatment of compound **138** or its reduced analogue **139** with **G1** in the presence of CSA afforded the desired products **140** and **141** with similar yields and selectivity in favor of the *Z* stereoisomer [126, 127]. Surprisingly, the selectivity was inverted when the metathesis was realized in the absence of Brønsted acid [128]. Finally, excellent *Z* selectivity was obtained with catalyst **142** developed by Schrock, Hoveyda et al. [129, 130].

The group of Dixon has reported the synthesis of natural product Manzamine A as well as related alkaloids Ircinol A and Ircinal A (for another synthesis of

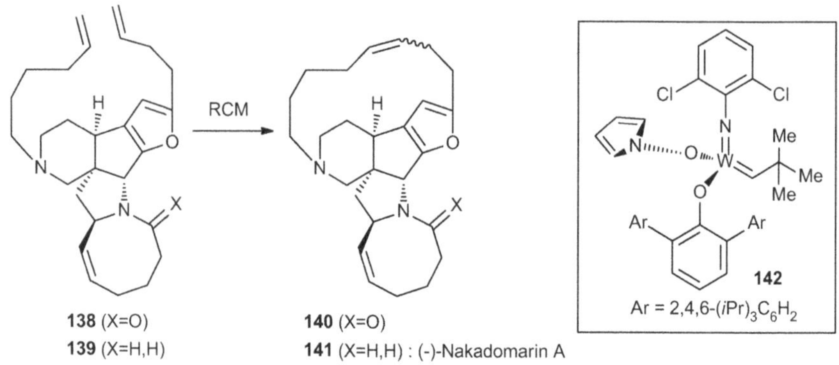

Substrate	Reaction conditions	Product	Yield	Z/E
138	2 equiv CSA, 25 mol % **G1**, DCM, reflux	**140**	65%	2/1
139	3 equiv CSA, 20 mol % **G1**, DCM, reflux	**141**	62%	63/37
138	40 mol % **G1**, DCM, reflux	**140**	66%	3/5
139	5 mol % **142,** Toluene, 22°C, 1 torr	**141**	63%	94/6

Scheme 51 Synthesis of (-)-Nakadomarin A

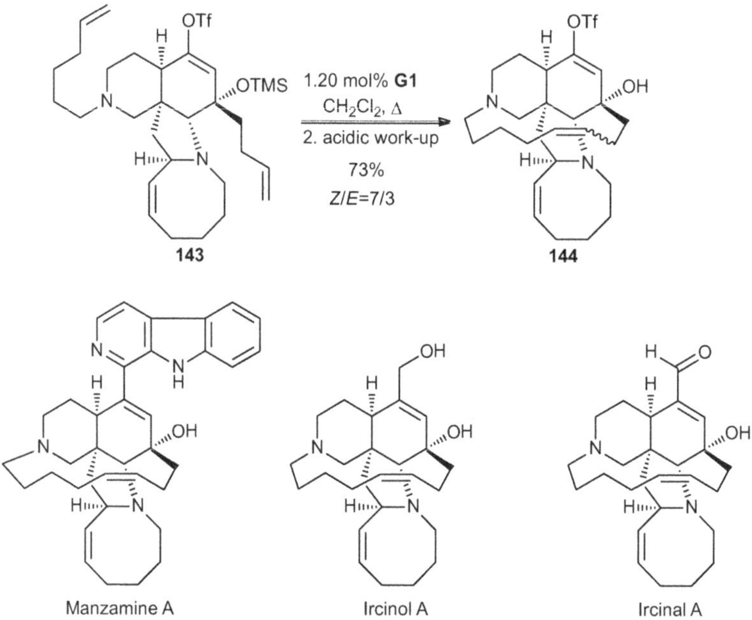

Scheme 52 Synthesis of alkaloids Manzanine A, Ircinol A and Ircinal A

Manzanine A see scheme 45). The key step was the RCM of intermediate **143**, which, after treatment with catalyst **G1**, afforded alcohol **144** in good yield and moderate selectivity (Scheme 52). The desired alkaloids were then obtained from **144** after one or two steps [131].

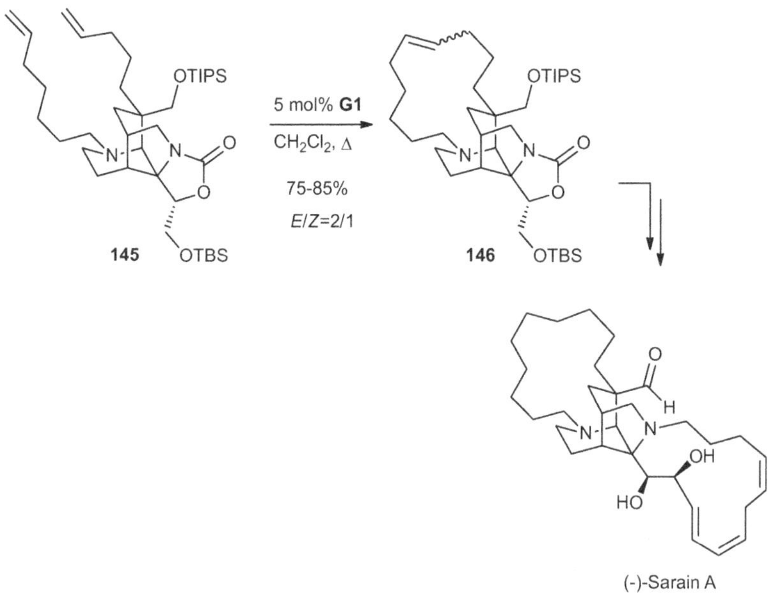

Scheme 53 Synthesis of the 13-membered macrocycle of (-)-Sarain A

Overman et al. reported the synthesis of (-)-Sarain A via RCM of **145** (Scheme 53). Compound **146** was obtained in good yield with **G1** catalyst. Interestingly, when the reaction was performed with 15 mol% of **G2** in the presence of acetic acid, desired product **146** was obtained in only 17% yield. This low yield was explained by the formation of a dimeric compound which is favored by catalyst **G2** [132].

Ring-closing alkyne metathesis (RCAM) is a powerful methodology for the synthesis of macrocyclic compounds. In addition, the reduction of the resulting alkyne can be perfectly stereocontrolled to afford the desired alkene configuration, which is not the case with classical RCM reactions (see, e.g., Schemes 51 and 52). This strategy has been applied in the total synthesis of racemic Haliclonacyclamine C [133]. Treatment of **147**, obtained from **134** (Scheme 50), with the catalytic system derived from in situ combination of complex **148** with Ph_3SiOH afforded, after reduction, the desired natural product in 50% yield for the two steps (Scheme 54).

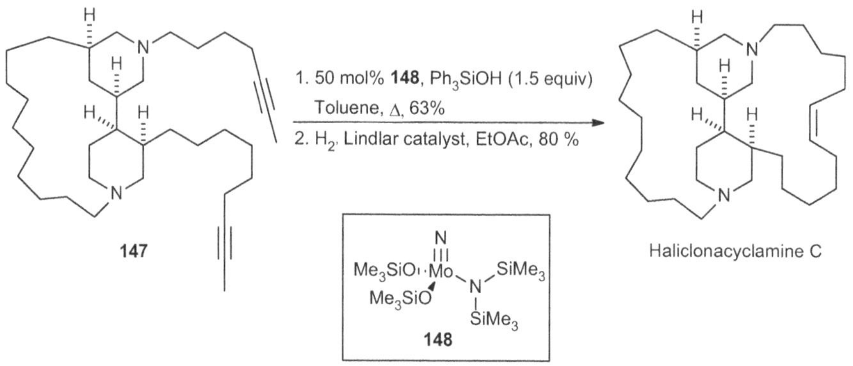

Scheme 54 Synthesis of Haliclonacyclamine C by way of RCAM

6 Tandem Reactions

Metathesis reactions are well adapted for tandem, domino, and multicomponent reactions. Indeed, if more than two double bonds are present in a given substrate, formation of multi-cycles in one operation is then possible. This strategy has been applied to the synthesis of polycyclic amines including the formation of macrocyclic molybdenum complexes by way of triple intramolecular RCM reactions [134]. Martin et al. have reported the total synthesis of racemic pseudotabersonine via a double RCM reaction as the key step [135]. Treatment of **149** with catalyst **HG2** afforded an inseparable mixture of diastereomers **150a** and **150b**. Regioselective reduction followed by deprotection of the TBS group led to **151a** and **151b** in 70% yield for the three steps (Scheme 55).

Lindsley et al. have recently reported an interesting methodology for the synthesis of natural product Stemaphylline and its analogues [136]. Double RCM of **152** resulted in a mixture of **154** and its epimer under a number of reaction conditions. In order to prevent epimerization, the authors applied relay Ring-closing metathesis (RRCM) strategy [137] (for examples of this methodology, see [138]). This methodology developed by Hoye and coworkers involves the incorporation of temporary ether to help the initiation of the catalytic cycle. RRCM of **153** proceeded with minor epimerization in the presence of **HG2** catalyst and led to the desired product **154** with good dr (10:1) (Scheme 56).

Enyne metathesis-based tandem processes have been used for the synthesis of polycyclic amines (for a review on enyne-metathesis tandem reactions, see [139]) [140, 141]. For example, a tandem dienyne metathesis has been applied for the total synthesis of ent-Lepadin F and G [141]. In presence of catalyst **G1** in 1,2-dichloroethane, **155** afforded intermediate **156** in excellent yield (Scheme 57).

Two pathways can be envisioned for this process (Scheme 58). The authors explained the favored formation of metal–carbene intermediate **157** by the higher reactivity of monosubstituted alkenes over disubstituted alkenes and by the plausible coordination of the allylic alcohol with the catalyst.

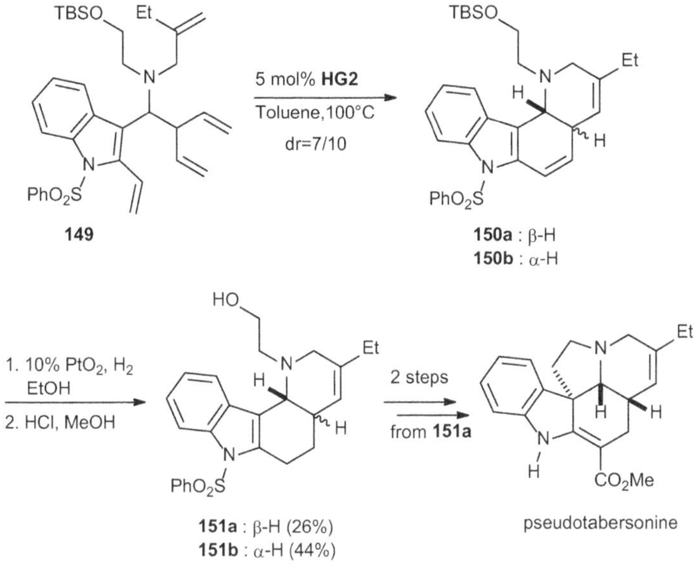

Scheme 55 Synthesis of pseudotabersonine by way of a double RCM reaction

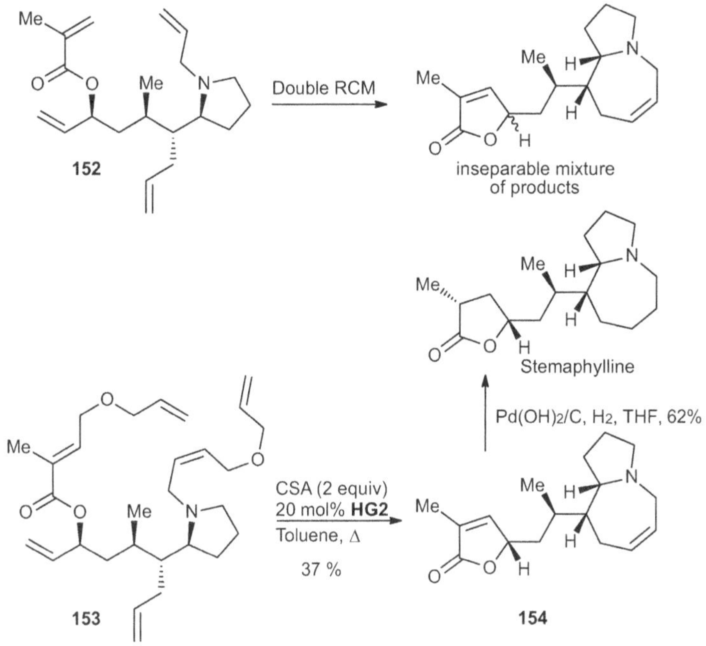

Scheme 56 Synthesis of Stemaphylline by way of relay ring-closing metathesis (RRCM)

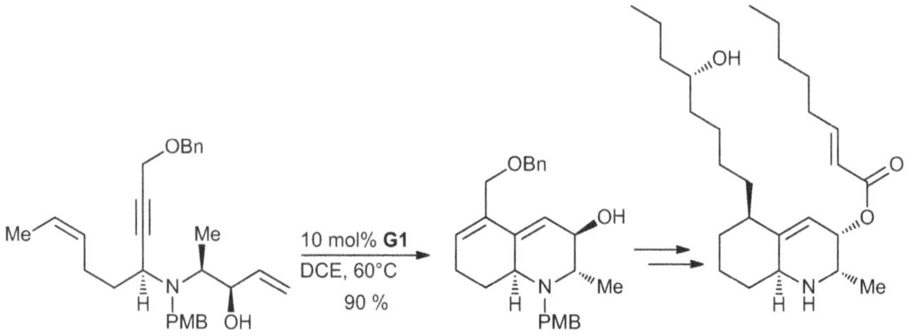

155 156 ent-Lepadin F

Scheme 57 Synthesis of ent-Lepadin F by tandem dienyne metathesis

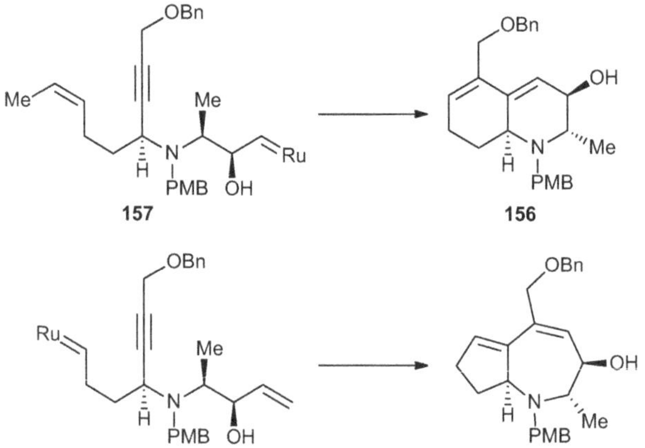

157 156
158 159

Scheme 58 The two possible pathways for the tandem dienyne metathesis of **155**

Chemists have taken advantage of the ability of ruthenium carbenes to catalyze various reactions including oxidation, isomerization, or cycloaddition to design elegant tandem or multicomponent processes as well as one-pot reactions [142]. For the synthesis of polyhydroxylated quinolizidine derivatives, a tandem RCM/dihydroxylation has been described ([143]). Catalyst **G2** can promote the RCM and subsequently the *syn*-dihydroxylation to afford bicyclic compounds **161** and **162** in 56% and 13% yield, respectively (Scheme 59). Thus, once the RCM was completed, $NaIO_4$–$CeCl_3$ was added to promote the oxidation step. In a related reaction, it was hypothesized that $CeCl_3$ reduces the high reactivity of the ruthenium-based oxidant RuO_4, allowing good selectivity and good yields ([143] and references cited). Interestingly, OsO_4-mediated dihydroxylation of the isolated RCM compound **163** gave rise to opposite diastereoselectivity.

Another example of tandem RCM/oxidation has been described for the synthesis of 2-quinolones [144]. RCM of compound **164** afforded **165**, which in the presence

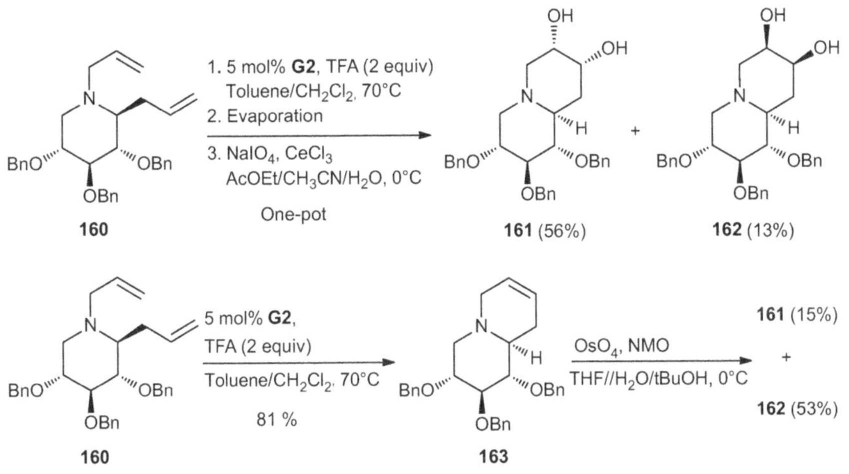

Scheme 59 Synthesis of polyhydroxylated quinolizidines by tandem RCM/*syn*-dihydroxylation

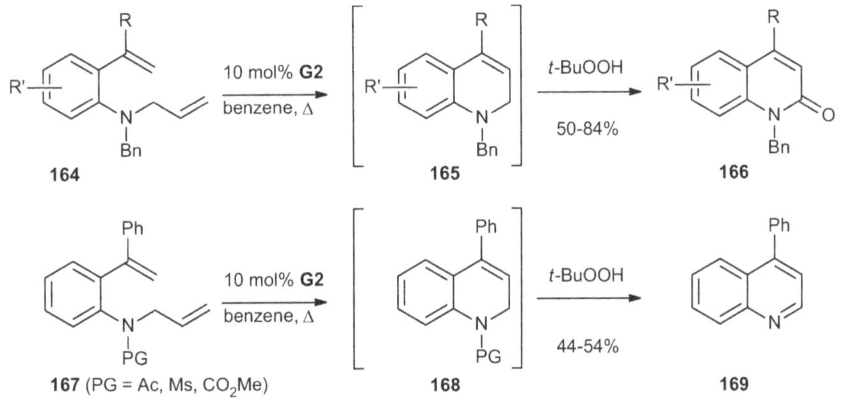

Scheme 60 Tandem RCM/oxidation for the synthesis of quinolinones **166** and quinolines **169**

of an oxidant, led to quinolinones **166** in good yields (Scheme 60). The best results were obtained with **G2** and *t*-BuOOH. It is noteworthy that oxygen can also promote the oxidation step but in lower yields. When nitrogen atom is deactivated, quinolines **169** were obtained under the same conditions. To explain the formation of **166**, the authors suggested that *t*-BuOOH converted **G2** to different ruthenium complexes which might catalyze the oxidation step.

Cascade RCM/Diels–Alder reactions have been reported for the synthesis of functionalized benzoxacines [145]. Treatment of enyne **170** with **G1** afforded diene **171**. Evaporation of the solvent and addition of suitable dienophiles led to compounds **172** and **173** in 73% and 35% yield, respectively (Scheme 61).

Nielsen et al. reported a ruthenium-catalyzed tandem RCM/isomerization/iminium cyclization which led to oxazabicyclooctanes **176** in moderate to good yields depending on the aromatic function in α-position to the nitrogen group

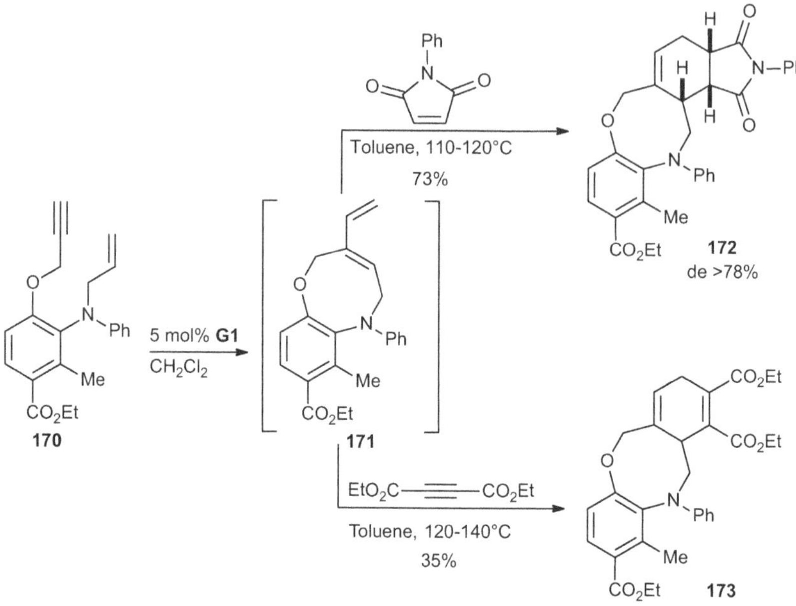

Scheme 61 Synthesis of benzoxacines by cascade RCM/Diels–Alder reaction

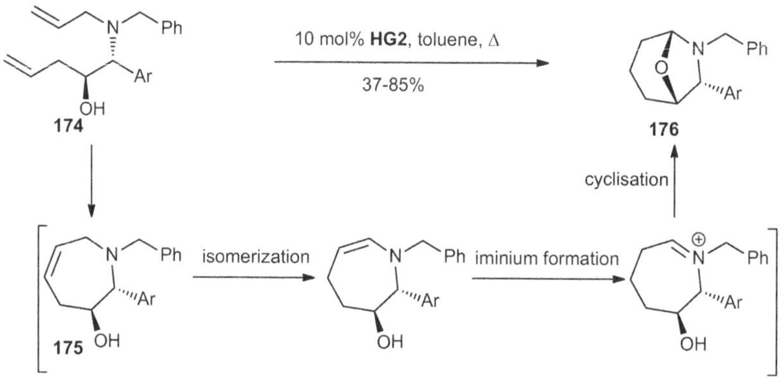

Scheme 62 Synthesis of oxazabicyclooctanes by tandem RCM/isomerization/*N*-alkyliminium cyclization

(Scheme 62) [146, 147]. Several catalysts have been screened and the best results were obtained by heating a toluene solution of **174** in the presence of **HG2** catalyst. This tandem process presumably involves the isomerization of the double bond in RCM product **175** to afford an *N*-alkyliminium. Finally, cyclization led to the desired bicyclic compound **176**. Depending on reaction conditions, compound **175** can be also isolated in modest to good yields. The authors showed that carefully purified **175** did not convert to **176** under thermal conditions whereas addition of **HG2** to **175** led to product **176**. This observation proved the importance of the Ru catalyst for the isomerization steps.

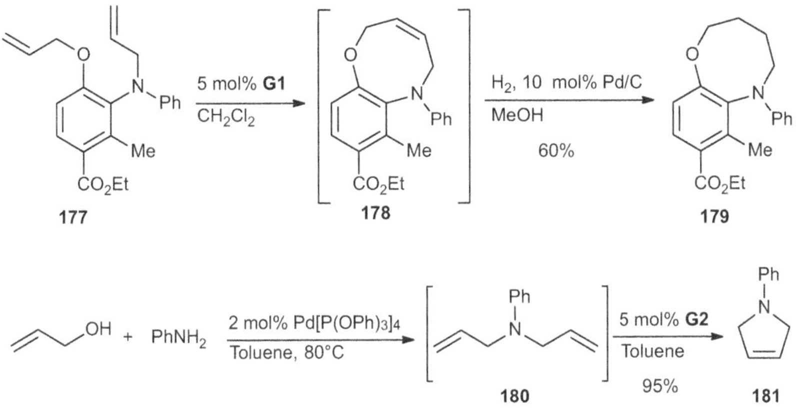

Scheme 63 Synthesis of cyclic amines by one-pot reactions combining Ru- and Pd-catalysis

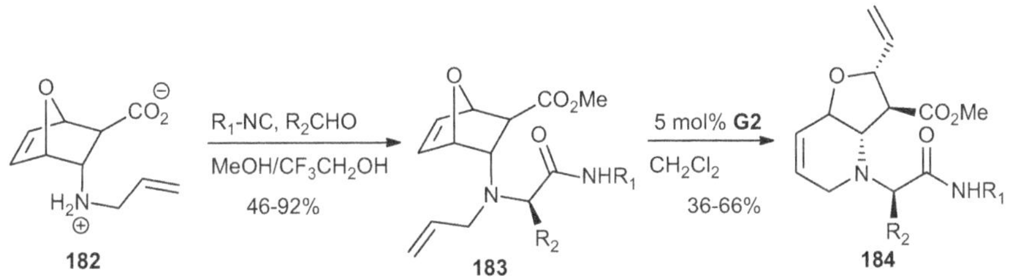

Scheme 64 Synthesis of hexahydrofuro[3,2-b]pyridine derivatives **184** by way of tandem ROM/RCM combined with the Ugi reaction

Ruthenium catalysts used for olefin metathesis are generally inert to almost all other metal-catalyzed reactions. Thus, one-pot reactions involving RCM and another catalyzed transformation have been applied for the synthesis of cyclic amines [32, 135, 145]. For example, benzoxacine **179** has been obtained from **177** by a one-pot RCM/hydrogenation sequence (Scheme 63) [145]. Sawadjoon and Samec reported a one-pot procedure for the synthesis of pyrroline **181** from allyl alcohol and aniline [32]. In this sequence, palladium-catalyzed diallylation of aniline affording **180** was followed by RCM in the same pot (Scheme 63).

Tandem ring-opening metathesis (ROM)/RCM combined with the Ugi reaction has been used for the synthesis of cyclic amines [148]. Thus, treatment of amide **183** with catalyst **G2** afforded bicyclic product **184** (Scheme 64).

Hoveyda and Schrock et al. reported the enantioselective synthesis of piperidines through asymmetric ROM/cross metathesis (CM) reactions, the tandem sequence being performed from cyclic amines [149, 150] (Scheme 65). For example, treatment of **185** with ten equivalents of the appropriate styrene derivative in the presence of chiral molybdenum catalyst **187** afforded piperidines **186** in good to excellent ee [149]. Application of the same reaction conditions to benzyl-protected azabicycle **188** showed that catalyst **187** was ineffective to promote acceptable enantioselectivity [150]. However, Ru catalyst **190** promoted the ROM-CM of **188**

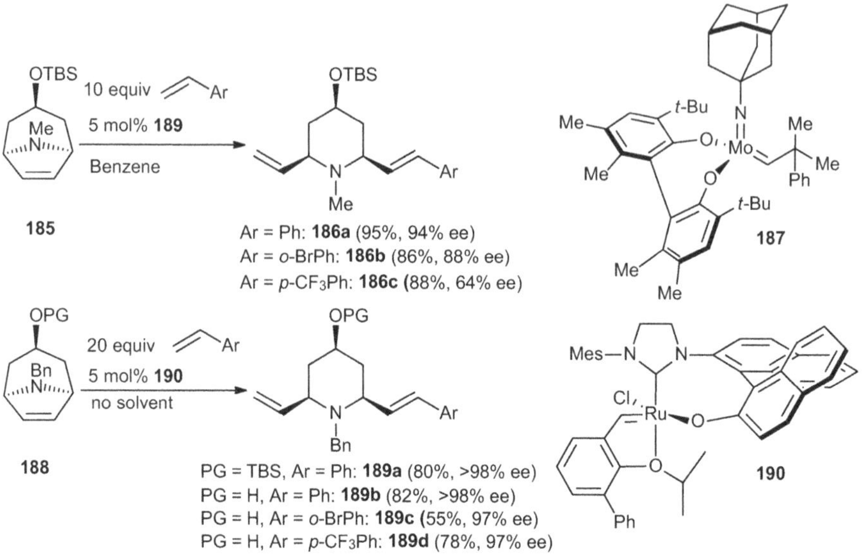

Scheme 65 Tandem asymmetric ROM-CM reactions of azabicycles **185** and **188**

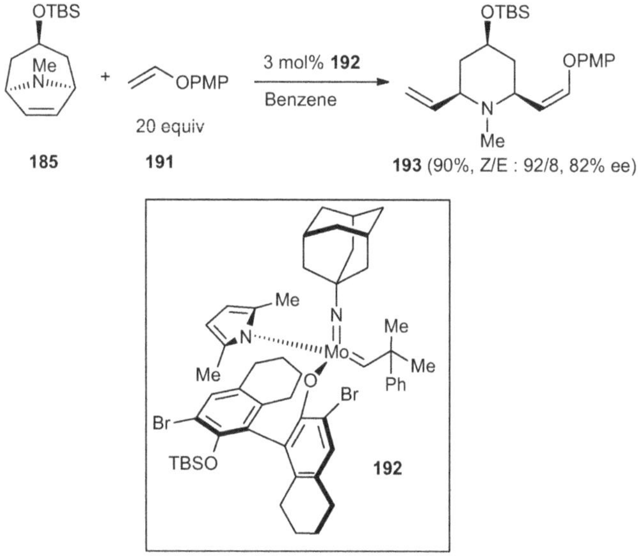

Scheme 66 Enantioselective synthesis of piperidine **193** by tandem asymmetric ROM-CM reaction

with excellent enantioselectivity. Furthermore, in contrast to catalyst **187**, **190** catalyzed the reaction with substrates bearing a free hydroxyl group as shown for the synthesis of compounds **189b–189d**.

The same authors applied a similar strategy for catalytic enantioselective ROM-CM involving enol ether **191** and azabicycle **185** (Scheme 66). The best catalyst found for this process was molybdenum complex **192**, the desired piperidine **193** being obtained in 90% yield and 82% ee [151].

7 Conclusion and Outlook

Since 2007 and our first review dealing with olefin metathesis of basic nitrogen-containing systems [13], numerous successful examples of direct metathesis with amine substrates have been reported in the literature. Consequently, the early "dogma" stating that efficient metathesis reactions are suppressed in the presence of basic amines and that such substrates must invariably be deactivated does not actually stand up to scrutiny. The success of a metathesis reaction often depends on the structure of the amine substrate. It is now well established that amines α-substituted with an electron-withdrawing group, arylamines, and sterically hindered amines are good substrates for metathesis reactions. In this context the use of hindered protecting groups can be an alternative to deactivation of the nitrogen atom by electron-withdrawing groups. However, many examples have shown that exceptions to these rules exist, as demonstrated by the successful metathesis of unhindered amines. In this context, the choice of olefin substituents, of the catalyst, or of the experimental conditions is of paramount importance. Indeed, introduction of a suitable substituent to the olefin (e.g., a phenyl group instead of hydrogen – Scheme 33) may enable successful metathesis reactions. Thus, in some cases, minor variations in substrate structure can make a strong difference in terms of reactivity. Subtle differences in the experimental conditions of a metathesis reaction can have also a great influence in its reactivity (temperature of the reaction, solvent, use of microwaves...). The choice of the catalyst is also crucial as successful metathesis reaction or inhibition of the catalytic process can take place depending on the catalyst. The most often used catalysts remain commercially available ruthenium catalysts **G1**, **G2**, **HG1,** and **HG2**. Although Schrock catalysts are more tolerant to amine-containing compounds, they are less tolerant to oxygen and moisture, which is the main limitation to their use. Alternatively, Brønsted or Lewis acid is also used as tools to deactivate the amine substrate and to block the formation of unproductive metallacycle intermediates by chelation of metal complexes. Although this method has been used for the synthesis of functionalized molecule including natural products, this strategy is not applicable to all substrates, in particular the acid-sensitive ones. Application of the metathesis reaction to complex nitrogen-containing molecules and recent development of tandem, multicomponent, or one-pot processes with amine substrates have further highlighted the power of this enabling methodology. In the future, development of more robust catalysts is likely to extend this power to more and more challenging substrates.

References

1. Fattorusso E, Taglialatela-Scafati O (2007) Modern alkaloids: structure, isolation, synthesis and biology. Wiley, Weinheim
2. Aniszewski T (2007) Alkaloids: secrets of life. Elsevier, Oxford
3. Compain P, Martin OR (2007) Iminosugar from synthesis to therapeutic applications. Wiley, Weinheim
4. Watson PS, Jiang B, Scott B (2000) Org Lett 2:3679–3681
5. Philips AJ, Abell AD (1999) Aldrichim Acta 32:75–89
6. Deiters A, Martin SF (2004) Chem Rev 104:2199–2238
7. Felpin FX, Lebreton J (2003) Eur J Org Chem 3693–3712
8. Chattopadhyay SK, Karmakar S, Biswas T, Majumdar KC, Rahaman H, Roy B (2007) Tetrahedron 63:3919–3952
9. Dragutan I, Dragutan V, Mitan C, Vosloo HCM, Delaude L, Demonceau A (2011) Beilstein J Org Chem 7:699–716
10. Dragutan I, Dragutan V, Demonceau A (2012) RSC Adv 2:719–736
11. Merino P, Tejero T, Greco G, Marca E, Delso I, Gómez-SanJuan A, Matute R (2012) Heterocycles 84:75–100
12. Lummiss JAM, Ireland BJ, Sommers JM, Fogg DE (2014) ChemCatChem 6:459–463
13. Compain P (2007) Adv Synth Catal 349:1829–1846
14. Yang Q, Xiao WJ, Yu Z (2005) Org Lett 7:871–874
15. Huo HH, Xia XE, Zhang HK, Huang PQ (2013) J Org Chem 78:455–465
16. Greene TW, Wuts PGM (2007) Protective groups in organic synthesis, 4th edn. Wiley, Hoboken
17. Lin NT, Ke YZ, Satyanarayana K, Huang SL, Lan YK, Yang HC, Luh TY (2013) Macromolecules 46:7173–7179
18. Xie M, Wang W, Ding L, Liu J, Yang D, Wei L, Zhang Y (2010) J Polym Sci Part A Polym Chem 48:380–388
19. Yang HC, Lee SL, Chen CH, Lin NT, Yang HC, Jin BY, Luh TY (2008) Chem Commun 6158–6160
20. Ranatunga S, Del Valle JR (2009) Tetrahedron Lett 50:2464–2466
21. Hoshi M, Kaneko O, Nakajima M, Arai S, Nishida A (2014) Org Lett 16:768–771
22. Davis FA, Song M, Qiu H, Chai J (2009) Org Biomol Chem 7:5067–5073
23. Delso I, Tejero T, Goti A, Merino P (2010) Tetrahedron 66:1220–1227
24. Kim IS, Qing RL, Dong GR, Kim YC, Hong YJ, Lee M, Chi KW, Oh JS, Jung YH (2010) Eur J Org Chem 1569–1573
25. Rajender A, Rao JP, Rao BV (2013) Eur J Org Chem 1749–1757
26. Chandrasekhar B, Rao BV, Rao KVM, Jagadeesh B (2009) Tetrahedron: Asymmetry 20:1217–1223
27. Jarvis SBD, Charette AB (2011) Org Lett 13:3830–3833
28. Iza A, Carillo L, Vicario JL, Badía D, Reyes E, Martínez JI (2010) Org Biomol Chem 8:2238–2244
29. Majumdar KC, Chakravorty S, Taher A (2008) Synth Commun 38:3159–3169
30. Balabani A, Hadjipavlou-Litina DJ, Litinas KE, Mainou M, Tsironi CC, Vronteli A (2011) Eur J Med Chem 46:5894–5901
31. Adak L, Chattopadhyay K, Ranu BC (2009) J Org Chem 74:3982–3985
32. Sawadjoon S, Samec JSM (2011) Org Biomol Chem 9:2548–2554
33. Shao M, Zheng L, Qiao W, Wang J, Wang J (2012) Adv Synth Catal 354:2743–2750
34. Déchamps I, Gomez Pardo D, Cossy J (2007) Tetrahedron 63:9082–9091
35. Déchamps I, Gomez Pardo D, Cossy J (2007) Arkivoc 38–45
36. Bates RW, Dewey MR (2009) Org Lett 11:3706–3708
37. Bates RW, Dewey MR, Tang CH, Safii SB, Hong Y, Hsieh JKH, Siah PS (2011) Synlett 2053–2055

38. Teichert JF, Feringa BL (2010) Synthesis 1200–1204
39. Woodward CP, Spiccia ND, Jackson WR, Robinson AJ (2011) Chem Commun 47:779–781
40. Gulajski L, Michrowska A, Naroznik J, Kaczmarska Z, Rupnicki L, Grela K (2008) ChemSusChem 1:103–109
41. Binder JB, Guzei IA, Raines RT (2007) Adv Synth Catal 349:395–404
42. Binder JB, Blank JJ, Raines RT (2007) Org Lett 9:4885–4888
43. Jordan JP, Grubbs RH (2007) Angew Chem Int Ed 46:5152–5155
44. Matsuo T, Yoshida T, Fujii A, Kawahara K, Hirota S (2013) Organometallics 32:5313–5319
45. Tomasek J, Schatz J (2013) Green Chem 15:2317–2338
46. Śledź P, Mauduit M, Grela K (2008) Chem Soc Rev 37:2433–2442
47. Miller II AL, Bowden NB (2007) Chem Commun 2051–2053
48. Selig P, Herdtweck E, Bach T (2009) Chem Eur J 15:3509–3525
49. Selig P, Bach T (2008) Angew Chem Int Ed 47:5082–5084
50. Zhang H, Curran DP (2011) J Am Chem Soc 133:10376–10378
51. Hayashi Y, Inagaki F, Mukai C (2011) Org Lett 13:1778–1780
52. Feldman KS, Antoline JF (2012) Org Lett 14:934–937
53. Feldman KS, Antoline JF (2013) Tetrahedron 69:1434–1445
54. Nidhiry JE, Prasad KR (2013) Tetrahedron 69:5525–5536
55. Sattely ES, Meek SJ, Malcolmson SJ, Schrock RR, Hoveyda A (2009) J Am Chem Soc 131:943–953
56. Deck JA, Martin SF (2010) Org Lett 12:2610–2613
57. Xu S, Unabara D, Uemura D, Arimoto H (2014) Chem Asian J 9:367–375
58. Tsukanov SV, Comins DL (2011) Angew Chem Int Ed 50:8626–8628
59. Race NJ, Bower JF (2013) Org Lett 15:4616–4619
60. Louvel J, Chemla F, Demont E, Ferreira F, Pérez-Luna A (2011) Org Lett 13:6452–6455
61. Louvel J, Botuha C, Chemla F, Demont E, Ferreira F, Pérez-Luna A (2010) Eur J Org Chem 2921–2926
62. Komatsu Y, Yoshida K, Ueda H, Tokuyama H (2013) Tetrahedron Lett 54:377–380
63. Watkin SV, Camp NP, Brown RCD (2013) Org Lett 15:4596–4599
64. Gálvez JA, Díaz de Villegas MD, Badorrey R, López-Ram-de-Víu P (2011) Org Biomol Chem 9:8155–8162
65. Pandey G, Kumar R, Banerjee P, Puranik VG (2011) Eur J Org Chem 4571–4587
66. Ansari AA, Vankar YD (2013) J Org Chem 78:9383–9395
67. Cheng X, Waters SP (2010) Org Lett 12:205–207
68. Allen CE, Chow CL, Caldwell JJ, Westwood IM, van Montfort RLM, Collins I (2013) Bioorg Med Chem 21:5707–5724
69. Biyikal M, Porta M, Roesky PW, Blechert S (2010) Adv Synth Catal 352:1870–1875
70. Jiangseubchatveera N, Bouillon ME, Liawruangrath B, Liawruangrath S, Nash RJ, Pyne SG (2013) Org Biomol Chem 11:3826–3833
71. Yoshida K, Kawagoe F, Hayashi K, Horiuci S, Imamoto T, Yanagisawa A (2009) Org Lett 11:515–518
72. Mailyan AK, Krylov IM, Bruneau C, Dixneuf PH, Osipov SN (2013) Eur J Org Chem 5353–5363
73. Bissember AC, Banwell MG (2009) Tetrahedron 65:8222–8230
74. Prasad KR, Nagaraju C (2013) Org Lett 15:2778–2781
75. Sun CS, Lin YS, Hou DR (2008) J Org Chem 73:6877–6880
76. Yoshida M, Maeyama Y, Shishido K (2012) Tetrahedron 68:9962–9972
77. Maity S, Zhu M, Shinabery RP, Zheng N (2012) Angew Chem Int Ed 51:222–226
78. Zhang Y, Silverman RB (2013) Tetrahedron Lett 54:573–575
79. Wang CH, Alluri S, Ganguly AK (2009) Tetrahedron Lett 50:1879–1881
80. Wang CH, Reilly J, Brand N, Schwartz S, Alluri S, Chan TM, Buevich AV, Ganguly AK (2010) Tetrahedron Lett 51:6213–6215
81. Wang CH, White AR, Schwartz SN, Alluri S, Cattabiani TM, Zhang LK, Chan TM, Buevich AV, Ganguly AK (2012) Tetrahedron 68:9750–9762

82. Hopkins CD, Malinakova HC (2007) Synthesis 3558–3566
83. Aslam NA, Babu SA, Sudha AJ, Yasuda M, Baba A (2013) Tetrahedron 69:6598–6611
84. Malcolmson SJ, Meek SJ, Sattely ES, Schrock RR, Hoveyda AH (2008) Nature 456:933–937
85. Fournier PA, Collins SK (2007) Organometallics 26:2945–2949
86. Jiang JK, Ghoreschi K, Deflorian F, Chen Z, Perreira M, Pesu M, Smith J, Nguyen DT, Liu EH, Leister W, Costanzi S, O'Shea JJ, Thomas CJ (2008) J Med Chem 51:8012–8018
87. Sançon J, Sweeney JB (2010) Synlett 664–666
88. Ren H, Wulff WD (2011) J Am Chem Soc 133:5656–5659
89. Fernandes RA, Nallasivam JL (2012) Org Biomol Chem 10:7789–7800
90. Chou SS, Zhang JW, Chen KH (2013) Tetrahedron 69:1499–1508
91. Kirkland TA, Lynn DM, Grubbs RH (1998) J Am Chem Soc 63:9904–9909
92. Lahiri R, Kokatla HP, Vankar YD (2011) Tetrahedron Lett 52:781–786
93. Piao F, Mishra MK, Jang DO (2012) Tetrahedron 68:7050–7055
94. Fu P, Snapper ML, Hoveyda AH (2008) J Am Chem Soc 130:5530–5541
95. Majumdar KC, Samanta S, Chattopadhyay B, Nandi RK (2010) Synthesis 863–869
96. Martínez-Estíbalez U, Sotomayor N, Lete E (2007) Tetrahedron Lett 48:2919–2922
97. Ramachary DB, Narayana VV (2011) Eur J Org Chem 3514–3522
98. Benedetti E, Lomazzi M, Tibiletti F, Goddart JP, Fensterbank L, Malacria M, Palmisano G, Penoni A (2012) Synthesis 3523–3533
99. Dieltiens N, Stevens CV, Masschelein K, Hennebel G, Van der Jeught S (2008) Tetrahedron 64:4295–4303
100. Hong SH, Day MW, Grubbs RH (2004) J Am Chem Soc 126:7414–7415
101. Vougioukalakis GC (2012) Chem Eur J 18:8868–8880
102. Brock EA, Davies SG, Lee JA, Roberts PM, Thomson JE (2013) Org Biomol Chem 11:3187–3202
103. Brock EA, Davies SG, Lee JA, Roberts PM, Thomson JE (2011) Org Lett 7:1594–1597
104. Lee JH, Shin S, Kang J, Lee S (2007) J Org Chem 72:7443–7446
105. Senter TJ, Schulte ML, Konkol LC, Wadzinski TE, Lindsley CW (2013) Tetrahedron Lett 54:1645–1648
106. Wang N, Zhang LH, Ye XS (2010) Org Biomol Chem 8:2639–2649
107. Kyle AF, Jakubec P, Cockfield DM, Cleator E, Skidmore J, Dixon DJ (2011) Chem Commun 47:10037–10039
108. Lowe JT et al (2012) J Org Chem 77:7187–7211
109. Lee HK, Him JH, Jung SH (2007) Tetrahedron 63:3321–3327
110. Davies SG, Fletcher AM, Roberts PM, Smith AD (2009) Tetrahedron 65:10192–10213
111. Nash A, Soheili A, Tambar UK (2013) Org Lett 15:4770–4773
112. Titanyuk ID, Beletskaya IP (2010) Russ J Org Chem 46:1277–1281
113. Vanlaer S, Voet A, Gielens C, De Maeyer M, Compernolle F (2009) Eur J Org Chem 643–654
114. Appukkutan P, Dehaen W, Van der Eycken E (2007) Chem Eur J 13:6452–6460
115. Kikuchi H, Horoiwa S, Kasahara R, Hariguchi N, Matsumoto M, Oshima Y (2014) Eur J Med Chem 76:10–19
116. Toma T, Kita Y, Fukuyama T (2010) J Am Chem Soc 132:10233–10235
117. Zhao Y, Hoveyda AH, Schrock RR (2011) Org Lett 13:784–787
118. Toya H, Okano K, Takasu K, Ihara M, Takahashi A, Tanaka H, Tokuyama H (2010) Org Lett 12:5196–5199
119. Hintermann L, Schmitz M, Englert U (2007) Angew Chem Int Ed 46:5164–5167
120. Cheng HY, Hou DR (2007) Tetrahedron 63:3000–3005
121. Kobayashi Y, Sano K, Saigo K (2009) Tetrahedron Lett 50:799–801
122. William AD et al (2011) J Med Chem 55:169–196
123. Jiang Y, Zhu XZ, Chen CF (2010) Chem Eur J 16:14285–14289
124. Zhao J, Bane S, Snyder JP, Hu H, Mukherjee K, Slebdonik C, Kingston DGI (2011) Bioorg Med Chem 19:7664–7678

125. Smith BJ, Qu T, Mulder M, Noetzel MJ, Lindsley CW, Sulikowski GA (2010) Tetrahedron 66:4805–4810
126. Cheng B, Wu F, Yang X, Zhou Y, Wan X, Zhai H (2011) Chem Eur J 17:12569–12572
127. Jakubec P, Cockfield DM, Dixon DJ (2009) J Am Chem Soc 131:16632–16633
128. Young IS, Kerr MA (2007) J Am Chem Soc 129:1465–1469
129. Yu M, Wang C, Kyle AF, Jakubec P, Dixon DJ, Schrock RR, Hoveyda AH (2011) Nature 479:88–93
130. Wang C, Yu M, Kyle AF, Jakubec P, Dixon DJ, Schrock RR, Hoveyda AH (2013) Chem Eur J 19:2726–2740
131. Jakubec P, Hawkins A, Felzmann W, Dixon DJ (2012) J Am Chem Soc 134:17482–17485
132. Becker MH, Chua P, Downham R, Douglas CJ, Garg NK, Hiebert S, Jaroch S, Matsuoka RT, Middleton JA, Ng FW, Overman LE (2007) J Am Chem Soc 129:11987–12002
133. Smith BJ, Sulikowski GA (2010) Angew Chem Int Ed 49:1599–1602
134. Cain MF, Forrest WP, Peryshkov DV, Schrock RR, Müller P (2013) J Am Chem Soc 135:15338–15341
135. Cheng B, Sunderhaus JD, Martin SF (2010) Org Lett 12:3622–3625
136. Schulte ML, Turlington ML, Phatak SS, Harp JM, Stauffer SR, Lindsley CW (2013) Chem Eur J 19:11847–11852
137. Hoye TR, Jeffrey CS, Tennakoon MA, Wang J, Zhao J (2004) J Am Chem Soc 126:10210–10211
138. Wallace DJ (2005) Angew Chem Int Ed 44:1912–1915
139. Li J, Lee D (2011) Eur J Org Chem 4269–4287
140. González-Gómez A, Domíngez G, Pérez-Castells J (2009) Tetrahedron 65:3378–3391
141. Niethe A, Fischer D, Blechert S (2008) J Org Chem 73:3088–3093
142. Alcaide B, Almendros P, Luna A (2009) Chem Rev 109:3817–3858
143. Malik M, Witkowski G, Ceborska M, Jarosz S (2013) Org Lett 15:6214–6217
144. Kato H, Ishigame T, Oshima N, Hoshiya N, Shimawaki K, Arisawa M, Shuto S (2011) Adv Synth Catal 353:2676–2680
145. Ramachary DB, Narayana VV, Prasad MS, Ramakumar K (2009) Org Biomol Chem 7:3372–3378
146. Ascic E, Jensen JF, Nielsen TE (2011) Angew Chem Int Ed 40:5188–5191
147. Ascic E, Le Quement ST, Ishoey M, Daugard M, Nielsen TE (2012) ACS Comb Sci 14:253–257
148. Sonaglia L, Banfi L, Riva R, Basso A (2012) Tetrahedron Lett 53:6516–6518
149. Cortez GA, Schrock RR, Hoveyda AH (2007) Angew Chem Int Ed 46:4534–4538
150. Cortez GA, Baxter CA, Schrock RR, Hoveyda AH (2007) Org Lett 9:2871–2874
151. Yu M, Ibrahem I, Hasegawa M, Schrock RR, Hoveyda AH (2012) J Am Chem Soc 134:2788–2799

Top Heterocycl Chem (2017) 47: 155–190
DOI: 10.1007/7081_2015_144

Published online: 16 June 2015

Synthesis of Lactams by Metathesis Reactions

Guillaume Vincent

Abstract The use of metathesis for the synthesis of lactams has become very important in the last decade. This reaction allows the synthesis of very diverse lactam-containing compounds: from simple building blocks to refined macro-lactams or polycyclic systems. The arsenal of metathesis reactions ranges from ring-closing metathesis (RCM) to ring-rearrangement metathesis (RRM) to ring-closing enyne metathesis (RCEYM) or cross metathesis (CM). The strategic deployment of such reactions for the synthesis of lactams will be described in this chapter, with a discussion on factors that govern the metathesis reaction.

Keywords Alkene • Alkyne • Amide • Cross metathesis • Lactam • Ring-closing enyne metathesis • Ring-closing metathesis • Ring-rearrangement metathesis

Contents

G. Vincent (✉)
Université Paris-Sud and CNRS, Institut de Chimie Moléculaire et des Matériaux d'Orsay (ICMMO), Equipe Méthodologie, Synthèse et Molécules Thérapeutiques (MS&MT), Bat. 410, 15 rue Georges Clemenceau, 91405 Orsay, France
e-mail: guillaume.vincent@u-psud.fr

1 Introduction

Lactams are important heterocycles that can be found in natural products or biological active compounds. They can also be used as synthetic intermediates such as in the synthesis of polymers [1]. Recent new methodologies for the direct functionalization of amides have increased the interest of lactams as molecular platform for the elaboration of complex structures [2]. Several strategies are known to achieve the formation of lactams [1]. Among them is the metathesis, which has emerged as one of the most powerful reactions for the preparation of lactams in the last decade. In addition, the unsaturation result of the metathesis step offers a straightforward entry to further functionalization. The diversity of metathesis reactions (RCM, CM, RCEYM, RRM) and the robustness of ruthenium catalysts offer various opportunities to the synthetic chemist in the design of synthetic routes toward lactams.

2 Ring-Closing Metathesis (RCM) of Olefins

The ring-closing metathesis (RCM) of alkenes has been by far more widely employed than enyne metathesis (RCEYM) or ring-rearrangement metathesis (RRM) to access lactams mainly because of its retro-synthetic simplicity, high yields, and the easy access to the precursors [3].

2.1 *From Acrylamides*

Acrylamide precursor **3** of small and medium lactams **4** could be easily obtained by a peptide-type coupling between acrylic acid derivatives **1** and primary or secondary amines **2** bearing a terminal alkene (Scheme 1). The RCM process could be performed in the next step, thus delivering in a very concise manner α,β-unsaturated lactams **4**. One of the key factors for a successful RCM is the control of the equilibrium between the *trans*- and *cis*-forms of the amide **3**.

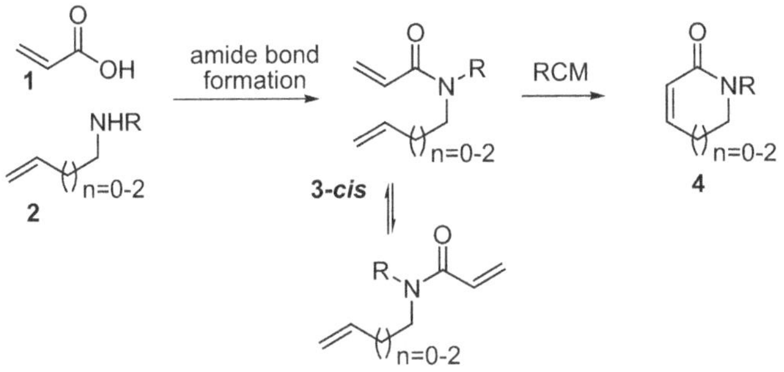

3-trans

Scheme 1 General process for the synthesis of α,β-unsaturated lactams by RCM from acrylamide precursors

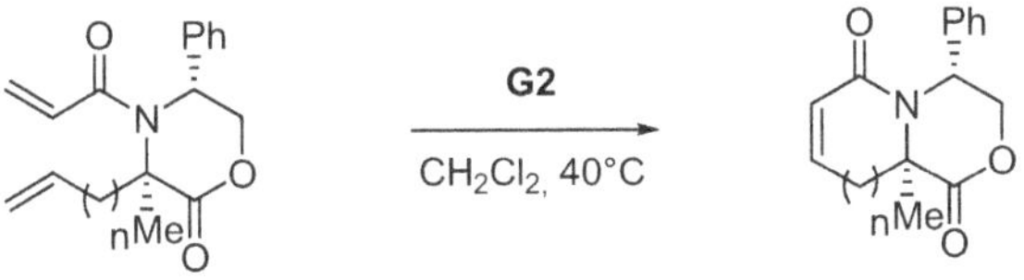

5a, n=0
5b, n=1
5c, n=2
5d, n=3

6a, n=0, 0%
6b, n=1, 92%
6c, n=2, 93%
6d, n=3, 20%

Scheme 2 Formation of five- to eight-membered lactams by RCM

2.1.1 Size of the Lactams Formed

The formation of five- to seven-membered rings by ring-closing metathesis is generally a really favored process. However, the formation of five-membered lactams by ring-closing metathesis is moderately reported, potentially because of the important strain of the ring formed. In contrast, the synthesis of six- and seven-membered lactams by RCM is abundantly reported.

The comparative yields of formation of five-, six-, seven-, and eight-membered lactams **6a–d** by Fustero et al. from precursors **5a–d** highlight this fact (Scheme 2) [4]. The six- and seven-membered rings **6b** and **6c** were obtained in very high yields with **G2** catalyst, while the corresponding five-membered ring **6a** was not obtained. The formation of eight-membered lactams is also a difficult task since only 20% of **6d** was formed.

Five-Membered Lactams

The lactam **6a** could not be obtained in the study of Fustero et al. (Scheme 2), probably because of the congested structure of the diene **5a** and/or the important strain of the bicyclic target. Fortunately, formations of five-membered lactams are

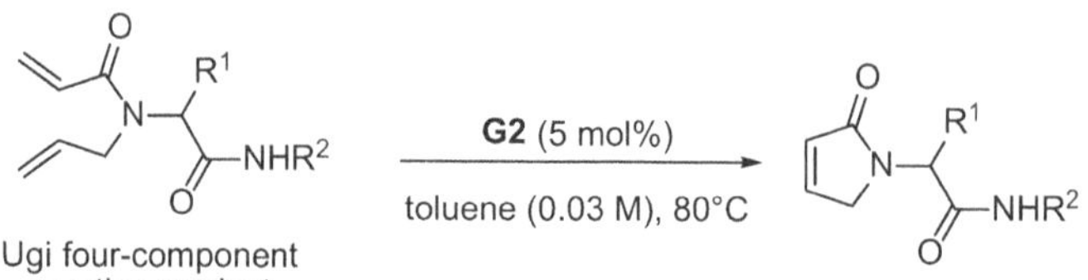

Ugi four-component reaction product
7a, R^1=H, R^2=Bn
7b, R^1=H, R^2=4-MeOBn
7c, R^1=H, R^2=2,4-diMeOBn
7d, R^1=H, R^2=*t*-Bu
7e, R^1=Et, R^2=Bn
7f, R^1=Et, R^2=4-MeOBn
7g, R^1=Et, R^2=2,4-diMeOBn

8a, R^1=H, R^2=Bn, 50%
8b, R^1=H, R^2=4-MeOBn, 63%
8c, R^1=H, R^2=2,4-diMeOBn, 64%
8d, R^1=H, R^2=*t*-Bu, 51%
8e, R^1=Et, R^2=Bn, 60%
8f, R^1=Et, R^2=4-MeOBn, 66%
8g, R^1=Et, R^2=2,4-diMeOBn, 61%

Scheme 3 Formation of γ-lactams by RCM from Ugi four-component products

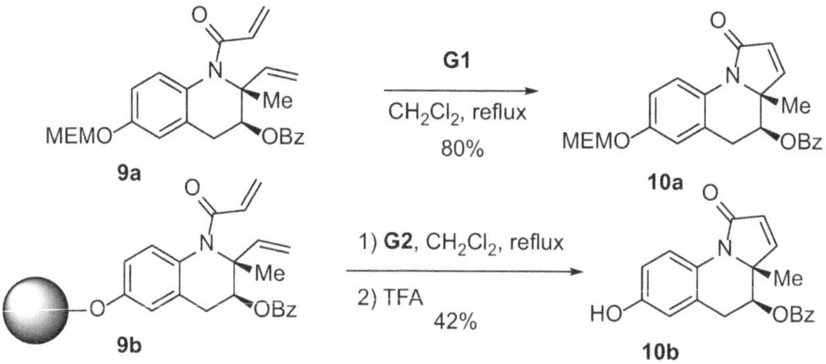

Scheme 4 Formation of pyrrolo[1,2-*a*]quinoline derivatives in solution and solid phase

reported in the literature by RCM from more suitable precursors such as in the following examples.

Keum et al. were able to transform the products of Ugi four-component products **7a–g** into γ-lactams **8a–g** through RCM (Scheme 3) [5]. While the reaction was unsuccessful with **G1** and **G2** catalysts in refluxing dichloromethane, the authors managed to find operative conditions for the RCM. A toluene solution of **G2** catalyst (5 mol%) was added slowly to a toluene-diluted solution (0.03 M) of dienes **7** at 80°C.

During the synthesis of pyrrolo[1,2-*a*]quinoline derivative **10a**, Khadem et al. performed the RCM of precursor **9a** with **G1** in refluxing dichloromethane with 80 % yield (Scheme 4) [6]. Interestingly, this reaction could be transposed to solid phase. The tetrahydroquinoline **9b** was anchored on a Wang resin, and the RCM was performed with the more active **G2** catalyst in refluxing dichloromethane. Cleavage of the product from the resin was accomplished with TFA, affording **10b** in 42% over two steps.

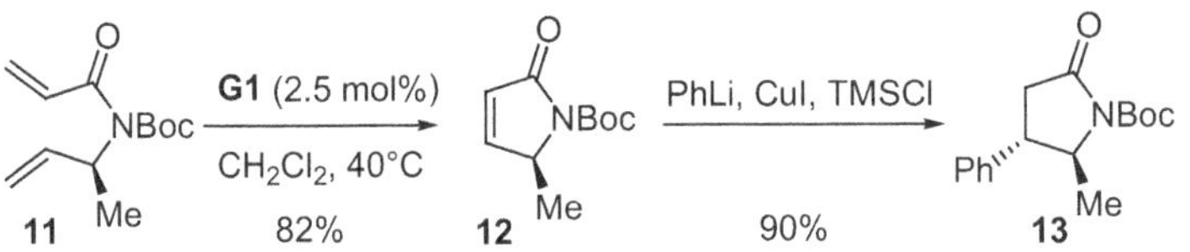

Scheme 5 RCM followed by conjugate additions

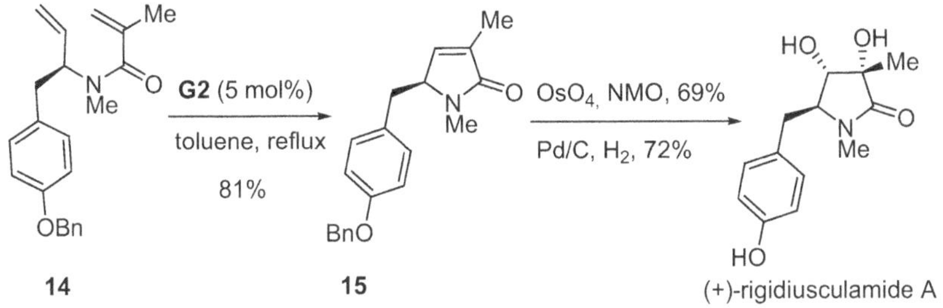

Scheme 6 RCM followed by dihydroxylation; total synthesis of (+)-rigiusculamide A

Beside hydrogenation, the double bond formed by RCM could be advantageously used for further functionalization. Helmchen et al. reported the diastereoselective conjugate addition of organocopper compounds to γ-lactams **12** produced by RCM with **G1** to yield *trans*-disubstituted lactam **13** (Scheme 5) [7].

The double bond formed by RCM could be readily transformed into a diol. During the total synthesis of (+)-rigidiusculamide A by Krishna and Reddy, diene **14** obtained from tyrosine and methacrylic acid was refluxed in toluene with **G2** catalyst (Scheme 6) [8]. The five-membered lactam **15** formed was subjected to dihydroxylation with the OsO_4/NMO system in order to form diastereoselectively the *cis*-diol of (+)-rigidiusculamide A.

One-pot isomerization of the double bond of **20** formed by RCM from **17** with **Zhan-1B** catalyst could generate the iminium intermediate **21**, which could react in a Mannich reaction with indole **16** to enantioselectively yield pyrrolidin-2-one **18** in the presence of a chiral phosphoric acid **19** as reported by You et al. (Scheme 7) [9].

Six-Membered Lactams

The general strategy implying the coupling of amines with acrylic acid derivatives followed by RCM was widely described for the synthesis of δ-lactams [10]. The synthesis of a series of δ-substituted six-membered lactams **23a–f** was described by Fiorelli and Savoia with **G2** (Scheme 8) from **22a** to **22f**. Alkyl, phenyl, naphthyl, and heterocyclic substituents such as furyl and thienyl were tolerated in α of the nitrogen. In the case of pyridyl **22f**, which contains a basic nitrogen, the use of TFA

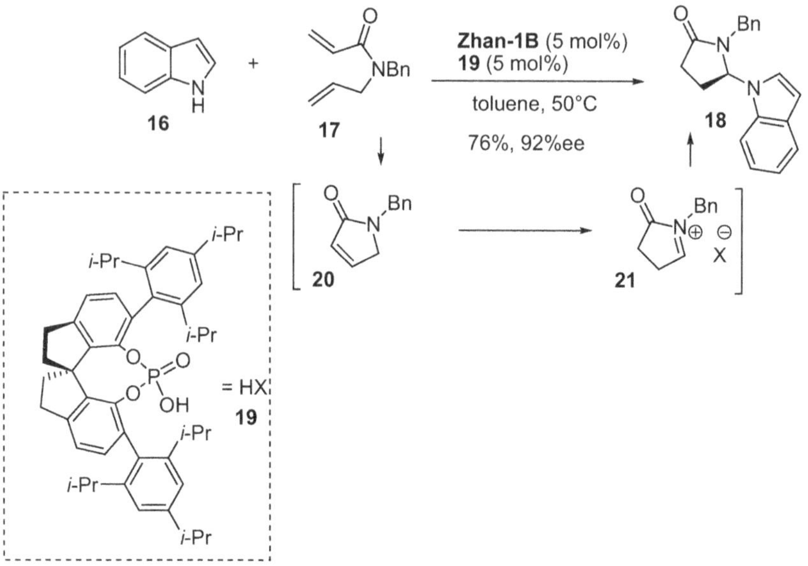

Scheme 7 RCM followed by isomerization and Mannich reaction

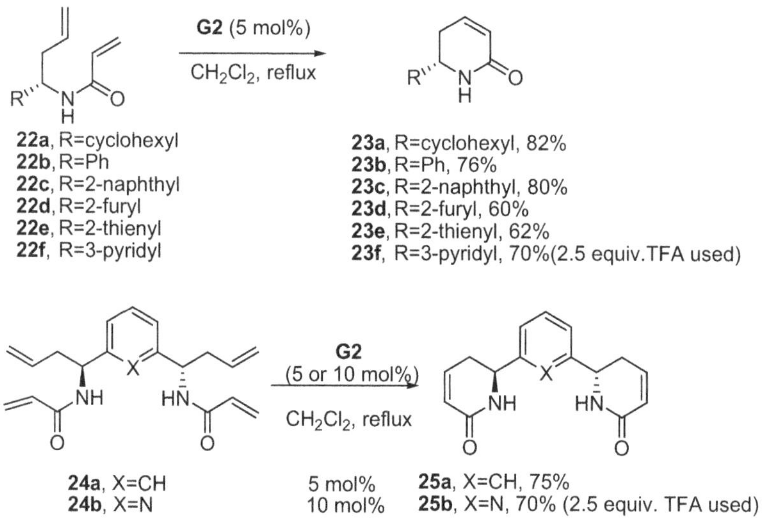

Scheme 8 Synthesis of δ-substituted δ-lactams by RCM

was required to achieve the protonation of the nitrogen and therefore prevents it to sequester the catalyst. C2-symmetric lactams **25a,b** could also be accessed by double RCM in good yields. With the pyridyl linker **24b**, addition of TFA was necessary.

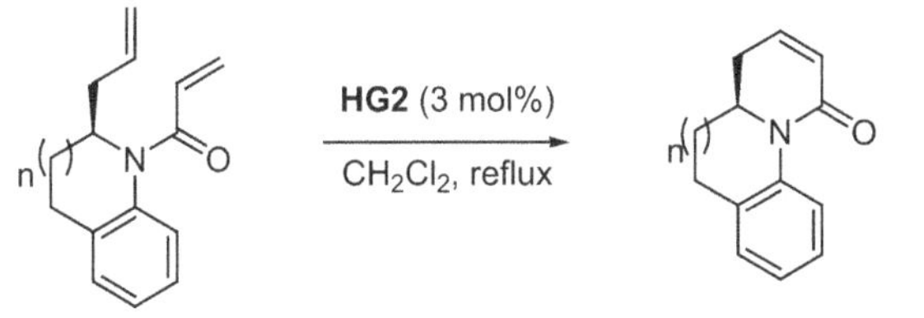

26a, n=1
26b, n=2

27a, n=1, 83%
27b, n=2, 79%

Scheme 9 Synthesis of fused δ-lactams by RCM

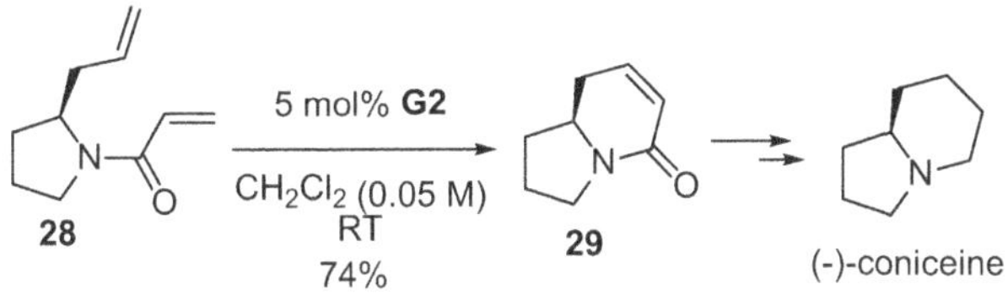

28 **29**

Scheme 10 Formal synthesis of (–)-coniceine

Additional synthesis of δ-substituted δ-lactams was reported by Yus et al. (Scheme 9) [11]. The fused tricycles **27a,b** were obtained from tetrahydroquinoline **26a,b** with **HG2** catalyst.

RCM of acrylate-containing dienes substituted at the α-position of the nitrogen has been employed for the total synthesis of several piperidine-containing alkaloids. The formation of the lactam is generally followed by reductions of the double bond and the carbonyl.

Chang et al. reported the formal synthesis of (–)-coniceine, one of the structurally simplest indolizidine alkaloids [12]. Amide **28** was obtained by coupling of acryloyl chloride with a secondary amine derived from proline. The metathesis reaction with 5 mol% of **G2** catalyst provided unsaturated lactam **29**, a known intermediate of the synthesis of (–)-coniceine (Scheme 10).

Yadav et al. accomplished the total synthesis of (–)-sedridine, (–)-halosaline, and (+)-norallosedamine through the RCM of 1,3-aminoalcohol derivatives **30a–c** (Scheme 11) [13, 14].

The synthesis of the δ-lactam core **33** of awajanomycin by Pritchard and Wilden featured an RCM with γ,δ-substitutions [15]. The metathesis reaction was performed by dropwise addition of **HG2** catalyst to a dichloromethane solution of **32** at reflux (Scheme 12).

Seven-Membered Lactams

The formation of seven-membered rings from acrylamide lactams is also reported.

Jenkins et al. described the synthesis of dihydroazepinones as templates for polyhydroxylated sugar analogs for evaluation as glycosidase inhibitors

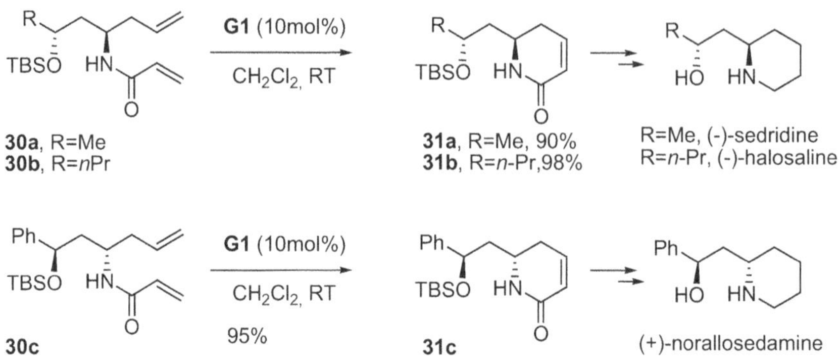

Scheme 11 Total synthesis of (–)-sedridine, (–)-halosaline, and (+)-norallosedamine

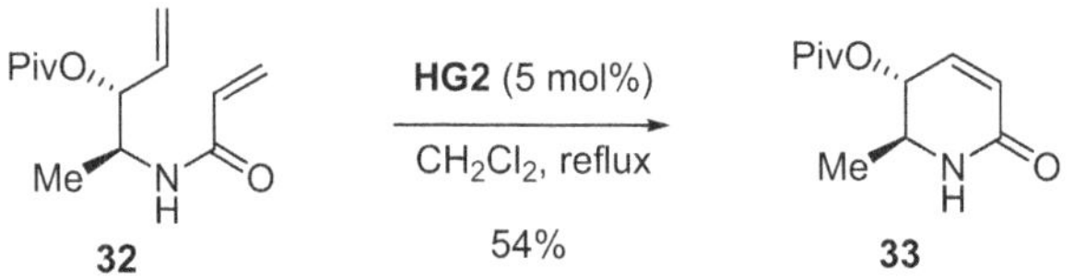

Scheme 12 Synthesis of a γ,δ-substituted δ-lactams by RCM

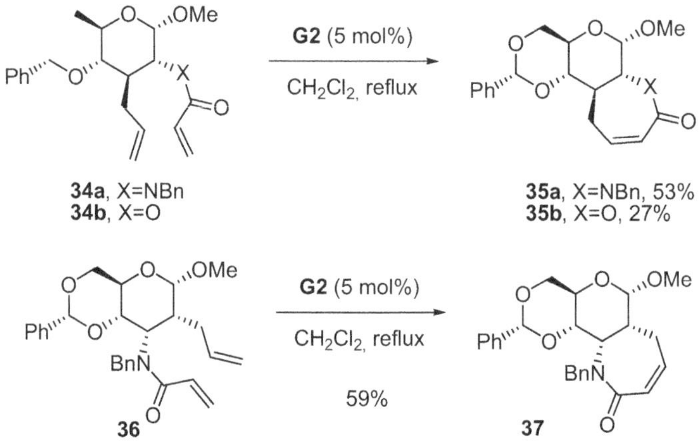

Scheme 13 Synthesis of seven-membered lactams by RCM

(Scheme 13) [16]. In this aim, carbohydrate-derived diene **34a** was subjected to RCM with **G2** catalyst in refluxing dichloromethane. The desired lactam **35a** was obtained in a satisfactory 53% yield. It should be noted that the corresponding lactone **35b** was obtained in only 27% with large amounts of dimeric products. It is supposed that the amide carbonyl group interferes less with the metathesis catalyst than the ester carbonyl group, and the double bond of the acrylamide is less electron deficient than the acrylate. The regioisomeric dihydroazepinone **37** was obtained from **36** under the same conditions.

In the goal of accessing indoline-derived natural product-like scaffolds, Arya et al. described the synthesis of seven-membered lactams from bicyclic acrylamides (Scheme 14) [17]. The reaction proceeded very easily at room temperature with **G2** catalyst in high yields. Both diastereoisomers **38a** and **38b** successfully delivered the desired lactams **39a,b**. This process was adapted to solid-phase synthesis. The indoline **38c** was loaded on a polystyrene resin with a spacer containing a silyloxy group. The RCM reaction was slow and required 50 mol% of **G2** but delivered the seven-membered lactam **39c**. Cleavage from the support was achieved with HF in pyridine.

Benzazepindiones **41a–d** were obtained by Couture et al. from the unusual RCM of styrene-derived *N*-benzoylacrylamides **40a–d** bearing no saturated C–C bonds (Scheme 15) [18]. Refluxing styrenes **40a–d** with 5 to 10 mol% of **G1** catalyst in

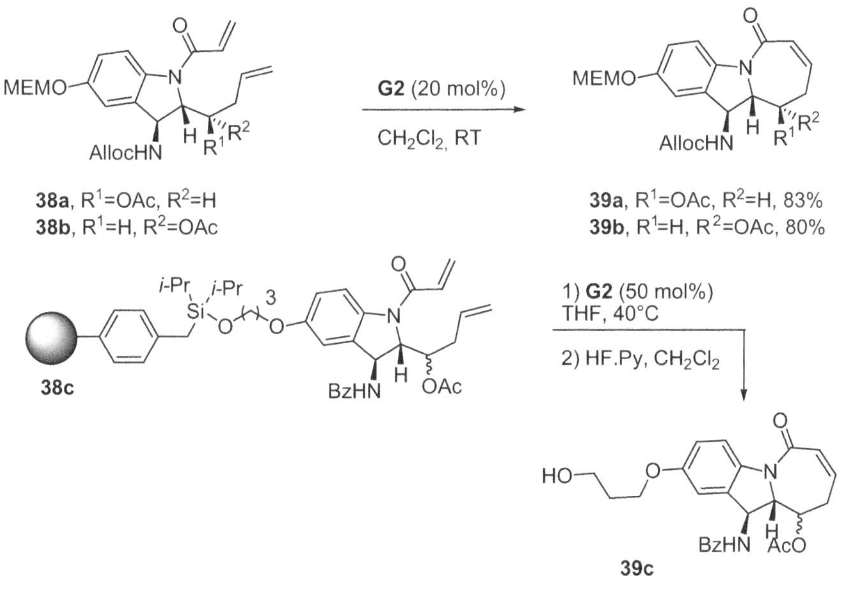

Scheme 14 Synthesis of seven-membered lactams derived from indolines by RCM

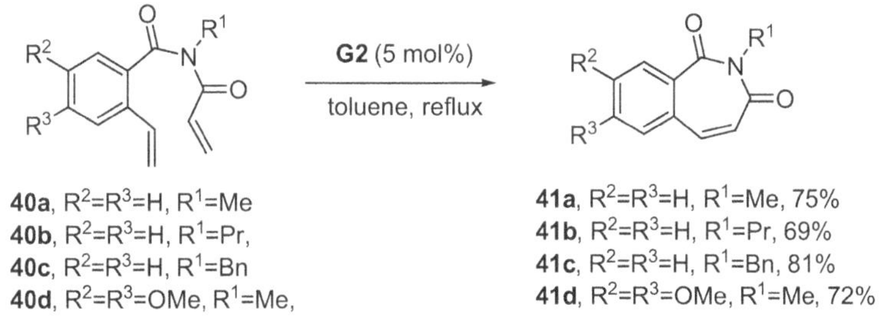

Scheme 15 Synthesis of benzazepindiones from *N*-benzoylacrylamide derivatives

dichloromethane during 24 h only led to 30–35% of lactam **41a–d**. Eventually, it was found that yields were increased to 69–81% with 5 mol% of **G2** catalyst in refluxing toluene.

Eight-Membered Lactams and More

The formation of eight-membered lactams via RCM of unsaturated conjugated amides is less encountered. The tetrahydroquinoline **42** was subjected to RCM with **G2** in refluxing dichloromethane, and the polycyclic compound **43** having a medium-sized lactam was isolated by Arya et al. (Scheme 16) [19]. This process could be transposed to solid-phase synthesis.

Enamides, which are isomers of acrylamides, could also be engaged in RCM reactions. During the total synthesis of (–)-paliurine E by Evano et al., the RCM of substituted enamide **44a** could deliver the 14-membered macrocycle **45a** of the natural product in 49 % with two additions of 10 mol% of **G2** catalyst in refluxing dichloroethane (Scheme 17) [20]. Subtle changes of the diene precursor showed

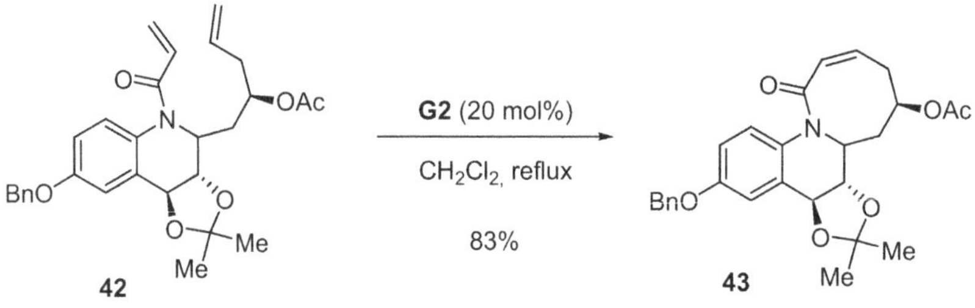

Scheme 16 Synthesis of eight-membered lactams derived from tetrahydroisoquinolines by RCM

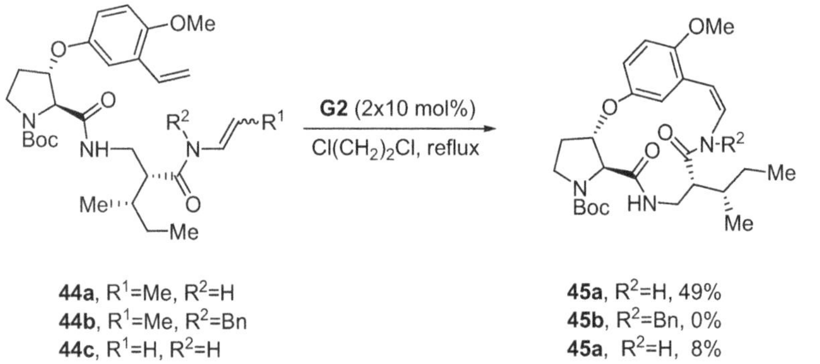

Scheme 17 RCM of enamides for the synthesis of a 14-membered lactam

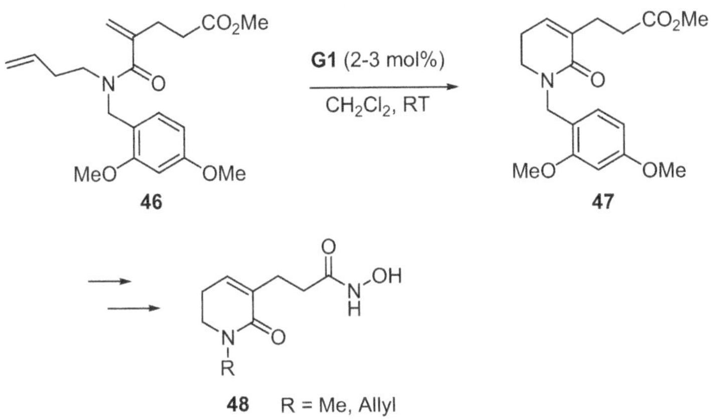

Scheme 18 RCM from α-alkylacrylamide

dramatic drops in yields. The addition of a benzyl on the nitrogen of the enamide **44b** did not lead to any RCM product, probably because the presence of the benzyl group changes the conformation of the enamide. Monosubstituted enamide **44c** led only to 8% of macrocycle **45a**, because **44c** was found to be thermally much less stable than disubstituted enamide **44a**.

2.1.2 Substitution Effects

This part will highlight cases in which the substitution of the diene precursor has an impact on the reactivity or cases of interesting synthetic application of substituted dienes.

Methacrylate derivatives could be easily assembled with amines to give α-substituted acrylamides.

In this context, inhibitors of histone deacetylase **48** (HDAC) containing a propionic hydroxamate appended to a δ-lactam have been obtained by the RCM of methacrylamide derivative **46** by Han et al. (Scheme 18) [21]. The RCM went smoothly with only 2–3 mol% of **G1** in dichloromethane at RT.

The RCM of α-aminoacrylamides allows the straightforward access to conformationally restricted peptidomimetics. In this context, the synthesis of bicyclic lactams **50a–c** was investigated by Manzoni et al. (Scheme 19) [22]. The RCM of the dehydro-α-aminoamide **49a**, in which the nitrogen is protected with an acetyl group, was carried out with 5 mol% of **G2** at RT in dichloromethane and yielded **50a**. Switching the acetyl group on the nitrogen to a Cbz did not lead to the expected product **50b** but only to dimers of the starting material. Steric and electronic factors from the Cbz group could be invoked. The formation of the seven-membered lactam required a higher catalyst loading, a higher temperature, and a higher dilution to avoid the formation of dimers. All starting material was consumed, but only 53% of **50c** was obtained along with 14% of six-membered

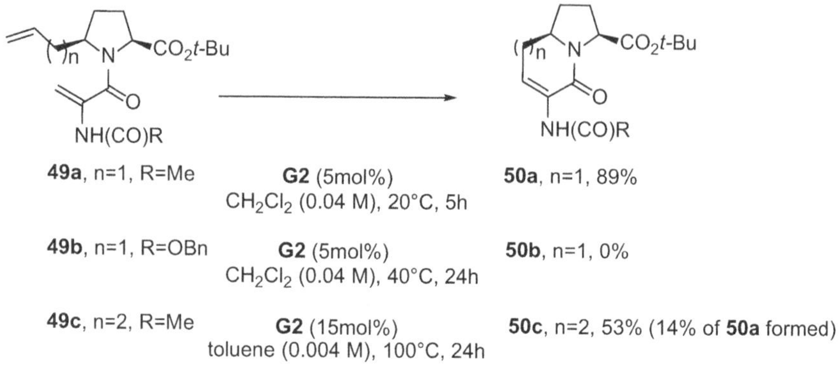

Scheme 19 RCM from α-aminoacrylamides

50a, which resulted from isomerization of the monosubstituted terminal olefin into a disubstituted olefin prior to RCM.

When studying the formation of monocyclic lactams from similar vinyl-α-amino acids, Nan et al. uncovered the substituent effect of the nitrogen acrylamide (Scheme 20) [23]. RCM of the unprotected amide **51** was unsuccessful, and the authors subsequently realized that the presence of the bulky dimethoxybenzyl on the nitrogen of **53a** allowed them to obtain successfully the γ-lactam **54a** with **G2** at RT. This result could be rationalized as follows. The lack of reactivity of the *N*-H amide **51** might be due to the fact that the *trans*-conformer **51-*trans*** is preferred over the *cis* **51-*cis*** in order to minimize steric repulsions between ally and enamide substituents. When the nitrogen of the amide is substituted by a bulky group, the *cis*-form **53-*cis*** would be preferred over the *trans*-form **53-*trans*** to avoid steric interactions between the dimethoxybenzyl and the enamide. The corresponding six- and seven-membered lactams **54b,c** were also synthesized with the corresponding *N*-substituted dienes. The former scaffold is of importance in drug discovery and natural product chemistry.

Haufe et al. demonstrated that the presence of a fluorine atom at the α-position of the acrylamide had some impact on the reactivity of dienes **55a–c** in RCM (Scheme 21) [24]. The five-membered lactam **56a** was obtained in only 10 min at RT in the presence of 2 mol% of **G2** catalyst, while 2 h at 80 °C was necessary to isolate a similar yield of the six-membered ring lactam **56b**. The seven-membered ring lactam **56c** was not obtained under those conditions; only dimerization by the non-fluorinated double bond occurred. The fact that the five-membered fluorinated lactam **56a** is easier to form than the six- or seven-membered lactams **56b,c** seems in contrast with the reactivity of unsubstituted acrylamides toward RCM, and it demonstrates the electronic effect of the fluorine on the RCM.

Unfortunately, Toueg and Prunet proved that acrylamides **57a,b** α-substituted by an acyl electron-withdrawing group were less suitable in RCM since diene **57a** did not lead to five-membered lactam **58a** but dimerized, while **57b** was converted in only 30 % into **58b** (Scheme 22) [25].

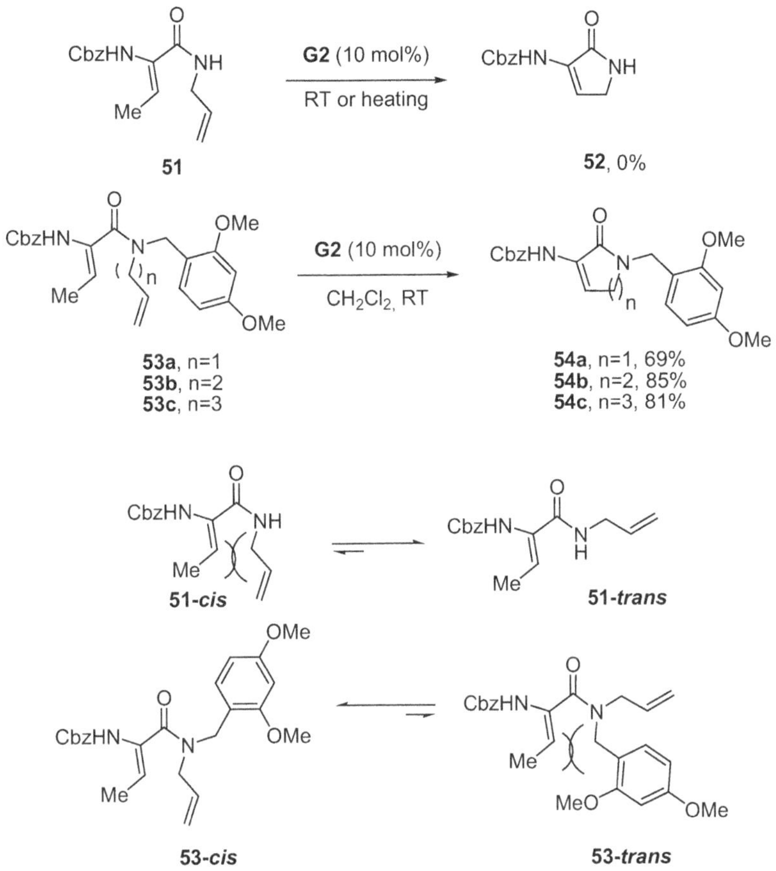

Scheme 20 RCM from α-aminoacrylamides; effect of the nitrogen substituent

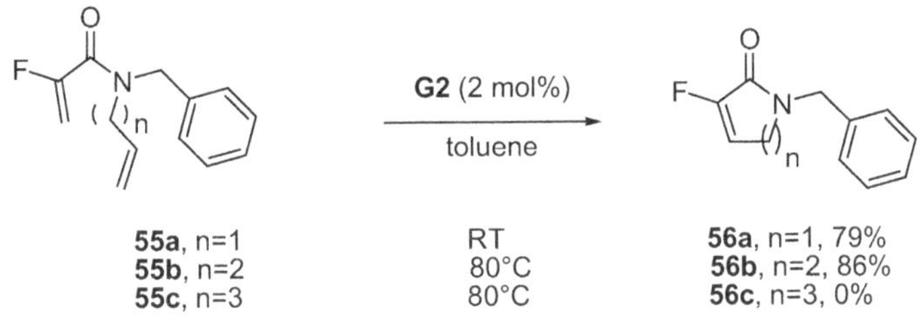

Scheme 21 RCM from α-fluoroacrylamides

The formation of *N*-hydroxylactams through RCM of *N*-alkoxyacrylamides is possible. Carotti et al. described the synthesis *N*-hydroxylactams by this strategy for fragment-based approach toward matrix metalloproteinase inhibitors (Scheme 23) [26]. Starting from suitable *O*-protected hydroxylamides, such as *N*-(*p*-

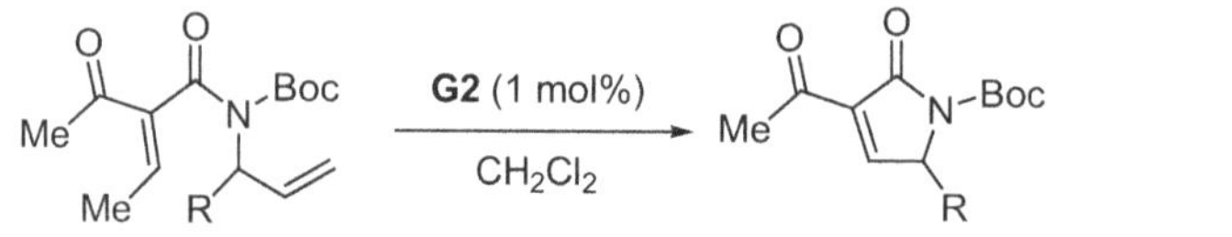

57a, R=H
57b, R=CH_2CH_2Ph

58a, R=H, 0% (dimer obtained)
58b, R=CH_2CH_2Ph, 30%

Scheme 22 RCM from α-acetylacrylamides

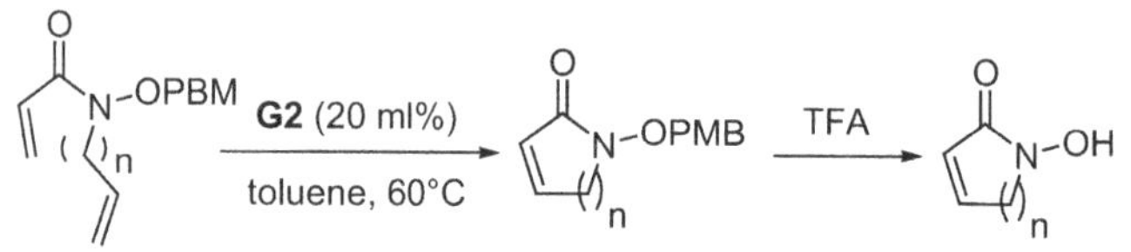

59a, n=1
59b, n=2

60a, n=1
60b, n=2

61a, n=1, 85%
61b, n=2, 97%

Scheme 23 RCM from N-alkoxyacrylamide; synthesis of *N*-hydroxylactams

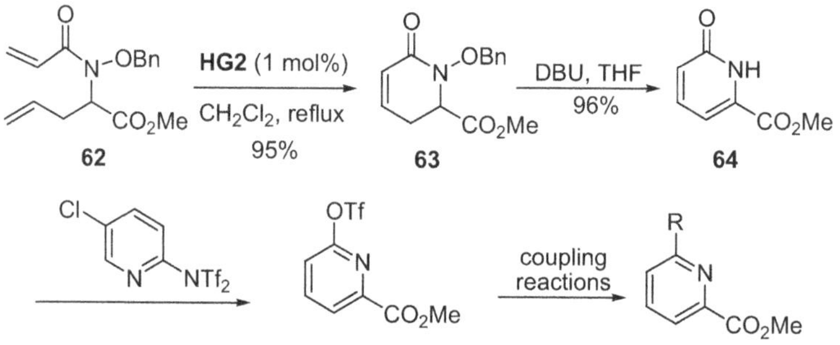

65

66

Scheme 24 RCM from an N-alkoxyacrylamide; synthesis of a 2-pyridone and 2-substituted pyridines

methoxybenzyl)hydroxylamide **59a,b**, the RCM with **G2** delivered *N*-(*p*-methoxybenzyl)hydroxylactams **60a,b**. The free hydroxylactams **61a,b** were then generated by deprotection in the presence of trifluoroacetic acid (TFA).

Alternatively, Donohoe et al. demonstrated that six-membered *N*-benzyloxy-hydroxylactam **63** obtained by RCM could be treated with DBU to eliminate benzyl alcohol and form 2-pyridone **64**, which could be further transformed into pyridine triflate **65** and functionalized with coupling reactions (Scheme 24) [27].

Complementary to *N*-hydroxylamides, the RCM of hydrazides could also lead to lactams. Couture et al. reported the synthesis of (–)-coniceine via the ring-closing metathesis of **68** (Scheme 25); **G1** catalyst was able to effect this transformation in refluxing dichloromethane [28]. In this strategy, the chiral non-racemic *N*-2-

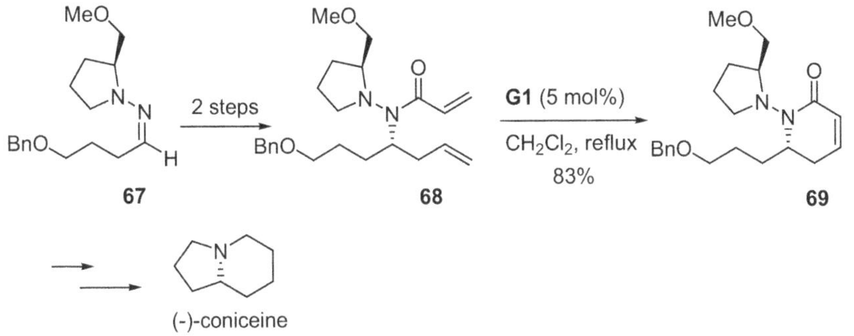

Scheme 25 RCM from hydrazides; total synthesis of (–)-coniceine

(methoxymethyl)pyrrolidine moiety served as a chiral auxiliary to form stereoselectively the stereogenic center of coniceine via the addition of an organometallic reagent to the corresponding hydrazone **67**.

2.1.3 Cleavage of Chiral Auxiliaries by RCM

It is also possible to release a chiral auxiliary during the formation of lactams by RCM. Spino et al. introduced *p*-menthane-3-carboxaldehyde **70** as a chiral template to create a new stereogenic center such as in alcohols **71** (Scheme 26) [29]. Conversion of alcohols **71** into a cyanate and transposition into isocyanate could lead to chiral *N*-allylamides **72a–c**. Concomitant to the formation of the desired lactams **73a–c**, the chiral auxiliary was released by RCM in the form of vinyl-menthane **74** which could be recycled into aldehyde **70** by ozonolysis. The metathesis step was achieved with **GN** or **G2** catalysts; the temperature of the reaction has to be increased in the presence of bulkier substituents in α of the nitrogen. Lactams **72b,c** bearing tetrasubstituted carbon centers were obtained. Since the olefin connected to the menthane template is sterically crowded, it is believed that the RCM is initiated at the acrylamide olefin. However, the methacrylamide function of **72d** was also unreactive toward any metathesis catalyst. Therefore, initiation of the RCM from both alkenes of the diene **72d** was not possible. To circumvent this problem, a relay ring-closing metathesis was deployed from **72e**. The initiation of the RCM would occur at the less sterically hindered olefin to generate carbene **75** which after release of cyclopentene would deliver carbene **76** and trigger the ring closing.

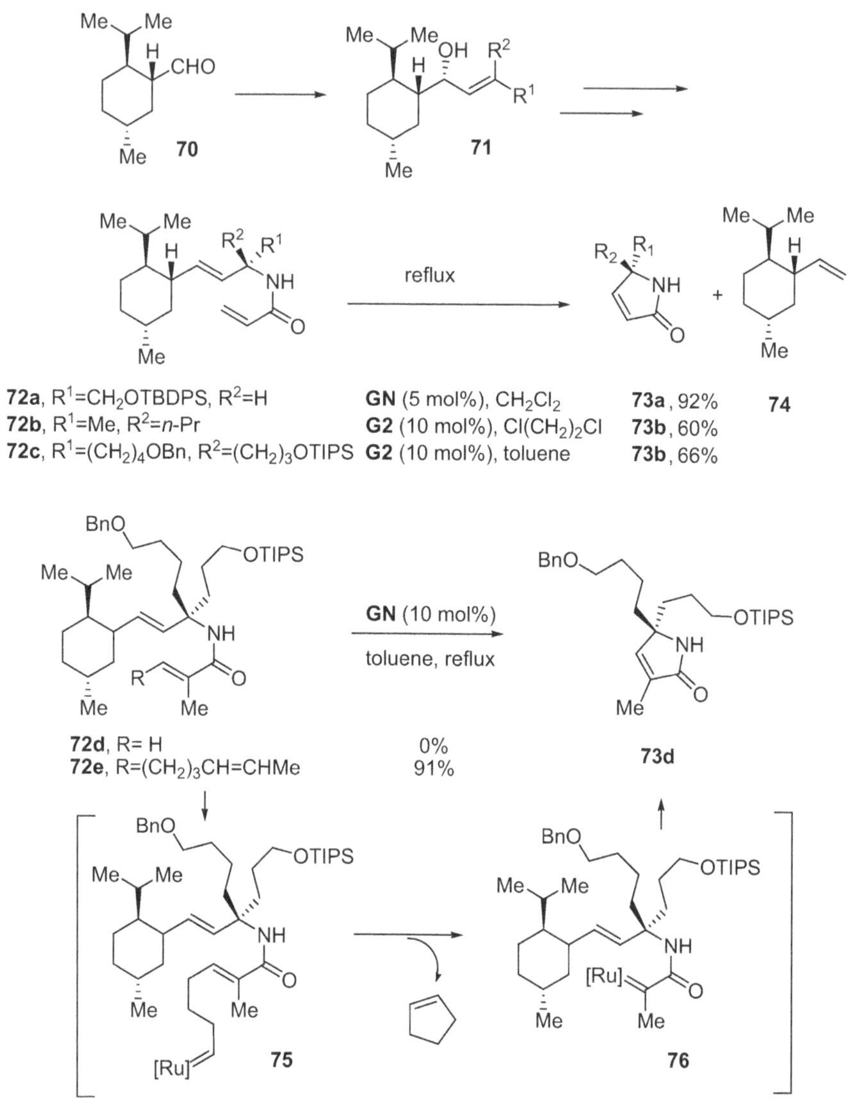

Scheme 26 Cleavage of a chiral auxiliary by RCM

2.2 *From Nonconjugated Amides*

2.2.1 Six- and Seven-Membered Lactams

Complementary to the cases of acrylamides, nonconjugated amides bearing dienes **79** could be easily assembled by the coupling of a carboxylic acid derivative **77** and an amine **78** and be engaged in RCM to generate lactams **80** (Scheme 27).

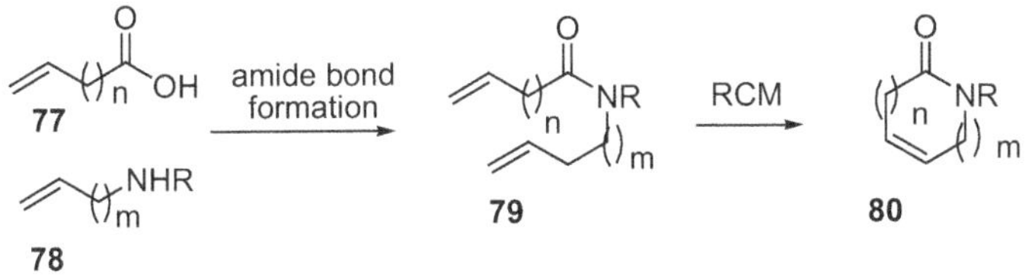

Scheme 27 General process for the synthesis of nonconjugated lactams by RCM

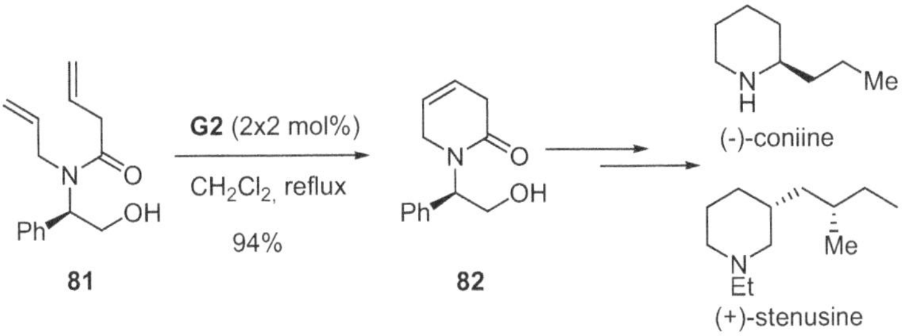

Scheme 28 Synthesis of (–)-coniine and (+)-stenusine by RCM

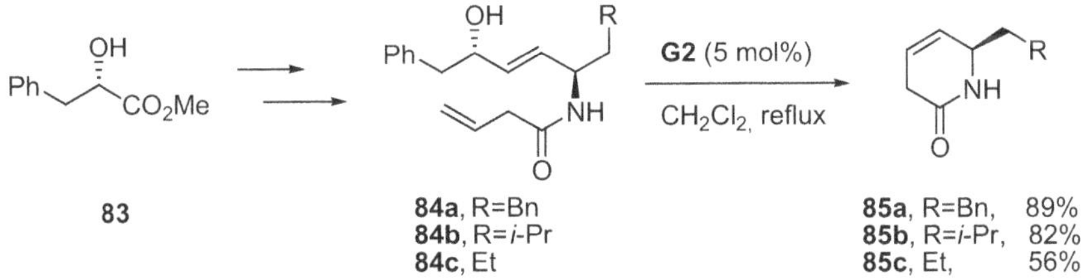

Scheme 29 Synthesis of δ-lactams and cleavage of the chiral auxiliary by RCM

Coldham et al. accomplished the formal syntheses of piperidine alkaloids via the RCM with **G2** of the chiral non-racemic diene **81** to form in an excellent yield the lactam **82**, which could be further transformed to a known common intermediate of (–)-coniine and (+)-stenusine (Scheme 28) [30].

For the stereocontrolled synthesis of δ-lactams, Sutherland et al. started from non-racemic (*S*)-phenyllactate **83** as inducer of chirality (Scheme 29) [31]. A series of reactions including a diastereoselective reduction of a ketone and an Overman [3,3]-sigmatropic rearrangement allowed to create diastereoselectively the new stereogenic center of amides **84a–c**. RCM with **G2** delivered expected lactams **85a–c** with concomitant cleavage of the chiral inducer.

The synthesis of seven-membered lactams is also described. Unfortunately, Kamimura et al. failed to achieve the RCM of unsubstituted diene-containing *N*-H or *N*-Boc amides **86a** and **88a** with **G1** catalyst (Scheme 30). In order to overcome this lack of formation of medium-sized lactams, the authors introduced a removable oxyoxazolidinone ring, which offers steric bias to induce the RCM [32]. Indeed, the bicyclic oxazoloazepine **91a** was obtained with **G1** catalyst. Removal of the RCM inducer was effected by hydrogenation and hydrolysis to yield seven-membered NH-lactam **87a**. The formation of oxazoloazecine **91b** required the use of **G2** catalyst instead of **G1**; opening of the oxyoxazolidinone and formation of the eight-membered NH-lactam **87b** were achieved with a lower yield.

While the formation of *N*-Boc lactam **89a** from diene **88a** was not effective with **G1** catalyst (Scheme 30), Hassan and Brown showed that it could be achieved with **G2** catalyst in high dilution (Scheme 31) [33]. The reaction in dichloromethane at 40 mM afforded mainly the dimeric macrolactam **92**. Performing the reaction at a

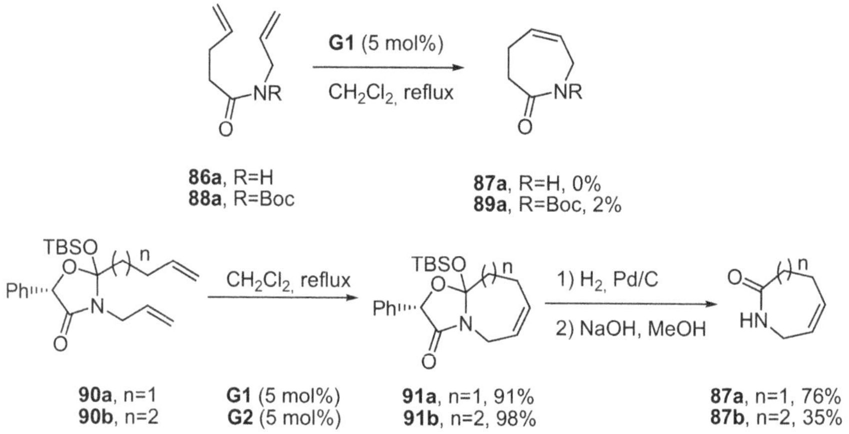

Scheme 30 Synthesis of NH-containing seven- and eight-membered lactams with assistance of a removable oxyoxazolidinone ring

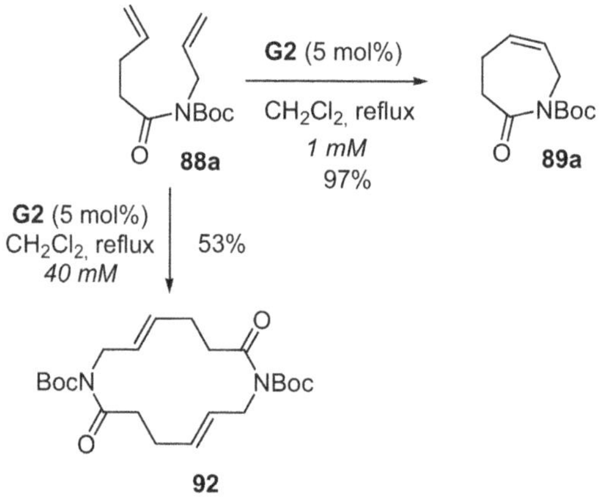

Scheme 31 Concentration effect in the synthesis of a *N*-Boc-containing seven-membered ring lactam

high dilution of 1 mM allowed the isolation of the desired *N*-Boc lactam **89a** in high yield.

Increase of the substitution of the diene precursor facilitates the RCM. The α-aminolactam **95** was thus obtained from **93** by Fuhshuku and Asano with **G1** catalyst at reflux of dichloromethane, followed by hydrogenation and removal of the Boc group by HCl (Scheme 32) [34].

During the synthesis of the orthogonally protected bis-amino acid **98**, Del Valle and Goodman deployed an RCM strategy from the dienic amide **96**, followed by opening of the N–(CO) bond of **97** (Scheme 33) [35].

The RCM of nonconjugated amides tolerates the introduction of functional groups. Helmchen et al. described the efficient synthesis of six- and seven-membered *N*-alkoxylactams **100a–c** with **G2** catalyst at 40–50°C from **99a** to **99c** (Scheme 34) [36].

Miller et al. described the synthesis of bicyclic alkoxylactams **101a–d** from β-lactam-containing dienes **100a–d** with **G2** catalyst (Scheme 35) [37].

In complement to the RCM of fluoroacrylamides(Scheme 21), Couve-Bonnaire et al. studied the synthesis of fluorolactams **103** from fluoroalkenes **102** (Scheme 36)

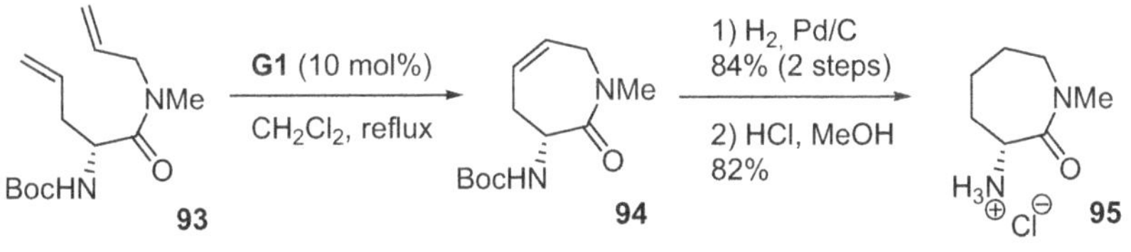

Scheme 32 Synthesis of a seven-membered α-aminolactam

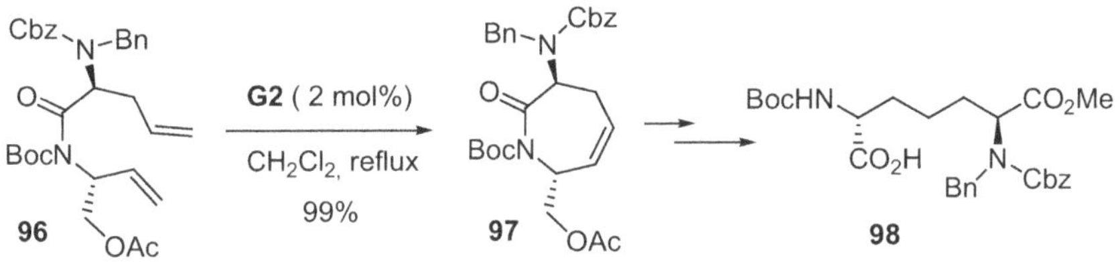

Scheme 33 Synthesis of an orthogonally protected bis-amino acid via RCM

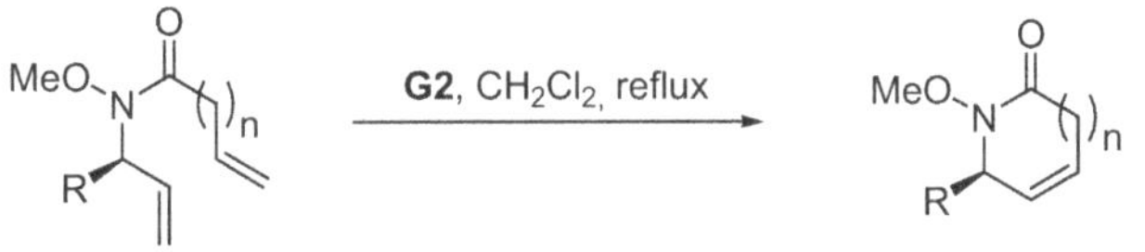

99a, n=1, R=Ph	(2 mol%) CH_2Cl_2, reflux	**100a**, n=1, R=Ph, 93%
99b, n=1, R=*n*-Hex	(2 mol%) toluene, 50°C	**100b**, n=1, R=*n*-Hex, 79%
99c, n=2, R=Ph	(4 mol%) toluene, 50°C	**100c**, n=2, R=Ph, 63%

Scheme 34 Synthesis of *N*-alkoxyamides by RCM

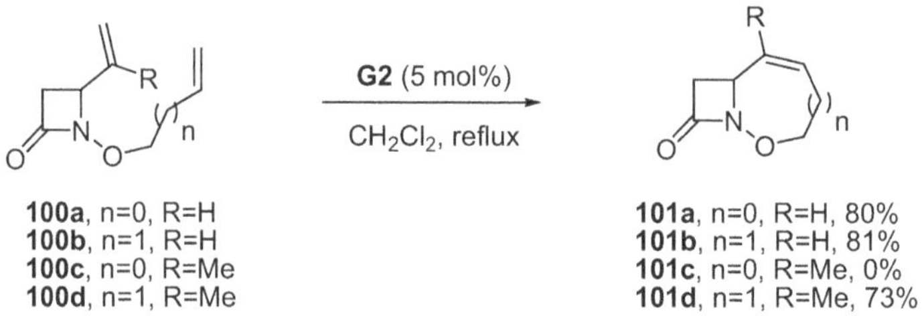

Scheme 35 Synthesis of *N*-alkoxyamides by RCM

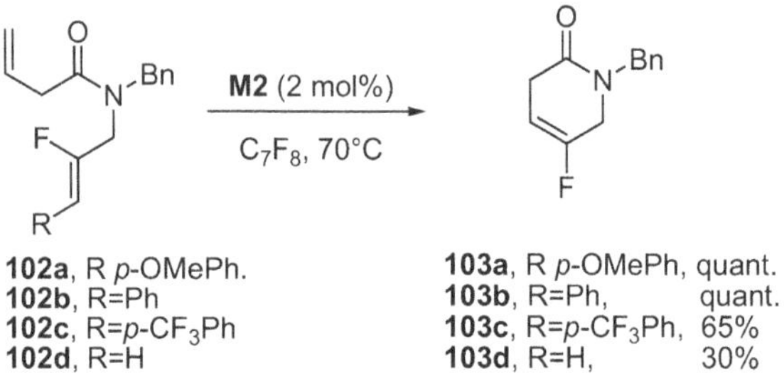

Scheme 36 RCM of fluorostyrenes

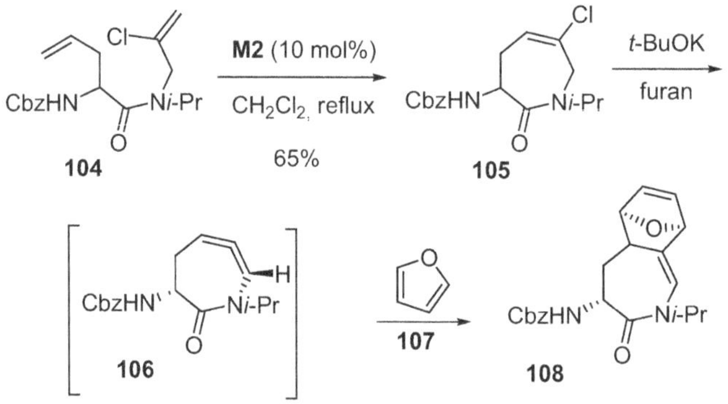

Scheme 37 RCM of chloroalkenes and cycloadditions

[38]. The authors discovered that the use of fluorotoluene as a solvent significantly enhanced the yield compare to toluene, and **M2** (2 mol %) proved to be the best catalyst at 70°C. Interestingly, trisubstituted fluorostyrene precursors **102a–c** were more efficient than the disubstituted fluoroalkene **102d**. This could be rationalized by the generation of a more stable and reactive Ru-arylidene active propagating catalyst species. The reaction with the trifluoro electron-withdrawing group on the fluorostyrene **102c** was less efficient than the unsubstituted fluorostyrene **102b**. The methoxy-containing fluorostyrene **102a** was the more reactive.

Chloroalkene **104** could also be employed in the formation of lactam **105** by RCM with **M2** catalyst (Scheme 37). Very importantly, Verniest et al. demonstrated

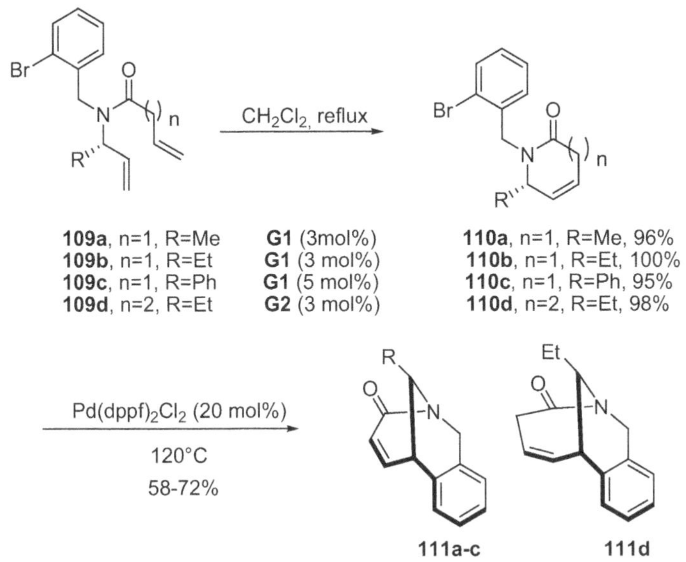

Scheme 38 Synthesis of bridged tricycles by RCM and Heck reactions

that deprotonation of the formed chlorovinyl lactam **105** could lead to a highly reactive allenamide **106**, which could be trapped in situ by a Diels–Alder cycloaddition with furan **107** to lead to **108** [39].

The Heck reaction of nonconjugated lactams obtained by RCM offers the possibility to generate bridged tricycles. Six- and seven-membered lactams **110a–d** substituted at the nitrogen by an *o*-bromobenzyl group were obtained respectively with **G1** or **G2** catalysts by Satyanarayana and Helmchen (Scheme 38) [40]. The Heck cyclization afforded diastereoselectively the tricyclic lactams **111a–d** that contained a nitrogen atom at one of the bridgehead positions.

2.2.2 Eight-Membered Lactams and More

Eight- and nine-membered ring lactams could be efficiently obtained by RCM if the dienic precursor is sufficiently substituted. One solution to favor the *cis*-conformation of the amide and therefore facilitate the RCM is to use the bulky 2,4-dimethoxybenzyl group on the nitrogen. Creighton, Reitz et al. could transform *N*-2,4-dimethoxybenzyl diene precursors **112a,b** into the eight-membered cyclic peptidomimetics **113a,b** [41] (Scheme 39), which could not be obtained from the corresponding *N*-H amide [42].

Formation of eight- and nine-membered lactams by RCM could also be favored if the diene precursor belongs to an adequately substituted cycle. Young et al. were able to synthesize bicyclic lactams **115a,b** as potential turn templates from dienes **114a,b** with **G1** catalyst (Scheme 40) [43].

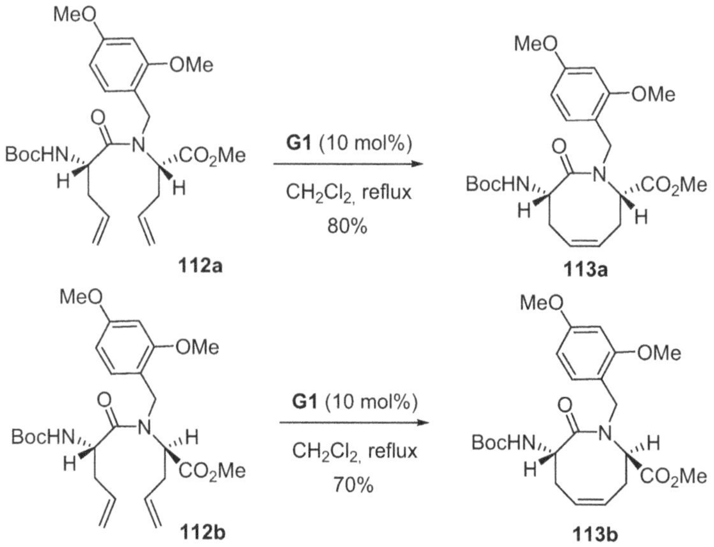

Scheme 39 Synthesis of a eight-membered lactam by RCM of a *N*-2,4-dimethoxybenzyl amide

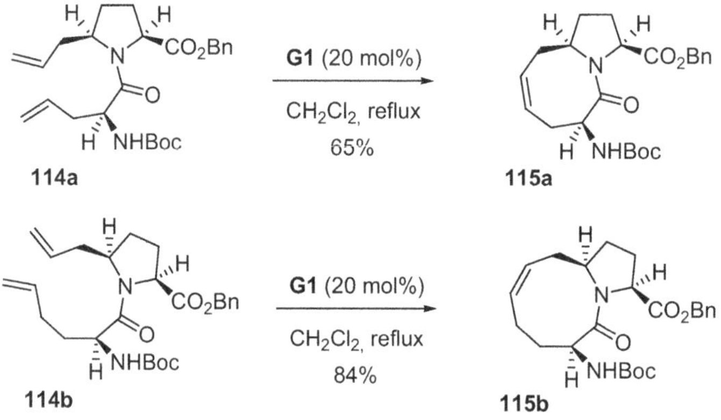

Scheme 40 Synthesis of eight- and nine-membered bicyclic lactams by RCM

Alternatively, to the 2,4-dimethoxybenzyl group, Lessene, Baell et al. introduced benzoylureas as removable *cis*-amide inducer to promote the formation of medium-sized lactams (Scheme 41) [44]. The RCM of amides **116a,b** failed with **G2** in various conditions. However, the benzoylureas **118a,b** adopt a closed *cis*-conformation maintained potentially by an intramolecular hydrogen bond. RCM with **G2** catalyst delivered efficiently the nine- and ten-membered lactams **119a,b**. Release of the inducer was achieved by removal of the Boc group and then intramolecular nucleophilic attack of the primary amine to the urea.

The use of a tether is an alternative strategy to induce the formation of medium-sized lactams by RCM lacking structural bias to facilitate the ring closure. Van Maarseveen et al. attached an allylamine and a carboxylic acid on a

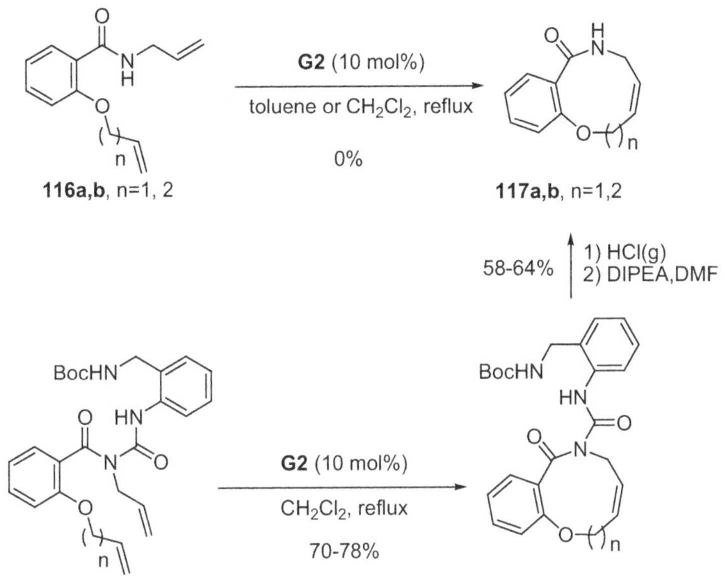

Scheme 41 Synthesis of a nine- and ten-membered lactams by RCM with a benzoylurea removable inducer

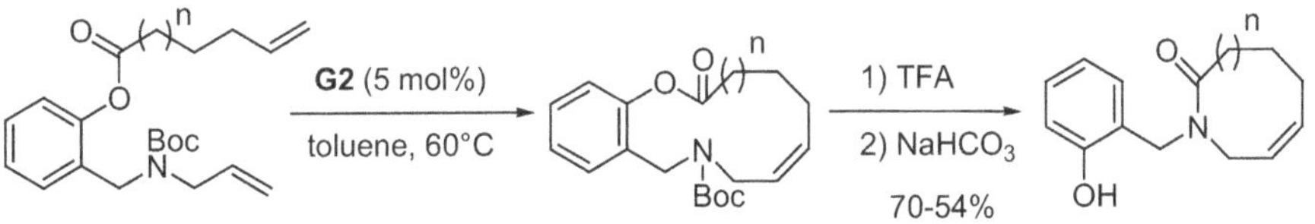

Scheme 42 Synthesis of eight- to ten-membered lactams using a salicylaldehyde-derived tether

salicylaldehyde-derived tethered **120a–c** to bring together the alkenes (Scheme 42) [45]. Macrocycles **121a–c** were obtained by RCM with **G2**, and transannular aminolysis led to the eight- to ten-membered lactams **122a–c**. Removal of the *o*-hydroxylbenzyl group could be achieved with sodium in ammonia.

Macrolactams can also be produced by RCM. During the synthesis of analogs of the naturally occurring macrolactone radicicol, Kitson, Moody et al. performed the synthesis of 14- to 16-membered lactams **124a–d** by metathesis with **G2** in moderate to good yields with various *E*/*Z* ratio of the olefin formed (Scheme 43) [46].

Finally, Szostak and Aubé were able to access challenging medium-sized bridged "twisted" amides via an RCM – transannular cyclization strategy (Scheme 44) [47]. The eight- to ten-membered lactams **126a–f** were obtained by RCM of the N-Ns-containing dienes **125a–f** with **HG2** catalyst in refluxing dichloroethane.

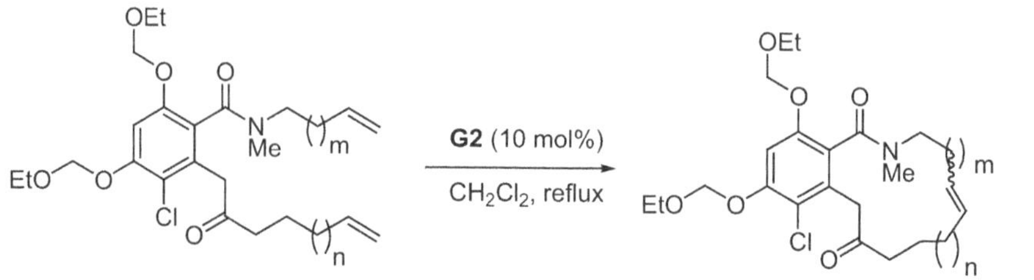

123a, n=2, m=1
123b, n=3, m=1
123c, n=4, m=1
123d, n=1, m=2,

124a, n=2, m=1, 51% (4:1E/Z)
124b, n=3, m=1, 39% (2:1E/Z)
124c, n=4, m=1, 34% (4:1E/Z)
124d, n=1, m=2, 68% (1:9E/Z)

Scheme 43 Synthesis of 14- to 16-membered lactams by RCM

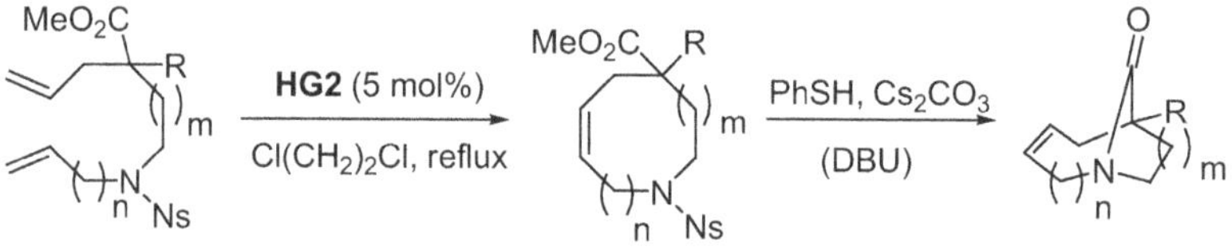

125a, n=1, m=1, R=CO_2Me
125b, n=1, m=2, R=CO_2Me
125c n=2, m=1, R=CO_2Me
125d n=2, m=2, R=CO_2Me
125e, n=3, m=1, R=CO_2Me
125f, n=1, m=3, R=Ph

126a, 90%
126b, 93%
126c, 94%
126d, 90%
126e, 92%
126f, 60%

127

Scheme 44 Synthesis of "twisted" amides by RCM and transannular lactamization

Removal of the nosyl group with thiophenol in basic condition was followed by lactamization to yield the desired bicyclic system **127** incorporating the nitrogen at a bridgehead position and the carbonyl on a bridge of only one carbon.

3 Cross Metathesis (CM)

Few syntheses of lactams incorporate a cross metathesis (CM) operation before the cyclization step.

3.1 CM and Lactamization

Vankar et al. disclosed the synthesis of sugar-derived spiroaminals using this general strategy (Scheme 45) [48]. Terminal alkene-containing azide **128** was subjected to CM with methyl acrylate in the presence of **G2** catalyst to achieve a one-carbon homologation to **129**. Reduction of the double bond and the azide followed by heating in MeOH delivered the desired spirolactam **130a** with noticeable epimerization at the anomeric position (**130b**).

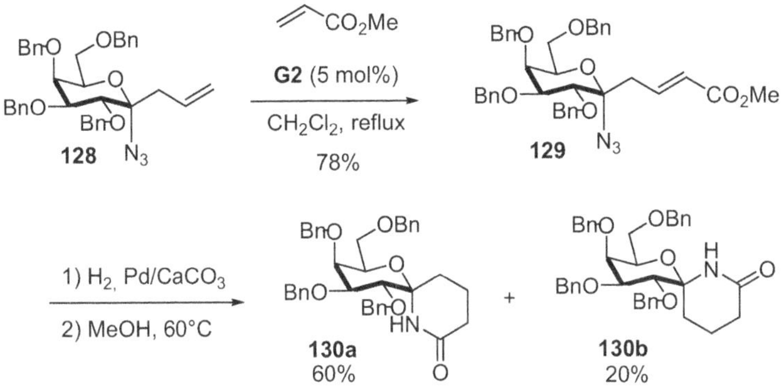

Scheme 45 Synthesis of a spirolactam by CM and reductive lactamization

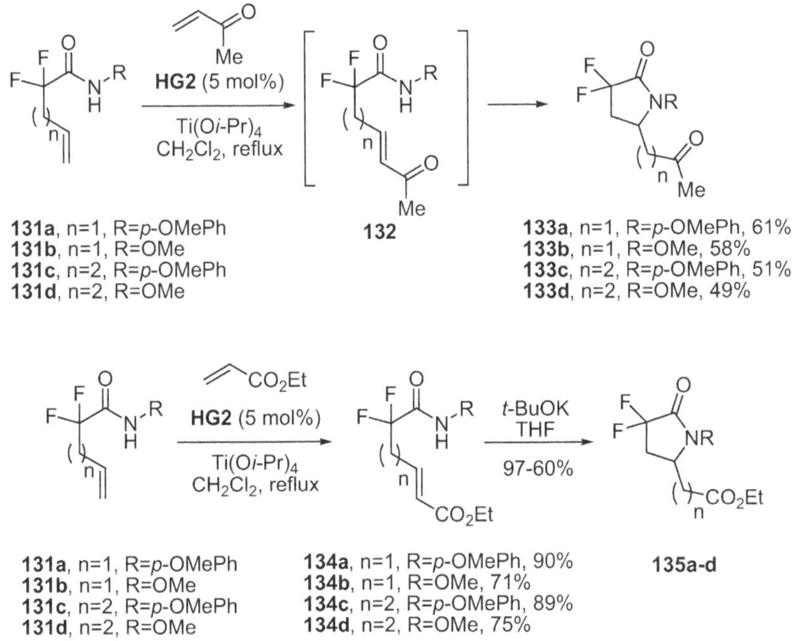

Scheme 46 Synthesis of lactams by CM and aza-Michael cyclization

3.2 CM and Aza-Michael Cyclization

A tandem CM – intramolecular aza-Michael addition – was reported by Fustero, del Pozo et al. for the synthesis of difluorinated lactams (Scheme 46) [49]. The metathesis of amides **131a–d** with methyl vinyl ketone in the presence of **HG2** catalyst and Ti(*Oi*-Pr)$_4$ was followed immediately by the aza-Michael reaction leading to the desired γ- and δ-lactams **133a–d**. When replacing methyl vinyl ketone by ethyl acrylate, the CM went smoothly but without spontaneous cyclization. The cross metathesis products **134a–d** were therefore isolated and treated with potassium *t*-butoxide to trigger the aza-Michael reaction and yielded lactams **135a–d**.

4 Ring-Closing Enyne Metathesis (RCEYM)

Beside the metathesis of alkenes, lactams could be obtained by metathesis of enynes.

4.1 From Acrylamides

Acrylamides **136a,b** proved to be reactive with **G1** catalyst to effect the RCEYM and deliver 4-vinyl-α,β-unsturated lactams **137a,b** as reported by Alper et al. [50] (Scheme 47). Ethylene gas was proven to be beneficial for the enyne metathesis.

4.2 From Nonconjugated Enynes

Nonconjugated enyne amides are also efficient precursors of lactams by metathesis. Arimitsu and Hammond achieved the synthesis of fluorinated δ-lactams **139a–c** by RCEYM of 1,7-enyne amides **138a–c** with **HG2** in the presence of ethylene (Scheme 48) [51]. Internal alkynes **138b,c** required higher temperatures than terminal alkyne **138a**. A domino metathesis could also be performed from the same enyne amide **138a**. A tandem RCEYM-CM occurred when styrene was added to the reaction mixture leading to 4-styrenyl-α,β-unsturated δ-lactam **140**. In the absence of ethylene gas, the yield was lower.

The synthesis of dienamide-containing lactams from the RCEYM of ene-ynamides was reported by Hsung et al. (Scheme 49) [52]. The six- and seven-membered lactams **142a,b** were obtained smoothly with **G2** at 80°C. The

Scheme 47 RCEYM of acrylamides

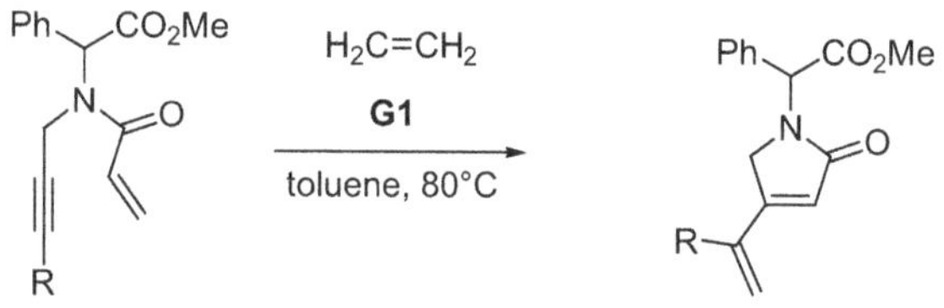

136a, R=H (7 mol%) **137a**, R=H, 76%
136b, R=Me (10 mol%) **137b**, R=Me, 75%

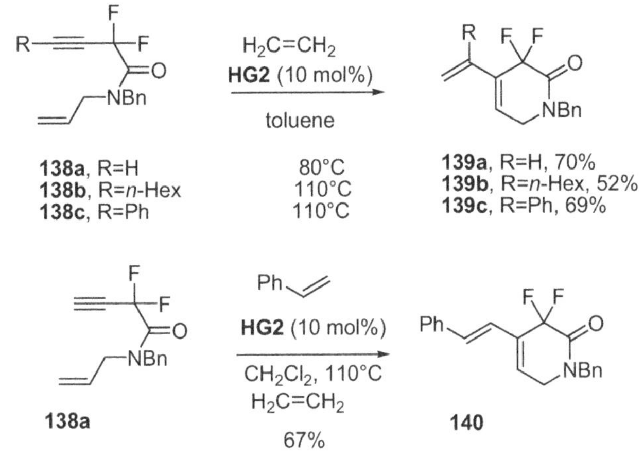

Scheme 48 RCEYM of 1,7-enyne amides

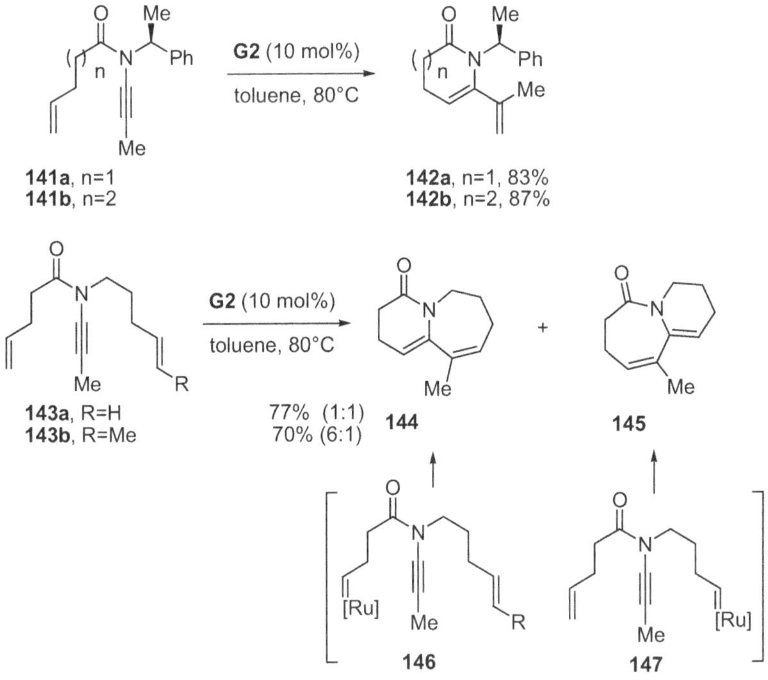

Scheme 49 RCEYM of ynamides

dienamide function thus obtained proved to be a suitable partner for Diels–Alder cycloaddition. Very interestingly, a domino metathesis was implemented from diene-ynamides **143a,b**; the RCEYM product could be engaged in a RCM of alkenes. Heating **143a** which contains two monosubstituted alkenes in the presence of **G2** resulted in the two bicyclic lactams **144** and **145** in a 1:1 ratio. It is assumed that the RCEYM is initiated at one of the olefins. The 1:1 ratio reflects that both monosubstituted double bonds in **143a** are equally reactive toward the catalyst and generate intermediates **146** and **147** in a 1:1 ratio. The use of a sterically differentiated diene such as **143b** allows to control the initiation of the RCEYM at the less substituted alkene and favors the formation of the bicyclic lactam **144**.

5 Ring-Rearrangement Metathesis (RRM)

Rearrangement by domino metathesis of strained rings obtained by cycloadditions has become recently a powerful tool, which indeed has been used in the synthesis of lactams (Scheme 50). Ring-opening metathesis (ROM) of cycloadduct **148**, favored by strain release, and subsequent RCM with an appended alkene and CM with an external alkene would result in the formation of bicyclic compound **149**. A N–(CO) functionality could be incorporated as X-Y or Y-Z in structures **148**, and therefore, lactams could be obtained by RRM.

5.1 *From All Carbon-Containing Bicycles*

One of the first examples of synthesis of lactams by RRM was achieved by Aubé et al. during the synthesis of alkaloid 251 F (Scheme 51). In fact, lactam **153** was created by an intramolecular Schmidt reaction of **152** after the RRM of the all carbon-containing framework of cycloadduct **150** [53].

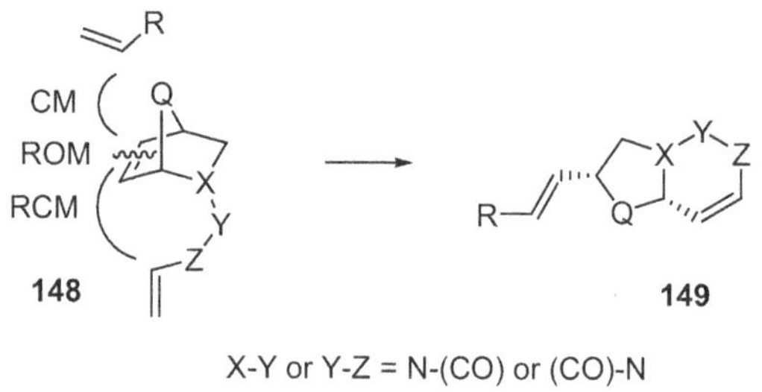

Scheme 50 Ring-rearrangement metathesis (RRM) for the synthesis of lactams

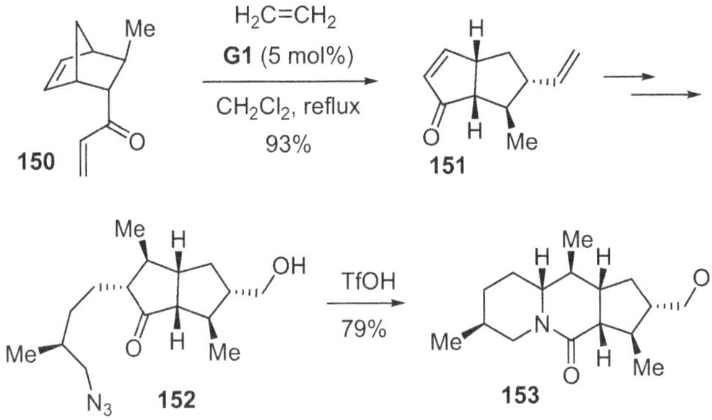

Scheme 51 Synthesis of a lactam by RRM and Schmidt reaction

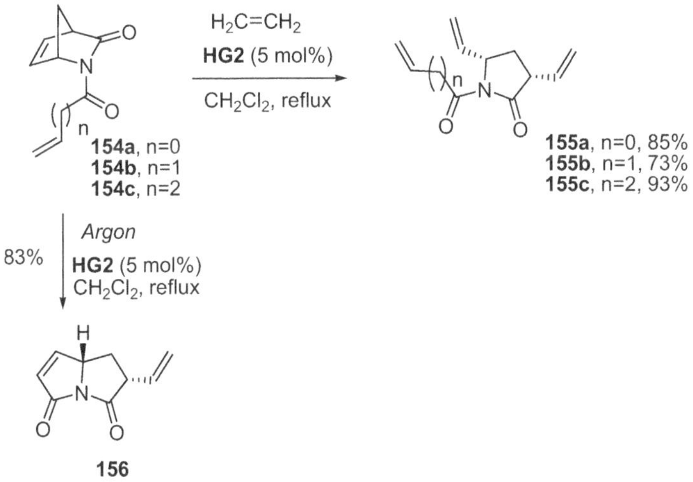

Scheme 52 RRM of azabicyclo[2.2.1]heptenones

5.2 *From Nitrogen-Containing Bicycles*

The RRM of bicyclic systems containing an amide function to bicyclic lactams was indeed described. Aljarilla and Plumet studied azabicyclo[2.2.1]heptenones **154a–c**, which contain an imide functionality with the nitrogen and one of the carbonyl on one of the bridge and the other carbonyl on the side chain (Scheme 52) [54]. RRM with **HG2** in the presence of ethylene, which is known to be beneficial to RRM reactions, only led to ring-opening products **155a–c** without ring closing. However, in the case of acrylamide **154a**, exclusion of ethylene in favor of an argon atmosphere allowed the RRM to take place and led to the desired 5,5-fused bicycle **156**.

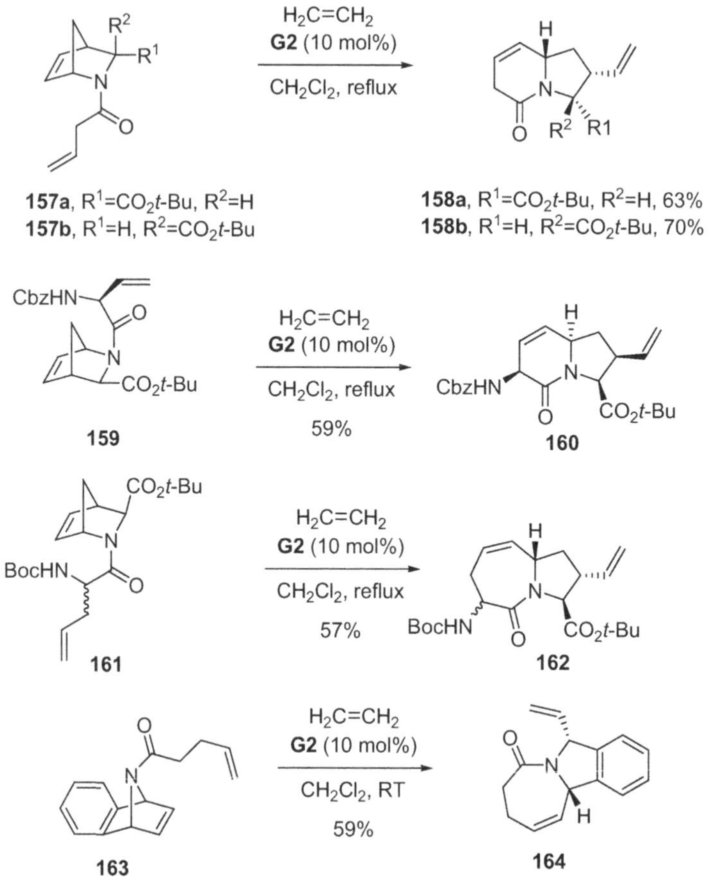

Scheme 53 RRM of azabicyclo[2.2.1]heptenes

Unfortunately, the corresponding 6,5- or 7,5-bicyclic imides were not obtained while all the starting material disappeared; ring-opening metathesis polymerization (ROMP) might have occurred.

The ring-rearrangement metathesis of azabicyclo[2.2.1]heptenes containing an amide functionality proved to be more favorable than the azabicyclo[2.2.1] heptenones. Maison et al. investigated the reaction of Diels–Alder cycloadducts **157a,b** in which the nitrogen of the amide is on one of the bridge and the carbonyl on the side chain (Scheme 53) [55]. Both *endo*-**157a** and *exo*-**157b** cycloadducts were converted into bicyclic 6,5-fused amides **158a** and **158b** in the presence of **G2** and ethylene atmosphere. Azabicyclo[2.2.1]heptenes **159** and **161**, derived respectively

Scheme 54 RRM of oxa-azabicyclo[2.2.1] heptenes and oxa-azabicyclo[2.2.2] octenes

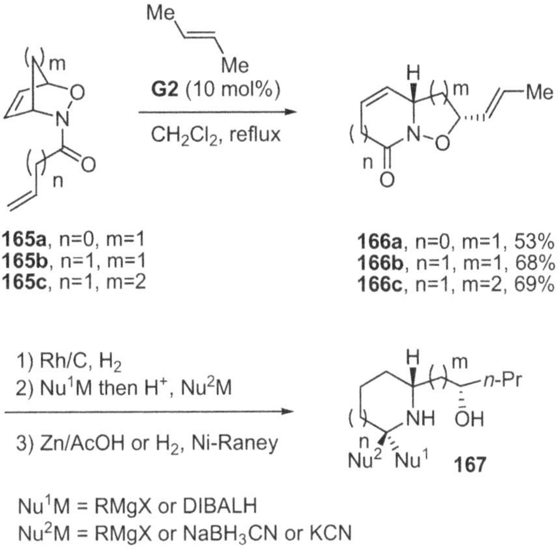

from vinylglycine and allylglycine, were subjected to the RRM conditions and led to 6,5- and 7,5-bicyclic dipeptide mimics **160** and **162**. Finally, amide **163** in which the nitrogen is on a one-atom bridge was also a suitable precursor of RRM, and tricyclic compound **164** could be obtained.

Complementary to azabicyclo[2.2.1]heptenes, Kouklovsky, Blanchard, Vincent et al. described the RRM of oxa-azabicyclo[2.2.1]heptenes and oxa-azabicyclo [2.2.2]octenes **165a–c** obtained by nitroso Diels–Alder cycloadditions (Scheme 54) [56, 57]. The RRM with **G2** in the presence of but-2-ene allowed the synthesis of 5,5- and 6,5- and 6,6-bicyclic *N*-alkoxyamides **166a–c**. Several external alkenes as partners of CM could also be employed. The successive additions of nucleophiles to this template allow the highly diastereoselective synthesis of piperidine scaffolds **167** found in natural products [58].

5.3 From Nitrogen-Containing Side Chains

The 2-aminonorbornene framework, in which the amide is incorporated in the side chain, showed a different trend in RRM. Nadany and Mckendrick were able to transform efficiently acrylamide **168a** into 6,5-bicyclic amide **169a** in the presence of **G2** under ethylene atmosphere (Scheme 55) [59]. Unfortunately, precursors **168b,c** with longer side chains did not react. Presumably, the formation of a nonreactive five- or six-membered ring carbene intermediate **170** might sequester

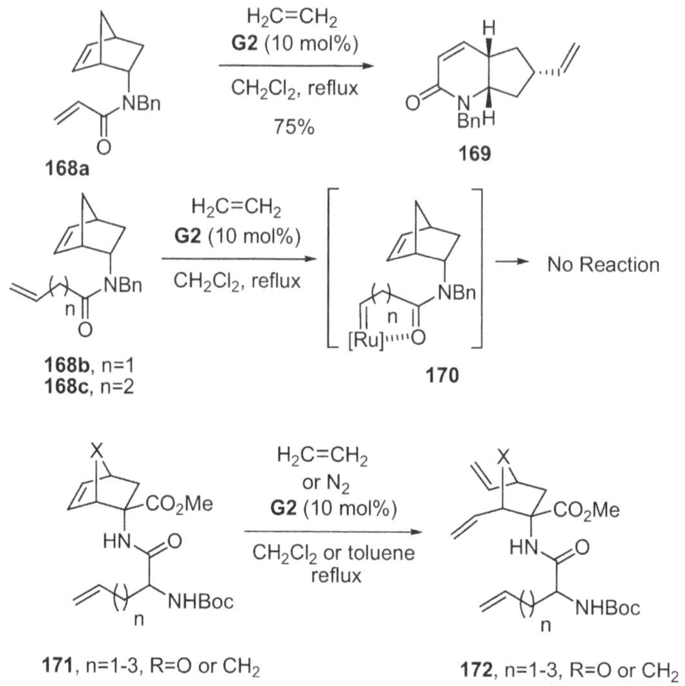

Scheme 55 RRM of norbornene-derived amides

the catalyst through chelation of ruthenium by the oxygen of the carbonyl. The precoordination of the amide with Lewis acids such as Ti(O*i*-Pr)$_4$ was not effective. It was also found by McKendrick, Blechert et al. that allyglycine-derived norbornenes **171** were also not suitable for RRM [60]. In contrast to **168b,c**, ROM occurred but not the RCM; this might be explained by the high preference of the *trans*-form of the amide **171**.

In the context of diversity-oriented synthesis, Spring et al. reported a RRM from norbornene-derived amides **173** similar to **168**, containing respectively an alkene and an alkyne on their two side chains (Scheme 56) [61]. Therefore, the domino metathesis sequence includes ROM, RCM, and RCEYM. A refined protocol was developed which includes first the reaction of norbornene **173a** with **G1** under ethylene atmosphere and microwave irradiation, followed by the addition of 10 mol % of both **G1** and **G2** catalysts. The tricyclic 7,5,6-lactam **174a** was obtained in 87% yield. *N*-ethylmaleimide could be added at the end of the reaction and after heating under microwave irradiation could lead to the pentacyclic compound **175**. A lower 36% yield of the 7,5,7-lactam **174b** was obtained along with 30% of the ROM-RCM lactam product **176**.

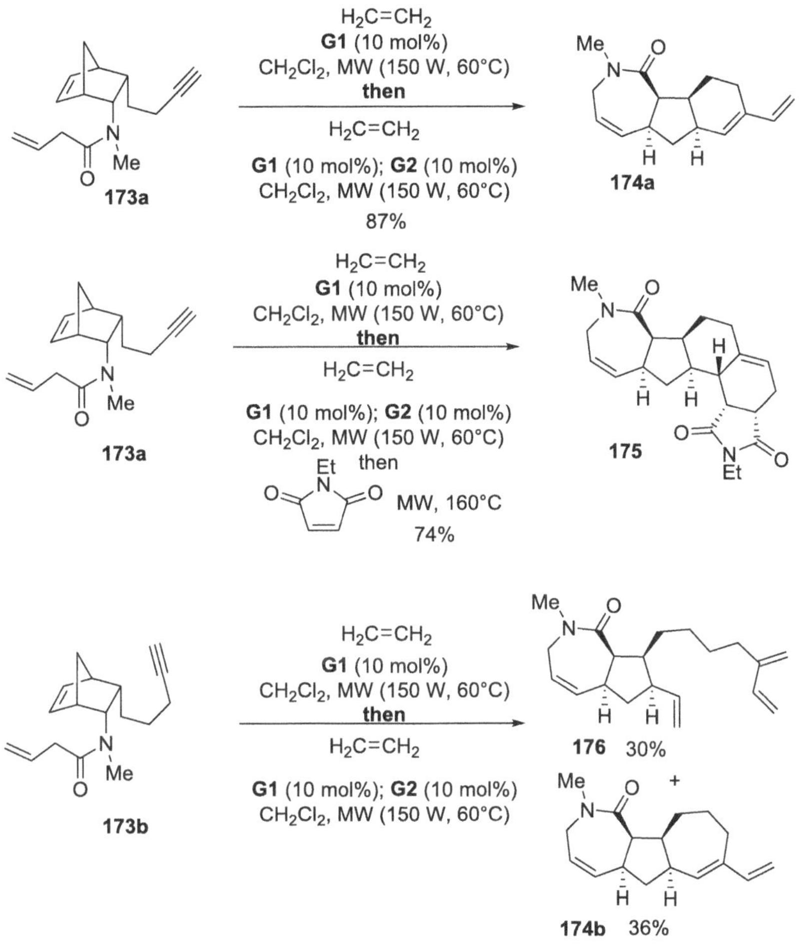

Scheme 56 RRM of norbornene-derived amides

The outcome of the metathesis of oxanorbornene derivatives **177** could be changed by subtle modifications of their structures as showed by Schreiber et al. (Scheme 57) [62]. Substrate **177a** in the presence of **G2** underwent RRM and delivered the highly complex fused tetracycle **178**, which contains two medium-sized lactams, a tetrahydrofuran, and a 12-membered macrolactam. In contrast, under the same conditions, the diastereoisomeric precursor **177b** led to the macrocycle **179** by RCM.

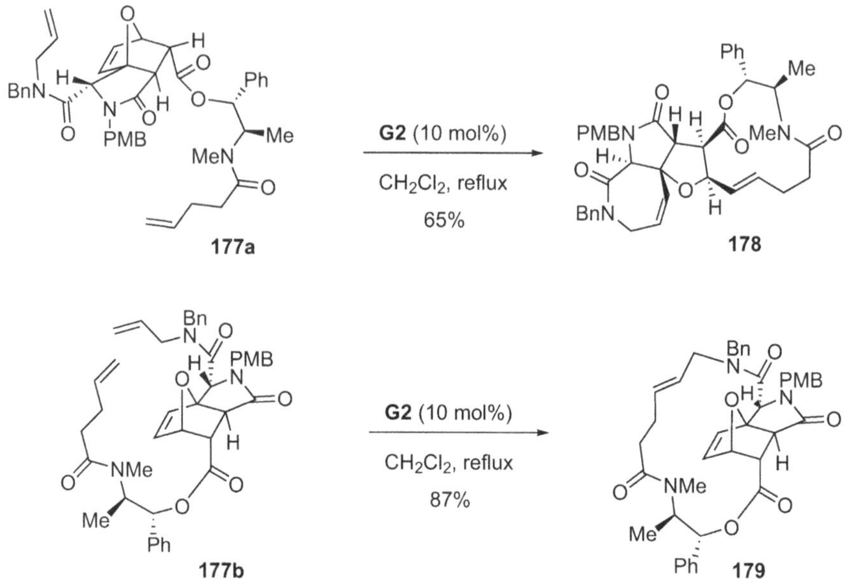

Scheme 57 Stereochemical control of metathesis of oxanorbornene derivatives

6 Conclusion

It is now well established that the synthesis of lactams has benefited from the development of metathesis and the first- and second-generation ruthenium catalysts **G1**, **G2**, or **HG2**. The ring-closing metathesis (RCM) of alkenes remains the most common metathesis reaction employed due to the simple access to acrylamides or nonconjugated amide precursors. The key feature for the success of the RCM is the access of the *cis*-form of the amide over the *trans* in order to bring together the reactive alkenes. Introduction of a suitable directing group on the nitrogen of the amide is of importance to favor the metathesis. Ring-closing enyne metathesis (RCEYM) is complementary since it affords lactams with a dienic system. A cross metathesis (CM) right before the lactam-forming step has also been employed. The ring-rearrangement metathesis (RRM) of bicycles allows the rapid generation of complexity and diversity in lactam synthesis.

References

1. Nubbemeyer U (2001) Top Curr Chem 216:125–196
2. Pace V, Holzer W (2013) Austr J Chem 66:507–512
3. Hassan HMA (2010) Chem Commun 46:9100–9106
4. Fustero S, Mateu N, Albert L, Acena JL (2009) J Org Chem 74:4429–4432

5. Ku IW, Kang SB, Keum G, Kim Y (2011) Bull Korean Chem Soc 32:3167–3170
6. Khadem S, Udachin KA, Arya P (2010) Synlett 199–202
7. Spiess S, Berthold C, Weihofen R, Helmchen G (2007) Org Biomol Chem 5:2357–2360
8. Krishna PR, Reddy BK (2013) Helv Chim Acta 96:1564–1569
9. Shi YC, Wang SG, Yin Q, You SL (2014) Org Chem Front 1:39–43
10. Fiorelli C, Savoia D (2007) J Org Chem 72:6022–6028
11. Sirvent JA, Foubelo F, Yus M (2014) J Org Chem 79:1356–1367
12. Park SH, Kang HJ, Ko S, Park S, Chang S (2001) Tetrahedron Asymm 12:2621–2624
13. Yadav JS, Jayasudhan Reddy Y, Addi Narayana Reddy P, Subba Reddy BV (2013) Org Lett 15:546–549
14. Subba Reddy BV, Ghanty S, Siva Senkar Reddy N, Jayasudhan Reddy Y, Yadav JS (2014) Synth Commun 44:1658–1663
15. Pritchard DR, Wilden JD (2010) Tetrahedron Lett 51:1819–1821
16. Laventine DL, Cullis PM, Garcia MD, Jenkins PR (2009) Tetrahedron Lett 50:3657–3660
17. Reddy PT, Quevillon S, Gan Z, Forbes N, Leek DM, Arya P (2006) J Comb Chem 8:856–871
18. Lebrun S, Couture A, Deniau E, Grandclaudon P (2011) Synthesis 669–673
19. Arya P, Couve-Bonnaire S, Durieux P, Laforce D, Kumar R, Leek DM (2004) J Comb Chem 6:735–745
20. Toumi M, Couty F, Evano G (2008) J Org Chem 73:1270–1281
21. Kim HM, Lee K, Park BW, Ryu DK, Kim K, Lee CW, Park SK, Han JW, Lee HY, Lee HY, Han G (2006) Bioorg Med Chem Lett 16:4068–4070
22. Manzoni L, Colombo M, Scolastico C (2004) Tetrahedron Lett 45:2623–2625
23. Chen Y, Zhang H, Nan F (2004) J Com Chem 6:684–687
24. Marhold M, Buer A, Hiemstra H, van Maarseveen JH, Haufe G (2004) Tetrahedron Lett 45: 57–60
25. Toueg J, Prunet J (2006) Synlett 2807–2811
26. Leonetti F, Muncipinto G, Stefanachi A, Nicolotti O, Cellamare S, Catto M, Pisanai L, Pellegrino G, Carotti A (2012) Tetrahedron Lett 53:4114–4116
27. Donohoe TJ, Fishlock LP, Procopiou PA (2008) Synthesis 2665–2667
28. Lebrun S, Couture A, Deniau E, Grandclaudon P (2008) Synthesis 2771–2776
29. Spino C, Boisvert L, Douville J, Roy S, Lauzon S, Minville J, Gagnon D, Beaumier F, Chabot C (2006) J Organomet Chem 691:5336–5355
30. Bensa D, Coldham I, Feinaugle P, Pathak RB, Butlin RJ (2008) Org Biomol Chem 6: 1410–1415
31. Daly M, Gill K, Sime M, Simpson GL, Sutherland A (2011) Org Biomol Chem 9:6761–6770
32. Kamimura A, Tanaka K, Hayashi T, Omata Y (2006) Tetrahedron Lett 47:3625–3627
33. Hassan HMA, Brown FK (2010) Chem Commun 46:3013–3015
34. Fuhshuku K, Asano Y (2012) Tetrahedron 68:6651–6655
35. Del Valle JR, Goodman M (2004) J Org Chem 69:8946–8948
36. Jäkel M, Qu J, Schnitzer T, Helmchen G (2013) Chem Eur J 19:16746–16755
37. Watson KD, Carosso S, Miller MJ (2013) Org Lett 15:358–361
38. Guérin D, Gaumont AC, Dez I, Mauduit M, Couve-Bonnaire S, Pannecoucke X (2014) ACS Catal 4:2374–2378
39. Schurgers B, Brigou B, Urbanczyk-Lipkowska Z, Tourwé D, Ballet S, De Proft F, Van Lommen G, Verniest G (2014) Org Lett 16:3712–3715
40. Satyanarayana G, Helmchen G (2014) Eur J Org Chem 2242–2252
41. Creighton CJ, Leo GC, Du Y, Reitz AB (2004) Bioorg Med Chem 12:4375–4385
42. Williams RM, Liu J (1998) J Org Chem 63:2130–2132
43. Duggan HME, Hitchcock PB, Young DW (2005) Org Biomol Chem 3:2287–2295
44. Brady RM, Khakham Y, Lessene G, Baell JB (2011) Org Biomol Chem 9:656–658
45. Bieraugel H, Jansen TP, Schoemaker HE, Hiemstra H, van Maarseveen JH (2002) Org Lett 4:2673–2674

46. Dutton BL, Kitson RRA, Parry-Morris S, Roe SM, Prodromou C, Moody CJ (2014) Org Biomol Chem 12:1328–1340
47. Szostak M, Aubé J (2009) Org Lett 11:3878–3881
48. Pal APJ, Kadigachalam P, Mallick A, Doddi VR, Vankar YD (2011) Org Biomol Chem 9: 809–819
49. Fustero S, Baez C, Sanchez-Rosello M, Asensio A, Miro J, del Pozo C (2012) Synthesis 44: 1863–1873
50. Yang Q, Lai YY, Xiao WJ, Alper H (2008) Tetrahedron Lett 49:7334–7336
51. Arimitsu S, Hammond GB (2010) Beilstein J Org Chem 6:48
52. Huang J, Xiong H, Hsung RP, Rameshkumar C, Mulder JA, Grebe TP (2002) Org Lett 4: 2417–2420
53. Wroleski A, Sahasrabudhe K, Aubé J (2002) J Am Chem Soc 124:9974–9975
54. Aljarilla A, Plumet J (2008) Eur J Org Chem 3984–3990
55. Buchert M, Meinke S, Prenzel AHGP, Deppermann N, Maison W (2006) Org Lett 8: 5553–5556
56. Calvet G, Blanchard N, Kouklovsky C (2007) Org Lett 9:1485–1488
57. Vincent G, Kouklovsky C (2011) Chem Eur J 17:2972–2980
58. Vincent G, Guillot R, Kouklovsky C (2011) Angew Chem Int Ed 50:1350–1353
59. Nadany AE, Mckendrick JE (2007) Synlett 1663–1666
60. Maechling S, Norman SE, McKendrick JE, Basra S, Koppner K, Blechert S (2006) Tetrahedron Lett 47:189–192
61. Spandl RJ, Rudyk H, Spring DR (2008) Chem Commun 3001–3003
62. Sello JK, Andreana PR, Lee D, Schreiber SL (2003) Org Lett 5:4125–4127

Top Heterocycl Chem (2017) 47: 191–244
DOI: 10.1007/7081_2015_142

Published online: 4 June 2015

Synthesis of Cyclic Peptides and Peptidomimetics by Metathesis Reactions

Dirk T.S. Rijkers

Abstract This review surveys developments in the field of ring-closing metathesis and cross-metathesis reactions applied to the synthesis of constrained amino acids, peptides, and peptidomimetics. Examples, in which metathesis is used as one of the synthetic tools to arrive at the desired peptide molecules as well as examples that describe in-depth optimized metathesis protocols, will be discussed. Currently, metathesis reactions are well-accepted synthetic tools within the field of peptide chemistry and provide peptides unprecedented properties like conformational, metabolic, and chemical stability and improved bioactivity.

Keywords Conformational constraints · Cross metathesis · Cyclic constraints · Depsipeptides · Dicarba peptides · Macrocyclization · Peptide-based therapeutics · Peptides and peptidomimetics · Rigidification · Ring-closing metathesis · Stapled peptides · Stitched peptides · Structure-inducing scaffolds · *Z*-Selective ring-closing metathesis · α-Helix inducing · α-Helix stabilization · β-Turn inducing

Contents

The original version of this chapter was revised: The missing 3D structure of Scheme 10 has been included now. The erratum to this chapter is available at: DOI 10.1007/7081_2016_203

D.T.S. Rijkers (✉)
Department of Pharmaceutical Sciences, Faculty of Science, Medicinal Chemistry and Chemical Biology, Utrecht Institute for Pharmaceutical Sciences (UIPS), Utrecht University, P.O. Box 80082, 3508 Utrecht, The Netherlands
e-mail: D.T.S.Rijkers@uu.nl

1 Introduction

Since the earliest scientific contributions of ruthenium-based metathesis reactions on amino acids and peptides, an enormous increase in synthesis routes and applications was realized. Metathesis on amino acids and peptides was regarded in first instance as rather exceptional, due to the large number of relatively polar functional groups; however, due to the development of more robust ruthenium catalysts [1–3], it became a more common and well-accepted tool in the synthesis of modified as well as constrained amino acids, peptides, and peptidomimetics [4–8]. This contribution does not aim to provide an extensive review of the current literature but rather emphasizes the most important and high-impact examples of metathesis reactions on peptides and peptidomimetics.

2 Application of Olefin Metathesis to Rigidify Amino Acids and Peptides

The ruthenium-based catalysts that have been most often used in the olefin metathesis reaction on peptides and peptidomimetics are abbreviated as follows: **G0** for the first well-defined catalyst [9, 10], **G1** for the first-generation Grubbs' catalyst [11], **G2** for the second-generation Grubbs' catalyst [12], and **HG2** for the Hoveyda-Grubbs' catalyst [13].

2.1 The First Applications of Olefin Metathesis in Peptide Synthesis

The first two seminal contributions to apply ruthenium-based olefin metathesis reactions to amino acids and peptides were reported in 1995. In order to synthesize conformationally restricted amino acids and peptides, Miller and Grubbs [14] reported the successful ring-closing metathesis of diene substrates **1**, **5**, and **7**, which were efficiently converted into their six-, seven-, and eight-membered cyclic amino acid derivatives **2**, **6**, and **8**, respectively, in yields varying between 50 and

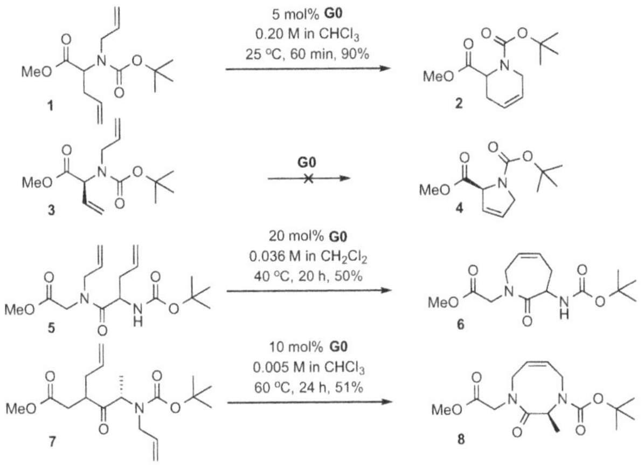

Scheme 1 Ring-closing metathesis to obtain cyclic amino acid residues

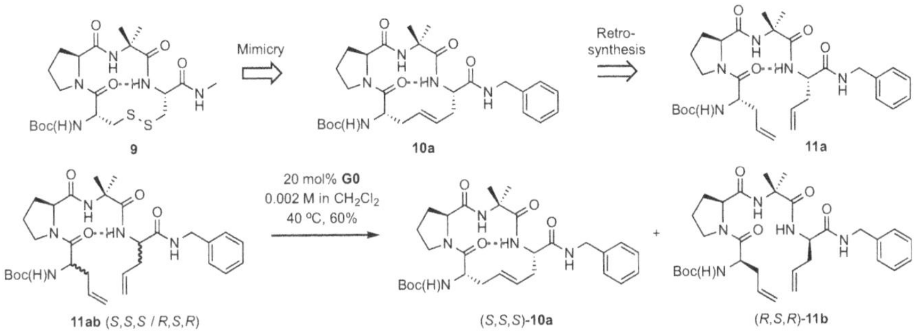

Scheme 2 Mimicking a covalently disulfide-stabilized β-turn by a carbon-carbon double bond, in red a putative hydrogen bond that facilitates ring closure of the "natural" configuration

90%, as shown in Scheme 1. However, under the same reaction conditions, the vinyl glycine derivative **3** did not cyclize into the five-membered dehydroproline derivative **4** (Scheme 1). It was assumed that the acidic α-proton at the chiral carbon atom resulted in a shift of the double bond with the ultimate result that this double bond was unreactive toward catalyst **G0** [15].

In the same paper [14], Miller and Grubbs reported the synthesis of a conformationally restricted tetrapeptide to mimic a disulfide-stabilized β-turn by replacing both cysteines by allylglycine derivatives (as shown in Scheme 2).

It turned out that the cyclization process is highly stereospecific since the (*S*,*S*,*S*)-diastereomer **11a** smoothly cyclizes into macrocyclic peptide **10a** (in 60% yield), while the (*R*,*S*,*R*)-diastereomer **11b** could be recovered unreacted. It is assumed that the observed stereospecificity is based on the hydrogen bond that facilitates ring closure of the (*S*,*S*,*S*)-diastereomer by preorganization, while it leads to a destabilizing transannular interaction in case of the other diastereomers making macrocyclization less likely to occur. In a follow-up paper [15], Grubbs and co-workers explored the scope of RCM on these tetrapeptide substrates by replacing the rigidifying (turn-inducing) sequence Pro-Aib by Pro-Tyr and finally by Leu-Leu, in which the conformationally constrained amino acids were absent, and found that these linear tetrapeptides could be smoothly converted into their cyclic congeners in a yield of 60, 80, and 60%, respectively [15]. These results clearly indicated that for the synthesis of cyclic tetrapeptides (as **10**) by RCM, there is no strict requirement to incorporate the Pro-Aib as a preorganization motif.

The second seminal contribution was reported by Clark and Ghadiri [16] in which they used olefin cross metathesis for the covalent stabilization of self-assembled dimers of cyclic octapeptides (as shown in Scheme 3).

They prepared a cyclic octapeptide *cyclo*[-L-Phe-D-N^{Me}Ala-L-Hag-D-N^{Me}-Ala)$_2$-] **12** that consisted of an even number of alternating L- and D-amino acid residues and specific *N*-methylated amide bond functionalities, while two homoallylglycine residues were present for covalent capture by olefin metathesis. Such peptides form cylindrical structures driven by hydrogen-bonded self-assembly [17–19], like dimers **13a,b**. Only in case of dimer **13b**, with the correct juxtaposition of the allylic side chains, cross metathesis would be successful, leading, after hydrogenation of the double bonds, to the kinetically stable tricyclic structure **14**, thereby shifting the equilibrium toward **13b**.

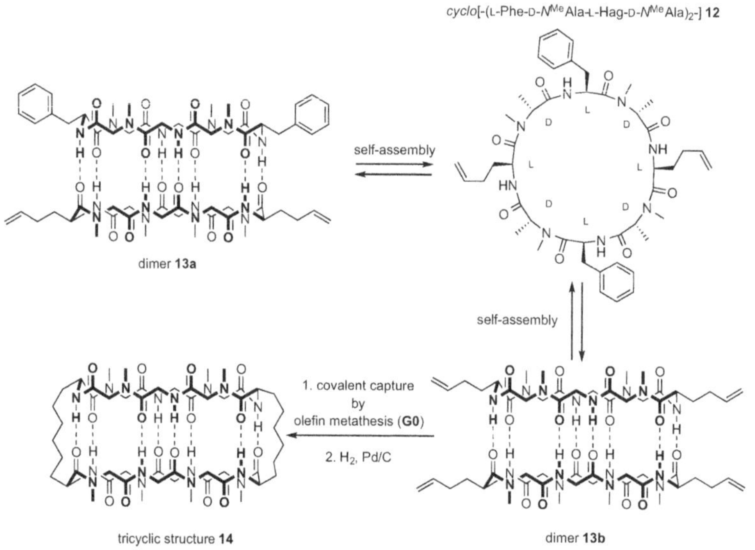

Scheme 3 The cyclic peptide **12** self-assembles by hydrogen bond interactions into an equal mixture of dimers **13a** and **13b**. Only **13b** is reactive toward metathesis and shifts the equilibrium to the stable tricyclic structure **14**. *Note*: for clarity side-chain functionalities are omitted

Scheme 4 Examples of conformationally restricted amino acids and peptidomimetics

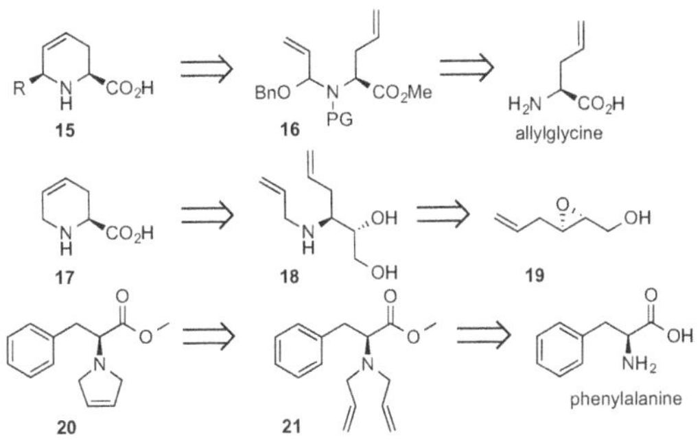

2.2 *Synthesis of Conformationally Restricted Amino Acids and Peptidomimetics*

Ring-closing metathesis is a versatile approach to the synthesis of enantiopure and conformationally restricted cyclic amino acid derivatives. The group of Rutjes reported an elegant combination of palladium-catalyzed *N*,*O*-acetal formation, ruthenium-catalyzed ring-closing metathesis, and *N*-sulfonyliminium ion-mediated carbon-carbon bond formation approach [20] to synthesize a series of enantiomerically pure 2,6-disubstituted unsaturated pipecolic acid derivates like **15**, as shown in Scheme 4.

Herein, allylglycine was the starting compound, and after conversion into its corresponding *N*-tosyl-protected methyl ester, and reaction with benzyl propadienyl ether in the presence of $Pd(OAc)_2$/dppp/Et_3N in acetonitrile as solvent, the desired *N*,*O*-acetal **16** was isolated in 85% yield. Diolefin **16** underwent ring-closure metathesis using the Grubbs Ru-benzylidene catalyst **G1** in an excellent yield. Functionalization of this scaffold by different R groups (to give molecular entity **15**) was performed with a variety of nucleophiles in the presence of $BF_3{\cdot}Et_2O$ as Lewis acid. In case of Et_3SiH (R=H) as the nucleophile, unsaturated pipecolic acid derivative **17** (baikiain [21]) was obtained in 88% yield [20]. Pipecolic acid derivatives are present in several natural products with important medicinal properties like the immunosuppressants FK506 [22] and rapamycin [23].

Alternative RCM-based syntheses of **17** have been reported by Riera and co-workers [24, 25]. They started with chiral epoxide **19** which underwent a nucleophilic ring opening using allylamine to afford aminodiol **18**, and only after protecting the secondary amine with a Boc group, RCM in the presence of **G1** was successful. Finally, the desired compound **17** (in its Boc-protected form) was isolated in 72% yield [24].

Pyrrolidines that contain a basic and nucleophilic nitrogen atom are rather difficult to prepare by ring-closing metathesis starting from diallylamines. These substrates require a proper deactivation by conversion into, among others, carbamates, e.g., by Boc protection as was the case with aminodiol **18**. To remedy this shortcoming, Yu and co-workers developed a versatile approach for an efficient RCM of chiral diallylamines by virtue of the presence of a strong Lewis acid

[26]. They found that diallylamine **21** (which was prepared from phenylalanine, Scheme 4) could be converted into pyrrolidine **20** (in 92% yield; an almost fourfold improvement) in the presence of **G2** as the RCM catalyst with 20 mol% $Ti(O^iPr)_4$ as the Lewis acid. This reaction was rather insensitive toward variations of the amino acid side-chain functionalities, since general yields varying between 78% and 91% were achieved, in case of alanine, leucine, aspartic acid, tryptophan, and glutamic acid, thus broadening the scope of this pyrrolidine scaffold synthesis.

2.3 *Secondary Structure-Inducing Peptide Mimetic Scaffolds Synthesized via Ring-Closing Olefin Metathesis*

The central tripeptide motif in Fig. 1 indicates the positions within the peptide backbone that have been used for modification with alkene-based moieties to arrive at cyclic covalent constraints to induce secondary structures in medium-sized peptides and peptidomimetics [27].

Type **I** scaffolds, as represented by **22**, employ the N^i to N^{i+1} connectivity pattern, and in case of *N*-allyl derivatives, Grubbs showed that RCM proceeded smoothly in the presence of **G0**, while **22** was isolated in 51% yield (R^i=CH_3 (alanine), R^{i+1}=H (glycine)) [15]. Reichwein and Liskamp broadened the scope of this reaction to amino acids other than glycine [28]; for example, Boc-N^{All}Phe-N^{All}Phe-OMe was cyclized into **22** (R^i=R^{i+1}=Bn (phenylalanine)) in the presence of **G1** in 1,1,2-trichloroethane at 115°C in 27% isolated yield. Moreover, in a follow-up paper, the same authors described the synthesis of a series of "amide to

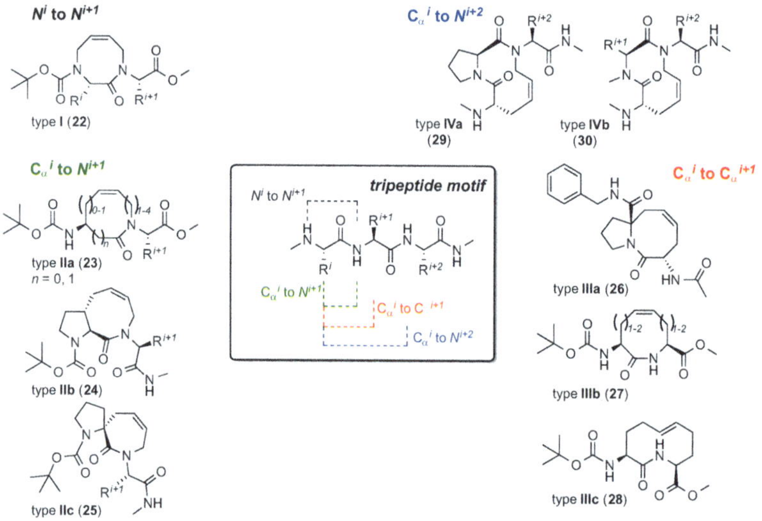

Fig. 1 Secondary structure-inducing peptide mimetic scaffolds synthesized via RCM

amide" cyclized peptides in order to define a set of rules for RCM (in the presence of Ru-catalyst **G1**) as function of the length of the alkene substituent versus the length/nature of the peptide sequence [29]. From their studies the following "rules" for cyclization of bis-*N*-alkenylated peptides by **G1**-mediated RCM could be inferred: cyclic dipeptides – involving two amide bonds – could be obtained starting from bis-*N*-allyl (or longer alkene) substituents. Cyclic tripeptides – involving three amide bonds – could be obtained starting from bis-*N*-4-pentenyl (or longer alkene) substituents. While cyclic tetra-, penta-, and hexapeptides – involving four, five, or six amide bonds, respectively – could be obtained starting from *N*-homoallyl (or longer alkene) substituents. In addition to this, the introduction of either proline or a single D-amino acid in the peptide sequence increased the overall yield of the cyclization reaction [29, 30]. These insights were used to develop the so-called rolling loop scan to probe the bioactive conformation of cyclic peptides [31].

Type **IIa** cyclic constraints, as represented by **23**, are known as Freidinger lactams [32] in which the $C_\alpha^{\,i}$ is connected to N^{i+1} leading to seven-membered macrocycles ($n = 0$). Their synthesis by ring-closing metathesis will result in dehydro lactams and was elaborated by Grubbs and co-workers [15] (**6**=**23** if $n = 0$, R^{i+1}=H, glycine) and Piscopio et al. (**23**, $n = 0$, R^{i+1}=Bn (phenylalanine)/^{i}Pr (valine)/Ph (phenylglycine)/$PhCH_2CH_2$ (homophenylalanine)) in yields varying between 16 and 62% [33]. The group of Gmeiner explored the synthetic accessibility of this type of constraints further by the preparation of eight-, nine-, and ten-membered macrocycles (**23**, $n = 0$) and by studying the conformational properties of these derivatives [34]. In addition to this, β-amino acids (**23**, $n = 1$) were also found to be versatile RCM substrates to access homo-Freidinger lactams [35].

Incorporation of proline as the R^i residue will lead to type **IIb** (represented by **24**) and eventually in spirocyclic type **IIc** (represented by **25**) fused lactams to tune the ψ^{i+1} dihedral angle to mimic the bioactive conformation of prolyl residues in medium-sized peptides [36]. Of special interest are type **IIc** cyclic constraints since they constitute potent type II β-turn-inducing molecular scaffolds [37]. In this context it is important to mention the contributions of the Moeller [38], respectively, Wagner [39] and Schmalz groups since they showed that stereo- and regiospecific functionalization of the proline ring with suitable alkene moieties is a powerful tool to define the conformation of the fused pyrrolidine ring as polyproline type II (PPII) helix mimic [40] or as an α-helix-nucleating motif [41] to stabilize the secondary structure of relatively small peptide sequences (vide infra).

Type **III** scaffolds, exemplified by structures like **26**–**28**, represent compounds in which $C_\alpha^{\,i}$ is connected to $C_\alpha^{\,i+1}$ resulting in eight- to ten-membered macrocycles (Fig. 1). Compound **26** (type **IIIa**) was synthesized by RCM and was a highly efficient proline-based type VI β-turn mimetic [42]. A series of eight- to ten-membered enantiomerically pure macrolactams (**27**, type **IIIb**) was synthesized by Lubell and co-workers [43] based on an earlier communication by Banfi et al. [44], who reported the synthesis of a racemic nine-membered lactam derivative via a combination of a Ugi 4-MCR [45] followed by RCM. Finally, compound **28** (type **IIIc**) was synthesized by Katzenellenbogen and co-workers [46] and represents a (3*S*,10*S*)-(6*E*)-2-azacyclodec-6-enone constraint as a dipeptide structural mimic (vide infra).

As the last example in this series, the C_α^{i} to N^{i+2} connectivity pattern is used in type **IV** constraints and is represented by structures **29** and **30**, as shown in Fig. 1. These molecular entities have been designed by the group of Gmeiner [47] with the idea to replace a hydrogen bond by a covalent alkene constraint in Asn-Pro/Asp-Pro turn motifs, thereby mimicking type I β-turn structures, and form important scaffolds to enable molecular recognition processes between pharmacologically relevant receptor-ligand interactions.

2.4 *Stabilization of α-Helices and β-Turns in Peptides*

For covalent stabilization of α-helices, the incorporation of disulfide and lactam bridges into peptide sequences belongs to the classical toolbox of the peptide chemist. The first application of RCM to stabilize helical peptides was described by Blackwell and Grubbs [48]. They used a hydrophobic heptapeptide, Boc-Val-Ala-Leu-Aib-Val-Ala-Leu-OMe, that was known to adopt an α-helical conformation both in the solid state and in solution. The alanine residues were replaced by serine (compound **31**) or homoserine *O*-allyl ethers to apply RCM for macrocyclization, as shown in Scheme 5. Precursor peptide **31** was treated with catalyst **G1** (20 mol% cat, 5 mM peptide in CH_2Cl_2 at 25°C for 3 to 4 h), to yield a 21-membered cyclic peptide **32** as a mixture of *E*/*Z* isomers (approximately 5:1) in 85% yield. Hydrogenation of the double bond resulted in an alkane-bridged peptide macrocycle **33** in an almost quantitative yield.

It turned out that the ellipticities as measured with circular dichroism did not differ much for the macrocyclic peptide compared to its linear precursor peptide. This was explained by the fact that diene **31** was highly preorganized in the organic solvent used for RCM, and cyclization did not result in a significant conformational change [49]. Although a noticeable helical stabilization was not observed, the

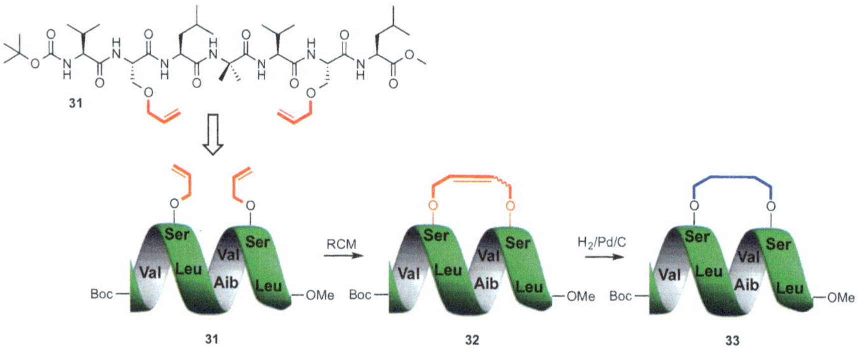

Scheme 5 The *O*-allyl ethers form the cyclic constraint to stabilize an α-helical conformation

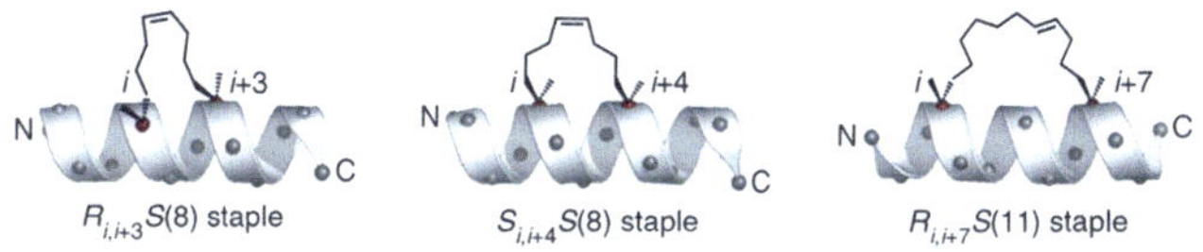

Fig. 2 Schematic representation of three stapled peptides [51]. The nomenclature $R_{i,i+3}S(8)$ refers to an eight-carbon metathesized cross-link with *R*-configuration at *i* and *S*-configuration at $i+3$ position [52]; $S_{i,i+4}S(8)$, eight-carbon tether with *S*-configuration at both *i* and $i+4$ positions; $R_{i,i+7}S(11)$ 11-carbon tether with *R*-configuration at *i* and *S*-configuration at $i+7$ position. Reprinted by permission from Macmillan Publishers Ltd. Nature Protocols (Nat Prot (2011) 6:761–771), © 2011. See [51]. *Note*: the double-bond geometry is a mixture of *E* and *Z*

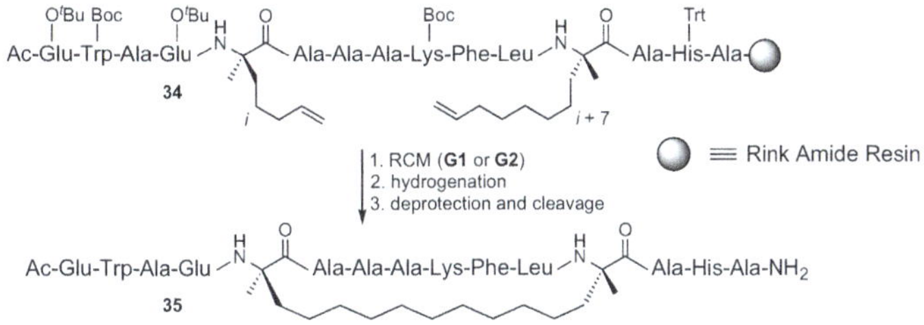

Scheme 6 Example of peptide stapling for stabilizing the α-helix through all-hydrocarbon cross-linking [50]

relatively easy incorporation of carbon-carbon bonds for rigidification of peptides and thereby also increasing their metabolic stability paved the way for a new era of constraining therapeutically relevant peptides.

Based on this concept, Verdine and co-workers screened a library of multiple configurations of all-hydrocarbon cross-links differing in position of attachment, stereochemistry, and cross-linker length [50]. This strategy was termed "peptide stapling" to stabilize the α-helical conformation of a peptide, while the "staple" consists of an all-hydrocarbon macrocyclic bridge connecting adjacent turns of the helix, flanked on each end by an α-methyl group, as shown in Fig. 2 [51]. The staple was incorporated into a peptide by the introduction of two appropriately α-amino-protected α-methyl-α-alkenyl amino acids via SPPS protocols; subsequently peptide macrocyclization was performed on the solid support by Ru-based RCM, followed by hydrogenation and treatment with acid to remove protecting groups and to cleave the peptide from the solid support (Scheme 6).

The peptide from the example shown in Scheme 6 represents the C-terminal sequence from RNAse A, and the original amino acid residues threonine at position *i* and alanine at position $i+7$ have been replaced by α-methyl-α-alkenyl amino acids as in the fully protected and resin-bound peptide **34**. This sequence was chosen since it adopts in aqueous solution a partial α-helical conformation, which allows to observe both increases and decreases in helical character. The unmodified RNAse A peptide is ~40% helical in water, while the RCM-precursor peptide is ~60% helical. Macrocyclic peptide **35**, with an 11-carbon cross-link, adopts in water a conformation that corresponds to 84% helical character and thus helix stabilization by 44% [50]. Besides the noticeable helix stabilization, peptide **35** was ~40-fold more resistant against trypsin digestion than its linear (unmetathesized) counterpart [50]. To address the need for a more in-depth understanding on how peptide stapling stabilizes peptide helices, all-atom Monte Carlo folding simulations were performed [53]. These insights are currently explored as tools in chemical biology or to improve the drug-like properties of therapeutically relevant stapled peptides [54–60]. A recent overview of hydrocarbon-stapled peptides, which highlights their use as biomedical research tool and prototype therapeutics, is recommended to the reader who is interested in more biological applications [61].

Recently, the next generation of hydrocarbon-stapled peptide systems was reported [62] in which two hydrocarbon staples are formed from a single amino acid precursor –*bis*-pentenylglycine(2-(amino)-2-(pent-4-enyl)hept-6-enoic acid)–, as shown in Scheme 7, thereby creating a spiro-macrocyclic connectivity pattern of the amino acid side chains which is known to be exceptionally rigidifying the bicyclic peptide. This approach of introducing contiguous hydrocarbon staples along one side of the α-helix has been coined "stitching" which provides peptides superior helix induction, thermal stability, resistance against proteolysis, and cell permeability properties relative to peptide stapling. Tetraolefin peptide **36** was synthesized by Fmoc/^{t}Bu SPPS (Scheme 7), with (*S*)-α-methyl-α-pentenylglycine on position *i*, *bis*-pentenylglycine on position *i*+4, and (*S*)-α-methyl-α-octenylglycine on position *i*+7. Ring-closing metathesis was performed on the solid support in the presence of **G1** (20 mol%) in 1,2-dichloroethane for 2 h at room temperature. RCM of **36** could lead to the formation of three bis-metathesized peptides, **37–39**. Based on model studies it turned out (see Table 1 in [62]) that intraresidue RCM (reaction ***a***; to form cyclononenylglycine) was disfavored as was the undesired reaction ***b***, suggesting that compound **37** would be highly unlikely. Reaction ***c*** on the other hand would

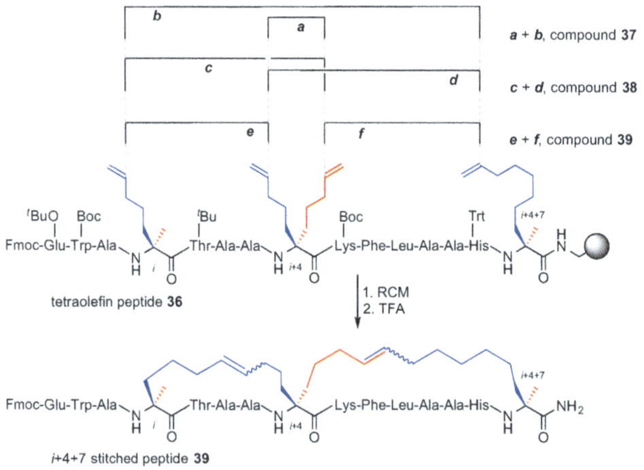

Scheme 7 Tetraolefin peptide **36** designed to undergo *bis*-RCM: peptide stitching [62]

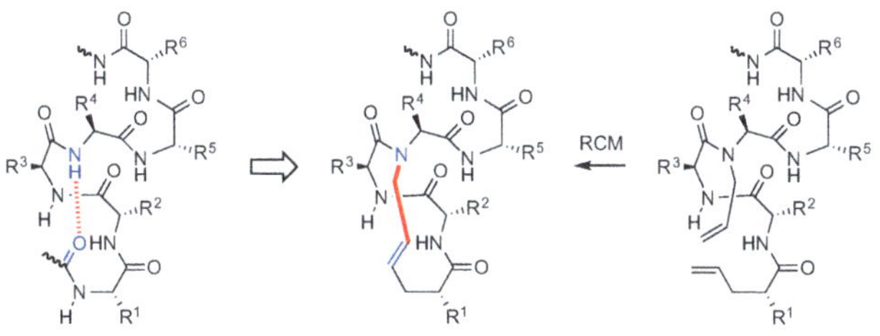

Scheme 8 α-Helix inducing by replacement of an i and $i+4$ hydrogen bond by covalent carbon-carbon constraint

lead to an $S_{i,i+4}R(8)$ staple (see Fig. 2), which is known to be highly disfavored [52, 53]. These model reactions also showed that reaction ***d*** is kinetically disfavored compared to ***f***, while reaction ***e*** and ***f*** lead to well-established $S_{i,i+4}S(8)$- and $R_{i,i+7}S$ (11) staples, respectively. Thus, there is a high degree of regio- and stereochemical control over the RCM reactions on these peptides to yield compound **39** rather than **38**.

Another approach for α-helix stabilization was elegantly pioneered by Arora and co-workers [63, 64]. In their strategy they replace an i and $i+4$ hydrogen bond (C=O···H-N) in an α-helix by a covalent constraint (C=X-Y-N) in which X and Y are carbon atoms introduced via olefin metathesis, as shown in Scheme 8.

Thus, α-helix sequence **40** is mimicked by its covalent congener **41**, which is synthesized from the linear *bis*-olefin precursor peptide **42** by ring-closing metathesis. This concept of α-helix stabilization was coined "hydrogen-bond surrogate" (HBS) and is a general and versatile strategy for generation of short α-helical peptide ligands to interact with therapeutically relevant receptors since the important amino acid side chains for binding interactions are still available [65]. Based on an X-ray crystal structure [66] of a short α-helix peptide (**43**; {HBS}Gln-Val-N^{All}Ala-Arg-Gln-Leu-Ala-Glu-Ile-Tyr-NH_2), it was found that this mimic indeed adopts an α-helical conformation with an RMS deviation of 0.74 Å upon superimposition of the backbone onto a model α-helical peptide, as shown in Fig. 3 [65, 66].

The synthesis of peptide **44** was carried out in solution, and RCM was performed with Hoveyda-Grubbs' catalyst (**HG2**) to afford (in its protected form) HBS α-helix **45** in 55% yield, as shown in Scheme 9. Finally, treatment with TFA removed the ^{t}Bu groups, and peptide **45** was studied by NMR and CD for structural analysis [63, 64]. For these peptides, a solid-phase synthesis approach was also developed (Scheme 8) [67]. Herein, Fmoc-Ile-N^{All}Ala-OH (indicated in red in peptide **46**) was used as a dipeptide building block for an efficient introduction of the N-allyl functionality. RCM was performed on the solid support with several Ru-based catalysts, since macrocyclization was far from trivial. Refluxing in CH_2Cl_2 in the presence of **G1** or **G2** did not yield significant amounts of the cyclic peptide. RCM in the presence of Ciba-Ruthenium [68] or Grubbs **G3** [69] (see insert Scheme 9) did also not exceed 20% cyclization yield; only in the presence of the Hoveyda-Grubbs' catalyst **HG2** in dichloroethane at 60°C with 25 mol% catalyst loading

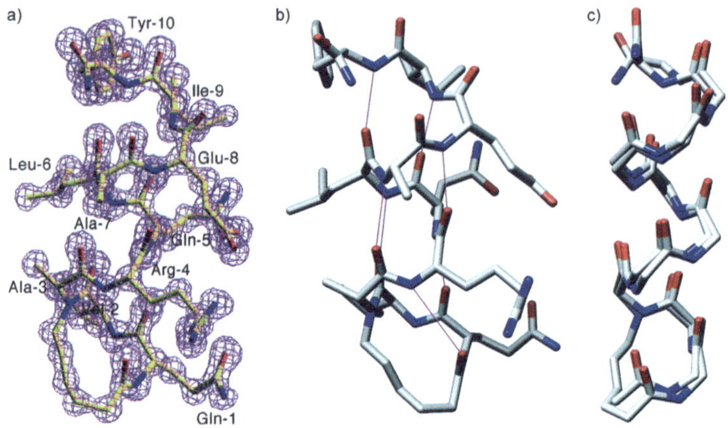

43

Fig. 3 (**a**) Crystal structure of the HBS α-helix with electron density superimposed onto the refined molecular model, (**b**) putative *i* and *i*+4 hydrogen bonds in crystal structure-derived molecular model of HBS helix, (**c**) overlay of crystal structure and a model of an idealized α-helix. Reprinted with permission from Acc Chem Res (2008) 41:1289–1300, © 2008 American Chemical Society. See [65]

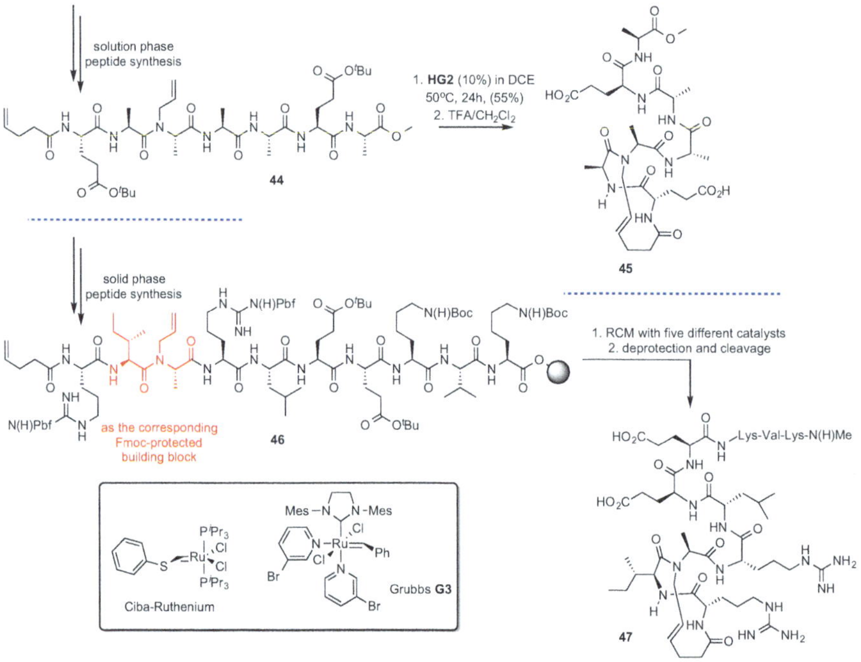

Scheme 9 Solution-phase synthesis and solid-phase synthesis of peptides **44** and **46** and HBS α-helices **45** and **51**

during 48 h, macrocyclization was efficient and cyclic peptide **47** could be isolated in approximately 90% yield [67].

The template-induced folding of short peptide sequences is a challenging approach to the design of folded peptides with a well-characterized secondary structure [70]. α-Helix-nucleating scaffolds are rather difficult to design; however, protein helices often contain proline residues especially in the N-cap region, e.g., at the N-terminus of the protein sequence. The Pro-Pro dipeptide segment forms a partial helical structure by two intramolecular hydrogen bonds [71] and might be used to nucleate helical folding [72, 73]. Based on Kemp's Pro-Pro derivative designed to function as an α-helix-inducing N-cap motif [74, 75], the groups of Kühne and Schmalz [41] designed a tricyclic diproline scaffold **48** for efficient α-helix induction in a linear peptide, as shown in Scheme 10.

Scaffold **48** is a rigidified diproline derivative by means of an ethylidene bridge and was synthesized from precursor **49** by ring-closing metathesis using **G2** in 59% yield. Dipeptide **49** was synthesized starting with two suitably protected vinylproline building blocks **50** (*cis*-4-vinylproline) and **51** (*trans*-5-vinylproline), as shown in Scheme 10. Scaffold **48** was protected at the secondary amine with an Fmoc group for application in Fmoc/^{t}Bu SPPS to afford template-peptide **52**. As a control peptide, the tricyclic Pro-Pro dipeptide was replaced by Pro-Pro, and both peptides were analyzed by circular dichroism on their helical content. It turned out that the Pro-Pro control peptide was random coiled (conformationally unordered), while peptide **52** adopted a high degree of α-helical character [41].

The same group employed a differently annulated tricyclic diproline scaffold [40] (compound **53**, as shown in Scheme 11) to induce a polyproline type II (PPII) helical conformation in short peptides to mimic proline-rich motif-recognition domain (PRD) ligands to interact with proteins containing proline-rich motifs

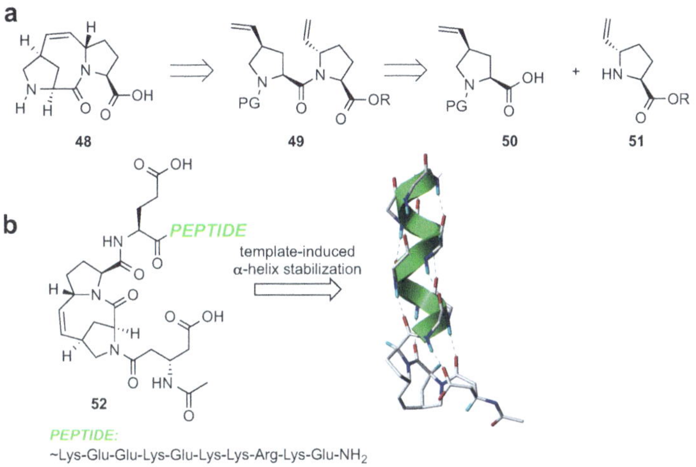

Scheme 10 (**a**) Retrosynthesis of tricyclic diproline scaffold **48**, (**b**) model of an N-capped α-helical peptide. Based on Angew Chem Int Ed (2013) 52:9539–9543. © Professor H. G. Schmalz, Universität zu Köln, Germany

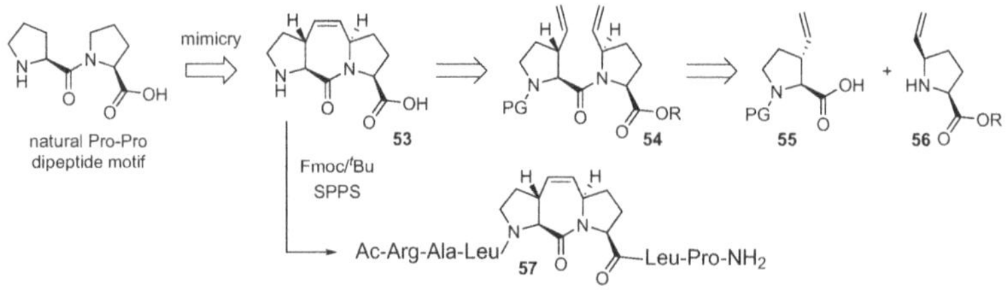

Scheme 11 Retrosynthesis of tricyclic diproline scaffold **53** to induce a polyproline type II (PPII) helical conformation in short peptides to mimic a proline-rich motif-recognizing domain (PRD) ligand

(PRMs). These interactions play an important role in nature and are involved in several biologically relevant processes, among others, tyrosine kinase receptor signaling, endocytosis, transcription, and splicing. Their studies showed that the dipeptide motif Pro-Pro is perfectly mimicked by scaffold **53** since replacing Pro-Pro in the natural ligand peptide (Ac-Arg-Ala-Leu-*Pro*-*Pro*-Leu-Pro-NH_2) by **53** to give **57** did not affect the binding affinity toward the target Fyn-SH3 as measured with isothermal titration calorimetry (ITC), the K_D being 18 ± 5 versus 62 ± 13 μM, respectively [40]. The retrosynthesis of tricyclic scaffold **53** is shown in Scheme 11. Ring-closing metathesis of **54** mediated by **G2** (5 mol%) in refluxing CH_2Cl_2 (15 h) gave **53** in 91% yield, while dipeptide **54** was obtained by peptide bond formation starting with *trans*-3-vinylproline (**55**) and *cis*-5-vinylproline (**56**) [40].

The synthesis of β-turn mimetics by ring-closing metathesis has already been described in Chapter 2.3 and is illustrated in Fig. 1. In the next section three examples will be discussed that make use of RCM-mediated β-turn mimetics to stabilize the conformation of the peptide, while in most cases this will also result in an increase of the biological activity of the rigidified peptidomimetic.

The first example refers to the work of Katzenellenbogen, as shown in Scheme 12 [46]. Herein, they mimicked the C-terminal tetrapeptide (~Phe-Gly-Leu-Met~) of Substance P (**58**, a bioactive undecapeptide that belongs to the tachykinins and plays important roles in pain sensation, inflammation, asthma, and vasodilatation [76]) by cyclic peptide **59**. By NMR studies [77] it was shown that Phe^8-Gly^9-Leu^{10}-Met^{11} adopted a β-turn, and to confirm this observation, it was envisioned that the truncated Substance P analog **59** might shed light on its bioactive conformation.

The versatile cyclic constraint **60** (indicated in red in compound **59**) allows selective functionalization with amino acids to design other tachykinin-selective receptor (ant)agonists. Its synthesis is relatively easily performed, starting with the properly protected (*S*)-2-aminohex-5-enoic acid to afford dipeptide **61**, which undergoes a smooth RCM in the presence of **G1** as the catalytic species to afford **60** in a good yield of 65% [46].

In the second example, ring-closing metathesis was applied to stabilize an octapeptide β-hairpin that exhibits enantioselective acylation activity, as shown in

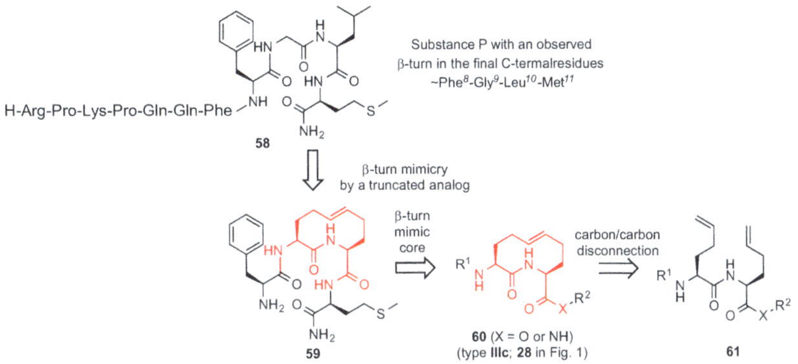

Scheme 12 The incorporation of a β-turn mimic in Substance P to define its bioactive conformation [46]

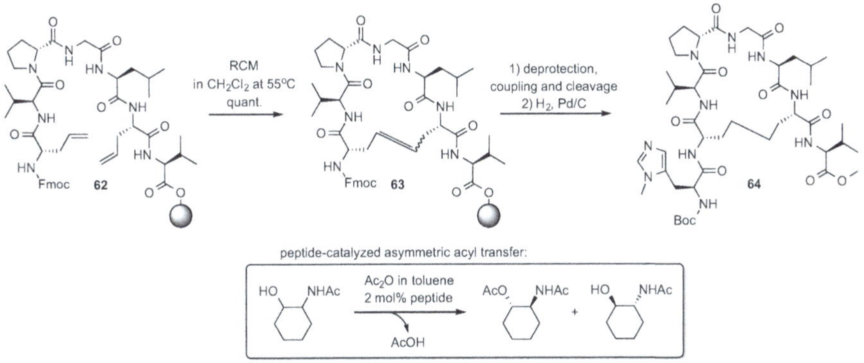

Scheme 13 Synthesis of rigidified peptide **64** as an asymmetric acyl transfer catalyst [78]

Scheme 13 [78]. The linear peptide, Boc-N^{Im}MeHis-*Val*-Val-D-Pro-Gly-Leu-*Val*-Val-OMe was found to be active as a catalyst in the asymmetric acyl transfer reaction with *rac*-2-acetylaminocyclohexanol as the substrate with a $k_{\mathrm{rel}} = 51$ (see insert Scheme 13) [78]. To be active as a catalyst, it should adopt a β-hairpin conformation. In a biomimetic approach, Miller and co-workers synthesized solid-phase-bound peptide **62**, in which two valine residues were replaced by allylglycine, and applied RCM mediated by **G1** to arrive at macrocyclic peptide **63** in quantitative yield. After Fmoc removal, Boc-N^{Im}MeHis-OH coupling, resin cleavage by transesterification, and catalytic hydrogenation, rigidified peptide **64** was obtained. It turned out that macrocyclic peptide **64** as well as its acyclic analog were less enantioselective than the original octapeptide in the kinetic resolution of *rac*-2-acetylaminocyclohexanol, $k_{\mathrm{rel}} = 12$ and 20, respectively [78]. This was explained by assuming "that substituents along the peptide backbone may be more important than covalent stabilization of a structural motif" [78].

The third example describes a general synthesis approach to apply ring-closing metathesis to resin-bound peptides [79]. In their approach, Schmiedeberg and

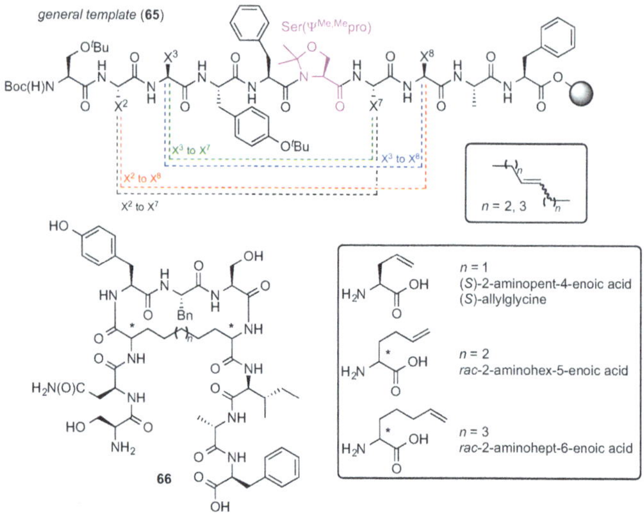

Scheme 14 Solid-phase RCM approach of backbone-protected peptides [79]

Kessler found that RCM was only successful by the introduction of a secondary structure disrupting reversible backbone protection, Ser($\psi^{Me,Me}$pro). Their decapeptide (H-Ser^1-Asn-Lys-Tyr-Phe^5-Ser-Asn-Ile-His-Trp^{10}-OH) was derived from the serine protease urokinase-type plasminogen activator (uPA) and was identified as a receptor-binding domain mimic. However, the C-terminal dipeptide ~ His-Trp ~ was replaced by ~ Ala-Phe ~ to remedy the chemical instability of the former dipeptide sequence against the reductive reaction conditions due to long reaction times. This modified peptide was used as the general cyclization template (**65**), as shown in Scheme 14.

Generally, the solid-phase synthesis of peptides that contain β-turn- or β-hairpin-like secondary structure as introduced by ring-closing metathesis often proceeds with good conversions. However, some peptides were found to be notoriously difficult to cyclize, independent of their ring size and stereochemical orientation, position, and number of carbon atoms of the olefinic amino acid residues. This failure against ring closure originates from the formation of inappropriate secondary structures that prevents the peptides from macrocyclization by RCM. As shown by Reichwein and Liskamp [28, 29], incorporation of *N*-alkenyl amino acids or proline residues will foster the efficiency of the cyclization reaction. Alternatively, reversible backbone protection by, among others, amide protection by either a 2,4-dimethoxybenzyl functionality [80] or *tert*-butyloxycarbonyl group [81] or by incorporation of pseudoprolines [82] is a versatile approach to overcome these difficulties during RCM-mediated cyclization reactions [79]. Schmiedeberg and Kessler [79] synthesized a series of uPA-derived peptides (general template, **65**) in which Ser^6 was replaced by pseudoproline Ser($\psi^{Me,Me}$pro) as introduced via building block Fmoc-Phe-Ser($\psi^{Me,Me}$pro)-OH [82], and the four combinations, Asn^2/Asn^7, Asn^2/Ile^8, Lys^3/Asn^7, and Lys^3/Ile^8, were substituted by the racemic amino acids

2-aminohex-5-enoic acid and 2-aminohept-6-enoic acid (see insert Scheme 14). RCM by **G1** proceeded smoothly, generally in yields varying between 56 and 84%. It should be noted, however, that incorporation of (*S*)-allylglycine as the olefinic amino acid was considered as not suitable, since too large amounts of unreacted peptide were found in the reaction mixture. After cleavage from the resin, the protected peptides with *cis/trans* double bonds were subjected to reduction, and remarkable differences in reactivity were found, strongly dependent on the number of carbon atoms spanning the bridge between the peptide backbone. As a representative example, cyclic peptide **66**, with an $X^3 \rightarrow X^7$ connectivity pattern was isolated in 8% ($n = 2$) and 6% ($n = 3$) yield, respectively, after reduction, deprotection, trituration, and HPLC purification [79].

2.5 Dicarba Analogs by RCM to Mimic Disulfide and Thioether Bridges

In their seminal contribution to RCM-based peptide cyclization [14, 15], Grubbs and co-workers described the successful mimicry of peptide disulfides by dicarba analogs, as shown in Scheme 2, compounds **9–11**, to stabilize β-turn conformational constraints. The disulfide bridge in peptides and proteins is an important entity to stabilize the secondary structure of peptide hormones. Since disulfides are rather sensitive toward reducing conditions, nucleophiles, and bases, several approaches have been developed to mimic the disulfide moiety by dicarba bridges. Thus, cystine **67**, can be replaced by an isosteric unit, like the all-carbon ~$(CH_2)_4$~ fragment, resulting in the non-reducible (2*S*,7*S*)-2,7-diaminooctanedioic acid ((*S*,*S*)-diaminosuberic acid) **68** (Scheme 15), which is easily accessible by olefin metathesis-based syntheses starting from suitably protected allylglycine derivatives [83–85].

One of the first applications to replace disulfide bonds in peptide hormones featuring the cyclization of bis-allylglycine-containing peptides was reported by Vederas and co-workers [86]. In their study, they synthesized a linear oxytocin-derived sequence **69** (in which $n = m = 1$, as shown in Scheme 16), in which the

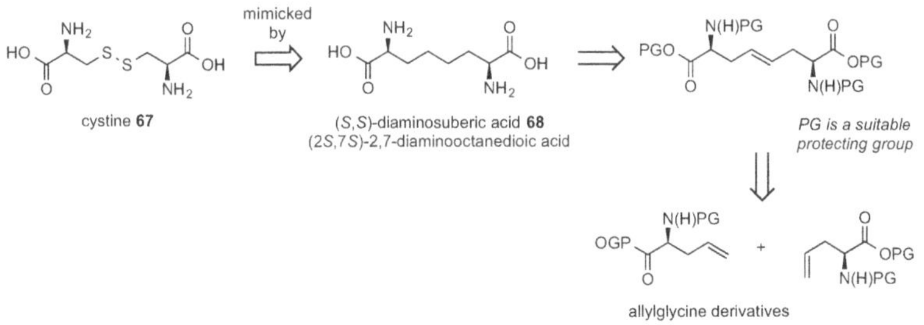

Scheme 15 Synthesis of (*S*,*S*)-diaminosuberic acid by metathesis approaches

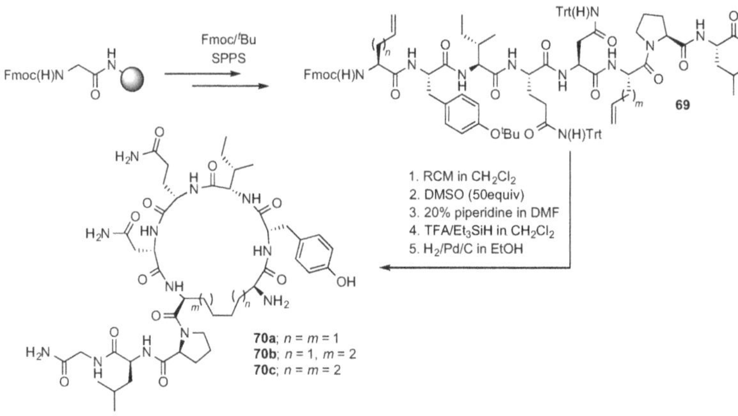

Scheme 16 Synthesis of dicarba oxytocin analogs using RCM on the solid support

cysteines were replaced by L-allylglycine residues. After cyclization, mediated by Ru-catalyst **G1**, the resin-bound peptide was treated with DMSO, an essential step to remove isomerization products and ruthenium-containing contaminants. Subsequent Fmoc removal, acidic deprotection and cleavage from the resin, and hydrogenation resulted in the isolation of oxytocin dicarba analog **70a** in an acceptable yield (8 to 28%). The higher cyclization yield was obtained when the second-generation Grubbs' catalyst **G2** was used. The biological activity of these oxytocin derivatives were tested, and the *cis*-intermediate was the most active oxytocin mimic ($EC_{50} = 38$ ng/mL), while the *trans*-intermediate ($EC_{50} = 242$ ng/mL) and the saturated analog **70a** ($EC_{50} = 338$ ng/mL) were two order of magnitude less active than native oxytocin ($EC_{50} = 2.7$ ng/mL) [86]. In a follow-up study by the same authors, the influence of the geometry of the double bond and the size of the peptide macrocycle were investigated for oxytocin (ant)agonistic activity to design long-lasting and commercially available oxytocin-derived peptide therapeutics for the initiation of parturition and treatment of preterm labor [87]. Their main conclusion was that 20-membered macrocycles, as in **70a**, were the most active, while the 21-membered oxytocin derivative **70b** was a very weak agonist ($EC_{50} = 1{,}400$ ng/mL), and the 22-membered congener **70c** was virtually inactive. However, half-lives of the dicarba analogs were found considerably increased (8 to 11 min longer) compared to their parent disulfide peptides.

Other examples of hydrocarbon bridges as isosteric replacements of disulfides in bioactive peptides have been reviewed [4, 6] and reported in the literature, among others, to arrive at redox-stable analogs of sunflower trypsin inhibitors [88], cyclic dermorphin tetrapeptides [89], cyclic enkephalin peptides [90, 91], octreotide- [92–94] and somatostatin-derived peptides [95], and an antimicrobial human β-defensin-1 derivative [96] and leucocin A analogs [97].

The regioselective formation of interlocked dicarba bridges as exemplified by the peptide toxin α-conotoxin Rg1A **71** (Scheme 17) required olefinic substrates with tunable reactivity against the ruthenium-based metathesis catalysts. Thus, the

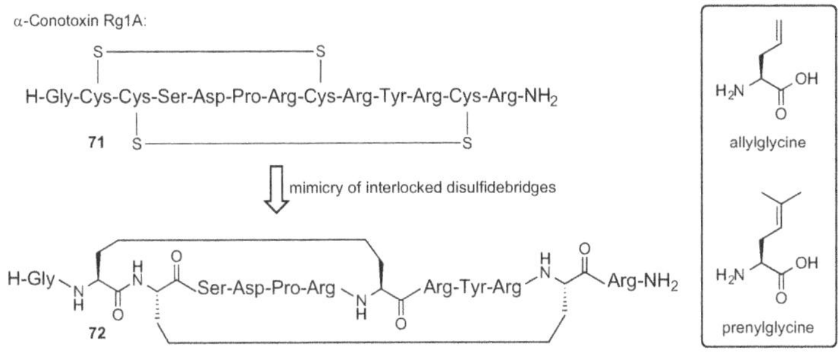

Scheme 17 α-Conotoxin Rg1A **71** and its interlocked dicarba analog **72**

incorporation of allylglycine (Alg) and prenylglycine (Pre) (insert Scheme 17), two metathesis-active olefinic amino acids, facilitates the regioselective formation of the intramolecular dicarba bridges in peptide **72**, as shown in Scheme 18 [98].

The linear peptide **73a** was synthesized on the solid support via well-established Fmoc/^{t}Bu protocols. Initially, ring-closing metathesis of resin-bound peptide **73a** that was performed in CH_2Cl_2 with Ru-catalyst **G2** (20 mol%) in the presence of 5% 0.4 M LiCl/DMF at 40°C for 40 h under conventional heating did not result in any conversion toward carbocycle **73b**. However, microwave irradiation of resin-bound peptide **73a** in CH_2Cl_2 with Ru-catalyst **G2** (20 mol%) in the presence of 5% 0.4 M LiCl/DMF at 100°C [99] resulted in complete ring closure to **73b** in only 2 h. Apparently, microwave irradiation is able to disrupt peptide aggregation and facilitates cyclization toward peptide carbocycles that cannot be obtained via conventional heating methods [100]. Gratifyingly, the prenyl side chains remain unaffected under these microwave-assisted reaction conditions. Selective hydrogenation of carbocycle **73b** with homogeneous Wilkinson's catalyst ([Rh(I)($PPh_3)_3Cl$]) and 80 psi H_2 in CH_2Cl_2/MeOH at room temperature resulted in macrocyclic resin-bound peptide **73c**. Subsequently, the prenyl moieties that resisted hydrogenation were activated by butenolysis by exposing construct **73c** to an atmosphere of *cis*-butene and Ru-catalyst **G2** (20 mol%) at 50°C for 72 h, and the formation of *bis*-crotylglycine-containing peptide construct **73d** was confirmed by mass spectrometry. Then, resin-bound peptide **73d** was subjected to microwave-assisted RCM under the same reaction conditions as for **73a**, which resulted in the correct bicyclic conotoxin framework **73e**. After hydrogenation and deprotection/cleavage, the saturated *bis*-dicarba conotoxin analog **72** was obtained [98].

Sulfide bridges (thioethers) as present in lanthionine **74** and 3-methyllanthionine **75** can also be mimicked by dicarba bridges, like derivatives **76** and **77**, as shown in Scheme 19.

Lanthionines are important conformational constraints to induce structural rigidity and to conserve the three-dimensional shape and bioactive conformation of an important class of antimicrobial peptides, called lantibiotics [101–105]. This family of peptide antibiotics is ribosomally synthesized by Gram-positive bacteria to act

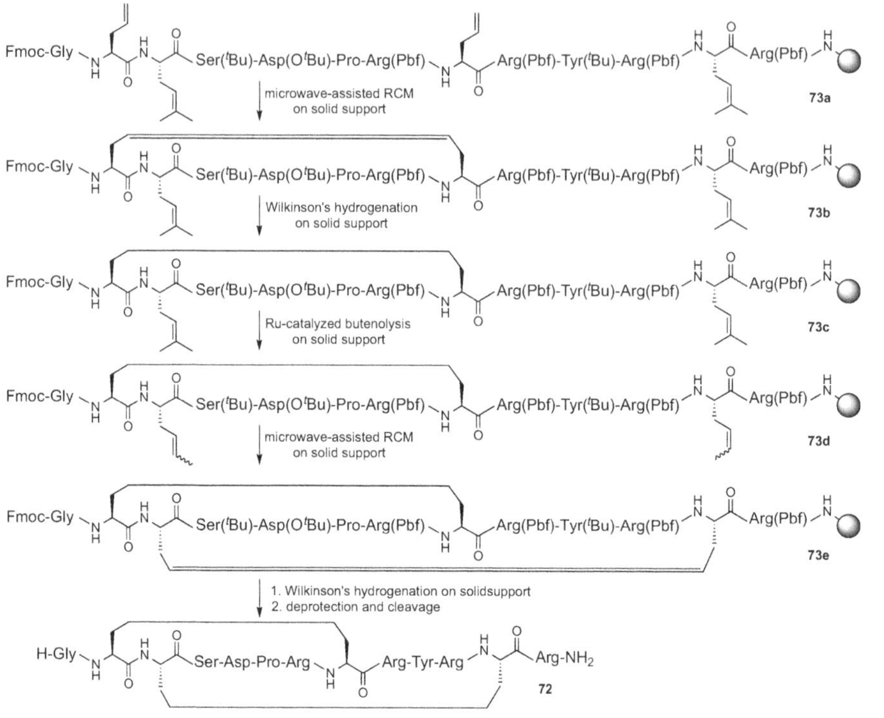

Scheme 18 Regioselective synthesis of interlocked dicarba bridges by tunable reactivity against RCM [98]

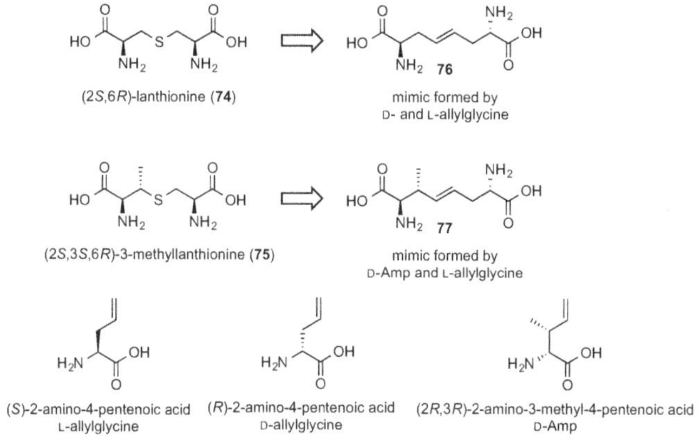

Scheme 19 Structures of the (2*S*,6*R*)-lanthionine and (2*S*,3*S*,6*R*)-3-methyllanthionine moieties and their corresponding alkene-bridged mimics formed by (*S*)-2-amino-4-pentenoic acid, (*R*)-2-amino-4-pentenoic acid, and (2*R*,3*R*)-2-amino-3-methyl-4-pentenoic acid; note: the stereochemistry of the stereogenic centers (mimic vs. lanthionine) is the same, and their *R*/*S* designations are opposite due to the CIP rules

against competing microorganisms and include nisin (**78**), subtilin, and lacticin 481. The lanthionine contains a sulfide bridge that is formed after a series of enzyme-catalyzed posttranslational modifications [106]. Although nisin is widely used as food preservative, it is unstable at basic pH and readily oxidizes and reacts with water or thiol-containing nucleophiles [107–109]. Therefore, we [110–116] and other researchers [117] initiated a program to find more redox-stable lantibiotics in which the thioether functionality is replaced by a two-carbon (dicarba) bridge, based on alkyne [113], alkene, or alkane moieties, which is conveniently introduced by ring-closing metathesis, and we successfully synthesized bioactive nisin AB [115]-, ABC [114, 116]-, and DE-fragment mimics [110–112] (**79**). The DE ring is of special interest since it contains an interlocking thioether bridge (as shown in Scheme 20) which is difficult to access synthetically [118–120].

The synthesis of bicyclic peptide **79** is shown in Scheme 21. Boc-L-Alg-OH was coupled to the solid support and subjected to a cross-metathesis reaction with Fmoc-D(or L)-Alg-OH in the presence of the second-generation Grubbs' catalyst to afford resin-bound alkene **80** [111]. Then, the Fmoc was removed and the amine was coupled to N_3-Ala-ONSu, in which the azide was used as a protecting group, orthogonally to the Boc functionality. Subsequently, the Boc group was removed by acid, followed by coupling of Fmoc-Asn(Trt)-OH and Fmoc-Alg-OH, respectively, and subsequent Fmoc removal of the allylglycine α-amino moiety. In the next step, macrolactamization between both allylglycine residues was performed in the presence of HATU/HOAt/DIPEA as coupling reagents in DMF during 16 h at room temperature. The correct side-chain to side-chain connectivity of ring E (a small amount of resin **81** was cleaved from the resin for analysis) was confirmed by NMR and mass spectrometry. After Staudinger reduction of the azide, Boc-Alg-OH was coupled, and the obtained monocyclic hexapeptide was cleaved from the resin, followed by RCM mediated by Ru-catalyst **G2** in solution. Bicyclic peptide **79** was

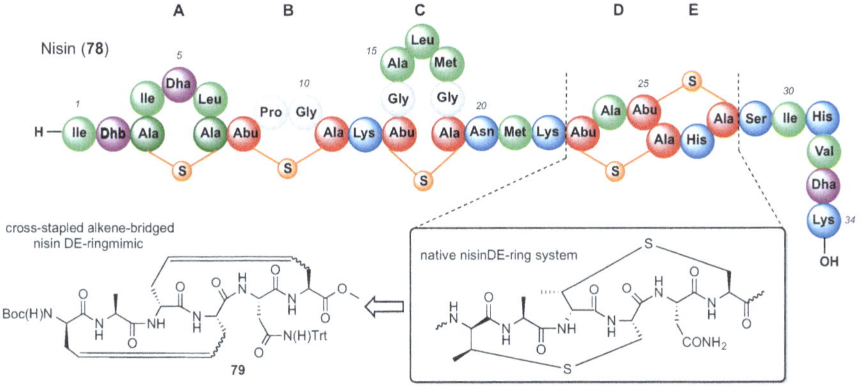

Scheme 20 Schematic representation of nisin (**78**); Dhb, *Z*-dehydrobutyrine; Dha, dehydroalanine; Abu, α-aminobutyric acid. The *insert* shows the structural formula of the DE ring, while **79** represents a cross-stapled alkene-bridged nisin DE-ring mimic

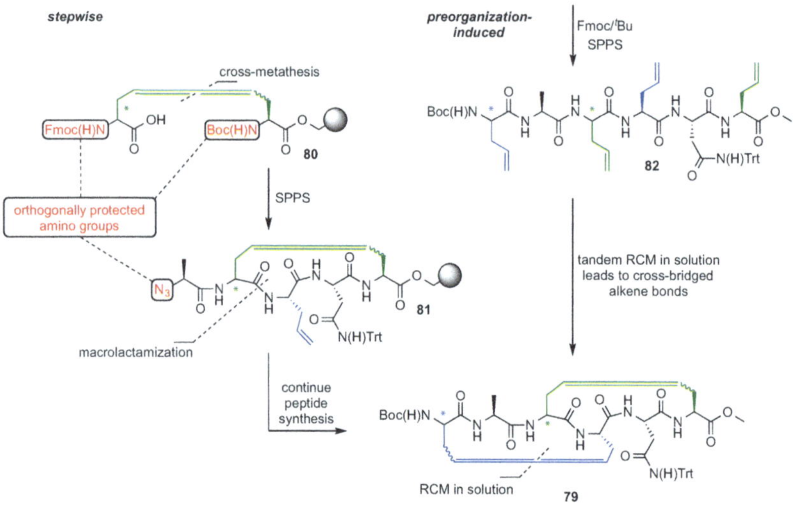

Scheme 21 Stepwise- and preorganization-induced synthesis of a crossed alkene DE-ring mimic [110–112]

obtained in 11% overall yield and characterized by NMR (^{1}H-500 MHz, COSY, TOCSY, ROESY) in combination with MS (LCES-TOF MS-MS) [111].

As an alternative synthesis route, linear hexapeptide **82** was assembled on the solid support and subsequently cleaved from the resin in its protected form. This peptide was treated with the second-generation Grubbs' catalyst in CH_2Cl_2 at 60°C for 2 h [110, 111]. The reaction mixture contained one compound, consisting as a mixture of the four geometrical isomers, which could be separated by preparative HPLC [112]. NMR and mass analyses indicated that only compounds were formed with the correct side-chain to side-chain connectivity pattern: $1 \rightarrow 4/3 \rightarrow 6$, this may hint to a certain degree of preorganization of the linear peptide. Ring-closing metathesis in the presence of the Hoveyda-Grubbs' catalyst gave an identical result [112], while treatment of peptide **82** with **G1** resulted in **79** and a monocyclic peptide with the $1 \rightarrow 4$ macrocycle in a ratio of 2:1 after 16 h of refluxing in CH_2Cl_2 [110, 111]. Dicarba AB- and ABC-ring mimics of nisin were able to recognize its natural substrate lipid II, although with a tenfold reduced affinity compared to its native counterpart [114, 115], while the DE-ring mimics were found to be devoid of membrane permeabilizing interactions [112, 116].

A multiple on-resin metathesis approach to introduce dicarba bridges as sulfide isosteres in lantibiotics was developed by Vederas and co-workers, as shown in Scheme 22 [117]. In their approach, resin-bound *bis*-allyl peptide **83** was treated with 20 mol% second-generation Grubbs' catalyst in refluxing CH_2Cl_2 for 12 h to give smoothly monocyclic peptide derivative **84**. Any ruthenium by-products were removed by a DMSO washing step. Peptide synthesis was continued, followed by a second RCM step (under the same reaction conditions as the first one) to give a single reaction product: bicyclic peptide **85**. SPPS was continued to give the

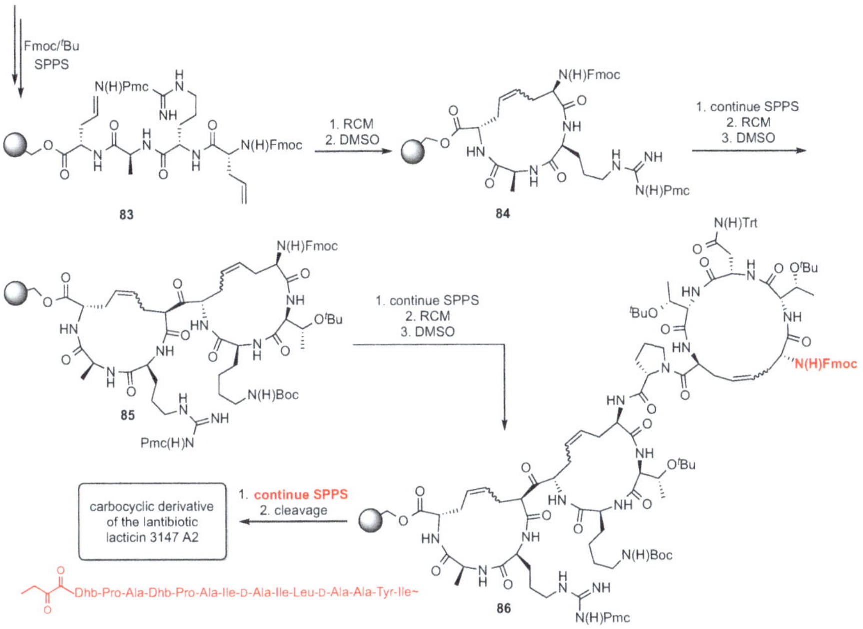

Scheme 22 Multiple on-resin ring-closing metathesis approach to form ring-expanded analogs of the lantibiotic lacticin 3147 A2 [117]

precursor to **86**; however, the last cyclization required an increase of catalyst loading to 50 mol%, and the reaction time was increased to 48 h, which ultimately gave a single reaction product. To exclude scrambling of the double bonds by ring-opening metathesis subsequently followed by RCM, a thorough LC MS-MS analysis was performed, which confirmed the absence of any crossover reaction products. The continuation of peptide synthesis by a combination of SPPS and fragment assembly in solution resulted in the successful synthesis of a carbon analog of lacticin 3147 A2. Unfortunately, the carbocyclic lacticin derivative was not active as an antimicrobial peptide at concentrations comparable to native lacticin, an indication that the mimic does not have the correct bioactive conformation to bind its target, the lacticin A1 complex, with lipid II.

2.6 *RCM of Peptides in Water*

Ring-closing metathesis of peptides, either in solution or on solid phase, is a well-established synthetic approach to arrive at cyclic peptides. Nevertheless, several peptide sequences are notoriously difficult to cyclize. It is assumed that a reduced solubility or aggregation phenomena on the solid support are responsible for these sluggish cyclization reactions. Semi- and unprotected peptides generally display

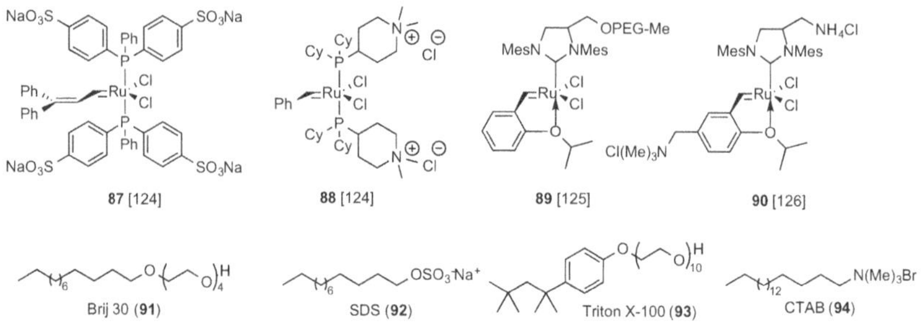

Fig. 4 Selected metathesis catalysts and surfactants used in aqueous olefin metathesis

high solubility in aqueous media due to the presence of polar functional groups, like amine (Lys, N-terminus), hydroxyl (Ser, Thr), carboxylic acid (Asp, Glu, C-terminus), amide (Asn, Gln, C-terminus), imidazole (His), and guanidine (Arg) moieties. Gratifyingly, these functional groups are well tolerated by the ruthenium-based metathesis catalysts. Therefore, it is assumed that aqueous olefin metathesis [121, 122] is a promising approach, not only for peptide synthesis but also in the field of protein chemistry and bioconjugation [123].

Aqueous olefin metathesis requires the development of water-soluble (pre)catalysts, and several have already been reported in the literature [124–126] (e.g., catalysts **87–90**, Fig. 4), while other approaches focus on the development of micellar environments (e.g., surfactants **91–94**, Fig. 4) to enhance transition-metal-catalyzed carbon-carbon formation reactions [127, 128].

The latter approach was applied by Vederas and co-workers to investigate aqueous ring-closing metathesis on two peptides that were difficult to cyclize on resin [129].

In their approach (shown in Scheme 23 and Table 1), two peptide derivatives were used as model systems, oxytocin (**95ae**) and crotalphine (**96ae**) [130], a peptide hormone that controls mammary and uterine smooth muscle contraction and an orally active analgesic, isolated from the venom of the rattlesnake *Crotalus durissus terrificus*, respectively. In case of the oxytocin derivatives, all on-resin cyclizations proceeded smoothly, except in case of **96e** (containing SAlg, an analog of *S*-allyl cysteine in which the position of the sulfur and alkene moiety is switched), while all crotalphine analogs were inert toward on-resin metathesis reaction conditions. Therefore, several detergents (shown in Fig. 4), e.g., sodium dodecyl sulfate (SDS, **92**), Triton X-100 (**93**), and cetyltrimethylammonium bromide (CTAB, **94**), were applied at critical micelle concentration with unprotected oxytocin and crotalphine analogs in water with **HG2** as the solubilized RCM catalyst (Scheme 23). All these attempts failed; however, some product formation in case where Xaa was pentenylglycine (Pgl, **b**), *O*-allylserine (Oas, **c**), and *S*-allylcysteine (Sac, **d**) could be identified by mass spectrometry, while product isolation was unsuccessful (Table 1).

Finally, an aqueous solution with 30% *tert*-BuOH as cosolvent with a high catalyst loading (50 equiv **HG2**) in the presence of a large excess of $MgCl_2$ (5000 equiv), to suppress ruthenium-peptide adduct formation, at 37°C for 24 h,

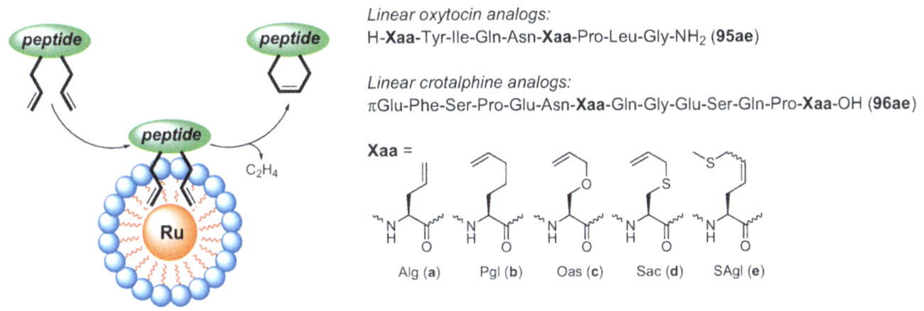

Scheme 23 Aqueous RCM of oxytocin and crotalphine analogs [129]

Table 1 Ring-closing metathesis of peptides in water [129]

	Alg	Pgl	Oas	Sac	SAlg
Oxytocin					
On solid support	✓	✓	✓	✓	✗
Micellar solution	✗	–[a]	–[a]	–[a]	n.d.
Aqueous solution	✗	<[b]	<[b]	78%[c,d]	✗
Crotalphine					
On solid support	✗	✗	✗	✗	✗
Micellar solution	✗	–[a]	–[a]	–[a]	n.d.
Aqueous solution	✗	<[b]	<[b]	63%[c]	✗

[a]Although the product was formed (only in 8.2 mM SDS), it could not be isolated
[b]Isolated yield <1% in H_2O/*tert*-BuOH 7:3 v/v with $MgCl_2$ (5,000 equiv)
[c]Isolated yield in H_2O/*tert*-BuOH 7:3 v/v with $MgCl_2$ (5,000 equiv)
[d]Isolated yield only with the α-amine protected by Fmoc
n.d. not determined
This sign (✓) indicates that product is formed; while this sign (✗) indicates that product is not formed

ultimately resulted in a successful RCM of **95d** and **96d**, since their cyclic congeners were isolated in 78% and 63% yield, respectively (Table 1).

The authors conclude that "the sulfur atom within *S*-allylcysteine drastically improved the efficiency of RCM compared to pentenylglycine and *O*-allylserine analogs" and that their method provides "an alternative to the on-resin cyclization of *S*-allylcysteine, which is particularly attractive in instances where on-resin cyclization is not possible" [129].

2.7 RCM of Peptide-Based Drug Molecules

Peptides have unprecedented biological activities regarding selectivity, specificity, and affinity. However, due to their polar character and many functional groups, their drug-like properties are not always optimal. Therefore, peptide lead molecules

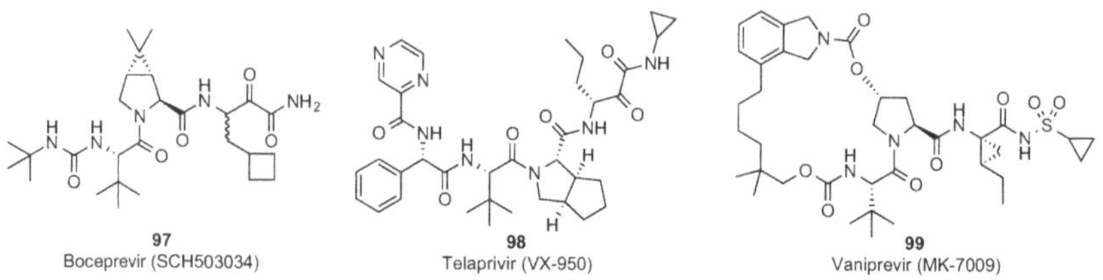

Fig. 5 HCV NS3/4a protease inhibitors

will be used to design peptidomimetics with improved properties like bioavailability and biostability. This approach has been shown to be very effective in case of the design of protease inhibitors [131]. The substrate-based inhibitor design led to the development of a series of highly active hepatitis C virus NS3/4a protease inhibitors [132, 133], like boceprevir (SCH503034, **97**) [134], telaprevir (VX-950, **98**) [135], and vaniprevir (MK-7009, **99**) [136], as shown in Fig. 5. The latter was prepared using an optimized ring-closing metathesis approach to ensure the supply of this molecule for ongoing clinical studies.

The primary challenges to develop a scalable and cost-effective synthesis for vaniprevir (MK-7009) **99** were to identify an efficient macrocyclization approach and versatile synthesis routes to the key intermediates required to assemble the cyclization precursor molecules. Macrolactamization and Heck-based couplings have been used to install the 20-membered macrocycle of vaniprevir in ten linear steps with 30% overall yield [137]. As an alternative, ring-closing metathesis was used for the highly efficient synthesis of vaniprevir in nine linear steps and 55% overall yield [138, 139]. This RCM approach employs a low catalyst loading (0.2 mol%) and a relatively high substrate concentration for ring-closing metathesis (0.13 M).

As shown in Scheme 24, there are three possible options for the construction of the macrocycle on the basis of different positions of the olefin metathesis: styryl-homoallyl (**102a**), allyl-allyl (**102b**), and homoallyl-vinyl (**102c**) diene precursors [138]. Based on an initial screening [137], ring-closing metathesis of **102a** was performed with M1 [140] (30 mol%) and gave **101a** in 91% yield, while RCM in the presence of Zhan-1B [141] (10 mol%) resulted in the isolation of **101a** in 98% yield. Despite the high yields, high catalyst loading and high dilution conditions (<3 mM) were required. When the catalyst loading was reduced or when the concentration was increased, lower yields were observed. Unfortunately, RCM of diene **102a** could not be improved by the variation of catalysts and reaction conditions. Moreover, catalyst removal after the cyclization reaction was challenging, and the catalysts' high costs and patent protection made this approach less amendable. Therefore, a more efficient ring-closing strategy was developed, and it was envisioned that the allyl-allyl diene **102b** would be a better cyclization precursor.

The first RCM reaction of diene **102b** was run by using 1 mol% of the second-generation Hoveyda-Grubbs' catalyst (**HG2**) under high dilution conditions, and

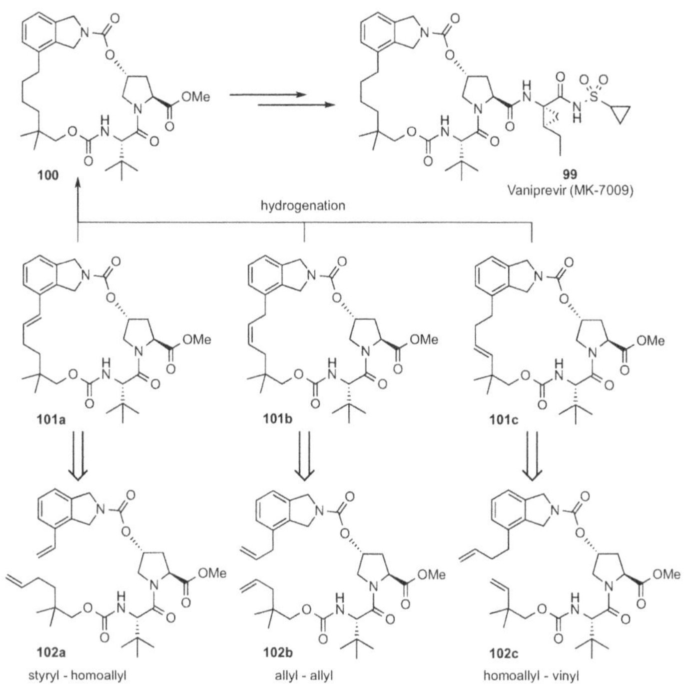

Scheme 24 Three possible options for the RCM route of the macrocyclic vaniprevir precursor **100**

macrocycle **100** was obtained in 57% yield, after hydrogenation. The yield was only marginally increased (to 67%) when the catalyst loading was increased to 5 mol%. However, the yield increased to 82% when 1 mol% of catalyst was added over 60 min at 60°C to a solution of diene **102b** in toluene (50 mL/g diene). The next step in the optimization process was to increase the concentration of the RCM reaction mixture. At 30 mL/g diene, the desired 20-membered macrocycle **101b** was obtained in only 61% yield, while a 19-membered side product was isolated in 9% yield. It was postulated that this side product was derived from a styryl isomer, which on its turn was generated from the isomerization of diene **102b**. This olefin isomerization reaction, due to the decomposition of the Ru catalyst [142], has been described in the literature, and vinyl formation can be effectively suppressed by the addition of quinone derivatives [143]. It was found that the addition of 2,6-dichloroquinone decreased the formation of the 19-membered ring to <1% without compromising the yield when the reaction was performed at 20 mL/g diene. After several optimizations, it was found that this RCM reaction gave 91% cyclization yield when the reaction was performed in toluene at 100°C at a concentration of 13.5 mL/g diene with a catalyst loading of 0.2 mol% by implementing a simultaneous slow addition of both catalyst and diene. Macrocycle **100** was conveniently obtained by hydrogenation of **101b** after crystallization from aqueous 2-propanol: 89% yield with >99% HPLC purity [138]. The third alternative route,

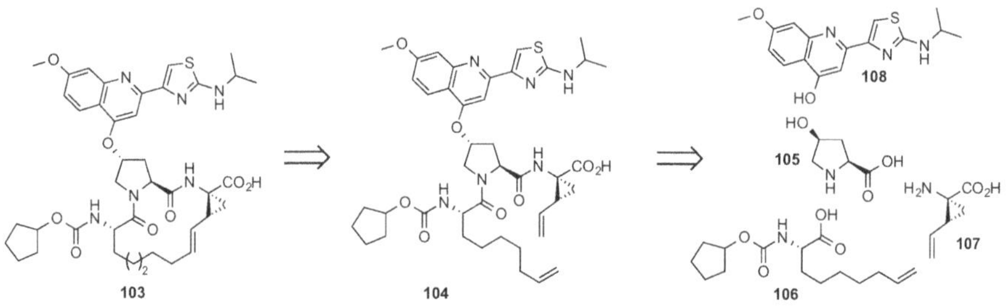

Scheme 25 Retrosynthesis of ciluprevir (BILN 2061) **103**, an NS3 HCV protease inhibitor [144]

the homoallyl-vinyl cyclization approach (**102c**), was not further explored by these authors, presumably by anticipating on the difficult cyclization due to the neopentylic olefin.

Another example of ring-closing metathesis to synthesize a peptide-derived drug molecule is given in Scheme 25, which describes the retrosynthesis of ciluprevir (BILN 2061) **103** [144], an NS3 protease inhibitor with proven antiviral activity in humans to treat hepatitis C infections [145].

The retrosynthesis of macrocycle **103** suggests that the 15-membered ring can be obtained from its linear precursor tripeptide **104** by ring-closing metathesis. This acyclic peptide can be efficiently assembled in a convergent manner in solution-phase peptide-coupling procedures by using four building blocks, including three unnatural amino acids: *trans*-(2*S*,4*R*)-hydroxyproline (**105**), N^{α}-cyclopentyloxy-carbonyl-(*S*)-2-amino-non-8-enoic acid (**106**), and (2*R*,3*S*)-3-vinyl-2-amino-2-cyclopropyl-carboxylic acid (**107**) and a 4-quinolinol moiety (**108**). Several ruthenium catalysts were used to optimize the ring-closing metathesis reaction of **104** [146]. The initial RCM reactions were performed with the first-generation Grubbs' catalyst **G1**. It turned out that the backbone conformations of each acyclic tripeptide, and in particular the ratio of the proline *cis*/*trans* conformers, were found to be crucial to the reaction rate, yield, and diastereomeric purity of the macrocyclic tripeptide(s). The absence of any conformational preference of the peptide backbone resulted in poor RCM yields (<15%), while the bulky Boc group was responsible for a certain degree of preorganization in a β-strand-like conformation that favored macrocyclization (~80% yield). Ring size was another important determinant for macrocyclization. Dienes with the C9 linker (as in **106**) incorporated were efficiently cyclized into the 15-membered ring structure, while the more constrained 13- and 14-membered macrocyclic peptides could only be obtained with an epimerized Cβ of the vinylcyclopropane moiety. Furthermore, RCM with **G1** in CH_2Cl_2 as solvent gave the highest yield (83%); however, the amount of epimerization was relatively high (13%) under these reaction conditions and de facto unacceptable, especially on multi-kilogram production scale. Changing the catalyst to **G2** resulted in a clean conversion into the macrocyclic peptide without any noticeable amount of epimerized product; unfortunately, under these conditions, the yield was significantly lower, 52% (**G2**) versus 83% (**G1**). RCM catalyzed by the first-generation Hoveyda's catalyst [147] was slower than in case of **G1**; however, the desired 15-membered macrocyclic peptide was obtained in 85% yield with an excellent purity of >99% as measured by chiral HPLC [146]. Besides that, Hoveyda's catalyst could be recovered from the crude reaction mixture and recycled [144].

2.8 RCM of Depsipeptides

Cyclodepsipeptides, peptides in which one or more amide bonds have been replaced by ester moieties, are valuable compounds with a diverse spectrum of high bioactivity, like antitumor, antiproliferative, cytotoxic, antimicrobial, and antifungal behavior. This class of highly potent peptides has been isolated from marine sponges, algae, and fermentation broths. Their intriguing bioactivity in combination with their complex chemical structure make cyclodepsipeptides challenging targets for total synthesis, not only for having access to larger quantities and other congeners but also to confirm the structural assignment and absolute configuration of the stereochemical centers. In the next section, four examples of relevant metathesis-based syntheses of cyclodepsipeptides will be discussed, e.g., cryptophycin-24 (arenastatin A) **109** [148]; (−)-spongidepsin (**111**) [149]; (−)-PF1163B (**113**) [150], as shown in Scheme 26; and lagunamide A (**116a**) [151], as shown in Scheme 27.

The cryptophycins, a unique class of 16-membered macrolides with antimitotic activity, have been isolated from the cyanobacteria *Nostoc* sp. They are currently considered to be promising lead structures in antitumor therapy [152]. A structural analog, cryptophycin-24 (**109**), was isolated from the marine sponge *Dysidea arenaria* [153]. Because of their interesting biological activity and intriguing molecular structure, several total syntheses of this cyclodepsipeptide and its

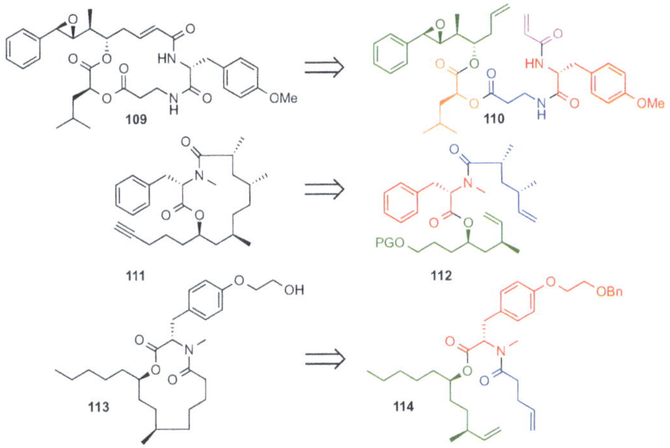

Scheme 26 Retrosynthesis of cyclodepsipeptides cryptophycin-24 **109** [148], (−)-spongidepsin (**111**) [149], and (−)-PF1163B (**113**) [150]

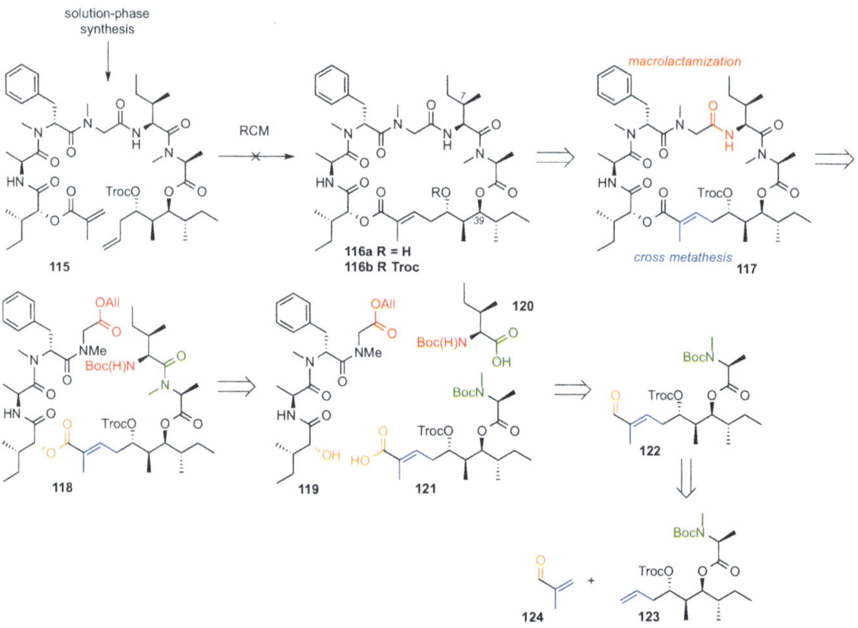

Scheme 27 Retrosynthesis of cyclodepsipeptide lagunamide A **116a** [156]. *Note*: compound **116a** represents the 7,39-*epi*-lagunamide A which corresponds to the originally proposed structure of lagunamide A

analogs have been reported [154, 155]. The first RCM-based approach was reported by Tripathy and Georg [148] and features the introduction of the β-epoxide prior to macrolide formation. Based on the retrosynthesis (Scheme 26), cryptophycin-24 (**109**) could be synthesized from diene **110** by ring-closing metathesis in the presence of the chemically reactive styrene epoxide moiety (the other required fragments are highlighted in color). Indeed, treatment of **110** with **G1** (10 mol%) in CH_2Cl_2 under reflux for 6 h resulted in a *trans*-selective macrocyclization, and cyclodepsipeptide **109** was obtained in 70% yield, and all spectroscopic data were in agreement with the reported values [155].

A second example describes the synthesis of (−)-spongidepsin **111** [149], featuring a ring-closing metathesis reaction to construct the 13-membered macrocycle (Scheme 26). (−)-Spongidepsin is a cyclodepsipeptide, originally isolated from the Vanuatu marine sponge *Spongia* sp., with potent antitumor activity [156]. Ring-closing metathesis of diene **112** in the presence of **G2** (20 mol%) in CH_2Cl_2 at 40°C resulted in the corresponding *Z*-lactam with 93% yield. After protecting group removal, hydrogenation of the double bond, and conversion of the alcohol into the aldehyde, the terminal acetylene was installed to obtain the title compound in 60% over the four last synthesis steps. Two other syntheses of (−)-spongidepsin, both applying RCM to assemble to macrocycle, had been reported two years earlier by Forsyth [157] and Ghosh [158], respectively. Noteworthy is the difference in reaction conditions while starting with a precursor in which PG was either 4-methoxybenzyl or *tert*-butyldiphenylsilyl, respectively. Forsyth described an efficient RCM reaction with **G2** (10 mol%) in toluene (110°C) during 20 min to obtain the macrocycle in 80% yield with a 10:1 *E*/*Z* ratio, while Ghosh reported an RCM reaction in the presence of **G2** (10 mol%) in CH_2Cl_2 (reflux) for 12 h affording the macrocycle as a single isomer (*Z*) in 93% yield. (−)-Spongidepsin synthesized by these approaches show identical spectroscopic data; however, different optical rotations were found. Follow-up papers that describe other RCM-based (formal) total syntheses of (−)-spongidepsin have been reported [159–162]; however, the issue of differences in the optical rotations was not further addressed.

Another example of an efficient RCM approach to obtain a 13-membered macrocycle deals with the synthesis of (−)-PF1163B (**113**) [150], a cyclodepsipeptide with antifungal properties. This class of bioactive peptides was isolated from the fermentation broth of *Penicillium* sp. [163, 164] and subsequently shown to be inhibitors of ERG25p to interfere with the ergosterol biosynthesis cascade. The 13-membered macrocycle incorporates an *O*-2-hydroxyethyl-*N*-methyl-L-tyrosine moiety, and a total synthesis starting with L-tyrosine and (*R*)-citronellol has been described in the literature [165]. This synthesis requires 13 steps, while the synthesis reported by Gesson [150] requires eight steps in which RCM of precursor **114** is the key step (Scheme 26). Ring-closing metathesis was attempted with different amounts of **G1** as the catalytic species. Unfortunately, only a limited degree of cyclization was found under these conditions. However, the use of **G2** led to better results: RCM of **114** (at a concentration of 2.1 to 3 mol/L) in refluxing $C_2H_4Cl_2$ in the presence of **G2** (12 mol%) gave the corresponding macrocycle in 60% yield as a mixture of (*E*)- and (*Z*)-diastereomers. Two other total syntheses of PF1163A have been described [166, 167] utilizing RCM as key reaction with a slightly increased yield of the macrocyclization (70 to 83%).

As a last example of a cyclodepsipeptide synthesis featuring a metathesis reaction, the synthesis of lagunamide A **116a** [151] will be discussed and is shown in Scheme 27.

Lagunamides are new cyclic depsipeptides isolated from the marine cyanobacterium *Lyngbya majuscula* collected from the western lagoon of Pulau Hantu Besar, Singapore [168, 169]. These compounds exhibit potent biological activities, such as antimicrobial, antimalarial, cytotoxic, and neurotoxic properties, which make these marine cyanobacterial compounds interesting lead compounds for further development as antitumor therapeutics. The originally assigned stereochemical configuration of lagunamide A (**116a**) [168] was revised by Dai et al. [170] and turned out to be the C7 and C39 epimer (Scheme 27). So far, the (*E*)-olefin in lagunamide A has been installed via the Horner-Wadsworth-Emmons olefination, while Huang et al. [151] designed a metathesis approach to introduce this double bond in its correct geometry. In first instance RCM was explored. Therefore, precursor **115** was prepared based on solution-phase peptide synthesis. Disappointingly, compound **115** did not undergo macrocyclization (to afford **116b**) in the presence of the second-generation Grubbs' catalyst **G2** (CH_2Cl_2 (1 mM) at reflux or toluene

(1 mM) at 60°C). Variation of the reaction conditions by different solvents and temperatures remained unsuccessful. As an alternative, macrolactamization of precursor **118** was envisioned, while the olefin was prepared by cross metathesis (CM) of alkene **123** with methacrylaldehyde (**124**). Gratifyingly, CM between compounds **123** and **124** proceeded smoothly in the presence of second-generation Grubbs' catalyst **G2** (refluxing CH_2Cl_2), since olefin **122** was obtained in 81% yield with a *E*/*Z* ratio of >99:1. Finally, via intermediates **119–121**, macrolactamization precursor **118** was synthesized, and after protecting group manipulation followed by a HATU/DIPEA-mediated macrolactamization, lagunamide A derivative **116a** was obtained in good yield. Thus, although RCM did not result in the desired macrocyclization, CM was a versatile alternative approach for RCM, since a series of different homoallyl alcohol derivatives, like **123**, could be obtained for the synthesis of lagunamide A analogs.

2.9 RCM for the Synthesis of Bioactive Peptides and Peptidomimetics

In this section a diverse set of examples that apply ring-closing metathesis and cross metathesis for the synthesis of bioactive peptides and peptidomimetics will be discussed. Herein, these examples focus on a large structural variation of the peptide derivatives, and special attention is paid to the reaction conditions of the RCM or CM protocols and to approaches to optimize the metathesis reaction.

The first example describes the solid-phase synthesis of peptide thioureas and thiazole-containing macrocycles (represented by **128**) obtained by Ru-catalyzed ring-closing metathesis [171], as shown in Scheme 28. This class of functionalized peptides is frequently encountered in biologically active derivatives, with antibacterial, antifungal, and antiviral properties.

Herein, Cohrt and Nielsen describe a versatile synthesis to obtain 15–17-membered thiazole-containing peptide macrocycles in all diastereomeric forms, and their developed "build/couple/pair" approach is well suited for the generation of larger libraries of said peptidomimetics. As a typical example, solid-phase-bound

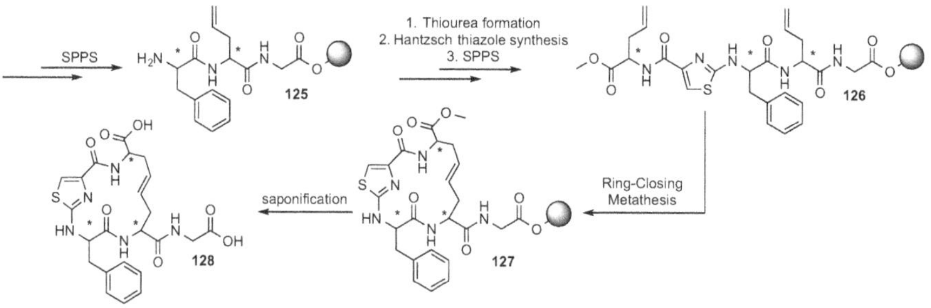

Scheme 28 Synthesis of thiazole-functionalized peptide macrocycles [171]

tripeptide H-Phe-Alg-Gly **125** was treated with *N,N′*-di-Boc-thiourea in the presence of Mukaiyama's reagent (2-chloro-1-methyl-pyridinium-iodide) and Et_3N as base in CH_2Cl_2 as the solvent, to afford the corresponding *N*-terminal α-thiourea intermediate in a yield >95%, after optimization of the addition/preactivation procedure. Subsequently, the Boc group could be smoothly removed by treatment of the peptide resin with 0.1 M $SnCl_4$ in CH_2Cl_2 for 2 × 30 min, and the free α-amine was then reacted with an α-halo ketone (e.g., $(HO_2C)COCH_2Br$, bromopyruvic acid) in $EtOH/H_2O$ 3:1 v/v at 60°C for 16 h to effect the Hantzsch thiazole synthesis in excellent yield and purity (after deprotection a small aliquot of resin). The resulting carboxylated thiazole readily underwent amide bond formation via on-resin activation, e.g., allylglycine methyl ester, to give resin-bound cyclization precursor **126**, as shown in Scheme 28. The ring-closing metathesis was performed by exposing diene **126** to 30 mol% catalyst in dry degassed CH_2Cl_2 at room temperature for 24 h. A total of nine ruthenium-alkylidene catalysts were tested, with the **G2** and **HG2** resulting in the highest conversions (67 to 95%). Further increasing the conversion could be accomplished by repeating the RCM reaction on the solid support once. Gratifyingly, this RCM protocol afforded access to 15-, 16-, and 17-membered macrocycles with up to three stereocenters in the ring, which all underwent cyclization, in excellent product yields and purity [171].

In the second example, macrocyclic analogs of angiotensin IV (H-Val-Tyr-Ile-His-Pro-Phe-OH) **129** have been synthesized according to a replacement/cyclization SAR study, to improve their chemical and metabolic stability, as potent inhibitors of insulin-regulated aminopeptidases, as shown in Scheme 29 [172]. In their study, Hallberg and co-workers designed cyclic analogs of the hexapeptide angiotensin IV (**129**) [172], which is known to play important roles in cognition processes, like memory and learning [173]. Angiotensin IV binds with high affinity to the insulin-regulated aminopeptidase (IRAP), an enzyme localized in the brain

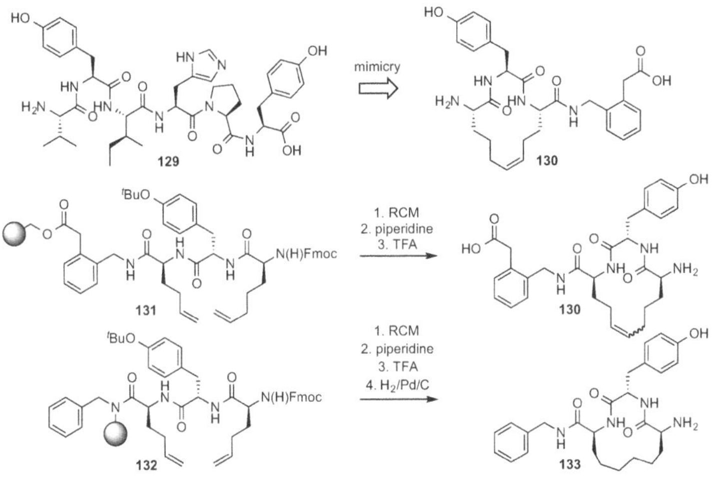

Scheme 29 Design and synthesis of angiotensin IV-derived macrocyclic peptidomimetics [172]

and is associated with the mentioned mental processes. Inhibitors of IRAP are promising novel lead compounds for the treatment of memory dysfunction, due to brain trauma, cerebral ischemia, Alzheimer's disease, and other age-related diseases.

Angiotensin IV (AngIV) is a linear hexapeptide, and to improve the drug-like character of this peptide, the group of Hallberg reported a disulfide scan which resulted in the discovery of efficient and potent cyclic AngIV analogs [174]. In the present study these disulfide peptides were converted into more stabile dicarba analogs by ring-closing metathesis. For this purpose, diene **131** was synthesized by SPPS and treated with an appropriate catalyst. For this purpose, **G2** and **HG2** (both 15 mol%) in 1,2-dichloroethane (DCE) under microwave irradiation at 150°C for 5 min were evaluated in this conversion, and after cleavage from the resin, complex mixtures of poorly separable compounds (like ring-contracted and double-bond migration products) were obtained, and only a 14-membered analog (with a shifted double bond compared to the peptide backbone) could be isolated in sufficient amounts. The addition of 1,4-dibenzoquinone (15 mol%) and a lower catalyst loading (7 mol% **HG2**) in combination with microwave heating at 120 or 140°C for 5 min in DCE resulted in the isolation of compound **130** as an *E*/*Z* mixture, in which the (*E*)-diastereoisomer was consistently the predominant 14-membered macrocycle (>60%). In another approach, benzylamide resin-bound diene **132** underwent RCM under the same reaction conditions as described for **131**. After Fmoc removal, cleavage from the resin and hydrogenation of the double bond, two products could be separated by HPLC, the desired 13-membered macrocycle **133** in a yield of only 2%, while the second product corresponding to a 12-membered ring-contracted derivative. Ring size was the most important factor toward the inhibitory activity against IRAP, since the following K_i values were found: AngIV 62 nM, macrocycle **130** 25 nM (14-membered ring), and macrocycle **133** 193 nM (13-membered ring) [172].

The third example describes the application of "Linked Amino Acid Mimetics" (LAAM): building blocks that provide a functional amino acid side chain in combination with a terminal olefin as a cyclization hinge via ring-closing metathesis [175]. Bornmann and co-workers synthesized two LAAM building blocks, amine **134** and carboxylic acid **135**, for the C- and N-terminal modification of a bioactive peptide sequence, respectively. In their communication, the tripeptide Arg-Gly-Asp (RGD) was chosen since it is known that cyclic RGD peptides (and especially *cyclo*[Arg-Gly-Asp-D-Phe-N^{Me}Val]) [176] exhibit high affinity toward the $\alpha_V\beta_3$ integrin [177], an important class of proteins that plays a role in angiogenesis. Precursor peptide **136** was synthesized on the solid support by Fmoc/^{t}Bu-based SPPS in which the C-terminal aspartic acid residue was anchored via its side chain to the solid support, while the α-carboxylate was coupled to amine **134** to install the C-terminal alkene moiety, as shown in Scheme 30. In turn, the α-amine of arginine was functionalized with acid **135**, and resin-bound diene **136** (0.82 mmol) was cyclized by ring-closing metathesis in the presence of **G2** (50 mol%) as the catalytic species in dichloroethane (25 mL) under microwave irradiation at 60°C for 40 h to yield resin-bound macrocycle **137**. After TFA

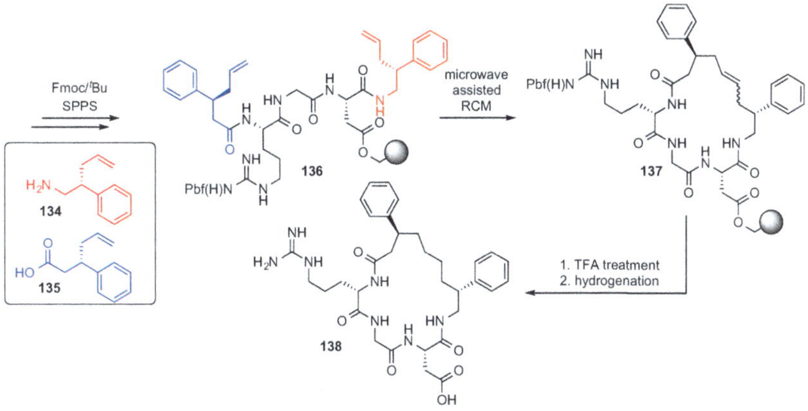

Scheme 30 Solid-phase synthesis of an RGD macrocycle based on Linked Amino Acid Mimetics [175]

treatment and hydrogenation, cyclic RGD peptide **138** was obtained in good yield (36%) [175].

Although the present paper does not provide any biological data regarding binding affinity of **138** toward the $\alpha_V\beta_3$ integrin, modeling suggests that the 19-membered macrocycle **138** is a representative mimic of the 15-membered *cyclo*[Arg-Gly-Asp-D-Phe-N^{Me}Val] peptide, since the RMSD was 0.8819 Å, as shown in Fig. 6.

In the fourth example, Liskamp and co-workers described the synthesis of a novel cyclic peptide MC4-ligand featuring ring-closing metathesis [178]. Based on a linear heptapeptide, Ac-Nle-Gly-Lys-D-Phe-Arg-Trp-Gly-NH_2, Ac-*Alg*-Gly-Lys (Boc)-D-Phe-Arg(Pbf)-Trp(Boc)-*Alg*-NH_2 (**139**) was synthesized by Fmoc/^{t}Bu SPPS and detached from the resin by ammonolysis to obtain the protected peptide amide (see Scheme 31).

RCM was performed in solution with a series of different solvents, reaction temperatures, and catalytic Ru species, as shown in Table 2. It has to be taken into account that precursor **139** is sparingly soluble in solvents commonly used in RCM reactions, while DMF was the only solvent in which the protected peptide reasonably dissolved (0.41 g peptide/5 mL solvent). Therefore, all cyclization reactions were carried out with 2.5 vol% DMF in the mentioned solvent (entries 1–8) at a final peptide concentration of 1.6 mM in the presence of 20 mol% catalyst under a continuous flow of dry N_2, and heating was performed in a conventional oil bath.

Remarkably, all cyclization attempts with **G2** as catalyst were unsuccessful, despite its higher activity and stability compared to **G1**. Cyclization entry 8 was performed for 1 h at 100°C, then an additional loading of 20 mol% of fresh catalyst was added, and RCM was extended for an additional 2 h at 40°C. This approach resulted in a separable mixture of linear and cyclic peptides as judged by LCES

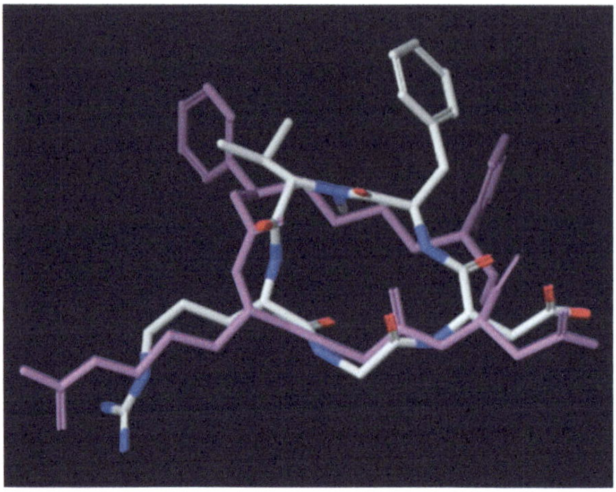

Fig. 6 Superimposition of the lowest energy structure of **138** (blue purple) with *cyclo*[Arg-Gly-Asp-D-Phe-N^{Me}Val] bound to $\alpha_V\beta_3$ integrin [177] (by-atom color). Reprinted from Tetrahedron Lett (2013) 54:5799–5801, Copyright (2013) with permission from Elsevier [175]

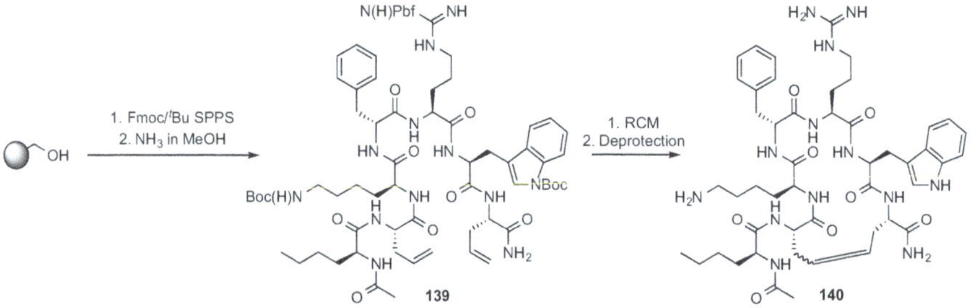

Scheme 31 SPPS and RCM for the synthesis of cyclic MC4-ligand **140** [178]

Table 2 Optimization of the RCM reaction of precursor **139** [178]

Entry	Solvent[a]	Catalyst	*T* (°C)	Result[b]
1	CH_2Cl_2	**G2**	40	Linear
2	DCE	**G2**	50	Linear
3	Toluene	**G2**	70	Linear
4	TCE	**G2**	70	Linear
5	TCE	**G2**	100	Linear
6	Toluene	**G1**	70	Linear
7	TCE	**G1**	100	Linear
8	TCE	**G1**	100 (1 h) → 40 (2 h)[c]	Linear/cyclic

[a]Solvents: 1,2-dichloroethane (DCE), 1,1,2-trichloroethane (TCE)
[b]As determined by LCES MS
[c]Second portion of **G1** (20 mol%) was added after 1 h

MS. After protecting group removal, cyclic peptide **140** could be isolated in 63% yield as an *E*/*Z* diastereomeric mixture.

The three peptides, Ac-Nle-Gly-Lys-D-Phe-Arg-Trp-Gly-NH_2 as the control, protected linear *bis*-allyl peptide **139**, and cyclic peptide **140**, were tested in a melanocortin receptor assay. Their K_i values for the MC4 receptor were 2.4 ± 0.3, 1.9 ± 1.0, and 2.3 ± 1.0 nM, respectively. However, the measured EC_{50} values were as follows: 1 ± 0.6, 0.21 ± 0.01, and 0.34 ± 0.02 nM, respectively. These data show a clear improvement with respect to the control peptide, while cyclization was apparently of less importance. In hindsight, the absence of the "cyclization effect" might be explained by an observation made by Derksen et al., in which they found that the non-covalent hydrophobic interactions of the allyl side chains afford the correct bioactive conformation of the peptide comparable to the cyclic peptide [97].

As the fifth example [179], RCM was applied to constrain and simplify caspofungin (**141**) as an antifungal peptide that belongs to the class of echinocandins that act by inhibiting the β-(1,3)-D-glucan synthase complex [180].

The total synthesis of caspofungin is quite challenging due to the many hydroxylated unnatural amino acids, like hydroxyproline, hydroxyornithine, *bis*-hydroxyhomotyrosine, and the presence of an ornithine hemiaminal. Mulder et al. [179] designed caspofungin mimic **142** in which the alkene moiety formed an alternative constraint for the hemiaminal, since the latter is known to be unstable at pH > 7. Precursor peptide **143** was synthesized on the solid support following Fmoc/^{t}Bu SPPS protocols. In this peptide, the C-terminal ornithine derivative was replaced by an allylglycine residue, while the N-terminus was extended by one allylglycine (Scheme 32). Ring-closing metathesis of **143** was performed according to the procedure of Robinson et al. [99], and for this purpose, resin-bound precursor peptide **143** was treated with **G2** (15 mol%) in CH_2Cl_2 containing 10 vol % 0.4 M

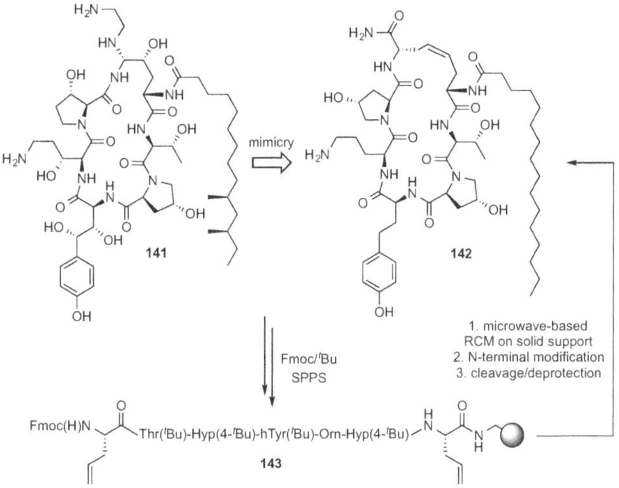

Scheme 32 Caspofungin (**141**) and its mimic **142** synthesized by RCM [179]

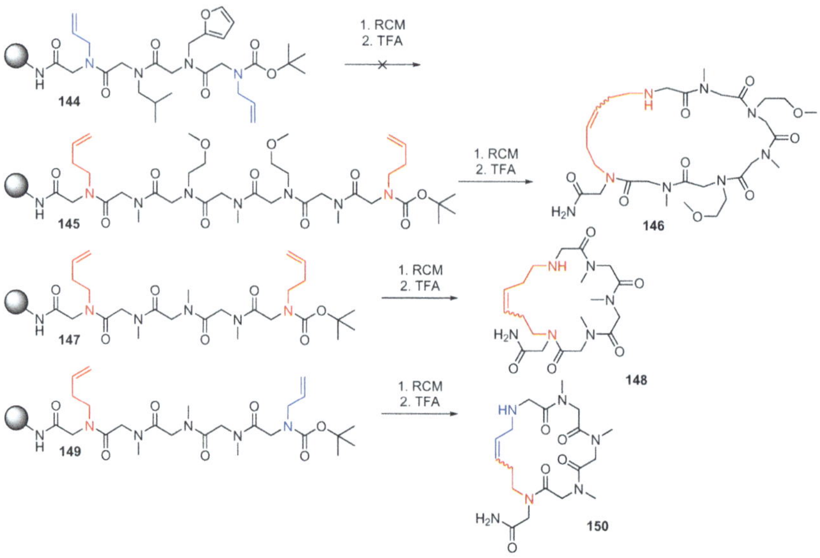

Scheme 33 RCM for the solid-phase synthesis of cyclic peptoids [181]

LiCl in DMA under microwave irradiation at 100°C for 75 min. Finally, macrocycle **142** was isolated in 11% overall yield as a mixture of *cis*/*trans* isomers. RCM reactions that were run without microwave irradiation were not successful, since treatment of the linear precursor peptide **143** with **G2** (20 mol%) in CH_2Cl_2 containing 10 vol % 0.4 M LiCl in DMA at 50°C for 24 h resulted in only trace amounts of cyclic peptide **142**. Unfortunately, these alkene-/alkane-bridged caspofungin mimics did not show any antifungal activity, up to 100 μg/mL (96 μM), mainly due to the increased ring size by two atoms to a 23-membered macrocycle.

The sixth example describes a solid-phase approach to synthesize cyclic peptoids via ring-closing metathesis to develop a tool for the easy access of cyclic peptoid libraries and other cyclic peptide-peptoid chimeras, as shown in Scheme 33 [181].

Peptoid tetramer **144** (containing allylamines) was subjected to RCM in the presence of **G1**, **G2**, or **HG2** (2 mol%) under various reaction conditions in common metathesis solvents as dichloromethane, 1,2-dichloroethane, and 1,2-dichlorobenzene. Ring-closing metathesis failed to give the cyclic 14-membered peptoid. However, the hexamer as well as the octamer underwent macrocyclization (to give a 20- and 23-membered macrocycle, respectively) only in the presence of **HG2** under microwave irradiation (300 W, 2 min) in 1,2-dichlorobenzene albeit in very low yield (10–20%). Since it is known that RCM efficiency strongly depends on the length of the olefinic amine [29, 182], heptamer **145** was synthesized containing 3-buten-1-amine moieties. This heptamer reacted with **G1**, **G2**, and **HG2** (2 mol%) under microwave irradiation (300 W, 2 min) in 1,2-dichlorobenzene, and after cleavage from the resin, the 25-membered macrocycle **146** was isolated in 10, 20, and 80% yield, respectively. Thus, **HG2** was selected in the macrocyclization of peptoid pentamer **147**, and under similar reaction conditions, 19-membered cyclic peptoid **148** was obtained in 70 to 85% yield. Based on MALDI-TOF analysis, dimerization products and linear peptoids were the major side products. To suppress these side reactions, **147** was treated with **HG2** in refluxing CH_2Cl_2 for 2 h, which resulted in a very clean RCM reaction, since **148** was obtained in 95% yield with <5% dimers present. In addition to this, pentamer **149** was synthesized with a combination of 3-buten-1-amine at the C-terminus and an allylamine at the N-terminus. RCM (**HG2** in refluxing CH_2Cl_2) resulted in the formation of 18-membered cyclic peptoid **150** in a yield of 60–70%. The reduced yield emphasized the importance of the correct length of the alkenyl moiety for being an optimal RCM substrate. Finally, a series of N-/C-terminally containing 3-buten-1-amino peptoids ranging from tetramers to heptamers with functionalized side chains like lysine and aspartic acid were synthesized and cyclized to give a whole range of 16- to 25-membered macrocyclic peptoids in yields ranging from 78 to 100% [181].

Finally, in the last example, six cyclic peptides (as shown in Fig. 7) will be discussed to emphasize the applicability of RCM to illustrate structural diversity within the peptide framework in relation to a wide variety in biological activities and biophysical properties.

Bicyclic pentapeptide **151** was designed as a CDE-ring mimic of the antibiotic glycopeptide vancomycin [183]. The central 4-hydroxyphenylglycine residue was found to be a suitable building block and was *bis*-allylated via a Stille coupling on the condition that the electron-donating hydroxyl functionality was converted into an electron-withdrawing moiety by acetylation. Both serine residues were introduced as their *O*-allylated congeners, and the linear pentapeptide with four allyl

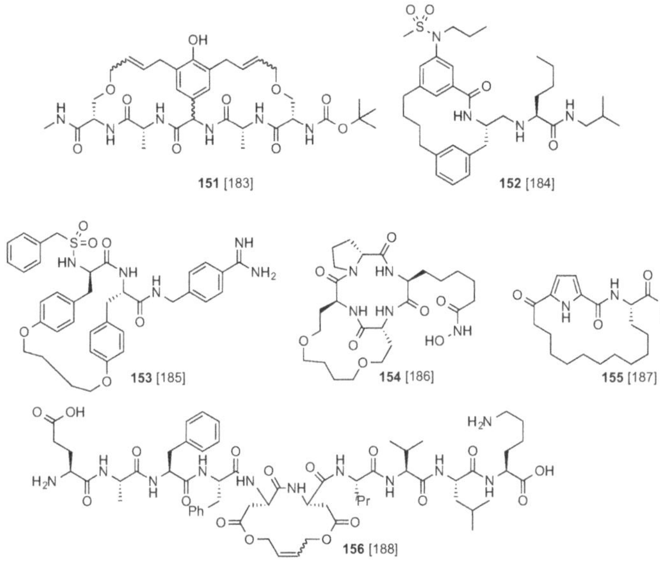

Fig. 7 Structural diversity of peptides in relation to a variety of biological activities

functionalities was treated with **G2** as the catalyst (100 mol%) to undergo tandem ring-closing metathesis in TCE/DMF 96:4 v/v (2.5 mM) at 80°C for 10 min, and bicycle **151** was obtained in 67% after column chromatography as a complex mixture of diastereomers [183].

Substituted isophthalamides emerged as early lead structures that were investigated for the inhibition of β-secretase, the enzyme responsible for releasing the neurotoxic peptide fragment $A\beta_{40-42}$ by processing the amyloid precursor protein. Targeting β-secretase is considered as an attractive therapeutic approach for the treatment and prevention of Alzheimer's disease. A macrocyclic active site-derived inhibitor of β-secretase, **152**, was designed and synthesized by covalently cross-linking the P1 and P3 side chains of the amino acid side chains that interact with the active site of the enzyme [184]. Allylation of the 3-aminobenzoic acid core was performed under Stille conditions, and ring-closing metathesis was carried out in the presence of **G2** (10 mol%) in CH_2Cl_2 (2.1 mM) at 50°C for 60 min resulting in the exclusive formation of the single (*E*)-regioisomer in high yield (84%). After hydrogenation of the double bond, macrocycle **152** displayed an IC_{50}-value of 4 nM against β-secretase with a 25% reduction of $A\beta_{40-42}$ levels in brain extracts from an in vivo mouse model [184].

A series of 20-membered macrocycles, like amidine **153**, was synthesized by Saupe and Steinmetzer to develop highly potent and selective plasmin inhibitors [185]. Herein, *O*-allyl protected tyrosine residues were ring-closed via **G1** (5 mol%) in refluxing CH_2Cl_2 for 6 h in a reasonable yield (65%). Cyclic amidine **153** was a potent plasmin inhibitor ($K_i = 1.1$ nM) with some affinity retained against plasma kallikrein ($K_i < 20$ nM), while proteases like thrombin, factor Xa, and activated protein C remained unaffected with K_i values of 510, 784, and 4,586 nM, respectively. This selectivity was explained by the so-called 99-loop residue, which is absent in plasmin, and the sterically less demanding glycine residue in case of plasma kallikrein, while bulky and sterically demanding residues (leucine, threonine, tyrosine) are present in the other proteases [185].

Histone deacetylases (HDACs) are a class of enzymes involved in the regulation of chromatin remodeling and gene transcription by deacetylation of acetylated lysine residues in the N-terminal amino acid sequences of histones. Malfunctioning of these enzymes is associated with carcinogenesis, and the development of HDAC inhibitors is a promising approach for novel anticancer drug molecules. Isoform-specific inhibitors of HDACs are often associated with less toxic side effect. To address this need, Islam et al. designed and synthesized a series of bicyclic tetrapeptides as selective HDAC inhibitors [186]. They found that bicyclic peptide **154** is almost tenfold more selective against HDAC-1 than HDAC-6. In a typical example, bicyclic peptide **154** was synthesized in solution, and the fully protected linear tripeptide was first cyclized by RCM via the *O*-allyl serine residues, followed by head-to-tail macrocyclization of the tetrapeptide with HATU/DIPEA as coupling reagents. Ring-closing metathesis in the presence of **G1** (5 mol%) in CH_2Cl_2 (7.1 mM) at room temperature during 48 h resulted in a monocyclic peptide in 52% yield, while macrolactamization was somewhat more efficient, 65% yield [186].

Protease substrates and inhibitors adopt a β-strand conformation upon binding toward the active site. Abell and co-workers [187] described a series of molecules, like **155**, in which a planar pyrrole has been used to replace a dipeptide sequence within a peptide backbone to generate a macrocycle that adopts a β-strand conformation. Macrocyclization was performed via RCM in the presence of **G2** (10 mol %) in refluxing CH_2Cl_2 (2.5 mL/mg acyclic diene) for 18 h. The yield was determined after hydrogenation of the newly formed double bond and was in the range between 28 (18-membered ring, **155**) and 56% (24-membered ring). After RCM, the subsequent reaction steps to arrive at compound **155** were ester saponification, amino acid alcohol ((*S*)-leucinol) coupling, and oxidation to the aldehyde. Macrocyclic inhibitor **155** was assayed against ovine *m*-calpain, human cathepsin L and S (cysteine proteases), and bovine α-chymotrypsin and human leukocyte elastase (serine proteases). Inhibitor **155** was highly potent with picomolar K_i values against cathepsin L and S (0.44 and 0.92 nM, respectively) and also potent against *m*-calpain (58 nM), while it had almost no activity against α-chymotrypsin and human leukocyte elastase, since here K_i values >10,000 and >2,500 have been found, respectively. Based on an X-ray structure, it was found that this class of inhibitors exactly adopts a β-strand conformation in which the pyrrole clearly defines the required geometry for optimal binding into the active site [187].

Amyloid peptides tend to self-assemble and form well-defined supramolecular assemblies, like sheets, fibrils, vesicles, and ribbons. Such assemblies can be used as bionanomaterials as drug-carrier and controlled-release vehicles for the targeted delivery of small drug molecules or peptide therapeutics. Hamley et al. report an approach in which they use olefin metathesis to design macrocyclic amyloid peptides that self-assemble into nanotubes that comprise wrapped β-sheets [188]. They synthesized a resin-bound protected decapeptide (Fmoc-Glu(O^{t}Bu)-Ala-Phe-Phe-Asp(OAll)-Asp(OAll)-Val-Val-Leu-Lys(Boc)-O-resin) in which the vicinal aspartic acid moieties were protected as allyl esters and used for RCM in the presence of **HG2** (11 mol%) in DCE (10 mL) at 60°C for 4 days. Macrocyclization resulted in two peptides, namely, **156** (intramolecularly cyclized peptide) and its corresponding dimer (intermolecularly cyclized peptide), while also peptide dimers were formed that underwent cross metathesis. In aqueous solution (0.5 wt%), acyclic peptide H-Glu-Ala-Phe-Phe-Asp(OAll)-Asp(OAll)-Val-Val-Leu-Lys-OH self-assembles into peptide nanotubes with a radius of 25 ± 4 nm, with tube walls that comprise 10% of the tube diameter. It turned out that the molecular organization is based on β-sheet packing, since a β-strand spacing of 0.47 nm and a β-sheet stacking distance of 1.06 nm were observed. Macrocyclic peptide **156** did not self-assemble in aqueous solution at 0.5 wt%; however, a higher concentration was required (2 wt%) at which β-sheet formation, identified as macroscopic fibrils, was observed. Peptide nanotubes, however, were not formed in contrast to the linear peptide. Apparently, the conformational constraints induced by ring-closing metathesis hinder nanotube formation; instead, fibrillar nanostructures as a different supramolecular morphology were preferentially formed. This study is a nice example which shows that ring-closing metathesis can be used as a "switch" to control the self-assembly of peptides and thereby to tune the supramolecular morphology of the aggregates [188].

2.10 Z-Selective RCM of Peptides

Olefin metathesis revolutionized organic synthesis by providing a powerful tool to introduce the carbon-carbon double bond in many biologically active molecules. Generally, metathesis reactions result in the predominant formation of *E*-alkenes, while the synthesis of the thermodynamically less stable *Z*- or *cis*-isomer is quite challenging. Approaches toward enantioselective metathesis were pioneered by Schrock and Hoveyda, and they reported in 2009 a catalyst-controlled approach for the synthesis of *Z*-olefins [189]. These "stereogenic-at-Mo" complexes were highly efficient in Z-selective cross metathesis as well as ring-closing metathesis for the stereoselective synthesis of antitumor agent KRN7000 [190] and the macrocyclic natural products epothilone C and nakadomarin A [191], respectively. Shortly thereafter, chiral *N*-heterocyclic carbene ruthenium complexes were reported as being *Z*-selective olefin metathesis catalysts [192–196]. Recently, the advances in the catalytic stereoselective olefin metathesis have been extensively reviewed by Hoveyda, and this overview is recommended to the reader for more information about this topic [197].

In a recent contribution, Grubbs and co-workers addressed *Z*-selective olefin metathesis on peptides [198]. Through the combined efforts of homodimerization, cross metathesis, and ring-closing metathesis, the authors have formulated guidelines to assess the influence of amino acids and peptides on catalyst activity and selectivity, and these principles were applied to carry out Z-selective metathesis on challenging substrates including peptides that comprise parallel β-sheets and on the stapling of α-helical peptides.

In first instance, the *i*, *i*+3 *Z*-selective RCM on Aib-containing peptides was studied by means of Boc-Ser(All)-Aib-Aib-Ser(All)-Aib-OMe (**157**) in which Ser (All) represents L-serine *O*-allyl residues. This peptide underwent *E*-selective RCM in the presence of **G2** (86% conversion; >20:1 *E*/*Z*, based on HPLC) and **HG2** (85% conversion; >20:1 *E*/*Z*) in DCE (5 mM) at 45°C for 3.5 h with 10 mol% catalyst loading [198] in accordance with earlier results to constrain the 3_{10}-helix in Aib-rich peptides [199]. Under similar conditions, the *Z*-selective ruthenium catalyst bearing *N*-adamantyl substituents and bidentate nitrato ligands **158** (Scheme 34) [195] did not result in macrocyclization even with a threefold increase of catalyst loading and extended reaction time. This result indicated that Aib residues at *i*+1 and *i*+2 control the *E*-selective macrocyclization by **G2** and **HG2**, while this conformational bias cannot be overcome by *Z*-selective catalysts **158** or **159** [198].

Z-Selective ring-closing metathesis on *i*, *i*+4 stapled peptides was explored by resin-bound peptide **160**, an α-helical peptide known to target the BCL-2 protein family in the regulation of apoptosis [54], Ac-Glu(O^{t}Bu)-Asp(O^{t}Bu)-Ile-Ile-Arg (Pbf)-Ile-S5*-Arg(Pbf)-Leu-Leu-S5*-Glu(O^{t}Bu)-Val-Gly-Asp(O^{t}Bu)-X-resin, in which S5* denotes (*S*)-2-(4-pentenyl)alanine. RCM of **160** in DCE (at 0.05 mM) to promote α-helicity in the presence of **158** or **159** (10 mol%) at 40°C for 2 h on Wang resin (X=O) resulted in a low conversion (~20–25%), while on MBHA resin (X=NH), the conversion was increased to 55–60% with greater than 85% *Z*-selectivity.

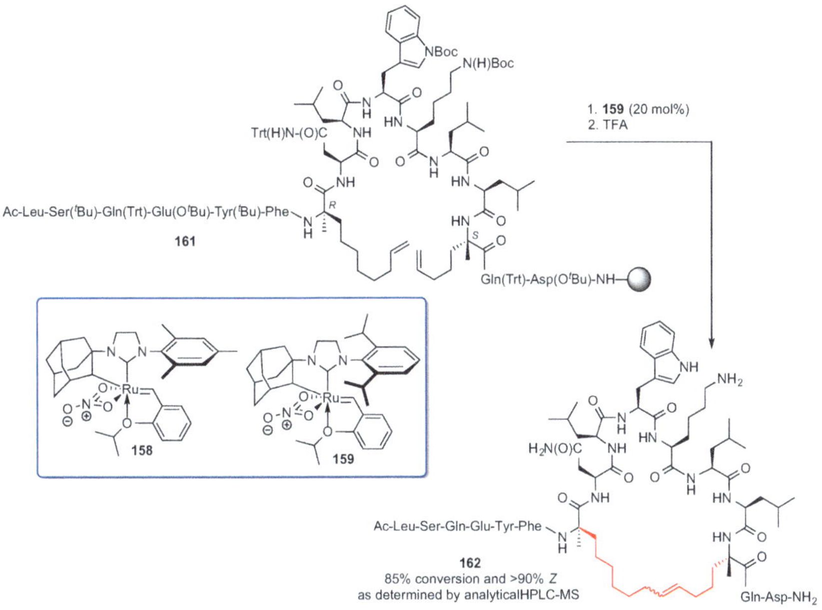

Scheme 34 *Z*-Selective ring-closing metathesis to obtain *i*, *i* + 7 stapled peptides [198]

Prolonging reaction time to four hours and two cycles of catalyst addition afforded the macrocyclic peptide in 80% yield with more than 90% *Z*-selectivity. To investigate the generality of their method, Grubbs and co-workers used resin-bound peptide **161** with two olefinic amino acids, R8* ((*R*)-2-(7-octenyl)alanine) on position *i* and S5* on position *i* + 7 which spans two turns of an α-helix, as shown in Scheme 34.

The incorporation of the combination R8* on position *i* and S5* on position *i* + 7 was previously found to facilitate RCM [50–53]. Under the optimized reaction conditions, solid-phase-bound peptide **161** in the presence of Ru-catalyst **159** (20 mol%) was transferred (after TFA treatment) into macrocyclic peptide **162** in high conversion (85%) and an excellent *Z*-selectivity greater than 90%. Thus, these results clearly showed that Ru catalysts can promote on-resin *Z*-selective ring-closing metathesis for the synthesis of stapled peptides [198].

3 Conclusion and Outlook

Due to the development of robust ruthenium-based metathesis catalysts that allow a wide range of polar groups within the olefinic substrates and that are compatible with diverse polar solvents and higher temperatures, ring-closing metathesis and cross metathesis became undisputable tools for the synthesis of peptides and peptidomimetics. Shortly after the seminal contributions of Grubbs and Ghadiri, who applied ring-closing metathesis for constraining dipeptides and cross metathesis to covalently cross-link self-assembled dimers of cyclic octapeptides, respectively, a plethora of metathesis reactions applied to peptide chemistry found their way in the scientific literature. The first applications of RCM in peptide chemistry dealt with alternatives for macrolactamization and disulfide bridge formation. Soon thereafter, RCM was applied to mimic and stabilize secondary structures in bioactive peptides, which culminated in the concept of peptide stapling as coined by Verdine. α-Helix stabilization and α-helix-inducing scaffolds became generally accessible, while pharmaceutically active peptides with improved conformational, chemical, and metabolic stability were realized by incorporating metathesis in their synthesis. Currently, *Z*-selective metathesis catalysts are in development, and the reported results are very promising. As discussed here, the bioactivity of a macrocyclic peptide may strongly depend on the geometry of the double bound. Tuning the *cis/trans* ratio of the newly double bond at will [200] will improve the efficiency of the synthesis as well as the purification of the desired peptides. In this context, ring-closing alkyne metathesis on peptides should be revitalized as an alternative approach to address this topic. Although many RCM syntheses with peptides have been described that did not require any optimization, catalyst screening, microwave irradiation, structure-inducing auxiliaries, chaotropic solvents, allyl sulfides as activated olefins, among others, have been used as a tool during optimization cycles for the efficient synthesis of peptide macrocycles. It is expected that ring-closing metathesis in aqueous solvents will usher in a new area of peptide macrocyclization, since unprotected peptides generally have higher solubility in water than (semi)protected peptides, while water-soluble ruthenium-based metathesis catalysts have been reported in the literature. To conclude this review, I will use the words of Hoveyda and Zhugralin:

> To consider catalytic olefin metathesis, or chemical synthesis, a consummated field would be akin to suggesting to Henry Ford that his model T was the be-all and end-all as far as automobiles are concerned [2].

Acknowledgments I would like to thank Professor H. G. Schmalz (Universität zu Köln, Germany) for providing me with a high-resolution picture of the model of an *N*-capped α-helical peptide as shown in Scheme 10. I gratefully acknowledge Professor Rob M. J. Liskamp (Utrecht University, the Netherlands, and University of Glasgow, United Kingdom) for being a mentor and teacher and for the stimulating discussions during the years of our fruitful collaboration.

References

1. Grubbs RH (2006) Olefin-metathesis catalysts for the preparation of molecules and materials (Nobel lecture). Angew Chem Int Ed 45:3760–3765
2. Hoveyda AH, Zhugralin AR (2007) The remarkable metal-catalyzed olefin metathesis reaction. Nature 450:243–251
3. Lozano-Vila AM, Monsaert S, Bajek A, Verpoort F (2010) Ruthenium-based olefin metathesis catalysts derived from alkynes. Chem Rev 110:4865–4909
4. Brik A (2008) Metathesis in peptides and peptidomimetics. Adv Synth Catal 350:1661–1675

5. Jacobsen Ø, Klaveness J, Rongved P (2010) Structural and pharmacological effects of ring-closing metathesis in peptides. Molecules 15:6638–6677
6. Pérez de Vega MJ, García-Aranda MI, González-Muñiz R (2011) A role for ring-closing metathesis in medicinal chemistry: mimicking secondary architectures in bioactive peptides. Med Res Rev 31:677–715
7. Cossy J, Arseniyades S, Meyer C (eds) (2010) Metathesis in natural product synthesis: strategies, substrates and catalysts. Wiley-VCH Verlag GmbH & Co. KGaA, Weinheim
8. Grela K (ed) (2014) Olefin metathesis: theory and practice. Wiley, Hoboken
9. Nguyen ST, Johnson LK, Grubbs RH, Ziller JW (1992) Ring-opening metathesis polymerization (ROMP) of norbornene by a group VIII carbine complex in protic media. J Am Chem Soc 114:3974–3975
10. Nguyen ST, Grubbs RH, Ziller JW (1993) Synthesis and activities of new single-component, ruthenium-based olefin metathesis catalysts. J Am Chem Soc 115:9858–9859
11. Schwab P, Grubbs RH, Ziller JW (1996) Synthesis and applications of $RuCl_2(=CHR')(PR_3)_2$: the influence of the alkylidene moiety on metathesis activity. J Am Chem Soc 118:100–110
12. Scholl M, Ding S, Lee CW, Grubbs RH (1999) Synthesis and activity of a new generation of ruthenium-based olefin metathesis catalysts coordinated with 1,3-dimesityl-4,5-dihydroimidazol-2-ylidene ligand. Org Lett 1:953–956
13. Garber SB, Kingsbury JS, Gray BL, Hoveyda AH (2000) Efficient and recyclable monomeric and dendritic Ru-based metathesis catalysts. J Am Chem Soc 122:8168–8179; Addition: (2001) J Am Chem Soc 123:3186
14. Miller SJ, Grubbs RH (1995) Synthesis of conformationally restricted amino acids and peptides employing olefin metathesis. J Am Chem Soc 117:5855–5856
15. Miller SJ, Blackwell HE, Grubbs RH (1996) Application of ring-closing metathesis to the synthesis of rigidified amino acids and peptides. J Am Chem Soc 118:9606–9614
16. Clark TD, Ghadiri MR (1995) Supramolecular design by covalent capture. Design of a peptide cylinder via hydrogen-bond-promoted intermolecular olefin metathesis. J Am Chem Soc 117:12364–12365
17. Ghadiri MR, Granja JR, Milligan RA, McRee DE, Khazanovich N (1993) Self-assembling organic nanotubes based on a cyclic peptide architecture. Nature 366:324–327
18. Ghadiri MR, Kobayashi K, Granja JR, Chadha RK, McRee DE (1995) The structural and thermodynamic basis for the formation of self-assembled peptide nanotubes. Angew Chem Int Ed Engl 34:93–95
19. Kobayashi K, Granja JR, Ghadiri MR (1995) β-Sheet peptide architecture: measuring the relative stability of parallel vs. antiparallel β-sheets. Angew Chem Int Ed Engl 34:95–98
20. Tjen KCMF, Kinderman SS, Schoemaker HE, Hiemstra H, Rutjes FPJT (2000) A ring-closing metathesis-mediated route to novel enantiopure conformationally restricted cyclic amino acids. Chem Commun 2000:699–700
21. King FE, King TJ, Warwick AJ (1950) The chemistry of extractives from hardwoods. Part III. Baikiain, and amino-acid present in *Baikiaea plurijuga*. J Chem Soc 3590–3597
22. Tanaka H, Kuroda A, Marusawa H, Hatanaka H, Kino T, Goto T, Hashimoto M (1987) Structure of FK506: a novel immunosuppressant isolated from *Streptomyces*. J Am Chem Soc 109:5031–5033
23. Swindells DCN, White PS, Findlay JA (1978) The X-ray crystal structure of rapamycin, $C_{51}H_{79}NO_{13}$. Can J Chem 56:2491–2492
24. Ginesta X, Pericàs MA, Riera A (2002) Straightforward entry to the pipecolic acid nucleus. Enantioselective synthesis of baikiain. Tetrahedron Lett 43:779–782
25. Alegret C, Riera A (2008) Enantioselective synthesis of indolizidine alkaloid *trans*-209D. J Org Chem 73:8661–8664
26. Yang Q, Xiao W-J, Yu Z (2005) Lewis acid assisted ring-closing metathesis of chiral diallylamines: an efficient approach to enantiopure pyrrolidine derivatives. Org Lett 7:871–874

27. Wakchaure S, Einsiedel J, Waibel R, Gmeiner P (2012) Conformationally restricted peptide mimetics by ring-closing olefin metathesis. Synthesis 44:2682–2694
28. Reichwein JF, Liskamp RMJ (2000) Synthesis of cyclic dipeptides by ring-closing metathesis. Eur J Org Chem 2000:2335–2344
29. Reichwein JF, Versluis C, Liskamp RMJ (2000) Synthesis of cyclic peptides by ring-closing metathesis. J Org Chem 65:6187–6195
30. Reichwein JF (2000) Synthesis of cyclic *N*-alkylated peptides by ring-closing metathesis. Dissertation, Utrecht University
31. Reichwein JF, Wels B, Kruijtzer JAW, Versluis C, Liskamp RMJ (1999) Rolling loop scan: an approach featuring ring-closing metathesis for generating libraries of peptides with molecular shapes mimicking bioactive conformations or local folding of peptides and proteins. Angew Chem Int Ed 38:3684–3687
32. Freidinger RM, Schwenk Perlow D, Veber DF (1982) Protected lactam-bridged dipeptides for use as conformational constraints in peptides. J Org Chem 47:104–109
33. Piscopio AD, Miller JF, Kock K (1999) Ring closing metathesis in organic synthesis: evolution of a high speed, solid phase method for the preparation of β-turn mimetics. Tetrahedron 55:8189–8198
34. Hoffmann T, Waibel R, Gmeiner P (2003) A general approach to dehydro-Freidinger lactams: ex-chiral pool synthesis and spectroscopic evaluation as potential reverse turn inducers. J Org Chem 68:62–69
35. Hoffmann T, Gmeiner P (2002) Short and efficient synthesis of homo-Freidinger lactams: an olefin metathesis approach towards conformationally restricted β-amino acid analogues. Synlett 2002:1014–1016
36. Einsiedel J, Lanig H, Waibel R, Gmeiner P (2007) Molecular building kit of fused-proline-derived peptide mimetics allowing specific adjustment of the dihedral ψ angle. J Org Chem 72:9102–9113
37. Bittermann H, Böckler F, Einsiedel J, Gmeiner P (2006) A highly practical RCM approach towards a molecular building kit of spirocyclic reverse turn mimics. Chem Eur J 12:6315–6322
38. Beal LM, Liu B, Chu W, Moeller KD (2000) Anodic amide oxidation/olefin metathesis strategies: developing a unified approach to the synthesis of bicyclic lactam peptidomimetics. Tetrahedron 56:10113–10125
39. Grossmith CE, Senia F, Wagner J (1999) Synthesis of novel unsaturated bicyclic lactams by ring-closing metathesis. Synlett 1999:1660–1662
40. Zaminer J, Brockmann C, Huy P, Opitz R, Reuter C, Beyermann M, Freund C, Müller M, Oschkinat H, Kühne R, Schmalz H-G (2010) Addressing protein-protein interactions with small molecules: a Pro-Pro dipeptide mimic with a PPII helix conformation as a module for the synthesis of PRD-binding ligands. Angew Chem Int Ed 49:7111–7115
41. Hack V, Reuter C, Opitz R, Schmieder P, Beyermann M, Neudörfl J-M, Kühne R, Schmalz H-G (2013) Efficient α-helix induction in a linear peptide chain by *N*-capping with a bridged-tricyclic diproline analogue. Angew Chem Int Ed 52:9539–9543
42. Hoffmann T, Lanig H, Wabel R, Gmeiner P (2001) Rational molecular design and EPC synthesis of a type VI β-turn inducing peptide mimetic. Angew Chem Int Ed 40:3361–3364
43. Kaul R, Surprenant S, Lubell WD (2005) Systematic study of the synthesis of macrocyclic dipeptide β-turn mimics possessing 8-, 9-, and 10-membered rings by ring-closing metathesis. J Org Chem 70:3838–3844; Addition: (2005) J Org Chem 70:4901–4902
44. Banfi L, Basso A, Guanti G, Riva R (2003) Application of tandem Ugi reaction/ring-closing metathesis in multicomponent synthesis of unsaturated nine-membered lactams. Tetrahedron Lett 44:7655–7658
45. Dömling A, Ugi I (2000) Multicomponent reactions with isocyanides. Angew Chem Int Ed 39:3168–3210
46. Fink BE, Kym PR, Katzenellenbogen JA (1998) Design, synthesis, and conformational analysis of a proposed type I β-turn mimic. J Am Chem Soc 120:4334–4344

47. Pinsker A, Einsiedel J, Härterich S, Waibel R, Gmeiner P (2011) A highly efficient type I β-turn mimetic simulating an Asx-Pro-turn-like structure. Org Lett 13:3502–3505
48. Blackwell HE, Grubbs RH (1998) Highly efficient synthesis of covalently cross-linked peptide helices by ring-closing metathesis. Angew Chem Int Ed 37:3281–3284
49. Blackwell HE, Sadowsky JD, Howard RJ, Sampson JN, Chao JA, Steinmetz WE, O'Leary DJ, Grubbs RH (2001) Ring-closing metathesis of olefinic peptides: design, synthesis, and structural characterization of macrocyclic helical peptides. J Org Chem 66:5291–5302
50. Schafmeister CE, Po J, Verdine GL (2000) An all-hydrocarbon cross-linking system for enhancing the helicity and metabolic stability of peptides. J Am Chem Soc 122:5891–5892
51. Kim Y-W, Grossmann TN, Verdine GL (2011) Synthesis of all-hydrogen stapled α-helical peptides by ring-closing olefin metathesis. Nat Prot 6:761–711
52. Kim Y-W, Kutchukian PS, Verdine GL (2010) Introduction of all-hydrocarbon i, i+3 staples into α-helices via ring-closing olefin metathesis. Org Lett 12:3046–3049
53. Kutchukian PS, Yang JS, Verdine GL, Shakhnovich EI (2009) All-atom model for stabilization of α-helical structure in peptides by hydrocarbon staples. J Am Chem Soc 131:4622–4627
54. Walensky LD, Kung AL, Escher I, Malia TJ, Barbuto S, Wright RD, Wagner G, Verdine GL, Korsmeyer SJ (2004) Activation of apoptosis in vivo by a hydrocarbon-stapled BH3 helix. Science 305:1466–1470
55. Walensky LD, Pitter K, Morash J, Oh KJ, Barbuto S, Fisher J, Smith E, Verdine GL, Korsmeyer SJ (2006) A stapled BID BH3 helix directly binds and activates BAX. Mol Cell 24:199–210
56. Bernal F, Tyler AF, Korsmeyer SJ, Walensky LD, Verdine GL (2007) Reactivation of the p53 tumor suppressor pathway by a stapled p53 peptide. J Am Chem Soc 129:2456–2457
57. Gavathiotis E, Suzuki M, Davis ML, Pitter K, Bird GH, Katz SG, Tu H-C, Kim H, Cheng EH-Y, Tjandra N, Walensky LD (2008) BAX activation is initiated at a novel interaction site. Nature 455:1076–1082
58. Stewart ML, Fire E, Keating AE, Walensky LD (2010) The MCL-1 BH3 helix is an exclusive MCL-1 inhibitor and apoptosis sensitizer. Nat Chem Biol 6:595–601
59. Bird GH, Madani N, Perry AF, Princiotto AM, Supko JG, He X, Gavathiotis E, Sodroski JG, Walensky LD (2010) Hydrocarbon double-stapling remedies the proteolytic instability of a lengthy peptide therapeutic. Proc Natl Acad Sci USA 107:14093–14098
60. Baek S, Kutchukian PS, Verdine GL, Huber R, Holak TA, Lee KW, Popowicz GM (2012) Structure of the stapled p53 peptide bound to Mdm2. J Am Chem Soc 134:103–106
61. Walensky LD, Bird GH (2014) Hydrocarbon-stapled peptides: principles, practice, and progress. J Med Chem 57:6275–6288
62. Hilinski GJ, Kim Y-W, Hong J, Kutchukian PS, Crenshaw CM, Berkovitch SS, Chang A, Ham S, Verdine GL (2014) Stitched α-helical peptides *via* bis ring-closing metathesis. J Am Chem Soc 136:12314–12322
63. Chapman RN, Dimartino G, Arora PS (2004) A highly stable short α-helix constrained by a main-chain hydrogen-bond surrogate. J Am Chem Soc 126:12252–12253
64. Wang D, Chen K, Kulp JL III, Arora PS (2006) Evaluation of biologically relevant short α-helices stabilized by a main-chain hydrogen-bond surrogate. J Am Chem Soc 128:9248–9256
65. Patgiri A, Jochim AL, Arora PS (2008) A hydrogen bond surrogate approach for stabilization of short peptide sequences in α-helical conformation. Acc Chem Res 41:1289–1300
66. Liu J, Wang D, Zheng Q, Lu M, Arora PS (2008) Atomic structure of a short α-helix stabilized by a main chain hydrogen-bond surrogate. J Am Chem Soc 130:4334–4337
67. Dimartino G, Wang D, Chapman RN, Arora PS (2005) Solid-phase synthesis of hydrogen-bond surrogate-derived α-helices. Org Lett 7:2389–2392
68. van der Schaaf P, Kolly R, Kirner H-J, Rime F, Mühlebach A, Hafner A (2000) Synthesis and reactivity of novel ruthenium carbine catalysts. X-ray structures of [$RuCl_2(=CHSC_6H_5)(PiPr_3)_2$] and [$RuCl_2(CHCH_2CH_2$-*C*,*N*-2-$C_5H_4N)(P^iPr_3)$]. J Organomet Chem 606:65–74

69. Love JA, Morgan JP, Trnka TM, Grubbs RH (2002) A practical and highly active ruthenium-based catalyst that effect the cross metathesis of acrylonitrile. Angew Chem Int Ed 41:4035–4037
70. Venkatraman J, Shankaramma SC, Balaram P (2001) Design of folded peptides. Chem Rev 101:3131–3152
71. Eichinger A, Beisel H-G, Jacob U, Huber R, Medrano F-J, Banbula A, Potempa J, Travis J, Bode W (1999) Crystal structure of gingipain R: an Arg-specific bacterial cysteine proteinase with a caspase-like fold. EMBO J 18:5453–5462
72. Venkatachalapathi YV, Balaram P (1979) An incipient 3_{10} helix in Piv-Pro-Pro-Ala-NHMe as a model for peptide folding. Nature 281:83–84
73. Rai R, Aravinda S, Kanagarajadurai K, Raghothama S, Shamala N, Balaram P (2006) Diproline templates as folding nuclei in designed peptides. Conformational analysis of synthetic peptide helices containing amino terminal Pro-Pro segments. J Am Chem Soc 128:7916–7928
74. Kemp DS, Boyd JG, Muendel CC (1991) The helical *s* constant for alanine in water derived from template-nucleated helices. Nature 353:451–454
75. Kemp DS, Curran TP, Boyd JG, Allen TJ (1991) Studies of the N-terminal templates for α-helix formation. Synthesis and conformational analysis of peptide conjugates of (2*S*,5*S*,8*S*,11*S*)-1-acetyl-1,4-diaza-3-keto-5-carboxy-10-thiatricyclo[2.8.1.$0^{4,8}$] -tridecane (Ac-Hel_1-OH). J Org Chem 56:6683–6697
76. Payan DG (1989) Neuropeptides and inflammation; the role of substance P. Ann Rev Med 40:341–352
77. Ward P, Ewan GB, Jordan CC, Ireland SJ, Hagan RM, Brown JR (1990) Potent and highly selective neurokinin antagonists. J Med Chem 33:1848–1851
78. Jarvo ER, Copeland GT, Papaioannou N, Bonitatebus PJ Jr, Miller SJ (1999) A Biomimetic approach to asymmetric acyl transfer catalysis. J Am Chem Soc 121:11638–11643
79. Schmiedeberg N, Kessler H (2002) Reversible backbone protection enables combinatorial solid-phase ring-closing metathesis reaction (RCM) in peptides. Org Lett 4:59–62
80. Creighton CJ, Reitz AB (2001) Synthesis of an eight-membered cyclic pseudo-dipeptide using ring closing metathesis. Org Lett 3:893–895
81. Goldring WPD, Hodder AS, Weiler L (1998) Synthesis of macrocyclic lactams and lactones via ring-closing olefin metathesis. Tetrahedron Lett 39:4955–4958
82. Wöhr T, Wahl F, Nefzi A, Rohwedder B, Sato T, Sun X, Mutter M (1996) Pseudo-prolines as a solubilizing structure-disrupting protection technique in peptide synthesis. J Am Chem Soc 118:9218–9227
83. Williams RM, Liu J (1998) Asymmetric synthesis of differentially protected 2,7-diaminosuberic acid, a ring-closure metathesis approach. J Org Chem 63:2130–2132
84. Gao Y, Lane-Bell P, Vederas JC (1998) Stereoselective synthesis of *meso*-2,6-diaminopimelic acid and its selectively protected derivatives. J Org Chem 63:2133–2143
85. Elaridi J, Patel J, Jackson WR, Robinson AJ (2006) Controlled synthesis of (S, S)-2,7-diaminosuberic acid: a method for regioselective construction of dicarba analogues of multicystine-containing peptides. J Org Chem 71:7538–7545
86. Stymiest JL, Mitchell BF, Wong S, Vederas JC (2003) Synthesis of biologically active dicarba analogues of the peptide hormone oxytocin using ring-closing metathesis. Org Lett 5:47–49
87. Stymiest JL, Mitchell BF, Wong S, Vederas JC (2005) Synthesis of oxytocin analogues with replacement of sulfur by carbon gives potent antagonists with increased stability. J Org Chem 70:7799–7809
88. Jiang S, Li P, Lee S-L, Lin CY, Long Y-Q, Johnson MD, Dickson RB, Roller PP (2007) Design and synthesis of redox stable analogues of sunflower trypsin inhibitors (SFTI-1) on solid support, potent inhibitors of matriptase. Org Let 9:9–12
89. Berezowska I, Chung NN, Lemieux C, Wilkes BC, Schiller PW (2006) Cyclic dermorphin tetrapeptide analogues obtained *via* ring-closing metathesis. Acta Biochim Pol 53:73–76

90. Berezowska I, Chung NN, Lemieux C, Wilkes BC, Schiller PW (2007) Dicarba analogues of the cyclic encephalin peptides H-Tyr-*c*[D-Cys-Gly-Phe-D(or L)-Cys]NH_2 retain high opioid activity. J Med Chem 50:1414–1417
91. Mollica A, Guardiani G, Davis P, Ma S-W, Porreca F, Lai J, Mannina L, Sobolev AP, Hruby VJ (2007) Synthesis of stable and potent δ/μ opioid peptides: analogues of H-Tyr-*c* [D-Cys-Gly-Phe-D-Cys]OH by ring-closing metathesis. J Med Chem 50:3138–3142
92. Whelan AN, Elaridi J, Mulder RJ, Robinson AJ, Jackson WR (2005) Metal-catalysed tandem metathesis-hydrogenation reactions for the synthesis of carba analogues of cyclic peptides. Can J Chem 83:875–881
93. D'Addona D, Carotenuto A, Novellino E, Piccand V, Reubi JC, Di Cianni A, Gori F, Papini AM, Ginanneschi M (2008) Novel sst_5-selective somatostatin dicarba-analogues: synthesis and conformation-affinity relationships. J Med Chem 51:512–520
94. Di Cianni A, Carotenuto A, Brancaccio D, Novellino E, Reubi JC, Beetschen K, Papini AM, Ginanneschi M (2010) Novel octreotide dicarba-analogues with high affinity and different selectivity for somatostatin receptors. J Med Chem 53:6188–6197
95. Martín-Gago P, Ramón R, Aragón E, Fernández-Carneado J, Martin-Malpartida M, Verdaguer X, López-Ruiz P, Colás B, Cortes MA, Ponsati B, Macias MJ, Riera A (2014) A tetrapeptide somatostatin dicarba-analog: synthesis, structural impact and biological activity. Bioorg Med Chem Lett 24:103–107
96. Heapy AM, Williams GM, Fraser JD, Brimble MA (2012) Synthesis of a dicarba analogues of human β-defensin-1 using a combined ring closing metathesis-native chemical ligation strategy. Org Lett 14:878–881
97. Derksen DJ, Stymiest JL, Vederas JC (2006) Antimicrobial leucocin analogues with a disulfide bridge replaced by a carbocycle or by noncovalent interactions of allyl glycine residues. J Am Chem Soc 128:14252–14253
98. Robinson AJ, van Lierop BJ, Garland RD, Teoh E, Elaridi J, Illesinghe JP, Jackson WR (2009) Regioselective formation of interlocked dicarba bridges in naturally occurring cyclic peptide toxins using olefin metathesis. Chem Commun 2009:4293–4295
99. Robinson AJ, Elaridi J, van Lierop BJ, Mujcinovic S, Jackson WR (2007) Microwave-assisted RCM for the synthesis of carbocyclic peptides. J Pept Sci 13:280–285
100. van Lierop BJ, Whelan AN, Andrikopoulos S, Mulder RJ, Jackson WR, Robinson AJ (2010) Methods for enhancing ring closing metathesis yield in peptides: synthesis of a dicarba human growth hormone fragment. Int J Pept Res Ther 16:133–144
101. van Kraaij C, de Vos WM, Siezen RJ, Kuipers OP (1999) Lantibiotics: biosynthesis, mode of action and applications. Nat Prod Rep 16:575–587
102. Chatterjee C, Paul M, Xie L, van der Donk WA (2005) Biosynthesis and mode of action of lantibiotics. Chem Rev 105:633–683
103. Nagao J-i, Asaduzzaman SM, Aso Y, Okuda K-i, Nakayama J, Sonomoto K (2006) Lantibiotics: insight and foresight for new paradigm. J Biosci Bioeng 102:139–149
104. Willey JM, van der Donk WA (2007) Lantibiotics: peptides of diverse structure and function. Annu Rev Microbiol 61:477–501
105. Ross AC, Vederas JC (2011) Fundamental functionality: recent developments in understanding the structure-activity relationships of lantibiotic peptides. J Antibiot 64:27–34
106. Knerr PJ, van der Donk WA (2012) Discovery, biosynthesis, and engineering of lantipeptides. Annu Rev Biochem 81:479–505
107. Liu W, Hansen JN (1990) Some chemical and physical properties of nisin, a small-protein antibiotic produced by *Lactococcus lactis*. Appl Environ Microbiol 56:2551–2558
108. Martin NI, Sprules T, Carpenter MR, Cotter PD, Hill C, Ross RP, Vederas JC (2004) Structural characterization of lacticin 3147, a two-peptide lantibiotic with synergistic activity. Biochemistry 43:3049–3056
109. Wilson-Stanford S, Kalli A, Hakansson K, Kastrantas J, Orugunty RS, Smith L (2009) Oxidation of lanthionines renders the lantibiotic nisin inactive. Appl Environ Microbiol 75:1381–1387

110. Ghalit N, Rijkers DTS, Kemmink J, Versluis C, Liskamp RMJ (2005) Pre-organization induced synthesis of a crossed alkene-bridged nisin Z DE-ring mimic by ring-closing metathesis. Chem Commun 2005:192–195
111. Ghalit N, Kemmink J, Hilbers HW, Versluis C, Rijkers DTS, Liskamp RMJ (2007) Step-wise and pre-organization induced synthesis of a crossed alkene-bridged nisin Z DE-ring mimic by ring-closing metathesis. Org Biomol Chem 5:924–934
112. Slootweg JC, Kemmink J, Liskamp RMJ, Rijkers DTS (2013) Synthesis and structural characterization of the individual diastereoisomers of a crossed-stapled alkene-bridged nisin DE-ring mimic. Org Biomol Chem 11:7486–7496
113. Ghalit N, Poot AJ, Fürstner A, Rijkers DTS, Liskamp RMJ (2005) Ring-closing alkyne metathesis approach toward the synthesis of alkyne mimic of thioether A-, B-, C-, and DE-ring systems of the lantibiotic nisin Z. Org Lett 7:2961–2964
114. Ghalit N, Reichwein JF, Hilbers HW, Breukink E, Rijkers DTS, Liskamp RMJ (2007) Synthesis of bicyclic alkene-/alkane-bridged nisin mimics by ring-closing metathesis and their biochemical evaluation as lipid II binders: toward the design of potential novel antibiotics. ChemBioChem 8:1540–1554
115. Harmsen RAG, Ghalit N, Kemmink J, Breukink E, Liskamp RMJ, Rijkers DTS (2014) A conformationally constrained fused tricyclic nisin AB-ring system mimic toward and improved pyrophosphate binder of lipid II. Tetrahedron 70:7691–7699
116. Slootweg JC, Peters N, Quarles van Ufford HC, Breukink E, Liskamp RMJ, Rijkers DTS (2014) Semi-synthesis of biologically active nisin hybrids composed of the native lanthionine ABC-fragment and a cross-stapled synthetic DE-fragment. Bioorg Med Chem 22:5345–5353
117. Pattabiraman VR, Stymiest JL, Derksen DJ, Martin NI, Vederas JC (2007) Multiple on-resin olefin metathesis to form ring-expanded analogues of the lantibiotic peptide, lacticin 3147 A2. Org Lett 9:699–702
118. Fukase K, Kitazawa M, Sano A, Shimbo K, Fujita H, Horimoto S, Wakamiya T, Shiba T (1988) Total synthesis of peptide antibiotic nisin. Tetrahedron Lett 29:795–798
119. Mothia B, Appleyard AN, Wadman S, Tabor AB (2011) Synthesis of peptides containing overlapping lanthionine bridges on the solid phase: an analogue of rings D and E of the lantibiotic nisin. Org Lett 13:4216–4219
120. Tabor AB (2011) The challenge of the lantibiotics: synthetic approaches to thioether-bridged peptides. Org Biomol Chem 9:7606–7628
121. Burtscher D, Grela K (2009) Aqueous olefin metathesis. Angew Chem Int Ed 48:442–454
122. Tomasek J, Schatz J (2013) Olefin metathesis in aqueous media. Green Chem 15:2317–2338
123. Lin YA, Chalker JM, Davis BG (2009) Olefin metathesis for site-selective protein modification. ChemBioChem 10:959–969
124. Lynn DM, Mohr B, Grubbs RH, Henling LM, Day MW (2000) Water-soluble ruthenium alkylidenes: synthesis, characterization, and application to olefin metathesis in protic solvents. J Am Chem Soc 122:6601–6609
125. Hong SH, Grubbs RH (2006) Highly active water-soluble olefin metathesis catalyst. J Am Chem Soc 128:3508–3509
126. Jordan JP, Grubbs RH (2007) Small-molecule N-heterocyclic-carbene-containing olefin-metathesis catalysts for use in water. Angew Chem Int Ed 46:5152–5155
127. Lipshutz BH, Aguinaldo GT, Ghorai S, Voigtritter K (2008) Olefin cross-metathesis reactions at room temperature using the nonionic amphiphile "PTS": just add water. Org Lett 10:1325–1328
128. Lipshutz BH, Ghorai S, Leong WWY, Taft BR, Krogstad DV (2011) Manipulating micellar environments for enhancing transition metal-catalyzed cross-couplings in water at room temperature. J Am Chem Soc 76:5061–5073
129. Cochrane SA, Huang Z, Vederas JC (2013) Investigation of the ring-closing metathesis of peptides in water. Org Biomol Chem 11:630–639

130. Konno K, Picolo G, Gutierrez VP, Brigatte P, Zambelli VO, Camargo ACM, Cury Y (2008) Crotalphine, a novel potent analgesic peptide from the venom of the South American rattlesnake *Crotalus durissus terrificus*. Peptides 29:1293–1304
131. Turk B (2006) Targeting proteases: successes, failures and future prospects. Nat Rev Drug Discovery 5:785–799
132. Njoroge FG, Chen KX, Shih N-Y, Piwinski JJ (2008) Challenges in modern drug discovery: a case study of boceprevir, an HCV protease inhibitor for the treatment of hepatitis C virus infection. Acc Chem Res 41:50–59
133. Tsantrizos YS (2008) Peptidomimetic therapeutic agents targeting the protease enzyme of the human immunodeficiency virus and hepatitis C virus. Acc Chem Res 41:1252–1263
134. Venkatram S, Bogen SL, Arasappan A, Bennett F, Chen K, Jao E, Liu Y-T, Lovey R, Hendrata S, Huang Y, Pan W, Parekh T, Pinto P, Popov V, Pike R, Ruan S, Santhaman B, Vibulbhan B, Wu W, Yang W, Kong J, Liang X, Wong J, Liu R, Butkiewicz N, Chase R, Hart A, Agrawal S, Ingravallo P, Pichardo J, Kong R, Baroudy B, Malcolm B, Guo Z, Prongay A, Madison V, Broske L, Cui X, Chenh K-C, Hsieh Y, Brisson J-M, Preluskey D, Korfmacher W, White R, Bogdanowich-Knipp S, Pavlovsky A, Bradley P, Saksena AK, Ganguly A, Piwinski J, Girijavallabhan V, Njoroge GJ (2006) Discovery of (1*R*,5*S*)-*N*-[3-amino-1-(cyclobutylmethyl)-2,3-dioxopropyl-3-[2(*S*)-[[[(1,1-dimethyl-ethyl)amino]carbonyl]amino]-3,3-dimethyl-1-oxobutyl]-6,6-dimethyl-3-azabicyclo[3.1.0]hexan-2(*S*)-carboxamide (SCH 503034), a selective, potent, orally bioavailable hepatitis C virus NS3 protease inhibitor: a potential therapeutic agent for the treatment of hepatitis C infection. J Med Chem 49:6074–6086
135. Perni RB, Almquist SJ, Byrn RA, Chandorkar G, Chaturvedi PR, Courtney LF, Decker CJ, Dinehart K, Gates CA, Harbeson SL, Heiser A, Kalkeri G, Kolaczkowski E, Lin K, Luong Y-P, Rao BG, Taylor WP, Thomson JA, Tung RD, Wei Y, Kwong AD, Lin C (2006) Preclinical profile of VX-950, a potent, selective, and orally bioavailable inhibitor of hepatitis C virus NS3-4A serine protease. Antimicrob Agents Chemother 50:899–909
136. McCauley JA, McIntyre CJ, Rudd MT, Nguygen KT, Romano JJ, Butcher JW, Gilbert KF, Bush KJ, Holloway MK, Swestock J, Wan B-L, Carroll SS, DiMuzio JM, Graham DJ, Ludmerer SW, Maso S-S, Stahlhut MW, Fandozzi CM, Trainor N, Olsen DB, Vacca JP, Liverton NJ (2010) Discovery of vaniprevir (MK-7009), a macrocyclic hepatitis C virus NS3/4a protease inhibitor. J Med Chem 53:2443–2463
137. Song ZJ, Tellers DM, Journet M, Kuethe JT, Lieberman D, Humphrey G, Zhang F, Peng Z, Waters MS, Zewge D, Nolting A, Zhao D, Reamer RA, Dormer PG, Belyk KM, Davies IW, Devine PN, Tschaen DM (2011) Synthesis of vaniprevir (MK-7009): lactamization to prepare a 22-membered macrocycle. J Org Chem 76:7804–7815; Correction: (2011) J Org Chem 76:9553
138. Kong J, C-y C, Balsells-Padros J, Cao Y, Dunn RF, Dolman SJ, Janey J, Li H, Zacuto MJ (2012) Synthesis of the HCV protease inhibitor vaniprevir (MK-7009) using ring-closing metathesis strategy. J Org Chem 77:3820–3828
139. C-y C (2012) Application of ring-closing metathesis strategy to the synthesis of vaniprevir (MK-7009), a 20-membered macrocyclic HCV protease inhibitor. Top Organomet Chem 42:135–150
140. Boeda F, Bantreil X, Clavier H, Nolan SP (2008) Ruthenium-indenylidene complexes: scope in cross-metathesis transformations. Adv Synth Catal 350:2959–2966
141. Cai Q, Zhao Z-A, You S-L (2009) Asymmetric construction of polycyclic indoles through olefin cross-metathesis/intramolecular Friedel-Crafts alkylation under sequential catalysis. Angew Chem Int Ed 48:7428–7431
142. Hong SH, Day MW, Grubbs RH (2004) Decomposition of a key intermediate in ruthenium-catalyzed olefin metathesis reaction. J Am Chem Soc 126:7414–7415
143. Hong SH, Sanders DP, Lee CW, Grubbs RH (2005) Prevention of undesirable isomerization during olefin metathesis. J Am Chem Soc 127:17160–17161

144. Faucher A-M, Bailey MD, Beaulieu PL, Brochu C, Duceppe J-S, Ferland J-M, Ghiro E, Gorys V, Halmos T, Kawai SH, Poirier M, Simoneau B, Tsantrizos YS, Llinàs-Brunet M (2004) Synthesis of BILN 2061, an HCV NS3 protease inhibitor with proven antiviral effect in humans. Org Lett 6:2901–2904
145. Llinàs-Brunet M, Bailey MD, Bolger G, Brochu C, Faucher A-M, Ferland JM, Garneau M, Ghiro E, Gorys V, Grand-Maître C, Halmos T, Lapeyre-Paquette N, Liard F, Poirier M, Rhéaume M, Tsantrizos YS, Lamarre D (2004) Structure-activity study on a novel series of macrocyclic inhibitors of the hepatitis C virus NS3 protease leading to the discovery of BILN 2061. J Med Chem 47:1605–1608
146. Poirier M, Aubry N, Boucher C, Ferland J-M, LaPlante S, Tsantrizos YS (2005) RCM of tripeptide dienes containing a chiral vinylcyclopropane moiety: impact of different Ru-based catalysts on the stereochemical integrity of the macrocyclic products. J Org Chem 70:10765–10773
147. Kingsbury JS, Harrity JPA, Bonitatebus PJ Jr, Hoveyda AH (1999) A recyclable Ru-based metathesis catalyst. J Am Chem Soc 121:791–799
148. Tripathy NK, Georg GI (2004) RCM approach for the total synthesis of cryptophycin-24 (Arenastatin A). Tetrahedron Lett 45:5309–5311
149. Ferrié L, Reymond S, Capdevielle P, Cossy J (2006) Total synthesis of (−)-spongidepsin. Org Lett 8:3441–3443
150. Bouazza F, Renoux B, Bachmann C, Gesson J-P (2003) Total synthesis and conformational analysis of the antifungal agent (−)-PF1163B. Org Lett 5:4049–4052
151. Huang W, Ren R-G, Dong H-Q, Wei B-G, Lin G-Q (2013) Diverse synthesis of marine cyclic depsipeptide lagunamide A and its analogues. J Org Chem 78:10747–10762
152. Eggen M, Georg GI (2002) The cryptophycins: their synthesis and anticancer activity. Med Res Rev 22:85–101
153. Kobayashi M, Aoki S, Ohyabu N, Kurosu M, Wang W, Kitagawa I (1994) Arenastatin A, a potent cytotoxic depsipeptide from the Okinawan marine sponge *Dysidea arenaria*. Tetrahedron Lett 35:7969–7972
154. Barrow RA, Hemscheidt T, Liang J, Paik S, Moore RE, Tius MA (1995) Total synthesis of cryptophycins. Revision of the structures of cryptophycin A and C. J Am Chem Soc 117:2479–2490
155. Golakoti T, Ogino J, Heltzel CE, Husebo L, Jensen CM, Larsen LK, Patterson GML, Moore RE, Mooberry SL, Corbett TH, Valeriote FA (1995) Structure determination, conformational analysis, chemical stability studies, and antitumor evaluation of the cryptophycins. Isolation of 18 new analogs from *Nostoc* sp. strain GCV 224. J Am Chem Soc 117:12030–12049
156. Grassia A, Bruno I, Debitus C, Marzocco S, Pinto A, Gomez-Paloma L, Riccio R (2001) Spongidepsin, a new cytotoxic macrolide from *Spongia* sp. Tetrahedron 57:6257–6260
157. Chen J, Forsyth CJ (2004) Total synthesis and structural assignment of spongidepsin through a stereodivergent ring-closing-metathesis strategy. Angew Chem Int Ed 43:2148–2152
158. Ghosh AK, Xu X (2004) Assignment of absolute stereochemistry and total synthesis of (−)-spongidepsin. Org Lett 6:2055–2058
159. Chandrasekhar S, Yaragorla SR, Sreelakshmi L (2007) Stereoselective formal total synthesis of the cyclodepsipeptide (−)-spongidepsin. Tetrahedron Lett 48:7339–7342
160. Chandrasekhar S, Yaragorla SR, Sreelakshmi L, Raji Reddy C (2008) Formal total synthesis of (−)-spongidepsin. Tetrahedron 64:5174–5183
161. Zhu G, Negishi E-i (2007) Fully reagent-controlled asymmetric synthesis of (−)-spongidepsin via the Zr-catalyzed asymmetric carboalumination of alkenes (ZACA reaction). Org Lett 9:2771–2774
162. Zhu Y, Loudet A, Burgess K (2010) An asymmetric hydrogenation route to (−)-spongidepsin. Org Lett 12:4392–4395
163. Nose H, Seki A, Yaguchi T, Hosoya A, Sasaki T, Hoshiko S, Shomura T (2000) PF1163A and B, new antifungal antibiotics produced by *Penicillium* sp. I. Taxonomy of producing strain, fermentation, isolation and biological activities. J Antibiot 53:33–37

164. Sasaki T, Nose H, Hosoya A, Yoshida S, Kawaguchi M, Watanabe T, Usui T, Ohtsuka Y, Shomura T (2000) PF1163A and B, new antifungal antibiotics produced by *Penicillium* sp. II. Physico-chemical properties and structure elucidation. J Antibiot 53:38–44
165. Tatsuta K, Takano S, Ikeda Y, Nakano S, Miyazaki S (1999) The total synthesis and absolute structure of antifungal antibiotics (−)-PF1163A and B. J Antibiot 52:1146–1151
166. Krishna PR, Srinivas P (2012) Total synthesis of the antifungal antibiotic PF1163A. Tetrahedron Asymmetry 23:769–774
167. Kumar H, Reddy AS, Reddy BVS (2014) The stereoselective total synthesis of PF1163A. Tetrahedron Lett 55:1519–1522
168. Tripathi A, Puddick J, Prinsep MR, Rottmann M, Tan LT (2010) Lagunamides A and B: cytotoxic and antimalarial cyclodepsipeptides from the marine cyanobacterium *Lyngbya majuscula*. J Nat Prod 73:1810–1814
169. Tripathi A, Puddick J, Prinsep MR, Rottmann M, Chan KP, Chen DY-K, Tan LT (2011) Lagunamide C, a cytotoxic cyclodepsipeptide from the marine cyanobacterium *Lyngbya majuscula*. Phytochemistry 72:2369–2375
170. Dai L, Lei H, Wang Z, Liu Y, Xu Z, Ye T (2012) Total synthesis and stereochemical revision of lagunamide A. Chem Commun 48:8697–8699
171. Cohrt AE, Nielsen TE (2014) Solid-phase synthesis of peptide thioureas and thiazole-containing macrocycles through Ru-catalyzed ring-closing metathesis. ACS Comb Sci 16:71–77
172. Anderson H, Demaegdt H, Johnsson A, Vauquelin G, Lindeberg G, Hallberg M, Erdélyi M, Karlén A, Hallberg A (2011) Potent macrocyclic inhibitors of insulin-regulated aminopeptidase (IRAP) by olefin ring-closing metathesis. J Med Chem 54:3779–3792
173. Wright JW, Harding JW (2009) The brain angiotensin IV/AT_4 receptor system as a new target for the treatment of Alzheimer's disease. Drug Dev Res 70:472–480
174. Anderson H, Demaegdt H, Vauquelin G, Lindeberg G, Karlén A, Hallberg M, Erdélyi M, Hallberg A (2010) Disulfide cyclized tripeptide analogues of angiotensin IV as potent and selective inhibitors of insulin-regulated aminopeptidase (IRAP). J Med Chem 53:8059–8071
175. Maxwell DS, Sun D, Peng Z, Martin DV, Bhanu Prasad BA, Bornmann WG (2013) Synthesis of a macrocycle based on linked amino acid mimetics (LAAM). Tetrahedron Lett 54:5799–5801
176. Dechantsreiter MA, Planker E, Mathä B, Lohof E, Hölzemann G, Jonczyk A, Goodman SL, Kessler H (1999) N-Methylated cyclic RGD peptides as highly active and selective $\alpha_V\beta_3$ integrin antagonists. J Med Chem 42:3033–3040
177. Xiong J-P, Stehle T, Zhang R, Joachimiak A, Frech M, Goodman SL, Arnaout MA (2002) Crystal structure of the extracellular segment of integrin $\alpha_V\beta_3$ in complex with an Arg-Gly-Asp ligand. Science 296:151–155
178. Wels B, Kruijtzer JAW, Garner K, Nijenhuis WAJ, Gispen WH, Adan RAH, Liskamp RMJ (2005) Synthesis of a novel potent cyclic MC4-ligand by ring-closing metathesis. Bioorg Med Chem 13:4221–4227
179. Mulder MPC, Kruijtzer JAW, Breukink EJ, Kemmink J, Pieters RJ, Liskamp RMJ (2011) Synthesis and evaluation of novel macrocyclic antifungal peptides. Bioorg Med Chem 19:6505–6517
180. Deresinski SC, Stevens DA (2003) Caspofungin. Clin Infect Dis 36:1445–1457
181. Khan SN, Kim A, Grubbs RH, Kwon Y-U (2011) Ring-closing metathesis approaches for the solid phase synthesis of cyclic peptoids. Org Lett 13:1582–1585
182. Kuhn KM, Champagne TM, Hong SH, Wei W-H, Nickel A, Lee CW, Virgil SC, Grubbs RH, Pederson RL (2010) Low catalyst loading in olefin metathesis: synthesis of nitrogen heterocycles by ring-closing metathesis. Org Lett 12:984–987
183. ten Brink HT, Rijkers DTS, Kemmink J, Hilbers HW, Liskamp RMJ (2004) Ring-closing metathesis for the synthesis of side chain knotted pentapeptides inspired by vancomycin. Org Biomol Chem 2:2658–2663

184. Stachel SJ, Coburn CA, Sankaranarayanan S, Price EA, Pietrak BL, Huang Q, Lineberger J, Espeseth AS, Jin L, Ellis J, Holloway MK, Munshi S, Allison T, Hazuda D, Somin AJ, Graham SL, Vacca JP (2006) Macrocyclic inhibitors of β-secretase: functional activity in an animal model. J Med Chem 49:6147–6140
185. Saupe SM, Steinmetzer T (2012) A new strategy for the development of highly potent and selective plasmin inhibitors. J Med Chem 55:1171–1180
186. Islam MN, Islam MS, Hoque MA, Kato T, Nishino N, Ito A, Yoshida M (2014) Bicyclic tetrapeptides as potent HDAC inhibitors: effect of aliphatic loop position and hydrophobicity on inhibitory activity. Bioorg Med Chem 22:3862–3870
187. Chua KCH, Pietsch M, Zhang X, Hautmann S, Chan HY, Bruning JB, Gütschow M, Abell AD (2014) Macrocyclic protease inhibitors with reduced peptide character. Angew Chem Int Ed 53:7828–7831
188. Hamley IW, Cheng G, Castelletto V, Handschin S, Mezzenga R (2012) Fibrillisation of ring-closed amyloid peptides. Chem Commun 48:3757–3759
189. Jiang AJ, Zhao Y, Schrock RR, Hoveyda AH (2009) Highly *Z*-selective metathesis homocoupling of terminal olefins. J Am Chem Soc 131:16630–16631
190. Meek SJ, O'Brien RV, Llaveria J, Schrock RR, Hoveyda AH (2011) Catalytic *Z*-selective olefin cross-metathesis for natural product synthesis. Nature 471:461–466
191. Yu M, Wang C, Kyle AF, Jakubec P, Dixon DJ, Schrock RR, Hoveyda AH (2011) Synthesis of macrocyclic natural products by catalyst-controlled stereoselective ring-closing metathesis. Nature 479:88–93
192. Endo K, Grubbs RH (2011) Chelated ruthenium catalysts for *Z*-selective olefin metathesis. J Am Chem Soc 133:8525–8527
193. Keitz BK, Endo K, Herbert MB, Grubbs RH (2011) *Z*-Selective homodimerization of terminal olefins with a ruthenium metathesis catalyst. J Am Chem Soc 133:9686–9688
194. Keitz BK, Endo K, Patel PR, Herbert MB, Grubbs RH (2012) Improved ruthenium catalysts for *Z*-selective olefin metathesis. J Am Chem Soc 134:693–699
195. Rosebrugh LE, Herbert MB, Marx VM, Keitz BK, Grubbs RH (2013) Highly active ruthenium metathesis catalysts exhibiting unprecedented activity and *Z*-selectivity. J Am Chem Soc 135:1276–1279
196. Occhipinti G, Hansen FR, Törnroos KW, Jensen VR (2013) Simple and highly *Z*-selective ruthenium-based olefin metathesis catalyst. J Am Chem Soc 135:3331–3334
197. Hoveyda AH (2014) Evolution of catalytic stereoselective olefin metathesis: from ancillary transformation to purveyor of stereochemical identity. J Org Chem 79:4763–4792
198. Mangold SL, O'Leary DJ, Grubbs RH (2014) *Z*-Selective olefin metathesis on peptides: investigation of side-chain influence, preorganization, and guidelines in substrate selection. J Am Chem Soc 136:12469–12478
199. Boal AK, Guryanov I, Moretto A, Crisma M, Lanni EL, Toniolo C, Grubbs RH, O'Leary DJ (2007) Facile and *E*-selective intramolecular ring-closing metathesis reactions in 310-helical peptides: a 3D structural study. J Am Chem Soc 129:6986–6987
200. Fürstner A (2013) Teaching metathesis "simple" stereochemistry. Science 341:1229713

Top Heterocycl Chem (2017) 47: 245–318
DOI: 10.1007/7081_2015_151

Published online: 26 August 2015

Synthesis of Macrocycles Other than Peptides by Metathesis

Ana Gradillas and Javier Pérez-Castells

Abstract Metathesis is one of the most efficient methodologies for the synthesis of large heterocycles. The wide variety of densely functionalized natural structures containing macrocycles and the interest of other nonnatural macrocycles as potential drugs or useful structures for supramolecular chemistry has triggered the synthetic efforts en route to these structures. This chapter is an overview of successful metathesis reactions, both diene ring-closing metathesis (RCM) and ring-closing alkyne metathesis (RCAM) as the key steps in the construction of large heterocycles (12 or more members excluding peptides). The focus will be made on the different conditions that may favor the metathesis process including substrate design, catalyst nature, and reaction conditions.

Keywords Alkyne metathesis · Macrocycles · Metathesis · Natural product synthesis

Contents

A. Gradillas and J. Pérez-Castells (✉)
Departamento de Química y Bioquímica, Facultad de Farmacia, Universidad CEU San Pablo, Urbanización Montepríncipe, 28668 Boadilla del Monte, Madrid, Spain
e-mail: jpercas@ceu.es

The impact of metathesis in modern synthetic chemistry is evidenced by the huge number of publications and reviews that have appeared in just a few years. Heterocycles of all sizes containing many functionalities are efficiently built using these methodologies. Among them, macrocycles have received great attention as they have always constituted a great synthetic challenge. This group includes many natural products with biological interest. Classical synthetic efforts toward macrocycles are many times inefficient and unpractical. Most early strategies were focused on the formation of carbon–heteroatom bonds in an attempt to apply the same methodologies as for small- and medium-sized rings. Thus, macrolactonizations, macrolactamizations, and macroaldolizations were extensively used although they were generally low yielding and side reactions were frequent. The macrocyclization encounters several problems including the competition between the desired intramolecular reaction and intermolecular processes that lead to polymers. In addition functional groups involved in the macrocyclization reaction like hydroxy, amino, or carboxy groups have to be differentiated from other polar functional groups at the time of the ring closure. Construction of these large rings by ring-closing metathesis (RCM) emerged some 15 years ago as a completely new methodology. Its main advantages were the compatibility with many functional groups and its high efficiency for almost every ring size. This is due, in part, to the concomitant loss of a volatile alkene which makes this reaction highly atom economical and entropically driven. In addition, the double bond formed is easily transformed into other functionalities. As the main problem, the stereochemical outcome of the RCM reaction is not completely predictable and depends on the reaction conditions and the substrate structure. More recently, the development of efficient catalysts able to perform alkyne metathesis has triggered the use of ring-closing alkyne metathesis (RCAM), followed by partial hydrogenation as a way to circumvent disadvantages of RCM.

Several reviews have covered the specific field of macrocyclic formation which includes an impressive amount of total syntheses of natural macrocycles. Among the numerous references, we will mention those that cover specifically, or in a particular section, the diene, ene–diene, and alkyne metathesis in the synthesis of macrocycles appeared since 2004 [1–12]. In this chapter we will provide an overview of successful synthesis of large heterocycles where the key step in the formation of the ring is a metathesis reaction. Most of the selected cases will deal with natural product total syntheses due to the special interest of natural macrocycles and the impressive solutions applied to the complex synthesis of these densely functionalized molecules. In the introductory section we will review the different conditions that may favor the metathesis process especially the recent efforts reported on the achievement of *Z*/*E* stereoselectivity. The examples described afterward are discussed in two sections. The first one is dedicated to natural product synthesis and is arranged according to families of natural products in order to compare the reaction conditions used with compounds bearing structural similarities. We have summarized the information regarding the metathesis step without details on other steps of the total syntheses. The last section is devoted to nonnatural macrocycle synthesis with the metathesis reaction as the key step. These include novel synthetic bioactive compounds and molecules designed as possible new materials or in supramolecular chemistry such as cyclophanes and rotaxanes.

1 Introduction

Macrocycles, defined as compounds containing at least one ring with 12 or more covalently connected atoms, constitute attractive targets for total synthesis. On one side there are a great number of natural products, many times from marine origin or found in bacteria or fungi, that have selective and potent biological activities (around 3% of all natural products described to date). Indeed, among the great number of applications of metathesis, the field of natural products containing heterocyclic macrocycles with varying ring sizes and functionalities is especially interesting and useful. On the other side large heterocycles have found interesting applications as possible new drugs and in material and supramolecular science, and their synthesis has been many times addressed using metathesis as the key step.

From the historical point of view, this story begins in 1980 with the disclosures by Villemin and Tsuji concerning non-stereoselective macrocyclic RCM reactions promoted by W-based complexes [13, 14]. In the late 1990s the RCM-based epothilone syntheses [15–18] and the synthesis of the 18-membered α,-β-unsaturated macrolide aspicilin [19] were the first total syntheses of natural macrocyclic products that used RCM to perform the ring closure. Since then, RCM has become one of the most utilized methodologies in macrocycle synthesis, and both ruthenium carbenic Grubbs' complexes and molybdenum Schrock's catalysts have been used. Molybdenum species like **S** are highly reactive but sensitive to water and oxygen and not compatible with several functionalities; thus they were less used than the other catalysts in macrocyclization [20]. However, new derivatives of Mo and W complexes able to give (*Z*)-selective metatheses have been described recently and triggered their use (vide infra). Most used catalysts in RCM-based synthesis of macrocycles have been to date ruthenium second-generation complexes **G2** and **HG2**. Although **G1** is cheap and easy to handle, it is thermally unstable and exhibits low reactivity, generally failing with substituted olefins (for reviews on ruthenium-based catalysts, see [21, 22]). Some derivatives of **G1** like **M1** improve the kinetic initiation and were used generally in early contributions. The second-generation catalysts **G2** and **HG2** give better results with substituted olefins. **HG2** has been modified by substitution of the aromatic ring, giving rise to a new family of catalysts including the possibility of anchoring the complex to a solid support.

Scheme 1 Macrocyclization via metathesis polymerization

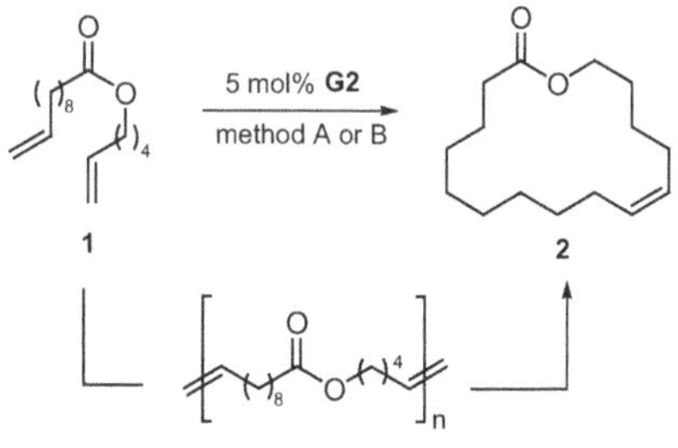

Not only is the choice of the catalyst crucial in these transformations but so are the solvent, temperature, concentration, and reaction time. In addition, subtle variations on the substrate structure lead to modifications in the conformational organization of the molecule which are critical for the success of the metathesis. In particular, the substitution pattern, the steric hindrance of the substituents in the substrate, the ring size to be formed, and the presence of coordinating heteroatoms have important influence on the results. One additional issue is the side reactions that may reduce the yield of the reaction. Mild conditions are highly desirable to avoid non-metathetic transformations both in the substrates and the products that may be catalyzed by decomposition species or by the catalysts itself. These include oxidations, hydrogenation of olefins, cycloaddition reactions, and double-bond isomerizations and migrations. Particularly important in macrocyclization is the competition with oligomerization. However, the reversible nature of the olefin metathesis makes oligomers common intermediates in RCM. In a recent example, an efficient macrocyclization strategy was reported and involved polymerization at high concentration of **1** followed by dilution and formation of macrocycle **2** (Scheme 1) [23].

The field of metathesis has reached such a mature state that scaled-up processes are now available, opening the use of these reactions at industrial level. Three main problems have to be overcome: the elimination of catalysts after the reaction, lowering of catalyst loading, and development of efficient reactions at high concentration as macrocycle ring closing generally needs high-dilution conditions to avoid side-oligomerization reactions.

Regarding the first challenge, solid-supported catalysts are a good alternative to avoid contamination of the products with metal traces which are especially important when synthesizing drug intermediates. Ruthenium-based metathesis catalysts supported on mesocellular siliceous foams showed high efficiency in macrocyclic RCM reactions. The immobilized catalysts **3–6** surpassed **HG2** catalyst in selectivity especially for the formation of greater than 20-membered rings (Fig. 1) [24].

Important improvements in the performance of RCM reactions were reported by the efficient volatilization of ethylene formed in the reaction. This methodology, specifically applied to macrolactones and macrolactams up to 16-membered, allowed catalyst loadings as low as 50–250 ppm to achieve near-quantitative

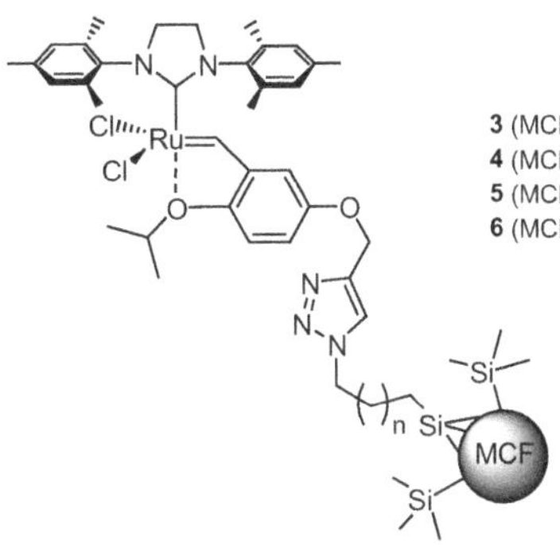

3 (MCF$_{24,\ 11nm}$, 0.29mmol/g, n = 1)
4 (MCF$_{25,\ 12nm}$, 0.13mmol/g, n = 1)
5 (MCF$_{24,\ 13nm}$, 0.27mmol/g, n = 5)
6 (MCF$_{30,\ 19nm}$, 0.17mmol/g, n = 1)

Fig. 1 A solid-supported ruthenium alkylidene catalyst

conversions. The procedure consisted of passing a continuous flow of argon through the reaction with toluene as solvent and **G2** or **HG2** as catalysts [25]. On the other hand, a "phase-separation" strategy can be utilized for controlling dilution effects at concentrations 150–500 times greater than those typically employed for macrocyclizations. The phase-separation strategy employs solvent mixtures of organic and poly(ethylene) glycol (PEG) solvents that form aggregates in solution, in which the macrocyclization substrate is preferentially solubilized. The catalyst system (**G2** and **HG2** complexes) is highly reactive in the organic solvent and inhibited in the PEG aggregate, so that slow diffusion of the substrate into the organic solvent "phase" mimics traditional slow addition strategies that are used for controlling dilution effects in macrocyclization reactions [26]. Other advances in scaling up procedures will be shown in particular examples especially regarding the synthesis of molecules entering clinical evaluation (*see* Sect. 4.1.)

2 Stereochemical Control

The large rings formed by metathesis are frequently obtained as (*E*/*Z*) mixtures where the (*E*)-isomer is dominant. The stereoselectivity is especially unclear for cycles going from 10- to 14-membered rings. In addition, the (*E*/*Z*) ratio of the product may be influenced by secondary isomerizations mediated by the metathesis complexes. In particular the many times desired (*Z*)-isomer is difficult to obtain in the synthesis of macrocycles as it is less energetically stable. In addition the reversible nature of catalytic olefin metathesis favors isomerization onto the (*E*)-isomer.

Since the first reports on RCM reactions directed to macrocycles, the absence of catalysts that reliably deliver kinetic control of stereoselectivity in the formation of large-ring olefins appeared as the major drawback. Dependence on substrate control does not typically lead to the selective formation of the targeted isomer, and many times the energy difference between the two isomeric alkenes is low and gives rise to mixtures.

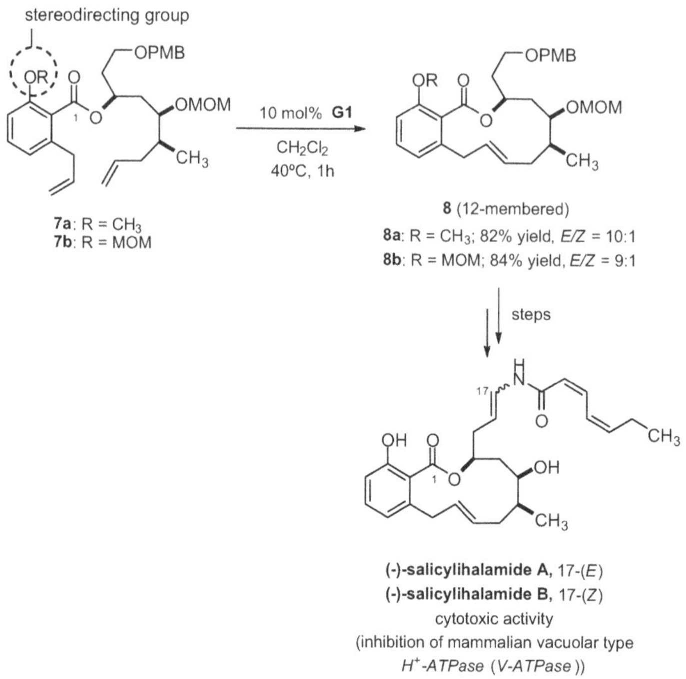

Scheme 2 Key step RCM in the formal synthesis of (−)-salicylihalamides A and B

A first solution to this problem has been the tuning up of precise conditions driving the delicate balance between kinetic and thermodynamic control, which is affected by solvent, catalyst, and temperature. In addition some authors have modified the substitution pattern of the substrate introducing steric demanding groups or coordinating heteroatoms which change the stereochemical outcome of the reaction. As a relatively early example, in the synthesis of salicylihalamides, a resorcynilic derivative, the (*E*/*Z*) outcome was strongly dependent on the remote protecting group situated at the aromatic ring. The careful study of the influence of remote substituents of RCM precursors and of the catalysts on the stereochemistry of the formed olefins was the subject of many efforts in order to obtain a good (*E*/*Z*) stereoselectivity. The remote phenolic hydroxy group favored the undesired (*Z*)-stereochemistry. If the phenolic hydroxy group was protected, the (*E*)-isomer became the major compound. The influence of the catalyst was also crucial as with catalysts **G1** a high (*E*) selectivity was observed. On the other hand, with the catalysts **G2** the equilibrium is reached, leading to a ratio (*E*/*Z*) 2:1 [27, 28]. With all this previous work in hand a recent contribution described a concise formal synthesis of (−)-salicylihalamides A and B in which the key step of the synthetic strategy includes a RCM reaction of **7a-b** using conditions developed by De Brabander (10 mol% of **G1**) to furnish the macrolides **8a-b** in 82–84% yield and with excellent (*E*)-selectivity (*E*/*Z* = 9–10:1) (Scheme 2) [29].

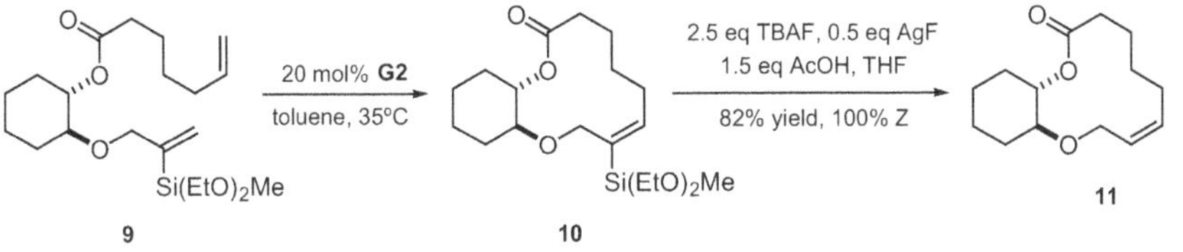

Scheme 3 Synthesis of a (*Z*)-olefin using a vinylsiloxane to achieve high (*E*)-selectivity and subsequent protodesilylation

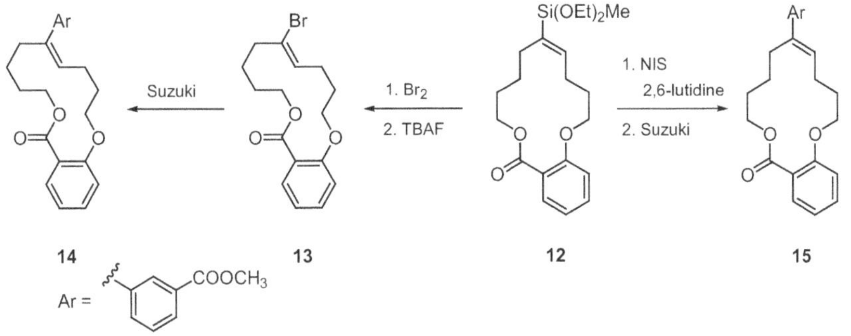

Scheme 4 Cross-coupling strategy to provide access to both (*E*)- and (*Z*)-trisubstituted alkenes

A second approach uses vinylsiloxanes (**9**) directly bonded to the double bond that give high (*E*)-selectivity and produce the desired (*Z*)-isomer **11** upon protodesilylation (Scheme 3) [30].

Alternatively, subsequent cross-coupling can provide access to both (*E*)- and (*Z*)-trisubstituted alkenes [31]. Conditions were developed to convert the alkenyl siloxane **12** into alkenyl halides with retention or inversion of configuration of the alkene geometry (**13**). The (*E*)- and (*Z*)-alkenyl halides, in turn, served as effective substrates for palladium-mediated cross-coupling reactions to generate trisubstituted alkenes **14** or **15** (Scheme 4).

Development of complexes that preferentially provide the (*Z*)-alkenes constituted a great challenge. (*Z*)-Selective catalysts must fulfill two significant requirements: they must facilitate the formation of the higher energy alkene isomer with high selectivity and not cause (*Z*)-to-*E* isomerization. In addition, catalysts need to be stable to decomposition, so that they survive the duration of the reaction in order to reach high turnover numbers (TONs). In the case of ruthenium complexes some decomposition products such as [Ru]–H species may promote unwanted side reactions such as olefin migration in both starting materials and products.

In 2009 Schrock and Hoveyda first reported a (*Z*)-selective metathesis catalysts and used it in a ring-opening/cross-metathesis (RO/CM) reaction. The catalyst was a stereogenic-at-Mo monoaryloxide-pyrrolide (MAP) complex [32]. Recently they have shown that the control of alkene geometry can be circumvented with Mo- or W-MAP complexes such as **18** in large macrocycle synthesis. As an example

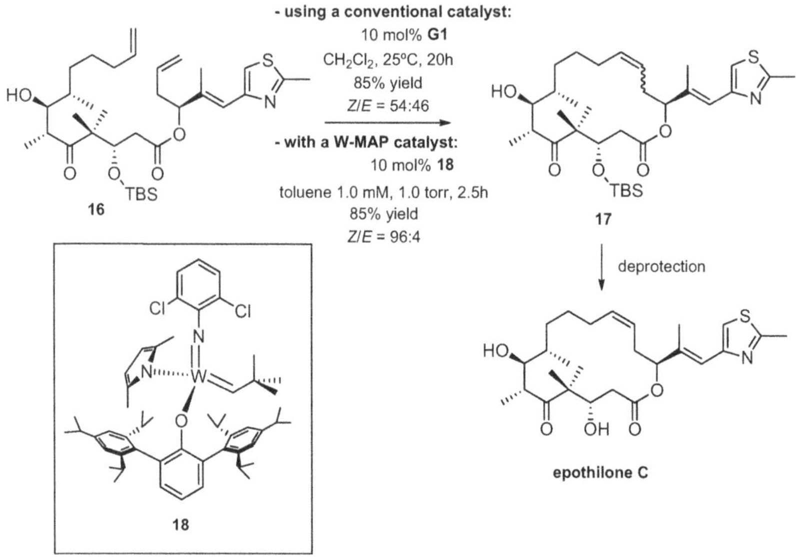

Scheme 5 RCM for the stereoselective total synthesis of epothilone C with a Ru- or a W-based complex

Scheme 5 shows the synthesis of epothilone C, comparing the result published by Nicolaou in 1997 with that achieved with these new complexes (Nicolaou synthesis: [33], Hoveyda synthesis: [34]).

The success of this methodology is based on the careful design of the catalyst. Both Mo and W are used, which is very advantageous as they are complementary. Thus, Mo-based alkylidenes exhibit higher activity, and in situations where the substrate dienes have a low degree of preorganization, it is the metal of choice. On the other hand, with relatively rigid starting materials, and in particular where post-RCM isomerization can be detrimental, the relatively less active W alkylidenes might emerge as more convenient catalysts. The latter complexes are more air and moisture stable than the Mo ones. They can be weighed in open air and reactions performed in a regular fume hood without the need for strict exclusion of air and moisture. As representative examples, Fig. 2 shows the syntheses of **19**, **20** and epilachnene with the best catalyst (**21** or **22**) used in each case and the results in terms of yields and selectivity [33, 34].

The (*Z*)-selectivity was explained in terms of steric interactions with the big aryloxide ligand which is much bulkier than the imido group. Thus, the bulky and freely rotating aryloxide (**I**→ **II**) can force the alkene to coordinate with the metal center so that the resulting metallacyclobutane substituents are oriented toward the smaller imido unit (**III**) (Scheme 6). Cycloreversion of **III** would furnish alkene complex **IV** and subsequent release of macrocyclic (*Z*) olefin produces methylidene complex **V**, which can react with another diene molecule to regenerate **I** [35].

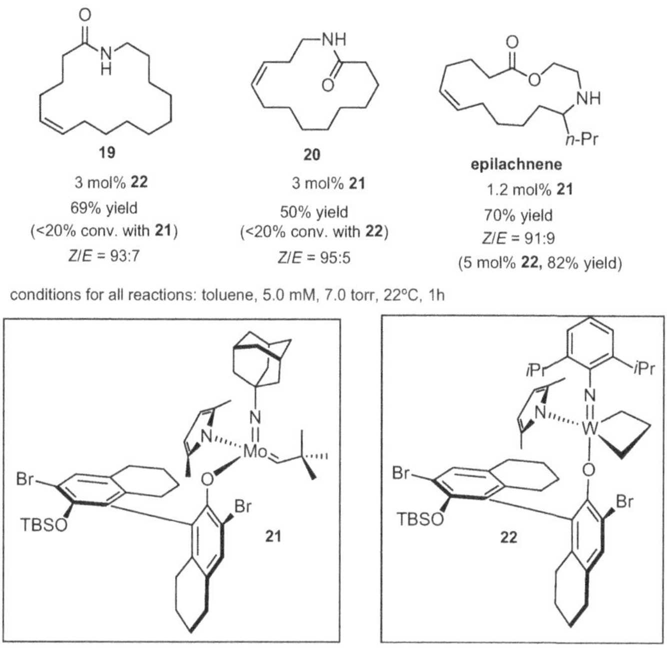

Fig. 2 A practical and (*Z*)-selective synthesis of epilachnene and related macrocycles

In another recent contribution, this group used the same type of MAP complexes to form a trisubstituted (*Z*)-alkene and access epothilones B and D. However this time they had to carefully modify the catalyst to achieve excellent conversions and a 91% (*Z*)-selectivity. They selected a Mo complex, due to inactivity of the parent W derivative, and improved (*Z*)-selectivity making aryloxides bigger and imido ligand smaller. The imido ligand was also changed to gain catalyst stability. Finally, complex **23** was able to convert **24** into **25** in 88% conversion (Scheme 7) [36].

The development of complementary ruthenium-based systems, and their use in (*Z*)-selective olefin metathesis, was first reported by Chen in 2010. They modified the first-generation ruthenium complexes with bulky sulfonates. Soon after, Grubbs and others published new ruthenium carbenes able to give (*Z*)-selective metathesis and used them in ROMP and CM. These new complexes, **26–29**, shown in Fig. 3, tend to be more tolerant to functionality than Mo- and W-MAP complexes and are easier to synthesize, but their structure–property relationships are less clear making their design more difficult (Complex **21**: [37], Complex **22**: [38], Complex **23**: [39], Complex **24**: [40]).

To date the only study of (*Z*)-selective macrocyclizations with ruthenium complexes has been reported by Grubbs using complex **29**. As shown in Scheme 8 they performed a systematic study of the scope of (*Z*)-selective macrocyclic RCM with respect to a broad range of ring sizes. The transformation was applied to the synthesis of a number of olfactory macrocycles (Scheme 8) [41].

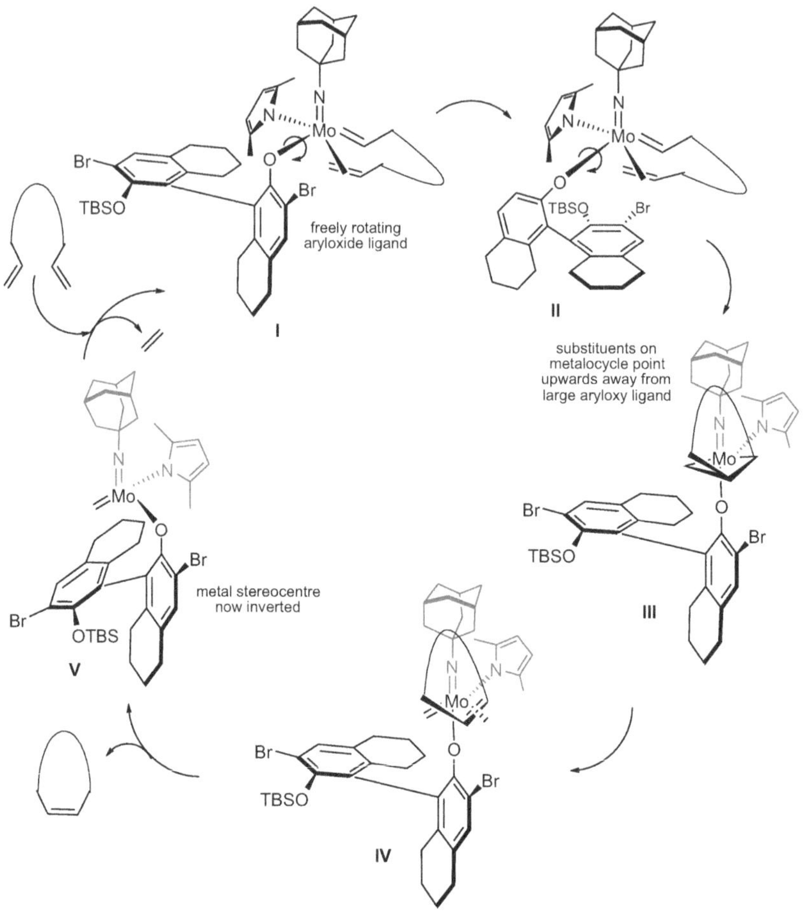

Scheme 6 Key structural features of MAP alkylidenes responsible for high (*Z*)-selectivity in macrocyclic RCM reactions

The explanation for this (*Z*)-selectivity has been suggested by Wang [42] for a CM case and is based on the association of the alkene in a *cis* manner to give **VII** in order to avoid steric interactions with the bulky mesitylene moiety. From this complex, formation of ruthenacyclobutane **VIII** and reductive elimination gives **IX** which would isomerize onto **VI** (Scheme 9).

A completely different way to control stereoselectivity in macrocyclizations performed using metathesis has emerged with the significant advances in catalyst design able to mediate in alkyne metathesis. Indeed, ring-closing alkyne metathesis (RCAM) followed by stereocontrolled partial hydrogenation of the formed macrocycloalkyne is becoming a useful way to construct macrocycles, especially when the elusive (*Z*)-stereochemistry is desired. In this context, complexes **32–36** are particularly noteworthy as they combine excellent reactivity with an

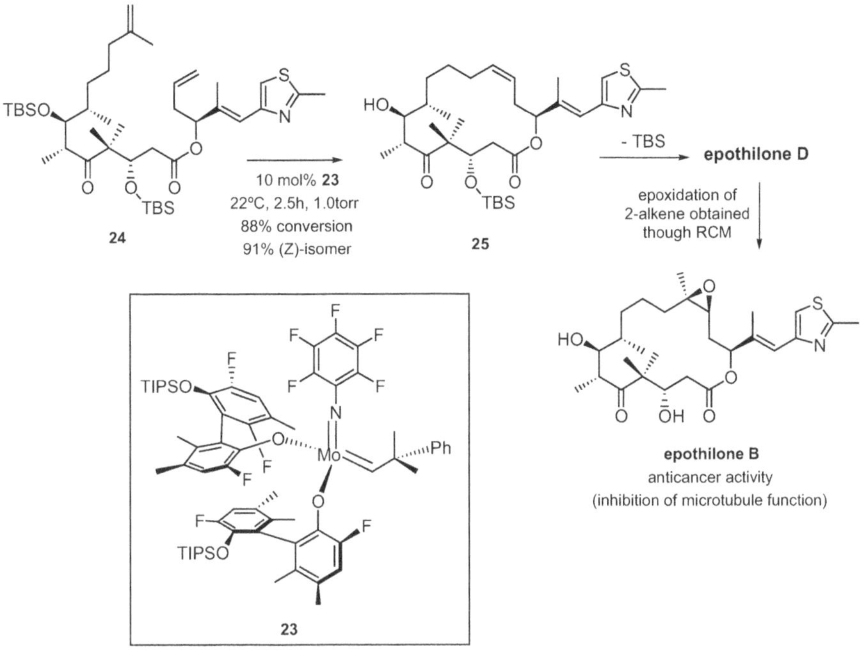

Scheme 7 MAP-catalyzed RCM for the stereoselective total syntheses of epothilone B

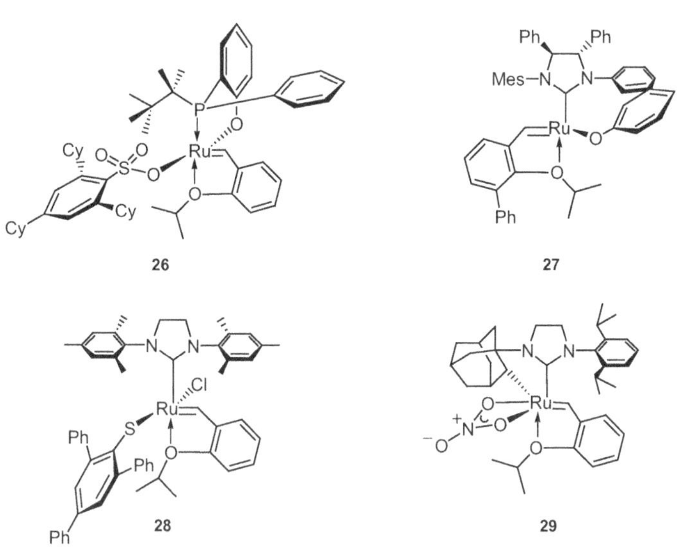

Fig. 3 Recently described ruthenium-based metathesis catalysts

outstanding functional group tolerance (Fig. 4). Their excellent application profile is ascribed to its triphenylsilanolate ligands. These impart a well-balanced level of

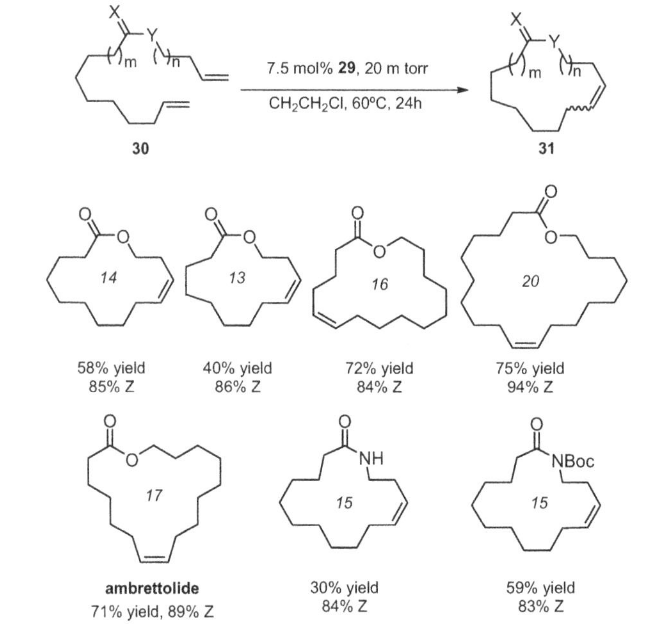

Scheme 8 (*Z*)-Selective macrocyclizations employing ruthenium catalyst **29**

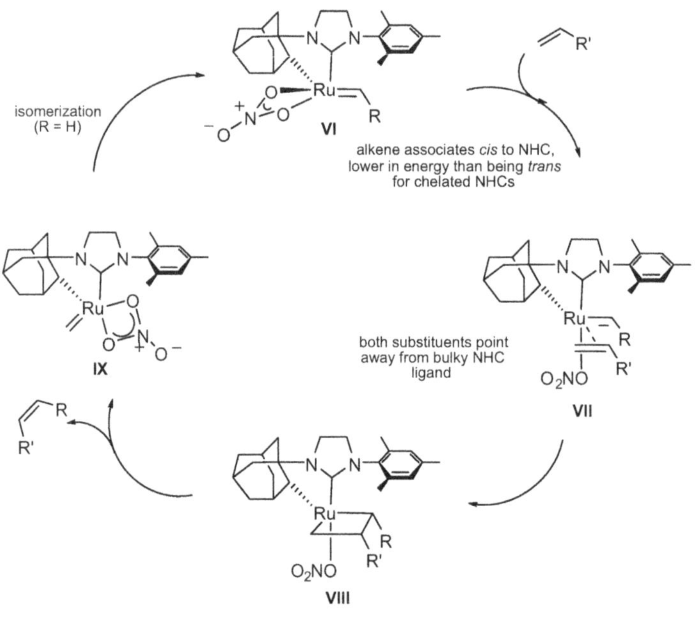

Scheme 9 A mechanistic explanation of the (*Z*)-selectivity observed with catalyst **29**

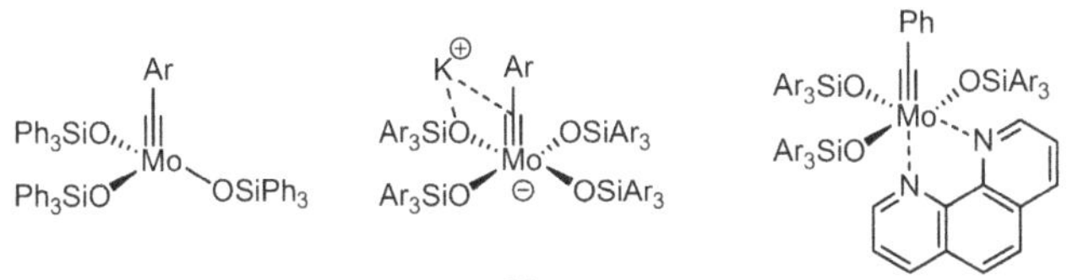

32: Ar = Ph
33: Ar = p-$CH_3OC_6H_4$
34
35: Ar = Ph
36: Ar = p-$CH_3OC_6H_4$

Fig. 4 Selected molybdenum-based catalysts for RCAM

Lewis acidity on the molybdenum center, which is high enough to ensure an outstanding catalytic performance, yet sufficiently tempered not to endanger acid-labile motifs. Examples using these new Mo-based catalysts will be shown in the following sections [43].

3 Natural and Closely Related Macrocyclic Heterocyclic Products Synthesized Using Metathesis as the Key Step

Millions of years of natural selection have rendered molecules with a good ability to interact with natural macromolecules. As differences in protein domains are not greatly significant in nature, these compounds interact efficiently, not only with their natural targets but with certain receptors of mammals, thus becoming useful for humans. The presence of macrocycles is particularly common in antitumorals, antibiotics, and antifungals [44, 45] and their use to access new leads is therefore highly desirable. They possess ring sizes going from 11- to 16-membered rings, with a special incidence for 14-membered rings. In addition they usually have complex structures, with a variety of chemical motifs dominated by macrocyclic lactones and ethers generally from polyketide origin. Some include aromatic rings in their structure like phenyl ethers, *ansa* cycles, and biphenyl structures. Nitrogen-containing macrocycles are very often lactams and have generally larger cycles reaching sometimes 30 members. Many of these are cyclopeptides, possessing three to six amino acids which are covered in [46] of this book. The total synthesis of all these compounds plays a crucial role in order to confirm their structure and to develop derivatives that can modulate the biological activity or the pharmacokinetic properties.

Next sections summarize recent efforts in natural macrocycle synthesis and close derivatives. Following a biogenetic classification, we will first address natural products from polyketide origin, with special impact of macrolides and macrolactams, and then the macrocycles issued from mevalonic pathway, followed by the amino acid metabolism (cyclodepsipeptides and alkaloids also treated in [47, 48] respectively) and finally the glycolipids.

3.1 *Macrocyclic Polyketides*

Macrocyclic natural products arising from polyketide origin comprise a huge number of structures. The majority of them have one or more lactone linkages (polyolides) and are generally referred to macrolides. They present diverse biological activities, so that they constitute ideal targets for synthetic studies. These compounds are isolated from bacteria or fungi and they are a large group that come from marine origin [49–51]. The classification of macrolides is not completely established. We will group the examples that follow according to families of natural products in increasing structural complexity. In addition to macrolides, macrolactams and examples of synthesis of ansamycins will be shown.

3.1.1 Monocyclic Macrolides: Macrocyclic Musk

Three classes of compounds, namely, nitroarenes, polycyclic benzenoids, and macrocyclic ketones and lactones, possess musky odors. The latter have been used for the longest time in fragrance formulations. Efficient synthetic methods that can be scaled up are necessary to decrease their cost of production. Fürstner's group reported the synthesis of several simple macrolides like recifeiolide, exaltolide, and muscopyridine in multigram scale using alkene RCM catalyzed by **G1** and other first-generation ruthenium catalysts (Fig. 5) [52]. The synthesis of exaltolide was improved recently and scaled up using only 1 mol% of **37** [53].

The group used RCAM for the stereoselective synthesis of other olfactory compounds, including ambrettolide and civetone [33, 34], and in their last contribution they published the synthesis of (*R*,*Z*)-muscenone [54]. In this work they improved catalyst loading by using new-generation **34** complex (Fig. 5), which impart exceptional reactivity as well as better functional group tolerance. The cyclization of diyne **38** proceeded with 1 mol% of the ate complex **34** in toluene at ambient temperature in 75% yield in only 30 min, when the reaction was performed in the presence of MS 5 Å (Scheme 7). Provided that subsequent hydrogenation proceeded in 94% yield, this result greatly improves the RCM approach from **39** as can be seen in Scheme 10.

3.1.2 Monocyclic Macrolides from Bacteria and Fungi and PKS Origin

Highly convergent stereoselective total syntheses of many macrolactones have been accomplished based on a strategy built around the construction of the macrocycle core using a RCM/deprotection or RCM/hydrogenation protocol. After early studies in which many conditions were tried, we have noted that most groups have used Grubbs second-generation complex **G2** as the catalyst and toluene or dichloromethane as solvents at room temperature or mild heating. Reaction time is usually variable. Figure 6 summarizes recent examples in which natural lactones

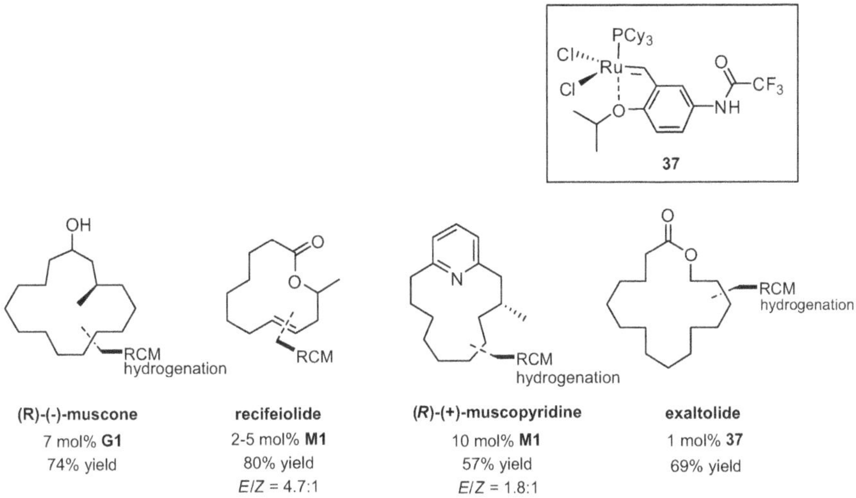

Fig. 5 Multigram-scale synthesis of macrocyclic musk using RCM

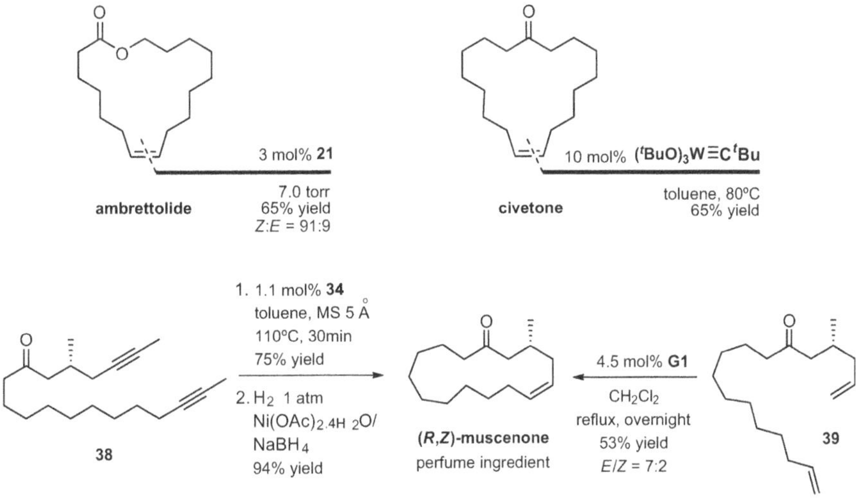

Scheme 10 RCAM-based synthesis of olfactory compounds

were built using diene RCM. The core motif of carolacton [55] and mycolactones [56] as well as two members of the methymicins, namely, pladienolides B and D, possessing antineoplastic activity, was described [57]. The lateral chain of carolacton was introduced via CM after the construction of the lactone ring. Finally, macrosphelide M, a 16-membered macrotriolide antibiotic having a trilactone backbone in its structure, was synthesized using RCM without observing side reactions arising from the second double bond present in the substrate [58].

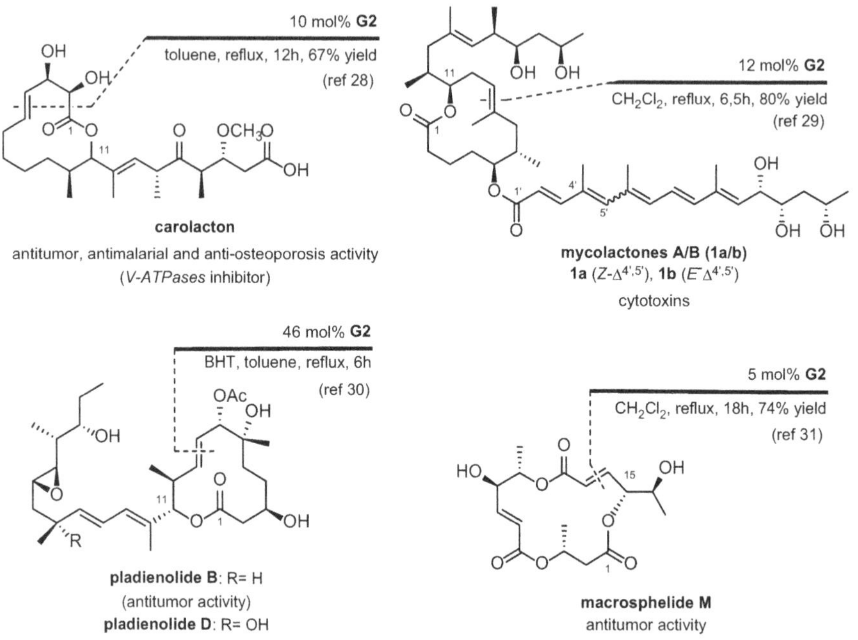

Fig. 6 Examples of natural lactones recently constructed using RCM

Very related structurally to the previous two compounds is lactimidomycin, a natural product with an important different challenge to overcome in its synthesis due to its polyenic nature. The problem is to discriminate between the two olefinic sites of the 1,3-diene substrate **40**. The reaction of the terminal double bond is desired, but activation of the internal double bond, which results in ring contraction, may become the dominant or even exclusive pathway. Fürstner's group disclosed a study in which a careful selection of catalyst and a modification of the substrates were used to produce macrocyclic 1,3-dienes with excellent yields and regio/stereoselectivity [59]. The authors found that the best catalyst choice was the more encumbered ruthenium carbene **41** [60]. This catalyst gave the desired product **42** in good yields, and only very small amounts of the 11-membered congener **43** were detected. A strategic positioning of a silyl group allowed protection of the internal alkene and gave stereoselectivity onto (*E*,*Z*)-configured macrocyclic 1,3-dienes (Scheme 11).

In other cases the competition between different alkenes has not been such a difficult problem. Migrastatin is a product which inhibits the migration of human tumor cells. This compound was synthesized via a RCM of the fully functionalized tetraene **44** (Scheme 12a). The reaction was highly chemoselective and proceeded in a totally (*E*)-selective fashion [61]. In another closely related example, the RCM reaction of **45** gave many side reaction products when using **G2** or **HG2** as the catalyst. However, the use of only 4 mol% of less reactive **G1** gave the desired

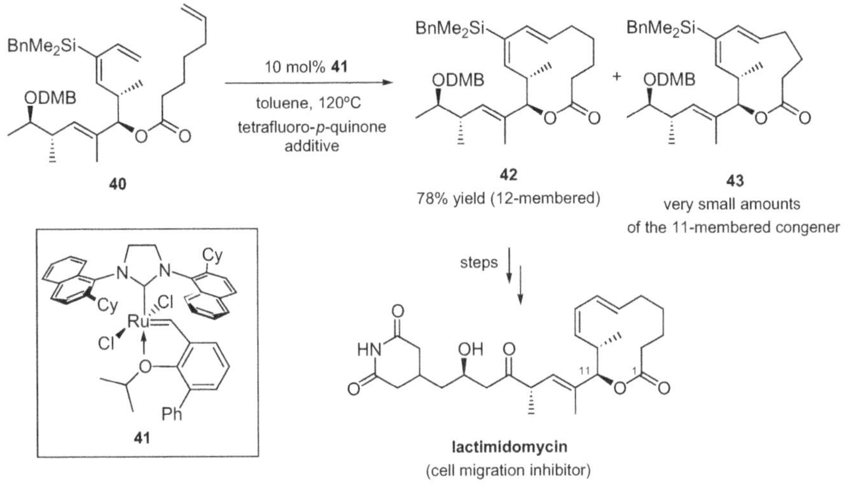

Scheme 11 Silyl-directed RCM for the synthesis of lactimidomycin

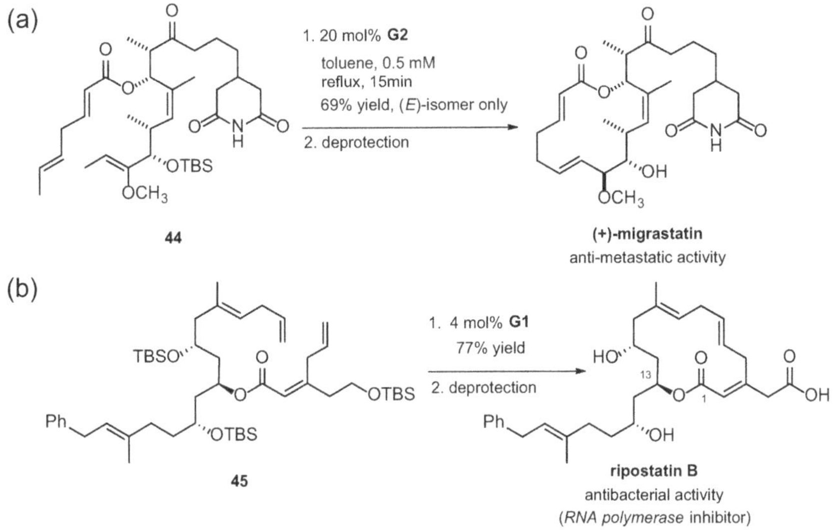

Scheme 12 Synthesis of migrastatin (**a**) and ripostatin B (**b**) using highly chemoselective and (*E*)-selective RCM

macrolactone in 77% yield as a single (*E*)-isomer in a much cleaner reaction. The resulting macrolactone was deprotected to obtain ripostatin B (Scheme 12b) [62].

Plecomacrolides are a large family of natural products typically featured with a 16- or 18-membered macrolactone possessing two conjugated diene units and a hemiacetal side chain. Plecomacrolides act as selective inhibitors of vacuolar H^+-ATPases (V-ATPases) and they are considered to have the potential for treatment of

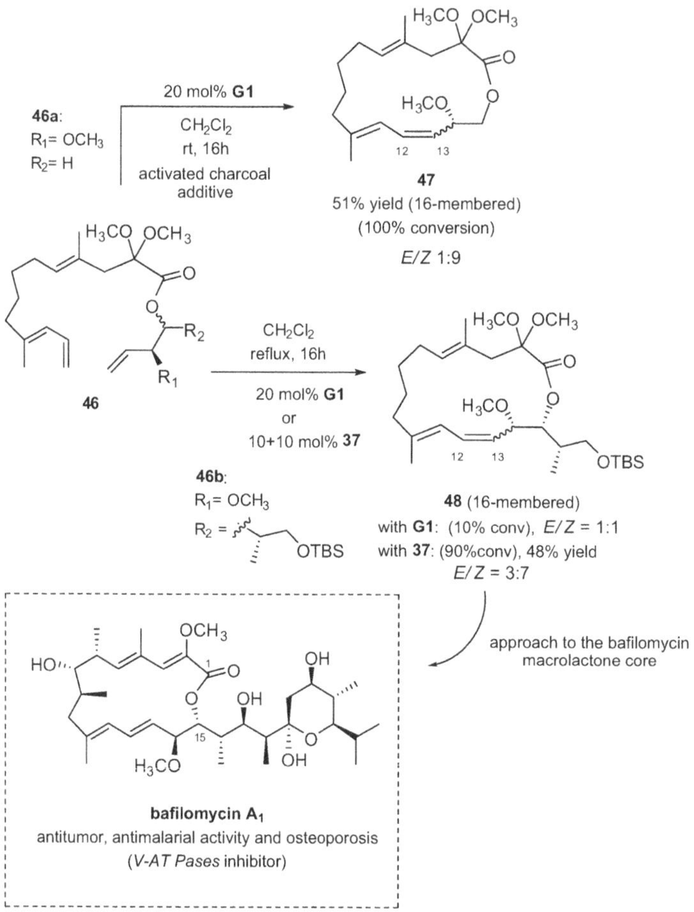

Scheme 13 (*E*/*Z*)-selectivity problems in the synthesis of the macrolactone core of bafilomycin A_1

postmenopausal osteoporosis (for a review on synthesis and activity of pleco-macrolide, see [63]). Recently the macrolactone core of a new promising member of the plecomacrolide, bafilomycin A_1, was synthesized, circumventing the regioselectivity problem of the RCM ene–diene reaction of substrates **46**. Different model esters of increasing complexity were subjected to the 1,3-diene-ene ring-closing metathesis reaction. The best promoter for the simplest esters **46a** was **G1**. However, for the most complex ester **46b**, a Hoveyda–Grubbs-type trifluoromethylamido-containing precatalyst (**37**; see Fig. 5 [64]) gave the best results. In all experiments, the 12-(*Z*)-configured isomer was obtained as the major product [65] (Scheme 13).

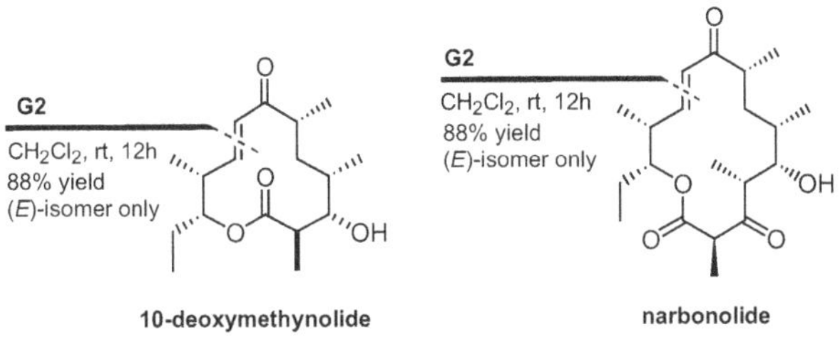

Fig. 7 Macrocycle formation by RCM of 10-deoxymethynolide and narbonolide

3.1.3 Monocyclic Macrolides: Ketolides

Antibiotic lactones with an additional keto group are called ketolides. For instance, 10-deoxymethynolide is the aglycone of the methymycin family of macrolides and was constructed via an exclusive *E*-selective RCM using **G2** in CH_2Cl_2 at rt. The same conditions were used for the synthesis of narbonolide, which possesses a 14-membered ring lactone instead of the 12-membered ring. Narbonolide is the aglycone of the pikromycin family of antibiotic macrolides (Fig. 7) [66].

3.1.4 Resorcylic Macrolides

The resorcylic acid lactones (RAL) are a family of naturally occurring homologous macrolides coming from different fungi. They have a typical chemical architecture including a benzo-fused macrolactone. Their biological activities are varied including antitumor, antibiotic, and antimalarial. However, the ability of these compounds to inhibit Hsp90 is outstanding and is the main cause of the great interest of many groups in these compounds.

The use of metathesis reactions for the synthesis of RALs was pioneered by Danishefsky in the late 1990s and was reviewed in 2007 by Winssinger (for a review on the structure, biology, and chemistry of RALs, see [67]). Radicicol and its derivatives were the first to be synthesized mainly through ene–diene RCM to avoid stereoselectivity problems. Scheme 14 summarizes these early syntheses. The aromatic ring was constructed from dienes of type **49** or alternatively through the so-called ynolide methodology which uses a stable cobalt complex like **50**. This approach avoids a competitive enyne metathesis and the bulky cobalt cluster brings into proximity the double bonds favoring the RCM. Oxidative metal decomplexation and a Diels–Alder reaction were used to elaborate the benzenic system [68, 69]. Other members of this family are aigialomycin D, zeranol, zeralenone, and pochonins, which share the basic skeleton consisting of a 14-membered ring lactone that includes a 1,2-phenylene. In search for biologically active compounds, many analogs have also been synthesized using metathesis methodologies (see [70]).

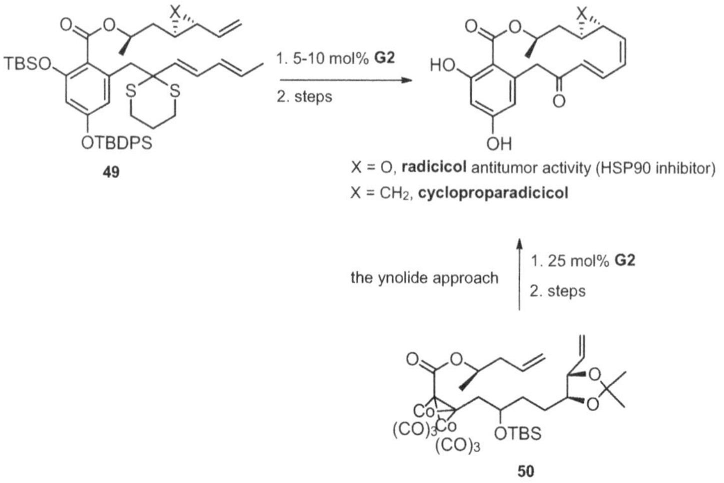

Scheme 14 RCM-based approaches for the synthesis of radicicol and cycloproparadicicol

Aigialomycin D has received recent synthetic attention as it possesses antimalarial and antitumor activity [71]. Two complementary approaches have used RCM to close the 14-membered ring of aigialomycin D building each of the two double bonds. When building the double bond which is closer to the aromatic ring (C1′-C2′), the main problem is a competitive reaction giving rise to diene **52** and the cyclohexene **53** in addition to the desired macrocycle **54** (Scheme 15a) [72]. The group of Barret reported recently the total synthesis of aigialomycin D, using for the macrocyclization a RCM (to form C7′-C8′) on triene precursors **55a-b** which did not require phenol protection (Scheme 15b). The key RCM was performed using **HG2** catalyst (15 mol%) at a concentration of 15×10^{-4} M. Triene **55b** was converted into the *trans*-macrocycle **56** in 83% yield with very little coproduction of cyclohexene **53** [73].

In a different way to avoid the formation of cyclohexenes, aigialomycin D was synthesized by RCM to build the C7′-C8′ bond, while the C1′-C2′ alkene was masked as a sulfone during formation of the macrocycle **58**. The macrocycle was completed via Ramberg–Bäcklund reaction which efficiently produced the C1′–C2′ (*E*)-alkene (Scheme 15c) [74].

Pochonins are an interesting group of resorcylic lactones isolated from *Pochonia chlamydosporia*. Pochonin C is active against HSV (herpes simplex virus), whereas pochonin D can adopt the so-called L-shape bioactive conformation against Hsp90. Much more structurally related to radicicol, pochonin A has both Hsp90 and HSV inhibition activity. Winssinger's group has disclosed the syntheses of this family of RALs. Ene–diene RCM of epoxide **59** was used for the synthesis of pochonin C (Scheme 16a) [75], and the synthesis of pochonins A [76] and D (Scheme 16b) [77] was achieved from **61** using in the three cases **G2** (5–10 mol%) in hot toluene. The synthesis of the purported *ent*-pochonin J was reported recently although the final product did not spectroscopically correlate to the initially disclosed natural product. Here the RCM was greatly efficient as the presence of the additional pyran ring

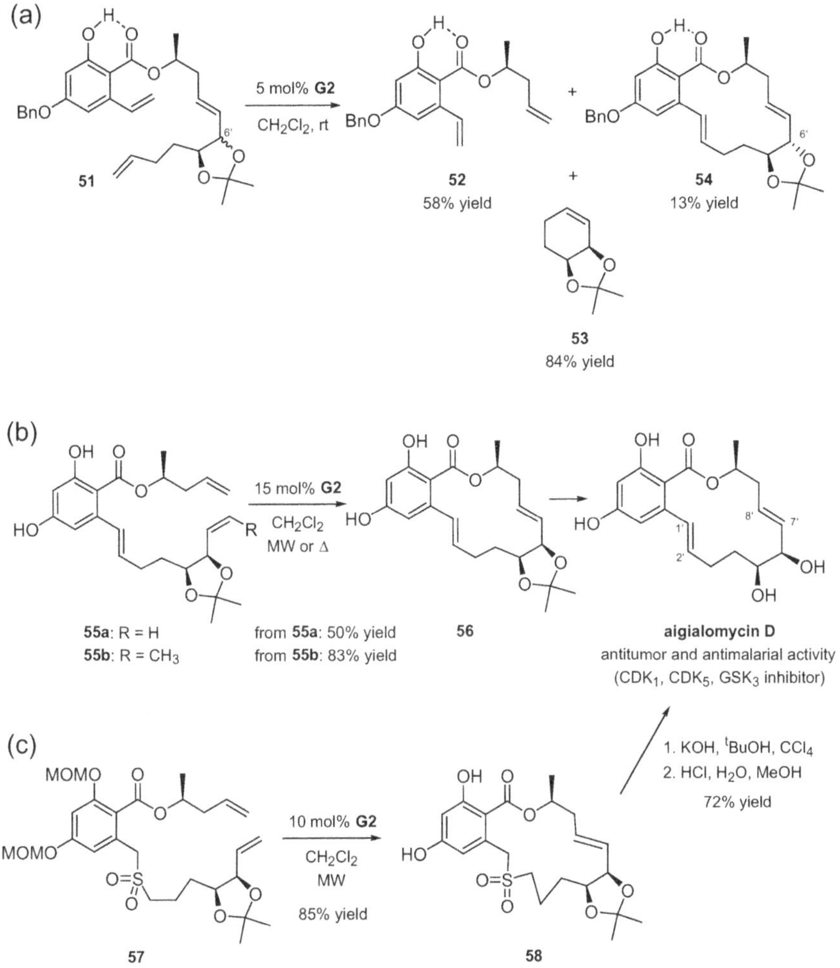

Scheme 15 Three different approaches to aigialomycin using RCM as the key step: (**a**) building of the C1′–C2′ double bond; (**b**) formation of C7′–C8′ bond from a triene precursor; (**c**) building of C7′–C8′ bond while the C1′–C2′ alkene is masked as a sulfone

helped to bring the starting material **63** into a reactive conformation (Scheme 16c) [78].

Other interesting members of the resorcylic macrolides are salicylates with a 12-membered ring benzolactone and a dienyl enamide side chain. They are potent inhibitors of the mammalian vacuolar ATPase and exert cytotoxicity in the NCI 60-cell lines (for a review on the chemistry and biology of salicylihalamide A and related compounds, see [79]). These natural products have been synthesized using metathesis methodologies as commented in the introductory section. Related to this group, Sch 351448, isolated from *Micromonospora*, is a dimer of salicylates featuring a 28-membered macrodiolide consisting of two identical hydroxy

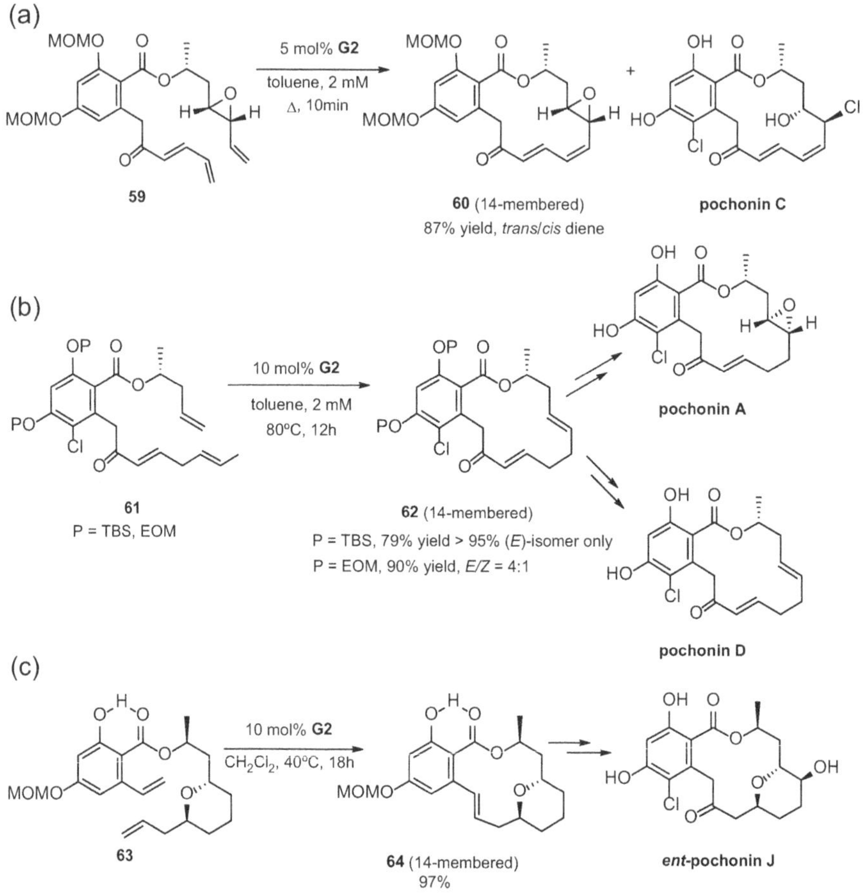

Scheme 16 Macrocycle formation by RCM in the synthesis of: (**a**) pochonin C; (**b**) pochonins A and D; (**c**) *ent*-pochonin J

carboxylic acid units. This product activates low-density lipoprotein receptor (LDL-R). In its synthesis, intramolecular olefin metathesis of **65** mediated by **G2** combined with hydrogenation gave the macrodiolide in 57% yield (Scheme 17) [80–83]. In a more recent synthesis the starting material was templated with a metal cation to improve the efficiency of the olefin metathesis which was raised up to 77% yield [84].

Other members of this family are apicularens, lobatamides A–F, and oximidines II and III, with an enamide side chain terminating by an *O*-methyl oxime. Early works have accounted for the synthesis of oximidines II [85] and III [86]. Another member of this group, cruentaren A which has similar biological activity as salicylihalamide, has been synthesized recently although following a different approach. The synthesis of cruentaren A illustrates a different way to circumvent the problem of achieving (*Z*)-selectivity in macrocyclizations. Three groups have published synthesis of this compound, all of them using the combination of alkyne

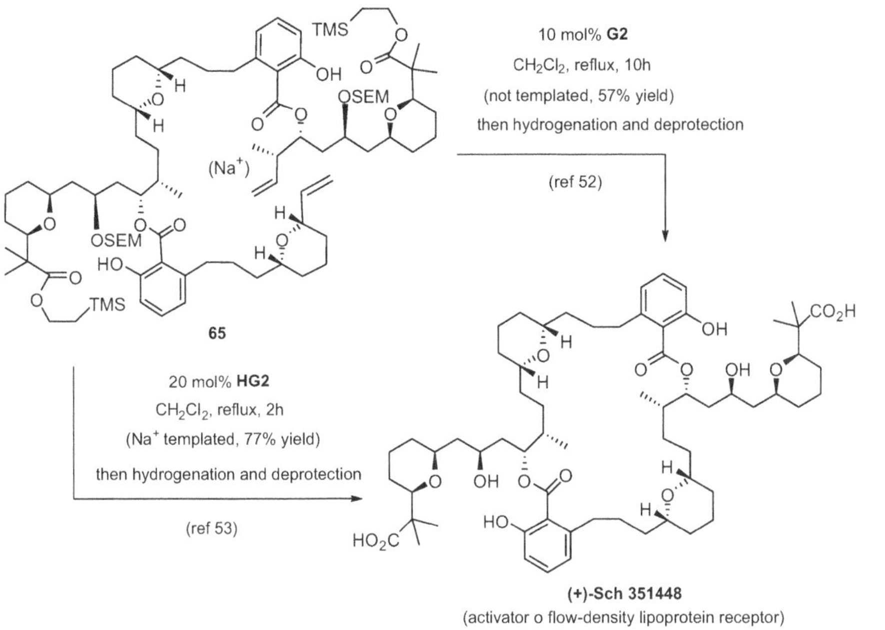

Scheme 17 A RCM–hydrogenation sequence for the construction of the 28-membered ring of macrodiolide (+)-Sch 351448

metathesis and palladium hydrogenation as the key step. The retrosynthetic strategies proposed by Maier [87, 88] and Fürstner [89, 90] have similar key disconnections. Both started with the derivatization of the aromatic unit and the macrolactone was elaborated from **66** to **67** using an esterification reaction followed by ring-closing alkyne metathesis of **68**. Lindlar hydrogenation of the resulting **69** afforded the unsaturated lactone with the desired (*Z*)-configuration. Barrett's group published later a variant in which the aromatic ring was constructed via late-stage cyclization and aromatization from starting compounds **70** and **71** [91]. Scheme 18 compares these three syntheses. As can be seen, the RCAM key step proceeded with excellent yields in the three variants with the use of different catalytic systems.

3.1.5 Marine Macrolides

Plants, animals, and microbes from the marine environment are a tremendous source of structurally diverse and bioactive natural products. Some of them have entered clinical trials as new classes of chemotherapeutic agents. Among them macrocyclic polyketides are frequent, and the synthesis of these promising molecules has been the subject of many studies. Structurally they are characterized by a highly oxygenated polyene backbone containing a macrocyclic lactone as a conformational constraint. Some are monocyclic but others present additional rings,

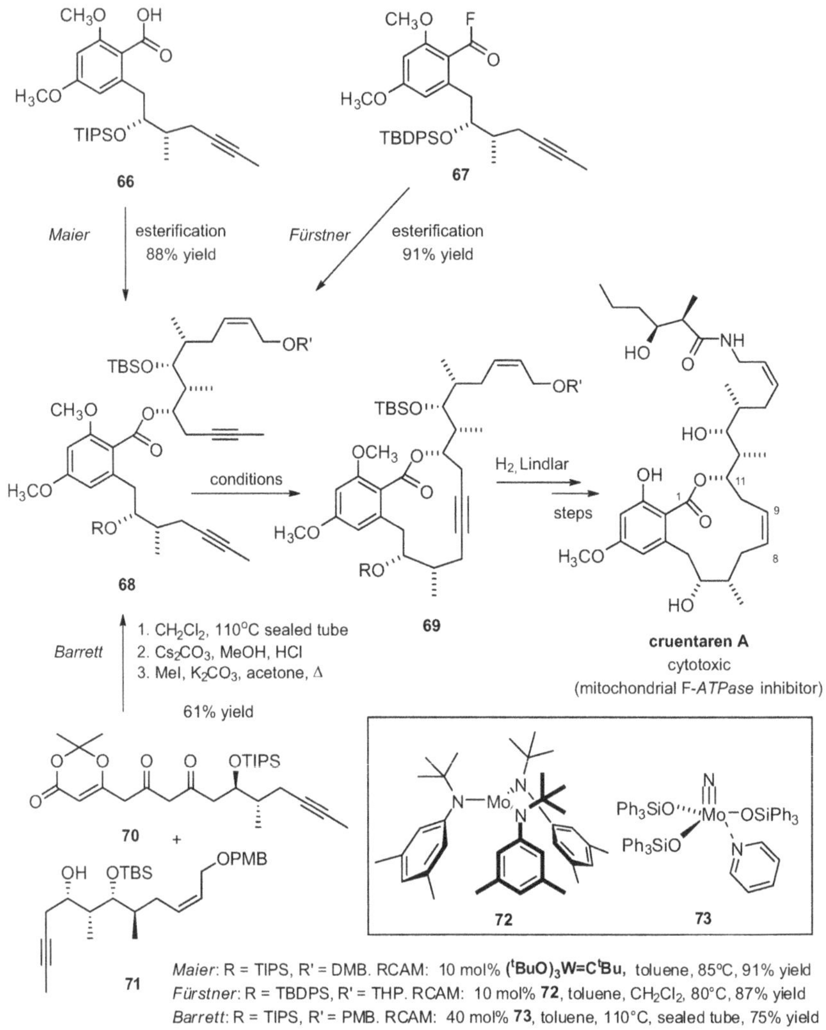

Scheme 18 RCAM used as the key step in three different syntheses of cruentaren A

generally one or two tetrahydrofuran or pyran rings. Many marine macrolides possess outstanding cell growth antiproliferative properties, therefore being promising lead compounds for the development of new antitumor chemotherapeutic agents. There is not a clear classification for this large family of molecules that are generally grouped following the organism from which they are isolated. We will follow a structural similarity organization starting with the monocyclic and less complex members.

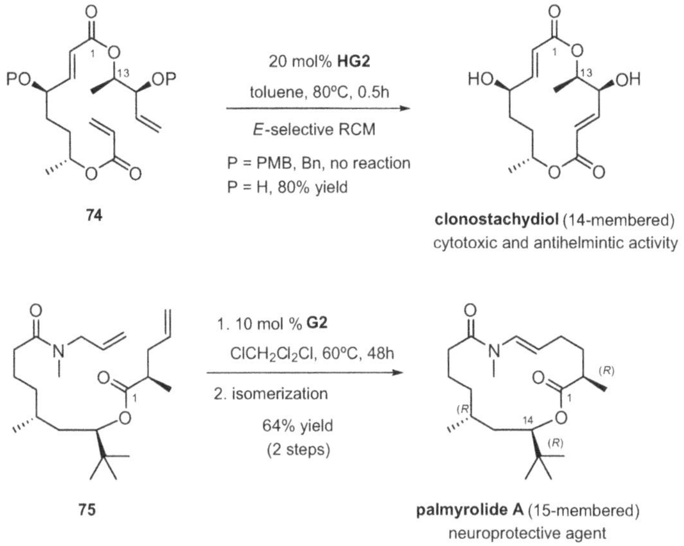

Scheme 19 Macrocycle formation by RCM in the syntheses of clonostachydiol and palmyrolide A

Monocyclic Marine Macrolides

Scheme 19 shows two relatively simple marine macrolides recently synthesized using diene RCM. Clonostachydiol is a 14-membered macrocyclic bislactone isolated from marine algae with cytotoxic and anthelmintic activities. RCM of the deprotected precursor **74** gave excellent results, while the parent protected substrate was unreactive under typical metathesis conditions [92]. On the other hand, palmyrolide A is a neuroprotective macrolide with an uncommon *N*-methyl enamide moiety. The targeted conjugated double bond was achieved using a combined RCM and isomerization of *N*-allylated tertiary amide **75** [93].

Palmerolide A (with a similar name but very different structure and origin) was isolated from *Synoicum adareanum* and possesses potent and selective activity against the melanoma cancer cell line UACC-62. To date, three total syntheses of palmerolide A have been reported, four groups disclosed formal syntheses, and various synthetic approaches have also been described [94]. Nicolaou's group reported the total syntheses of five stereoisomers of palmerolide A including the originally proposed and the revised [*ent*-(19-*epi*-20-*epi*)] structures. The RCM cyclization proceeded smoothly under mild conditions (Fig. 8) [95]. This synthesis resulted in the stereochemical reassignment of the configuration of the natural product at the C7, C10, and C11 positions from *R* to *S*. Among the seven available syntheses of **palmerolide A** core, not only Nicolaou and Chen but also Prassad [96] and Kaliappan [97] used RCM under similar conditions to form the C7–C8 bond.

Dictyostatin was isolated in tiny quantities from a marine sponge in 1994, showing a potent anticancer activity. It is one of the most potent microtubule-

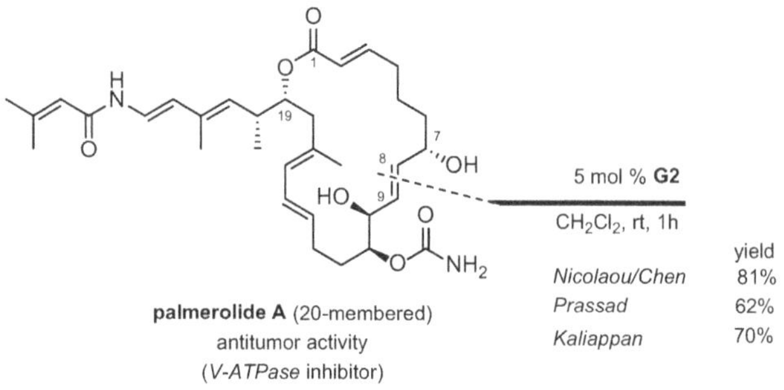

Fig. 8 Formation of the macrocycle of palmerolide A using RCM

stabilizing agents known, inducing the assembly of purified tubulin even more rapidly than paclitaxel. Several total syntheses of this natural product and some formal syntheses have appeared [98]. In 2010 Curran's group finalized its study on the total synthesis of dictyostatin and several derivatives. They developed three highly convergent approaches based on three fragments which were coupled, and in one of the syntheses, the final ring closing was performed using metathesis. However they encountered serious regioselective problems: when reacting substrate **76** under typical conditions, the only products were **78** and a contracted macrocycle **79** as a consequence of a relay RCM where the internal double bond reacted with the formed carbene **77** (Scheme 20a). In view of this result, they redesigned the starting material several times and finally found that **80** gave the desired 22-membered macrocycle (Scheme 20b). The RCM reaction was run with $Ti(OiPr)_4$ as additive, and the product was isolated in 65% yield. As initial reactions provided little or no conversion, they added the Lewis acid to prevent potential chelate formation of ruthenium carbene intermediates with the acetal oxygens [99].

Other polyene macrolides with a 24-membered ring were synthesized using RCM but without encountering the regioselectivity problems shown above (Fig. 9). Thus, an effective RCM was successfully used for the synthesis of the group of lejimalides containing no less than seven double bonds. These are a new group of cytotoxic compounds, possibly targeting V-ATPases. The RCM reaction gave the desired macrocycle in 55% yield and high *E* stereoselectivity [100]. The same group has recently improved the conditions of the RCM reaction in order to efficiently scale up the synthesis. Thus, in contrast to performing the ring-closing metathesis in CH_2Cl_2 at ambient temperature, it was found beneficial to run the reaction in toluene at 50°C under a constant stream of argon. This setup allowed the catalyst loading to be reduced from 20 to 10 mol%. Yields exceeded 70% of a single isomer [101]. The group has used these conditions to develop lejimalide–archazolid chimeras **81**, derivatives with the core of lejimalides and lateral chain of archazolid, a secondary metabolite derived from terrestrial bacteria [102]. Finally, Fig. 9 shows the conditions of a RCM used to build the core of another 24-membered marine macrolide, namely, dolabelide C with cytotoxic activity [103].

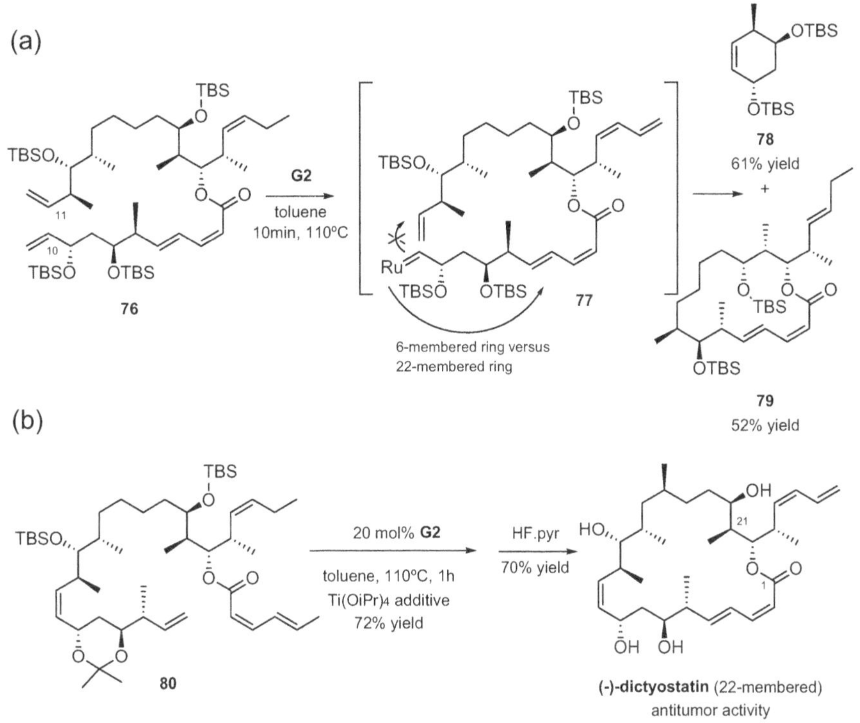

Scheme 20 Regioselective problems in the synthesis of dictyostatin: (**a**) relay RCM of **76** gave a contracted macrocycle; (**b**) substrate **80** produced the desired product

Amphidinolides are marine natural products, with outstanding cytotoxic effects. There are different subfamilies of these compounds possessing various macrocyclic ring sizes. In addition they may have one or two tetrahydrofuran or tetrahydropyran rings. Many groups have dedicated their synthetic efforts to these molecules due to their diverse functionality, stereochemical complexity, and low natural abundance. The total synthesis of more than 15 out of around 40 of the amphidinolides known have been described, some using RCM to build the ring.

In 2002 the first synthesis of one of these natural products, amphidinolide A, was disclosed. The key step was a RCM of **82** with 50 mol% of **G2**, which yielded the corresponding macrocycle in 35% yield. This result is the starting point in a huge effort of various groups that have produced more and more efficient syntheses in the last decade (Scheme 21) [104].

Continuing with monocyclic members, amphidinolides B, G, and H were constructed from epoxy-polyenes using a key RCM cyclization utilizing the **G2** catalyst. In Scheme 22 the synthesis of amphidinolide H reported by Fürstner in 2007 is shown. The desired macrocycle was obtained from **83** as a single (*E*)-isomer in 73% yield [105]. Below, the structures of amphidinolide G and B with the

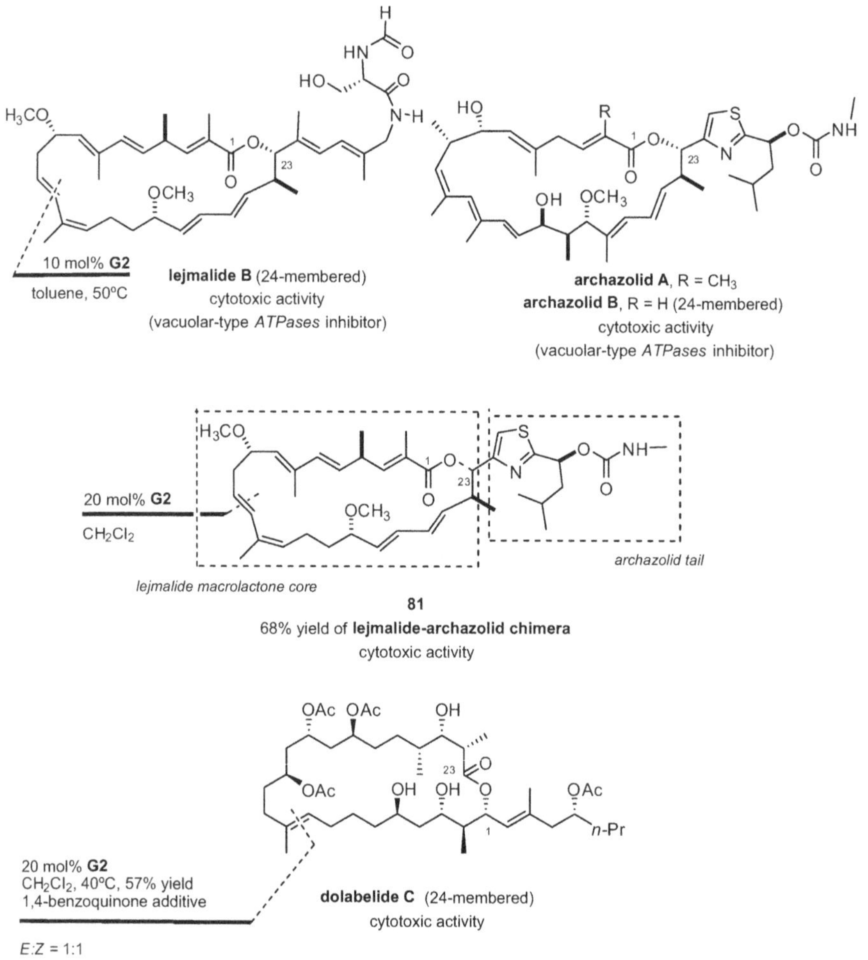

Fig. 9 Polyene macrolides with a 24-membered ring synthesized using RCM

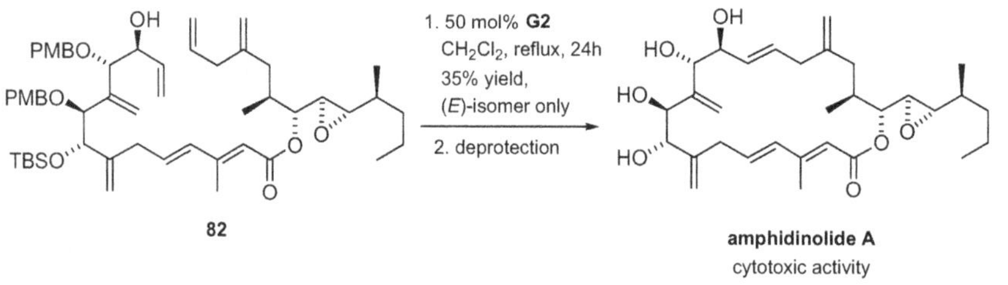

Scheme 21 First synthesis of an amphidinolide member using RCM

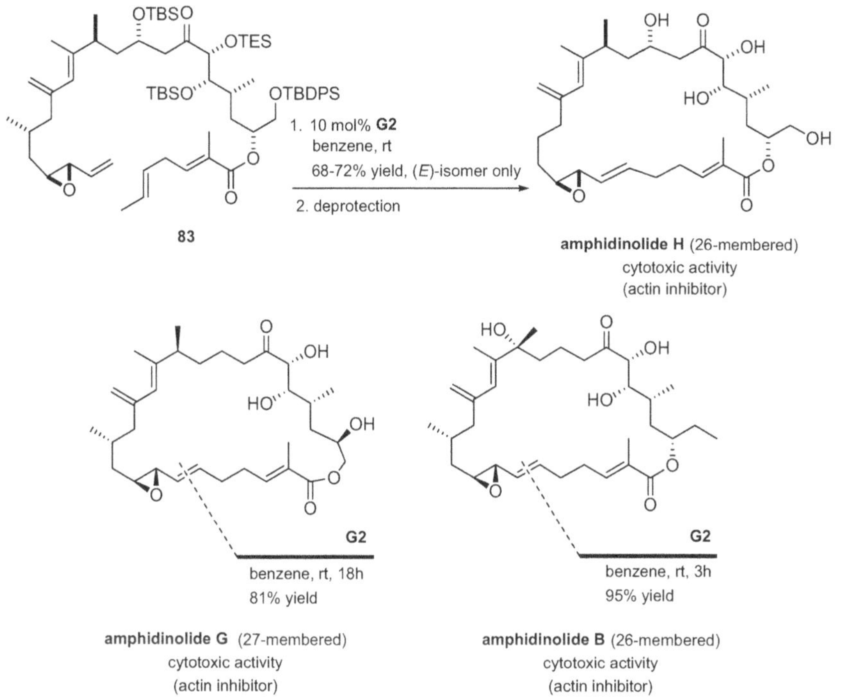

Scheme 22 RCM conditions used in the syntheses of three amphidinolides

conditions used for the RCM step and results are shown. The formal synthesis of the latter compound was reported recently using Fürstner conditions [106].

The amphidinium strain HYA024 was found to produce cytotoxic compounds iriomoteolides 1a–c (vide infra) and a rare 15-membered macrolide, iriomoteolide 3a. The synthesis of this macrolide was reported using a highly (*E*,*E*) stereoselective cross-metathesis/ring-closing metathesis sequence shown in Scheme 23. CM between compounds **84** and **85** afforded a mixture of products which gave a sluggish reaction under further metathesis conditions, with metatheses of the internal double bonds as predominant side reactions. However, the protection of the free hydroxy groups furnished a mixture of silyl derivatives **86** and **87** that underwent cleanly the desired RCM to give macrocycle **88** [107].

Marine Macrolides with Additional Fused Rings in the Core Structure

Amphidinolide T subfamily is characterized by having one tetrahydrofuran ring and by exhibiting interesting antineoplastic activity. They were first synthesized by Fürstner's group [108]. For the synthesis of amphidinolide T3, ring closure was performed to create C4–C5 bond from **89**, using 20 mol% **M1**, delivering cycloalkene **90** in 82% yield as an (*E*/*Z*) mixture of isomers in a ratio 2:1, which was further hydrogenated (Scheme 24a). Recently, Dai's group reported the total

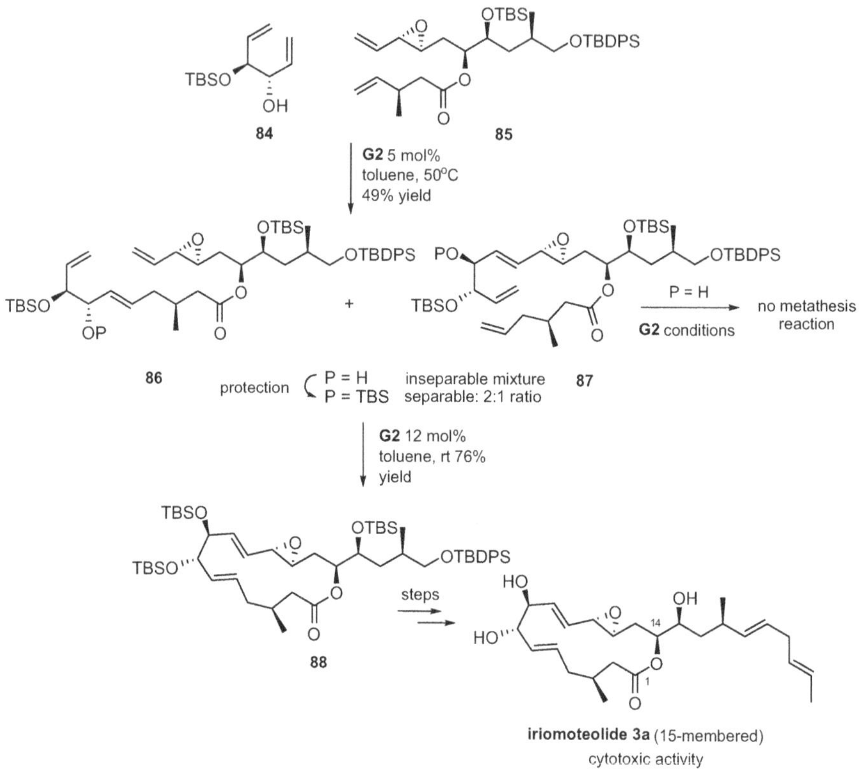

Scheme 23 Synthesis of iriomoteolide 3a using a highly (*E*,*E*)-selective cross-metathesis/ring-closing metathesis sequence

syntheses of amphidinolides T2 and T3. The methodology relayed on a RCM and asymmetric dihydroxylation (AD) strategy. This time the C12–C13 double bond was assembled via RCM in 80% yield and in 76:24 (*E*/*Z*) ratio using 5 mol% **G2** for the regioselective RCM reaction. No RCM products derived from the C16 methylene group in **91** were detected. As a whole, amphidinolide T3 was obtained in 8% overall yield from methyl (*S*)-lactate (Scheme 24b) [109, 110]. More recently, Clark's group has synthesized amphidinolides T1 and T3–T4 from a single common intermediate using metathesis methodology and macrolactonization for the synthesis [111].

The total syntheses of amphidinolide Y needed the formation of an (*E*)-trisubstituted alkene, which was achieved with the aid of remote substituent effect in the RCM. The cyclization of the C6 ketone **93** in refluxing CH_2Cl_2, using **G2** as the catalyst (up to 50 mol%), gave the macrocyclic core which was deprotected onto the natural product (Scheme 25a) [112]. For the synthesis of amphidinolide E, it was necessary to use **G1** to avoid the metathesis of an internal bond and decomposition of the starting polyene **94** as was observed when using **G2** (Scheme 25b)

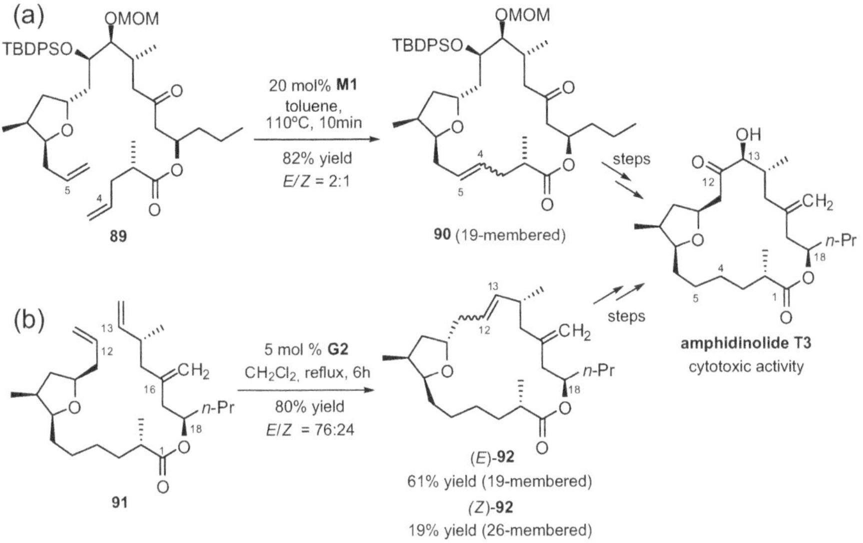

Scheme 24 (**a**) Fürstner's and (**b**) Dai's approaches to amphidinolide T3

[113]. The parent (−)-amphidinolide X was synthesized using RCM as the very last step (Scheme 25c) [114].

Amphidinolides O and P have one 6-membered ring bridged hemiacetal moiety and were synthesized from a common intermediate. The ring closure of the macrolide started from intermediates **96** and **97** which were coupled to afford the corresponding ester, and the ester was subjected to olefin metathesis to implement selectively the (*E*) C8–C9 double bond in 93% yield (**98**). Subsequent stereoselective epoxidation and formation of the required *exo*methylene groups gave the final products (Scheme 26) [115].

Iriomoteolides 1a–1c are recent additions to the amphidinolide family. Iriomoteolide 1a was reported to be cytotoxic toward human B-lymphocyte DG-75 cells and Epstein–Barr virus infected B-lymphocyte Raji cells. Several synthetic approaches to these two natural products all using RCM to form the macrocycle have appeared and been recently reviewed [116]. Horne and coworkers completed the first synthesis of the originally proposed structure of iriomoteolide-1a with spectra inconsistent with those of the natural product. The RCM was carried out on **99** using 10 mol% of **G2** to form the C6–C7 double bond giving a 73% yield of a 2.5:1 ratio of (*E*/*Z*) isomers (Scheme 27a) [117]. Xu and Xe's group focused their synthesis on a RCM directed to formation of the C15–C16 bond. Using **HG2** catalyst they transformed the substrate **100** into the intermediate macrocycle **101** in 51% yield albeit using 1 equiv. of the catalyst which was added in four portions (Scheme 27b) [118]. Finally, Yang's group prepared three diastereomers of iriomoteolides 1a–1c by using a RCM again to form the C15–C16 bond with a different substrate, **102**. The reaction gave exclusively the desired (*E*)-isomer **103** in 70% yield using 10 mol% of **G2** (Scheme 27c) [119].

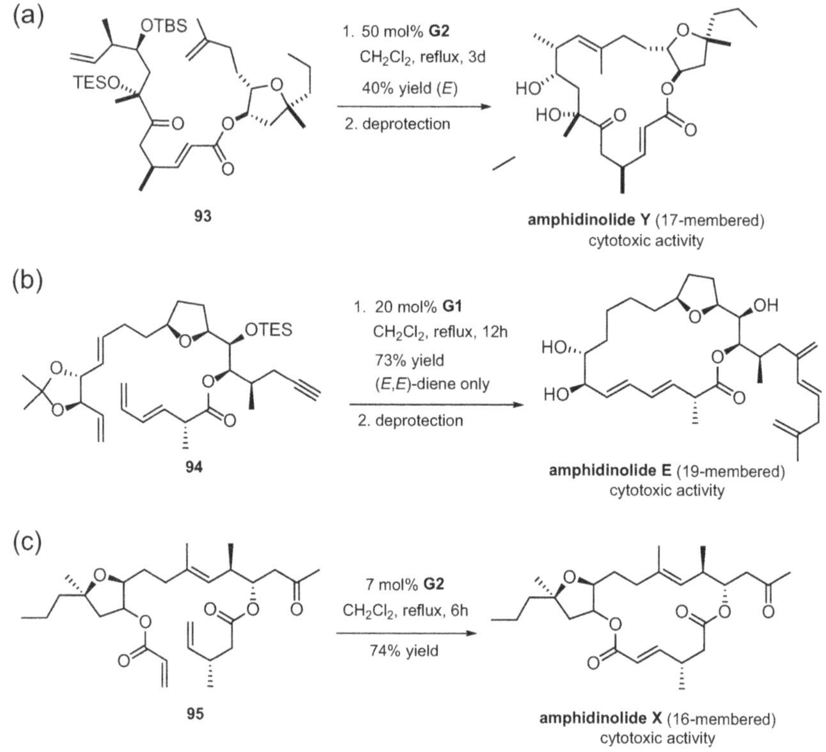

Scheme 25 Conditions used for the RCM step in the synthesis of: (**a**) amphidinolide Y; (**b**) amphidinolide E; (**c**) amphidinolide X

A number of marine natural 14-membered macrolides are known, which possess a tetrahydropyran ring very often forming part of a hemiacetal or acetal moiety. Aspergillides, neopeltolide, and lyngbyaloside are among these structures. All of them exhibit cytotoxicity pointing to a microtubule stabilization activity. The syntheses published to date share the RCM to form the C8–C9 bond as the way to construct the macrolide core. However it is interesting to comment on the differences in stereochemical outcome of the RCM. Thus, RCM of **104a** using 11 mol% of **HG2** in toluene (3 mM) at 140°C allowed isolation of **105** in 81% yield (Scheme 28a) [120]. The newly formed bond was exclusively *E* and **105** could be further transformed into 13-demethyllyngbyaloside. It appears that the C10 and C11 substituents affected the stereochemical outcome of the RCM as the stereoselectivity observed for RCM of **106** was opposite. This latter substrate was used to prepare the structurally related 14-membered macrolide neopeltolide. The exclusive product obtained in the RCM was (*Z*)-**107** (Scheme 28b) [121]. In these cases 1,4-benzoquinone was added to the RCM reaction to prevent undesired isomerizations [122]. In the synthesis of aspergillide A, both the first- and the second-generation ruthenium catalysts yielded (*Z*)-**109** as the sole reaction product, with **G1** giving higher yields than **G2** (99% vs 67%). As in this case the (*E*)-double bond

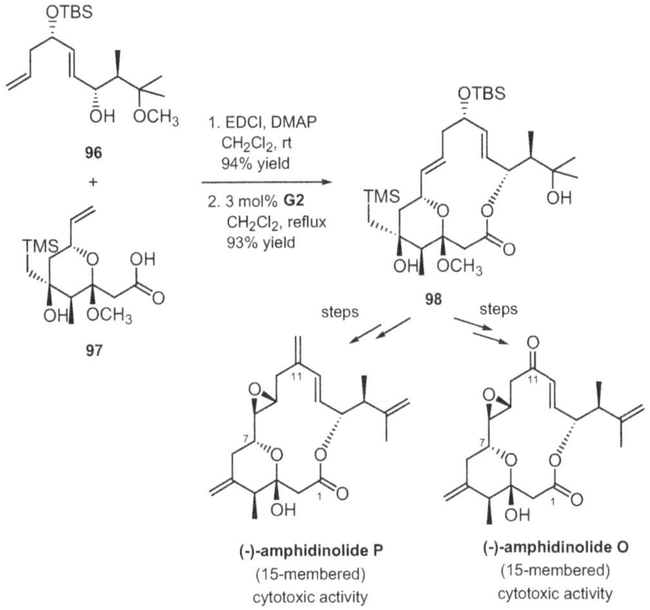

Scheme 26 Amphidinolides O and P synthesized from a common intermediate prepared using RCM

was present in the target product, a photochemical isomerization was performed to invert the stereochemistry of the double bond (Scheme 28c) [123].

Peloruside A is a cytotoxic polyketide isolated from a marine sponge which was found to have paclitaxel-like microtubule-stabilizing activity. Several groups have reported syntheses of analogs of peloruside A such as **110–112**. The synthetic routes encompassed a RCM key step. Figure 10 summarizes the conditions used for these reactions [124–127].

Zampanolide and dactylolide are two structurally close but stereochemically opposite macrolides that exhibit significant activity against a variety of tumor cell lines. The RCM cyclization step for the synthesis of these compounds was performed using **G2**. In the case of dactylolide, the desired macrolide was isolated in only 45% yield. Use of other catalysts or modification of the substrate in order to do a relay metathesis did not improve the result (Fig. 11) [128, 129].

RCAM was used to forge the macrocyclic skeleton of several amphidinolides and analogs. Scheme 29 summarizes these efforts. Thus, a sequence of ring-closing alkyne metathesis on **113** followed by an intermolecular enyne metathesis of the resulting cycloalkyne **114** with ethene gave the core macrocycle and set the vicinal *exo*methylene branches characteristic for the cytotoxic marine natural product amphidinolide V (Scheme 29a) [130, 131]. The flexibility of this methodology reported by Fürstner's group allowed the synthesis of several diastereomers and other analogs of the natural product. In a similar approach amphidinolide F was constructed using RCAM on compound **115**. Two possible scenarios were

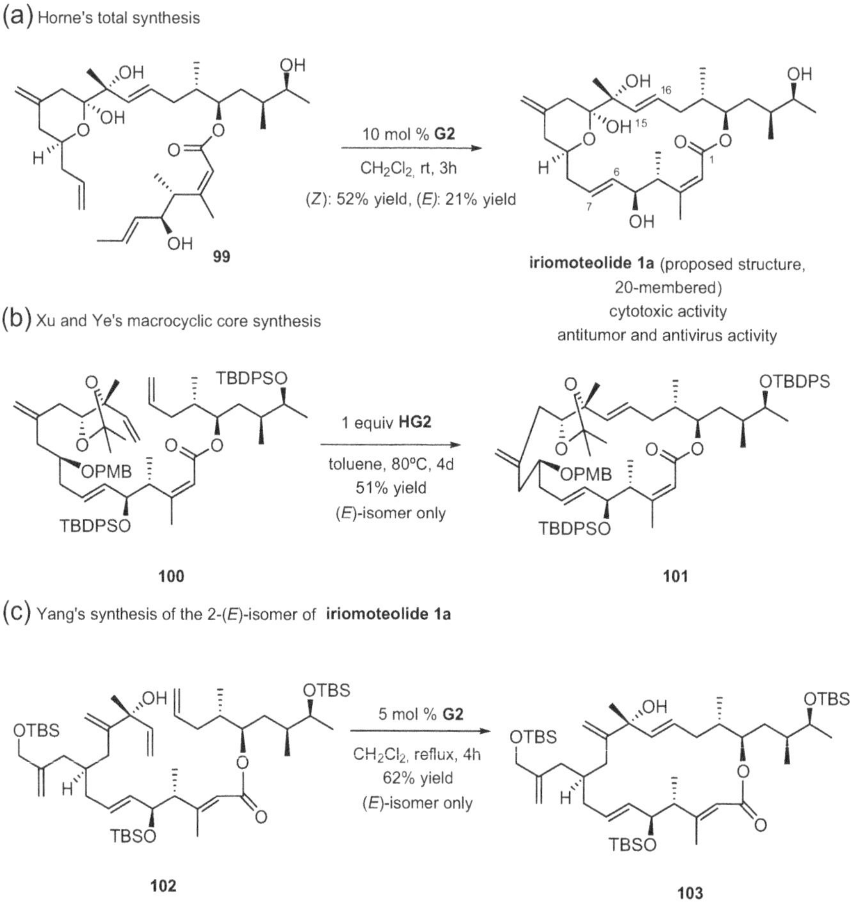

Scheme 27 Key steps in the preparation of iriomoteolide 1a: (**a**) RCM for the formation of the C6–C7 double bond; (**b**) and (**c**) RCM to build the C15–C16 bond from different substrates

explored: the ring closing of the fully protected diyne (using **72** as the catalysts; Scheme 18), followed by selective deprotection of the TES ether at C15, or inversion of the order and performing the RCAM step with a substrate already exhibiting the free OH group (using **33** as the catalyst; Fig. 4). Both alternatives were efficient and it was demonstrated that the free alcohol in **115** was tolerated by the catalyst giving **116** in good yield (Scheme 29b) [132]. Another contribution accounted for the synthesis of the core structure of the parent compound polycavernoside A, **118** using RCAM as depicted in Scheme 29c [133].

The marine oxylipins hybridalactone and the ecklonialactone family are unusual metabolites from polyketide origin that were recently synthesized using RCAM methodology. The macrocyclic scaffold of several members of this family was annulated to the cyclopentane head group by RCAM of **119–121** using the new generation of exceedingly tolerant molybdenum catalysts compatible with the

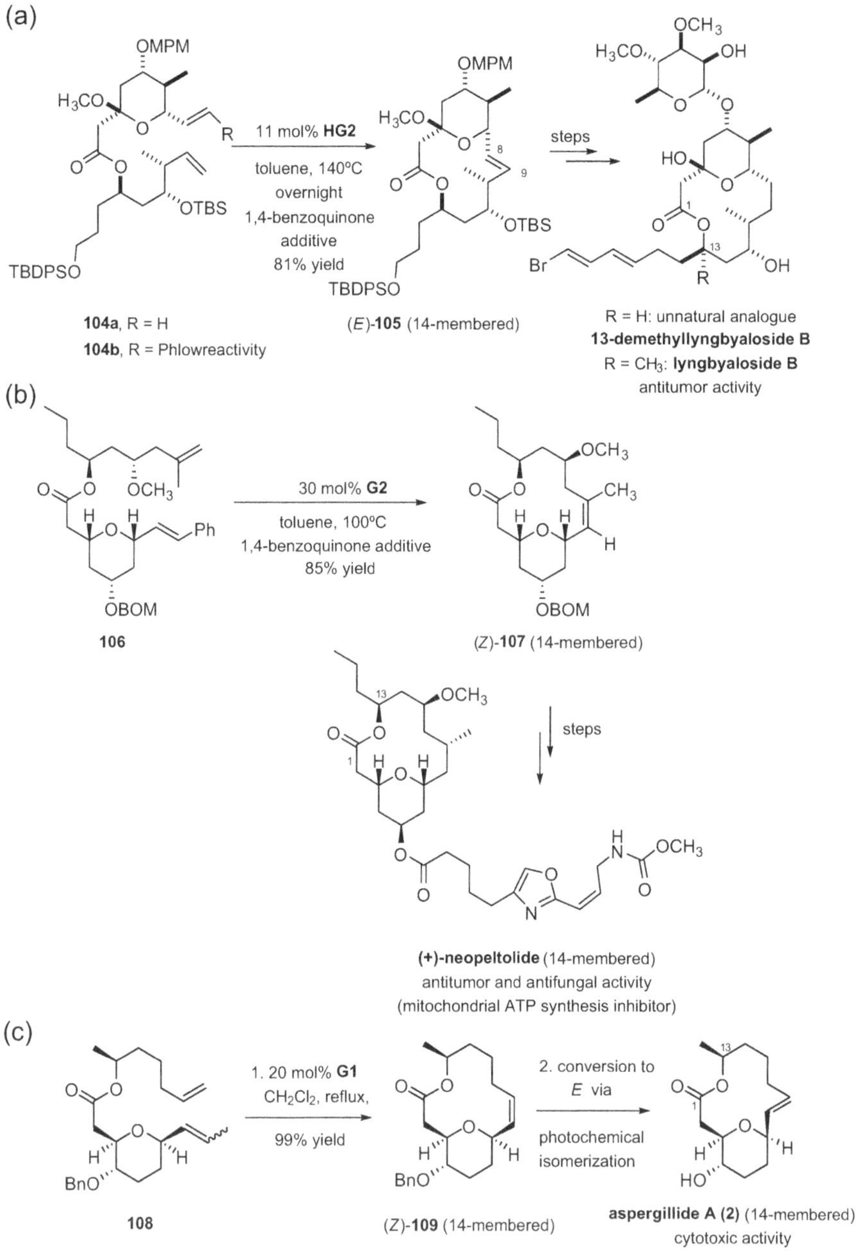

Scheme 28 Stereochemical outcomes of RCM in the syntheses of 14-membered macrolides: (**a**) total (*E*)-selectivity in the synthesis of **105,** a precursor of 13-demethyllyngbyaloside; (**b**) (*Z*)-stereoselectivity in the RCM of **106** used to prepare neopeltolide; (**c**) (*Z*)-selectivity in the preparation of **109** for the synthesis of aspergillide A

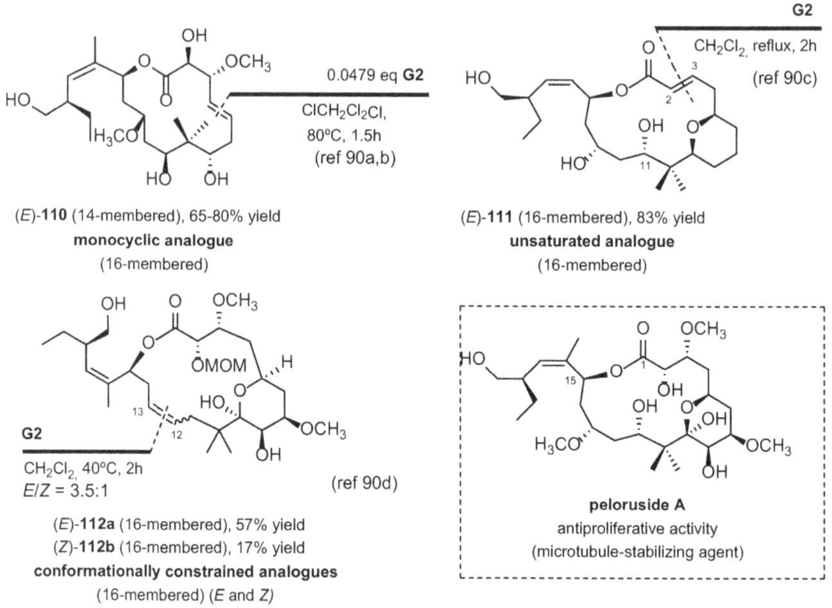

Fig. 10 Conditions for the RCM key step in the synthesis of peloruside A analogs

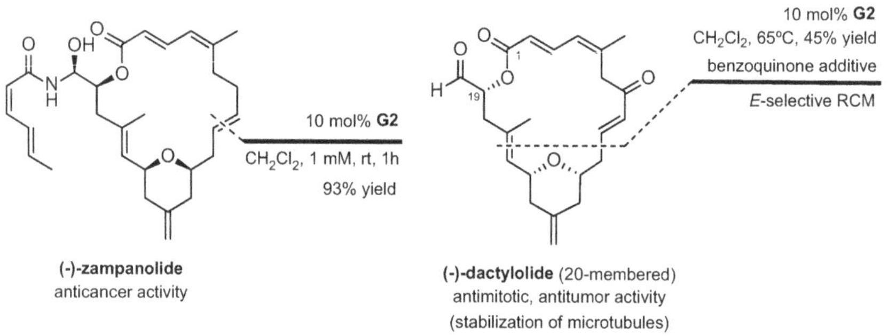

Fig. 11 Conditions for the RCM step in the synthesis of zampanolide and dactylolide

epoxy moiety. The ancillary triarylsilanolate ligands of **34** (Fig. 4) temper the Lewis acidity of the molybdenum center but are not sufficiently nucleophilic to engage in the opening of the fragile epoxide ring. A final semireduction of the cycloalkynes **122–124** formed in the RCAM step onto the required (*Z*)-alkene completed the total syntheses (Scheme 30) [134].

3.1.6 Epothilones

Application of the RCM to the synthesis of natural macrocycles started with approaches to epothilones. These well-known tubulin–polymerization agents are

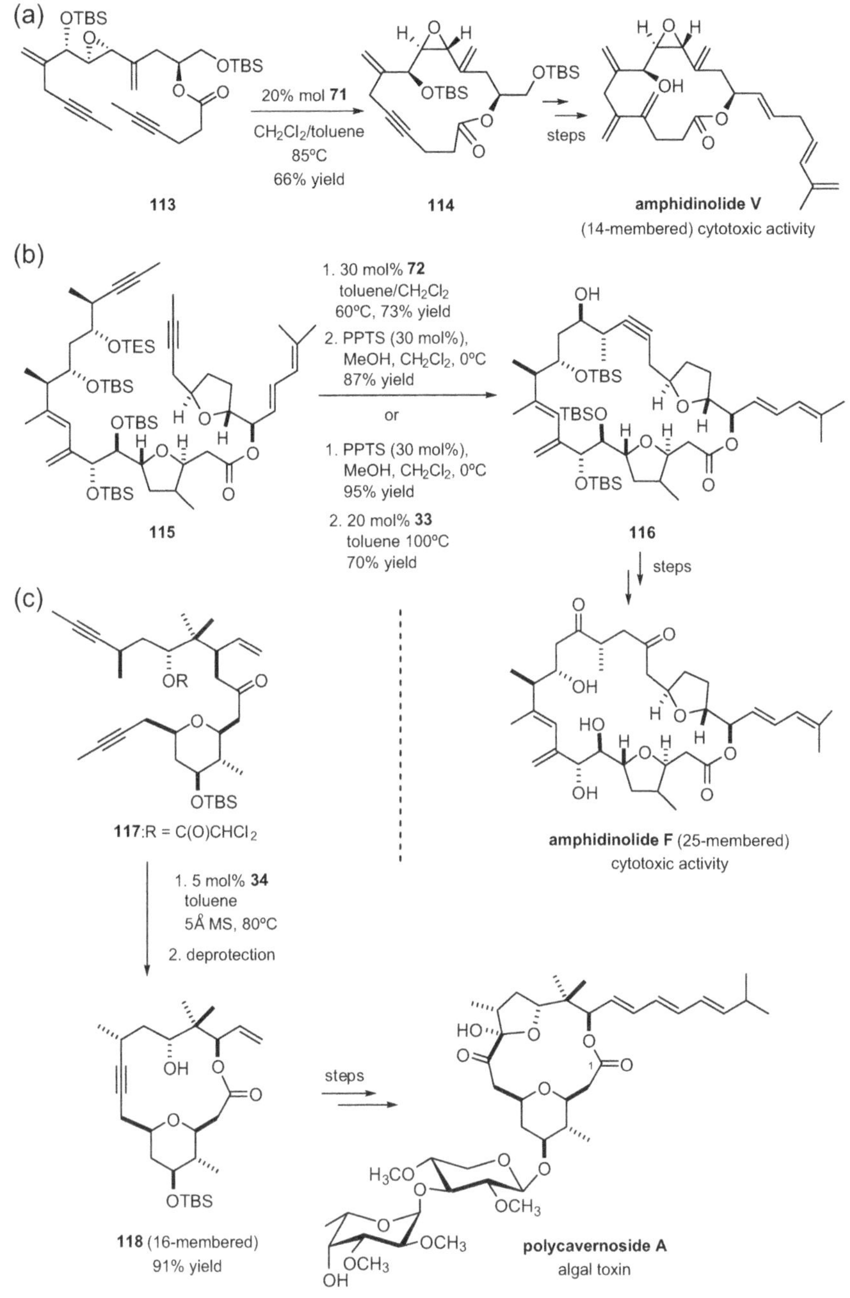

Scheme 29 Applications of RCAM to forge the macrocyclic skeleton of: (**a**) amphidinolide V; (**b**) amphidinolide F; (**c**) polycavernoside A

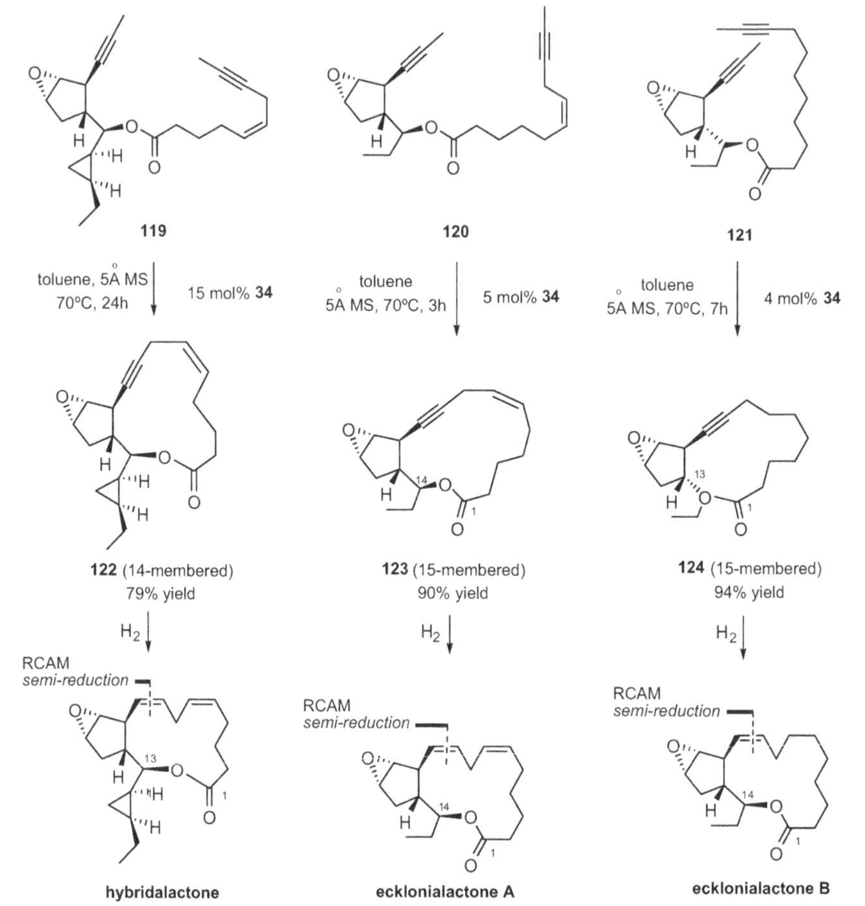

Scheme 30 Three marine oxylipins synthesized using RCAM methodology

able to maintain their cytotoxic potency against paclitaxel-resistant cell lines. Epothilones interrupt cell mitosis through interfering with the binding and functioning of tubulins. There is currently a good understanding of the conformational and binding issues involved in their interaction with tubulin (for reviews on epothilones, see [135–137]). An epothilone derivative, ixabepilone, received FDA approval for the treatment of metastatic breast cancer in 2007 [138].

The structure of epothilones consists of a 16-membered lactone with an additional 5-keto group and a lateral chain featuring a thiazole ring. The relative rigidity of the core lactone makes epothilones suitable for RCM methodologies which have been extensively used by Danishefsky and others (reviews: [16, 17]). Two strategies are possible: the (*Z*)-C12–C13 double-bond disconnection that implies a (*Z*)-selective RCM and the C9-C10 bond disconnection generally followed by hydrogenation as this is a single bond in the final structures. The first strategy, developed before the recent findings on new (*Z*)-selective catalysts, gave mixtures of (*E*/*Z*)

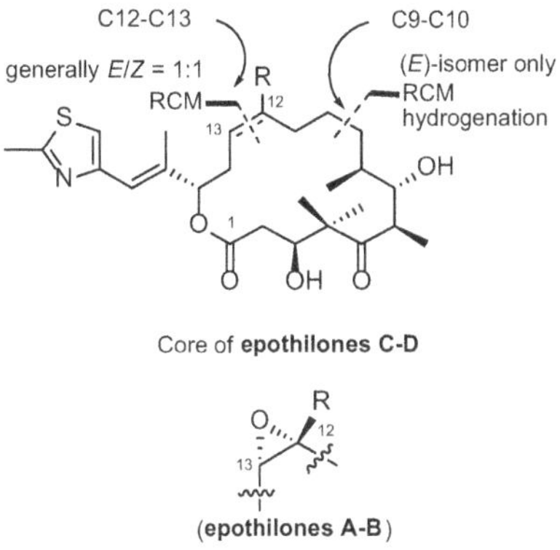

Fig. 12 Conditions for the two different RCM-based strategies used for the synthesis of epothilones

isomers which had to be separated. The second approach has the advantage of being compatible both with epoxides already situated at C12–C13 (for epothilones A–B) and with double bonds at this position (epothilones C, D diene-ene RCM) and gives exclusively the (*E*)-isomer which can be further hydrogenated. Yields are high using **G2** catalyst under typical reaction conditions (Fig. 12).

More recent synthetic efforts have been focused on the synthesis of potentially active epothilone derivatives. Numerous analogs retain potent in vitro activity especially 12,13-cyclopropyl analogs of EpoA and B. Cyclopropyl analogs are equally or even more potent than the epoxide-based parent compounds possibly because the epoxy ring has only a conformational role in the interaction with the target. In addition, replacement of the natural 12,13-epoxide moiety by a cyclopropane leads to better chemical and metabolic stability. An example of recent synthesis of these derivatives, 12,13-cyclopropyl-epothilone B and side-chain-modified derivatives, is shown in Scheme 31. A combination of a Yamaguchi esterification of alcohol **125** and acid **126** with RCM-based macrocyclization (**G2** catalyst in refluxing CH_2Cl_2) provided the macrolactones **127** as a 12:1 mixture of (*E*/*Z*) isomers [139].

The rhizoxin family is a group of natural polyketides isolated from fungi that can be considered closely related to epothilones because of their structural similarity. Indeed, they present a 16-membered lactone, a lateral chain bearing a heterocycle (oxazole in these cases) and epoxides in some members of the family. In addition, a fused six-membered lactone makes these compounds less flexible. Not surprisingly they are potent inhibitors of eukaryotic tubulin polymerization, exhibiting pronounced in vitro and in vivo antitumor activity. The only synthesis appeared to date that uses metathesis methodology is due to Altman's group. They have synthesized WF-1360F, a natural rhizoxin congener. In spite of extensive efforts, they were unable to close the macrocycle using diene RCM with a diverse number of alternative RCM substrates. The synthetic strategy was thus changed to ring-closing alkyne metathesis on **129**. The key reaction was catalyzed by Mo complex **35** (Fig. 4) giving **130** in 69% yield, which was further transformed into the target product (Scheme 32) [140].

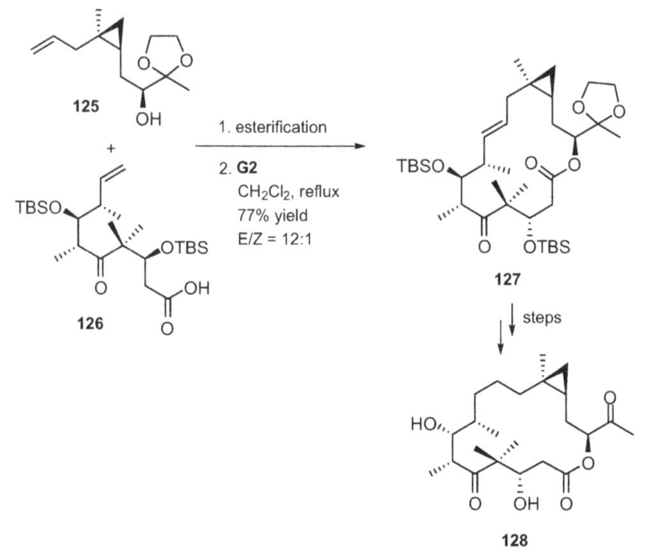

Scheme 31 Synthesis of 12,13-cyclopropyl analogs of EpoA and B

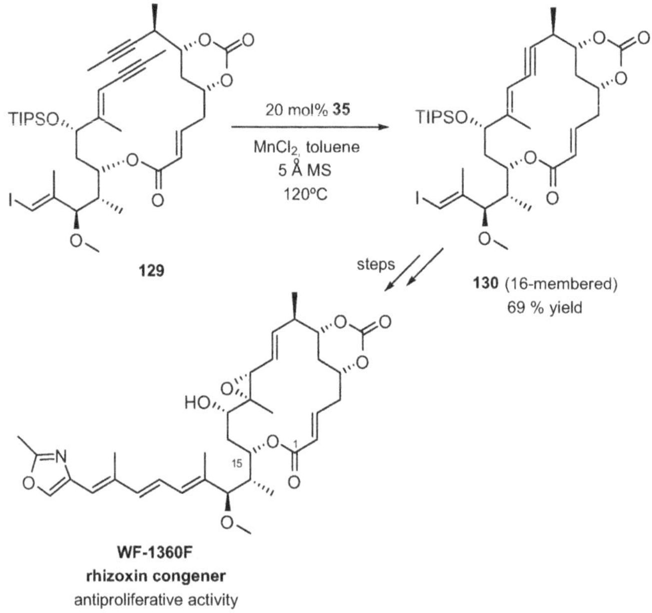

Scheme 32 Application of RCAM for the synthesis of a rhizoxin congener

3.1.7 Polyether Macrolides

With the improvement in synthetic efficiency of metathesis catalysts and their compatibility with many functional groups, greater synthetic challenges on more complex molecules have been undertaken recently by the synthetic community. Polyether macrolides can be treated as a separated group because they generally have less free hydroxyl groups but include several ethereal moieties generally cyclic and very often spiropolycyclic. Among them we encounter some of the most complex natural products synthesized to date. Most products are of marine origin and have been isolated in minute amounts. We present a selection of the most interesting examples reported recently.

Spirolide is a member of cyclic imine marine toxins that are selective inhibitors of nicotinic acetylcholine receptors (nAChRs). A study on the formation of the 23-membered macrocyclic framework en route to the spirolides was disclosed. The RCM reaction proceeded with low conversions in general. The best result was achieved with a compact substrate, **131**, which was subjected to optimized conditions for RCM (**G2** 30 mol%, 1,2-dichloroethane, 60°C), affording the macrocyclic product **132** (25% isolated yield) and recovered starting material (54% isolated yield; Scheme 33) [141].

Pectenotoxin is a powerfully toxic compound isolated from a scallop. The complex 34-membered lactone was constructed performing a RCM of triene **133**

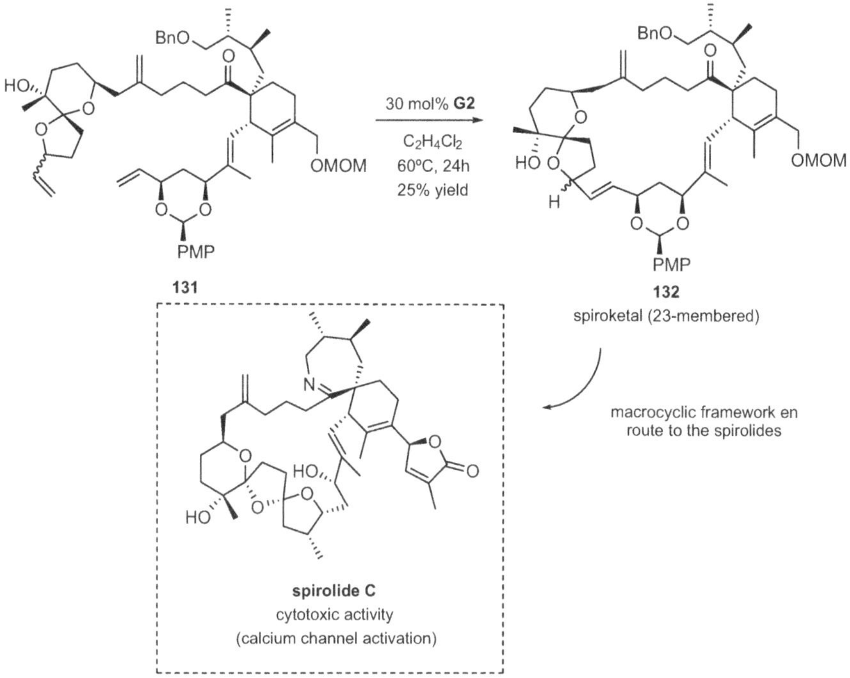

Scheme 33 Spirolide C, a marine toxin synthesized using a low-yielding RCM

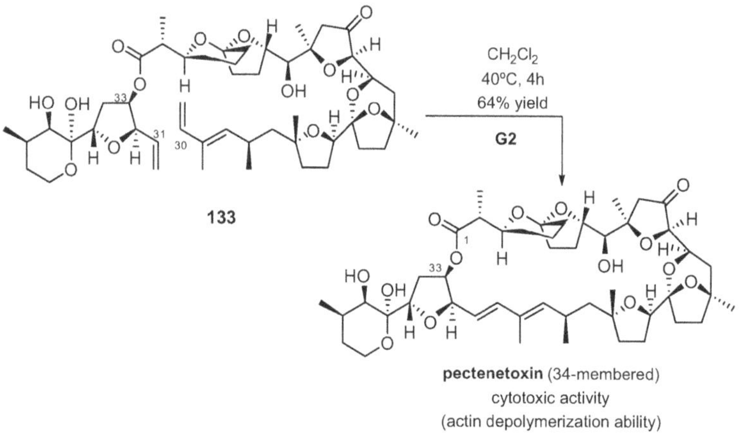

Scheme 34 The 34-membered lactone in pectenotoxin built using RCM

with 10 mol% **G2** (64% yield) without isomerization of the spiroacetal moiety (Scheme 34) [142].

In the total synthesis of pinnatoxin A, using **G2**, the bulky silyl protecting groups had to be removed, as they prevented RCM. Therefore, RCM macrocyclization of **134** gave **135**, which led to pinnatoxin A (Scheme 35) ([143]; for other RCM conditions in a recent synthesis, see [144]).

Bryostatins were isolated from extracts of a species of bryozoan, *Bugula neritina*. Bryostatins are potent modulators of protein kinase C. They are currently under investigation as anticancer agents and as a memory-enhancing agent. Few total syntheses have appeared to date due to the great complexity of these molecules. Trost's group is actually working on the total syntheses of these compounds and they selected RCM to construct the 21-membered lactone from densely functionalized substrates. Thus, they prepared compound **136** with the idea of performing a relay RCM. To avoid the undesired dimerization, a solution of **136** was added dropwise to a solution of **HG2** (18 mol%) in benzene. However, no relay-RCM product **137** was observed, but direct RCM products **138** were isolated as a mixture (*E*/*Z*) of isomers in 80% yield. Changing the catalysts to **G2** allowed an improvement in the stereoselectivity in favor of the (*E*)-isomer (Scheme 36) [145, 146].

The successful RCAM methodology developed by Fürstner has one of its most advanced applications in the total synthesis of spirastrellolide F methyl ester. The elaborated substrate **139** cyclized smoothly to cycloalkyne **140** when exposed to catalytic amounts of the molybdenum alkylidyne complex **34** (Fig. 4), which was used because of its efficiency and functional group compatibility. Product **140** was formed in 87% yield. From **140**, gold catalysis was used to construct the spiroketal moiety from the alkyne (Scheme 37) [147].

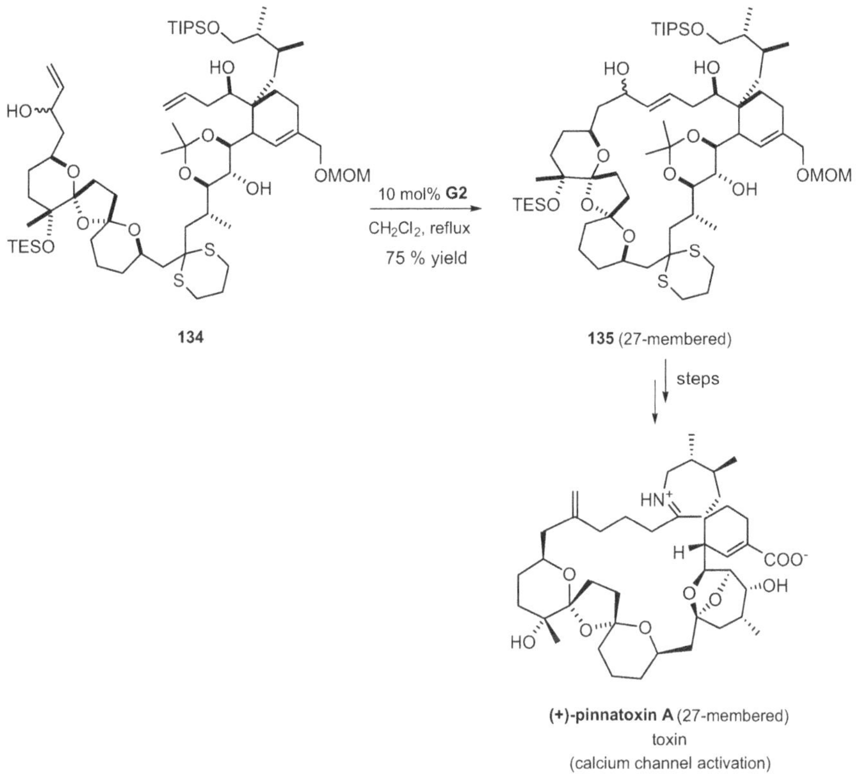

Scheme 35 Total synthesis of pinnatoxin A using a **G2**-catalyzed RCM

3.1.8 Polyketide Macrolactams

Pharmacologically active macrolactams from polyketide origin are highly functionalized and isolated from different sources. Macrolactamization is very often an excellent way to construct these macrocycles but some groups have applied metathesis protocols to achieve this target. Recent examples of biologically active macrolactams synthesized by metathesis follow [148].

A new class of ansamycin compounds, including a (*E*,*E*,*E*)-triene within a 21-membered ring lactam and aromatic cores, was described in recent years. In particular, cytotrienins and trienomycins are important antitumoral compounds. Synthesis of cytotrienins A–D was disclosed by Panek's group and involved a challenging diene–diene RCM step on polyenic substrate **141**. In the reaction of **141** with **G2**, activation of the internal bond of one of the 1,3-dienes gave a 19-membered ring cycle. Fortunately, the desired activation of the terminal olefins by the catalyst was achieved using the less reactive catalyst **G1** in refluxing CH_2Cl_2. These conditions afforded the expected (*E*,*E*,*E*)-triene with excellent selectivity and yield (Scheme 38a) [149]. A synthesis of cyclotrienin A, appeared later, used the same RCM approach on a more elaborated polyene bearing the cyclopropyl-

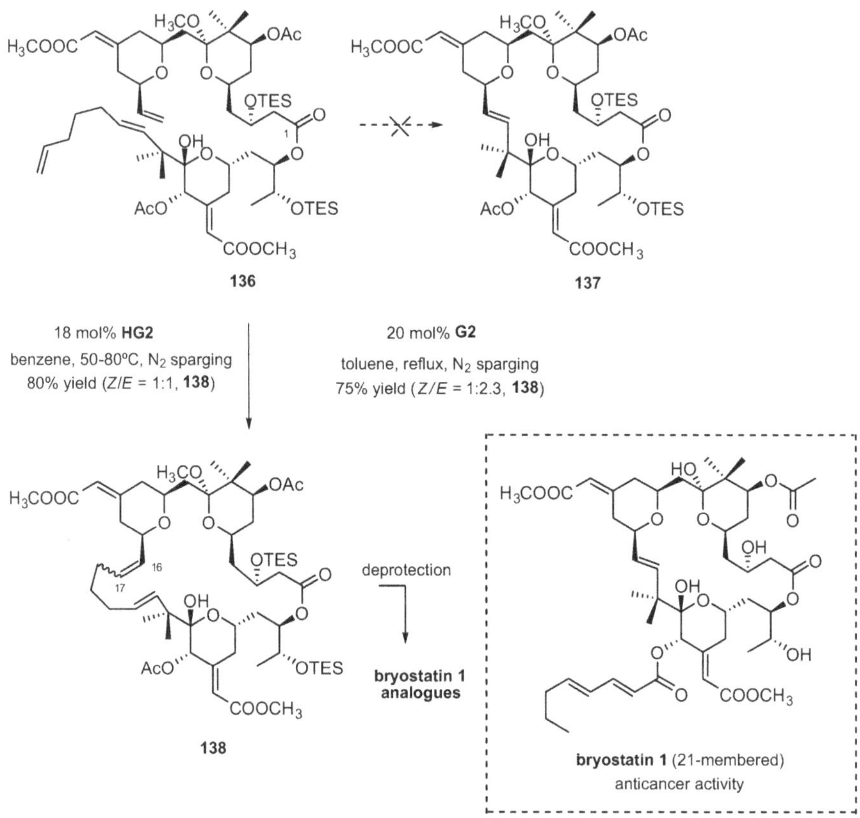

Scheme 36 The 21-membered lactone in bryostatins prepared using a non-stereoselective RCM

containing lateral chain. The metathesis furnished the desired product albeit in 39% yield under **G1** catalysis [150]. More recently, the synthesis of trienomycins A and F was disclosed with a similar strategy. As reported, with **G1**, **G2**, and **HG2**, RCM occurred to form substantial quantities of undesired diene products (30–40%). The indenylidene analog **M1** was more selective for triene formation, although modest yields were obtained, perhaps due to the lack of a conformational-biasing element in the form of a C19 substituent in **142** (Scheme 38b) [151].

Two more examples of polyenic macrolactam synthesis account for the difficulties in achieving regioselectivity in diene–diene RCM. After exploring reaction conditions for the RCM of **143** with various ruthenium carbene catalysts, **Grela** catalyst in the presence of *p*-methoxyphenol (PMPOH) gave incednam in 17% overall yield after deprotection of the TES groups in the resulting cyclic product. Incednam is the aglycon of the 24-membered macrolactam glycoside antibiotic incednine (Scheme 39a) [152]. For the synthesis of protected vicenilactam, on the other hand, the desired polyene **145** was obtained using **G1** with *p*-quinone as

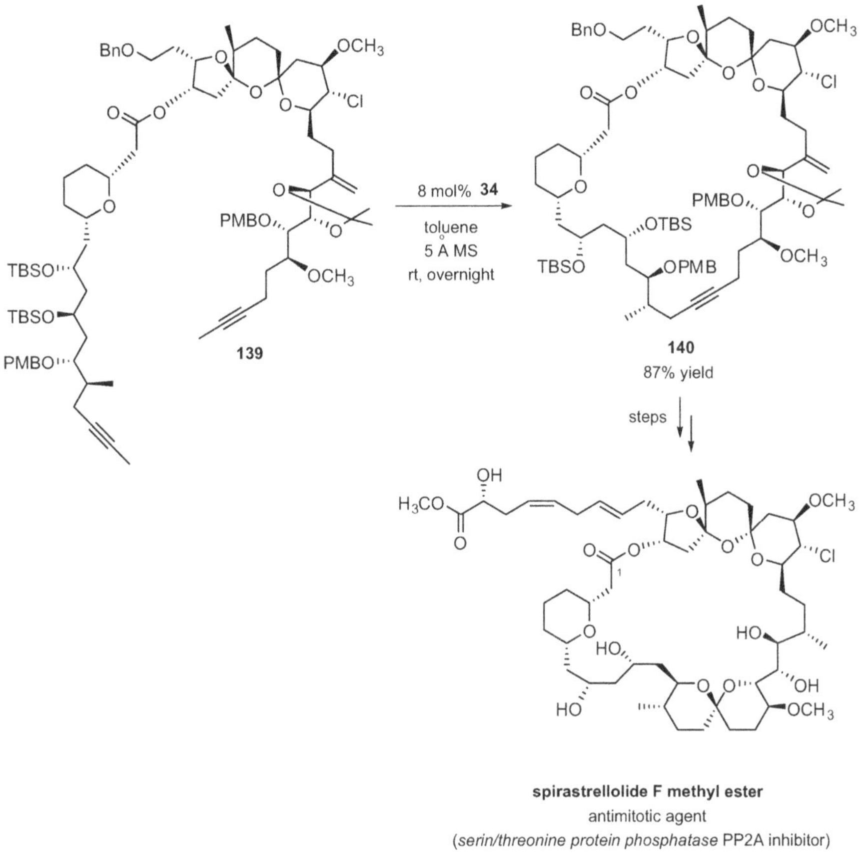

Scheme 37 RCAM methodology applied for the total synthesis of spirastrellolide F methyl ester

additive. A 5:1 mixture of (*E*/*Z*) isomers was obtained in 65% yield (Scheme 39b) [153].

3.1.9 Ansamycins

Kendomycin discloses a unique molecular architecture with an ansa-polyketide chain, a pentasubstituted tetrahydropyran ring and a *p*-quinone-methide chromophore. Due to the diverse pharmacological qualities, several total and partial syntheses have appeared. In two cases RCM was selected to construct the macrocycle [154]. In 2006, a total synthesis of kendomycin was achieved constructing the C13–C14 bond via RCM. It was interesting to note that the presence of a keto group at C19 precluded the metathesis step while a hydroxyl group (**146**) allowed the formation of **147** in 57% yield using **G2** as the catalyst. Unfortunately the exclusive product obtained in all conditions tested was invariably

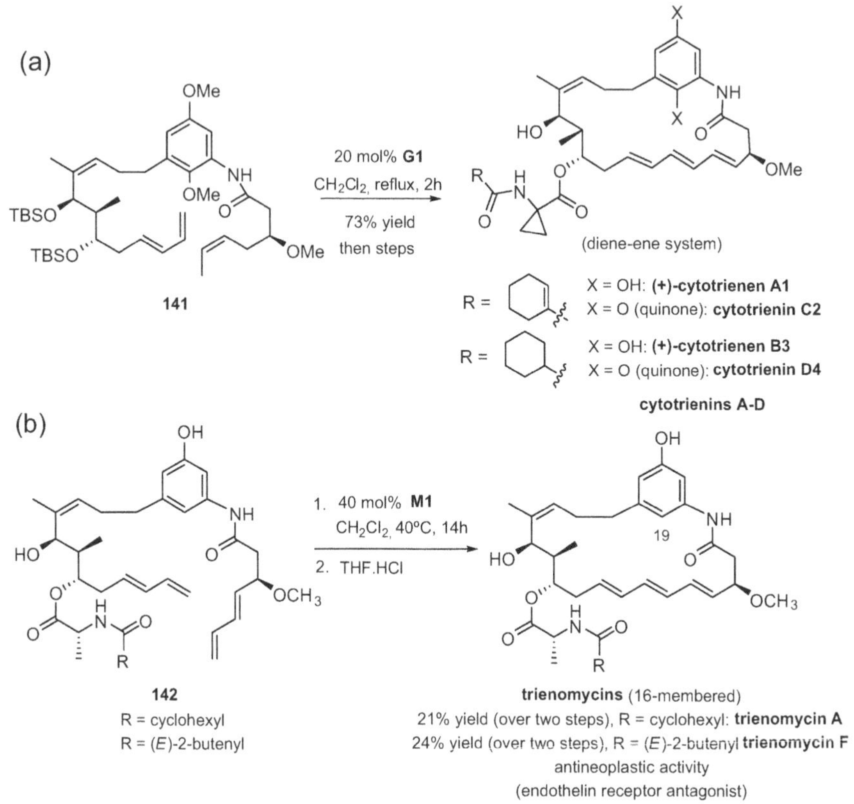

Scheme 38 A selective RCM for the synthesis of (*E*,*E*,*E*)-triene-containing ansamycin compounds: (**a**) synthesis of cytotrienins A–D; (**b**) synthesis of trienomycins A and F

the undesired (*Z*)-isomer, which had to be isomerized in four steps (Scheme 40a) [155]. More recently, the macrocyclic core of kendomycin was built by RCM to form the C10–C11 bond, which was subsequently reduced. The metathesis was effected on substrate **148** under identical conditions affording the desired macrocyle in excellent yield and almost exclusively as the (*E*)-isomer **149** (15:1). The success of this RCM came after bad results with other less elaborated starting materials with keto or hydroxyl groups at C5 (Scheme 40b) [156].

In view of the problematic construction of C13–C14 bond through RCM, Fürstner envisaged using ring-closing alkyne metathesis (RCAM) as noteworthy alternative. In addition the new molybdenum catalysts had proved to be very tolerant to delicate functionalities and they expected good results with elaborated substrate bearing the benzofurane and the phenolic hydroxyl group. Indeed, they prepared substrate **150** and reacted it with the latest generation of alkyne metathesis catalysts **33** (Fig. 4) obtaining **151** in excellent yield under notably mild conditions (Scheme 40c). This cycloalkyne was engaged in a gold-catalyzed

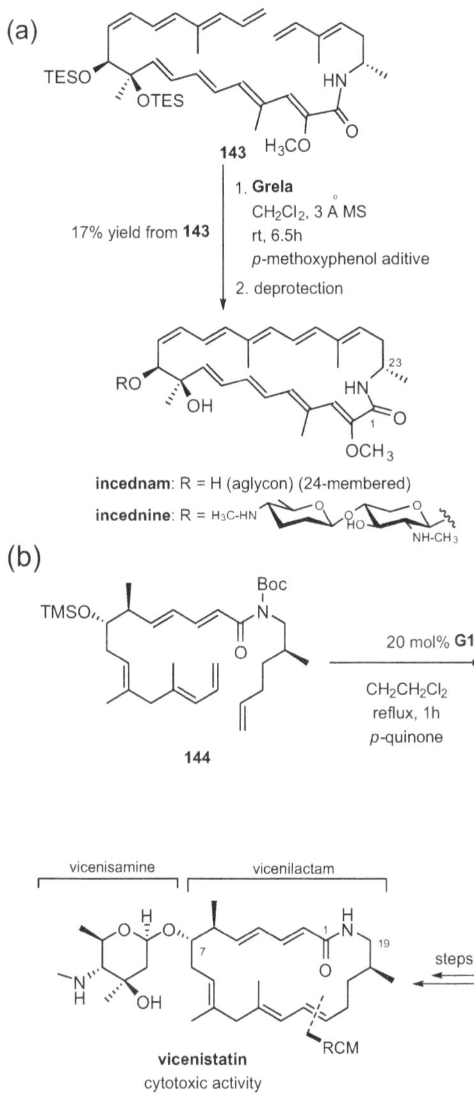

Scheme 39 Diene-RCM for the synthesis of polyene containing macrocycles: (**a**) synthesis of glycoside antibiotic incednine; (**b**) synthesis of protected vicenilactam

hydroalkoxylation, which led to a benzofuran derivative that had already previously served as a late-stage intermediate en route to kendomycin [157].

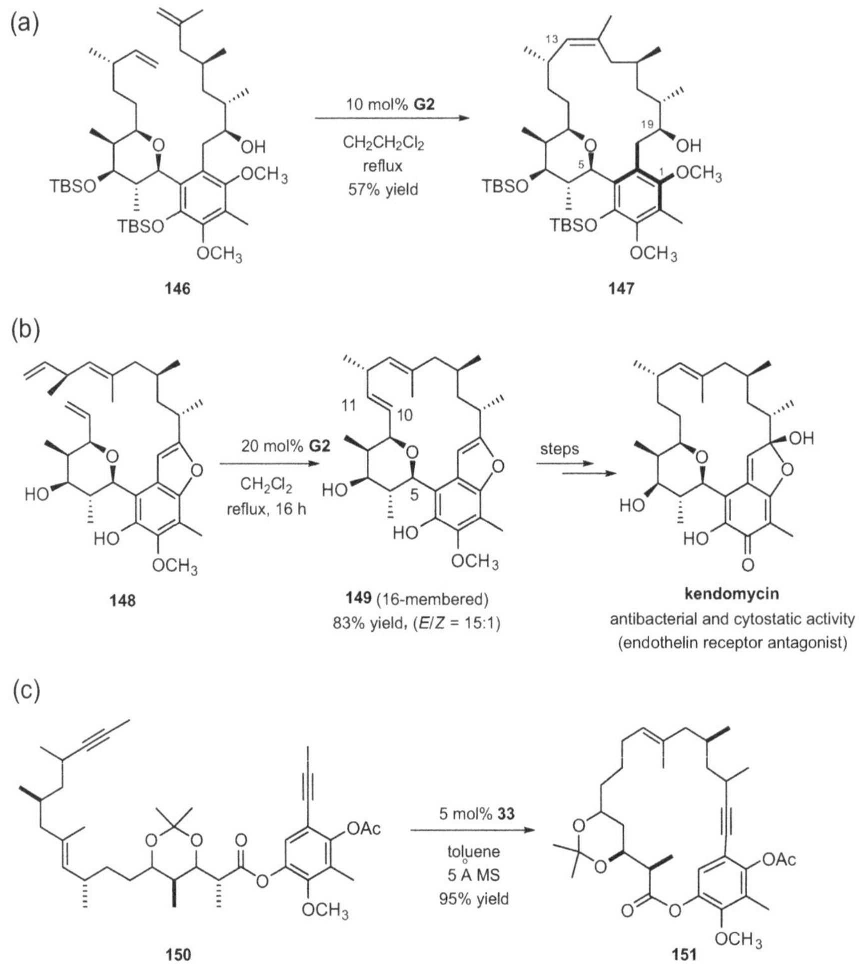

Scheme 40 Construction of the macrocyclic core of kendomycin through three different strategies: (**a**) building of the C13–C14 bond via RCM which gives the undesired Z-isomer; (**b**) macrocyclization by RCM/reduction to form the C10–C11 bond; (**c**) construction of the C13–C14 bond trough RCAM methodology

3.2 Metabolites from Mevalonate Origin: Terpenoids

Terpenoids constituted by four mevalonate units are among the most abundant macrocycles in nature. Some present oxygen bridges in their structure and thus fall into the heterocycle group. Scheme 41 summarizes the examples in which RCM was used for the cyclization step.

Among the two main possible strategies to construct the macrocyclic diterpene tonantzitlolone by using a RCM, when trying to build the double bond at C1–C2,

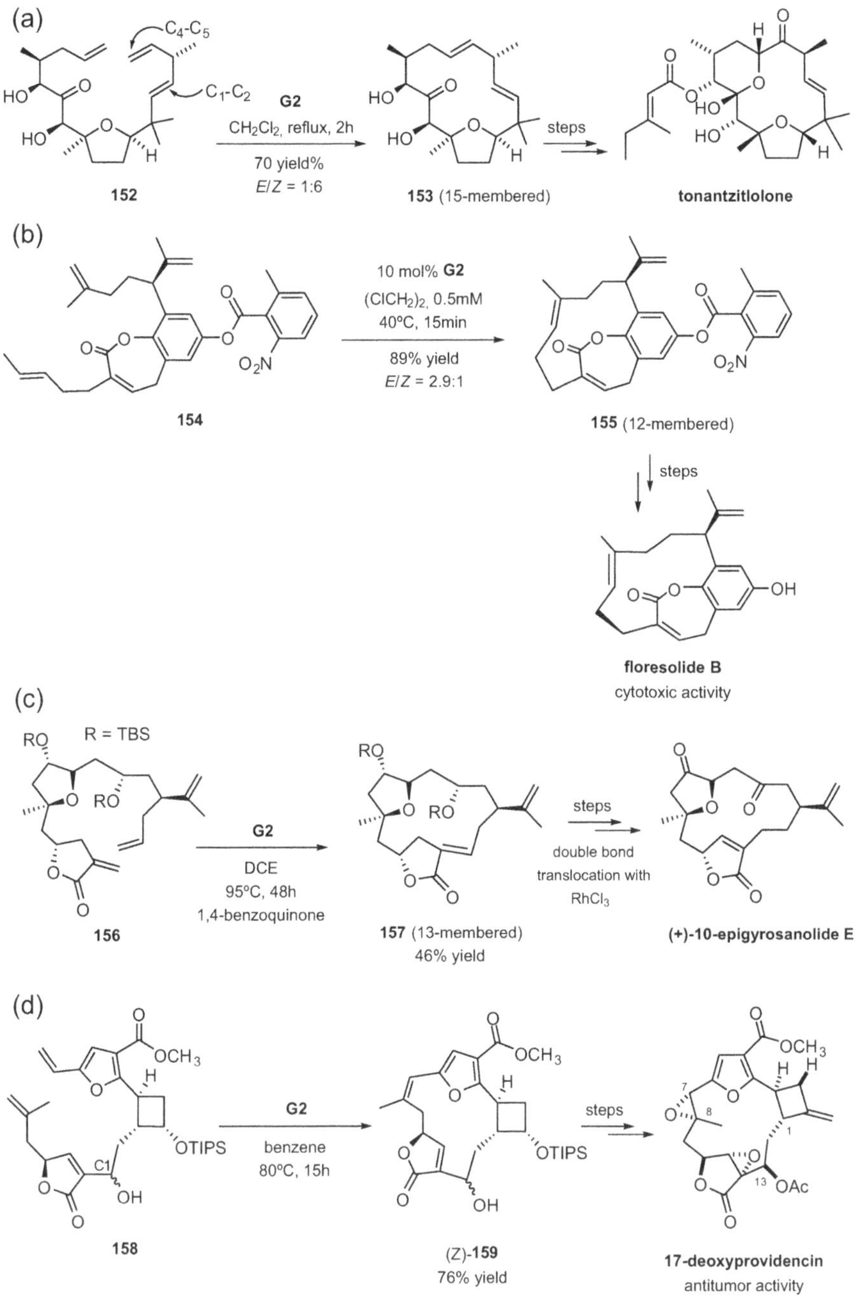

Scheme 41 Recent examples of terpenoid synthesis using RCM: (**a**) construction of the macrocyclic diterpene tonantzitlolone; (**b**) synthesis of floresolide B; (**c**) synthesis of (+)-epigyrosanolide E; (**d**) synthesis of 17-deoxyprovidencin

no final product was formed, probably due to steric congestion created by the quaternary allylic position [158]. On the contrary the macrocycle **153** was obtained from **152** with high *E* selectivity due to less congested olefins (Scheme 41a).

The second example accounts for the synthesis of floresolide B, a recently discovered natural product that exhibits cytotoxicity against KB tumor cells. This product contains an aromatic ring connected to a [12]-metacyclophane and a 7-membered ring lactone, which was installed previously to the metathesis. This rendered a rather constrained substrate, **154**, favoring the metathesis process, which proceeded in high yield albeit low (*E*/*Z*) selectivity (Scheme 41b) [159].

For the synthesis of (+)-epigyrosanolide E, a representative member of soft coral norcembranolides, the crucial RCM reaction was successfully carried out in the presence of **G2** catalyst and 1,4-benzoquinone providing macrocycle **157**. Formation of endocyclic olefin isomers from the starting material was a serious problem in the absence of 1,4-benzoquinone. Double-bond shift in **157** proceeded smoothly in the presence of rhodium chloride, which also effected global TBS deprotection (Scheme 41c) [160].

Furanocembranoids are a family of complex marine natural products isolated from a Caribbean sea plume. Their biosynthesis is unknown but probably come from the mevalonic pathway. In the synthesis of 17-deoxyprovidencin, large-scale RCM was performed on a diastereomeric mixture (at C1), which resulted in the formation of the readily separable *Z* olefins **159** (Scheme 41d) [161].

3.3 Natural Products from Amino Acid Origin

3.3.1 Cyclodepsipeptides

Jun et al. [162] in this book covers the synthesis of cyclic peptides by RCM and includes examples of cyclodepsipeptides. These are interesting macrocycles which contain proteinogenic and non-proteinogenic amino acids with one or more peptidic bonds substituted by a lactone moiety. In order to complete the panorama of macrocyclizations using RCM, we show here the synthesis of (+)-jasplakinolide (jaspamide) and chondramine C. These marine natural products are known for stabilizing F-actin fibers and for a wide variety of other biological activities. Jasplakinolide has shown potent cytotoxicity toward various cancer cell lines as well as fungicidal, insecticidal, anthelmintic, and antimalarial activity. The first synthesis of these two natural products in which metathesis was used to construct the cycle appeared in 2010. The authors showed the feasibility of using this methodology under typical conditions although yields were not completely satisfactory. Metathesis of substrate **160** proceeded smoothly (52% yield as (*E*)-isomer) toward the precursor of chondramine C. However, in the case of jaspamide, a combined yield of only 28% was achieved in the metathesis step although close derivatives could be obtained in much better yields (Scheme 42) [163, 164].

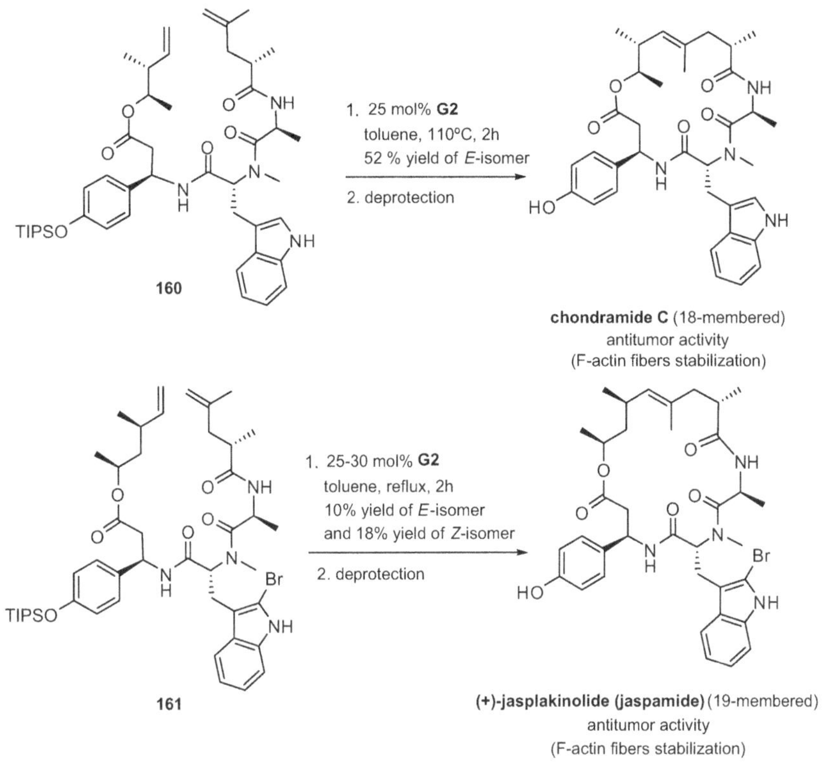

Scheme 42 Construction of the macrocycles in cyclodepsipeptides chondramine and jasplakinolide by RCM

3.3.2 Alkaloids

Alkaloids, generally from plant origin but also found in animals, present a rich variety of biological activities. This has stimulated many organic synthetic groups, and these compounds very often present macrocyclic structures. The main problem from a metathesis point of view is the ability of the nitrogen to coordinate to metal-alkylidene complexes and to interfere unproductively with the catalytic activity. In [46], devoted to metathesis reactions en route to cyclic amines, the ways to circumvent this problem are explained including examples of macrocyclic amines. Therefore we will just summarize herein additional cases to complete the view of the metathesis protocols applied to macrocycles. Figure 13 outlines examples of structurally similar alkaloids. The substrates selected for the key metathesis reactions leading to these macrocycles present great conformational flexibility. However, after careful selection of conditions, the syntheses were finalized successfully.

Marineosin A is a novel macrocyclic spiroaminal alkaloid. After surveying a number of RCM catalysts and conditions, the 13-membered macrocycle was built in 84% yield using **G1** (30 mol%) under dilute conditions, as **G2** provided primarily

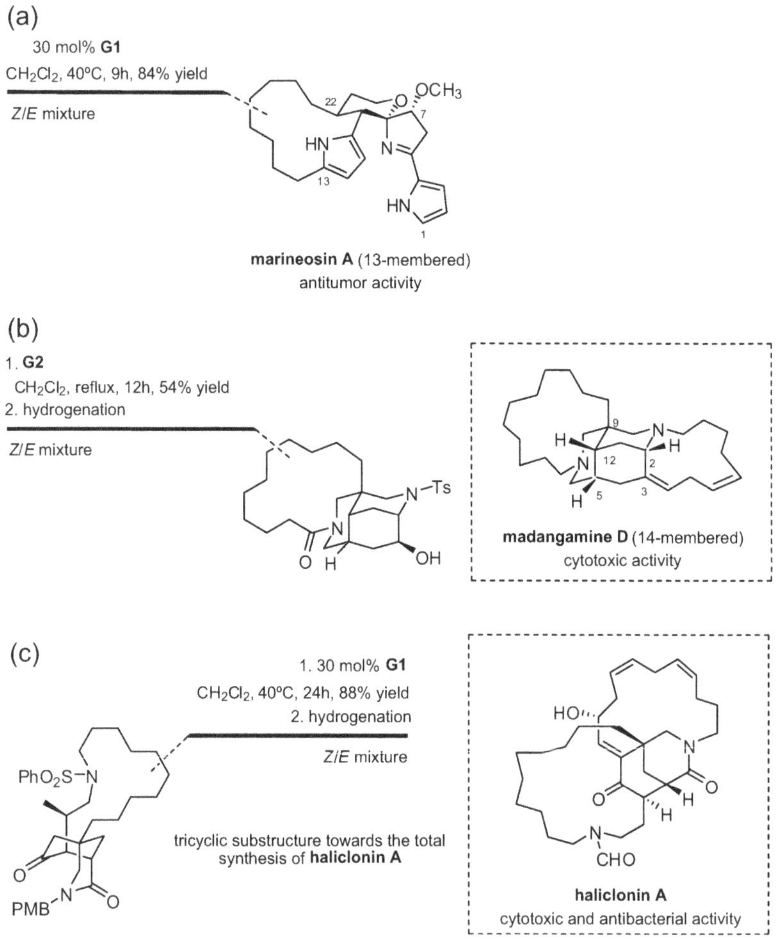

Fig. 13 Conditions of the RCM key step for the synthesis of alkaloids: (**a**) marineosin A; (**b**) madangamines; (**c**) haliclonin A

cross-metathesis products. Subsequent hydrogenation was performed (Fig. 13a) [165]. Madangamines are a group of bioactive marine sponge alkaloids with a challenging diazatricyclic core and two peripheral macrocyclic rings. Using **G2** under high-dilution conditions, RCM satisfactorily gave the desired macrocycle as a (*Z*/*E*) mixture of alkenes which was further hydrogenated (Fig. 13b) [166]. A related antimicrobial alkaloid, haliclonin A, which exhibited antibacterial activity against diverse microbial strains, was synthesized using a similar RCM/hydrogenation strategy (Fig. 13c) [167].

We have also selected the very recent synthesis of (−)-lythranidine, a piperidine-containing alkaloid, to show the applicability of the RCAM/reduction methodology to this group of natural products. A ring-closing alkyne metathesis (RCAM) of a propargyl alcohol derivative was envisaged although this type of

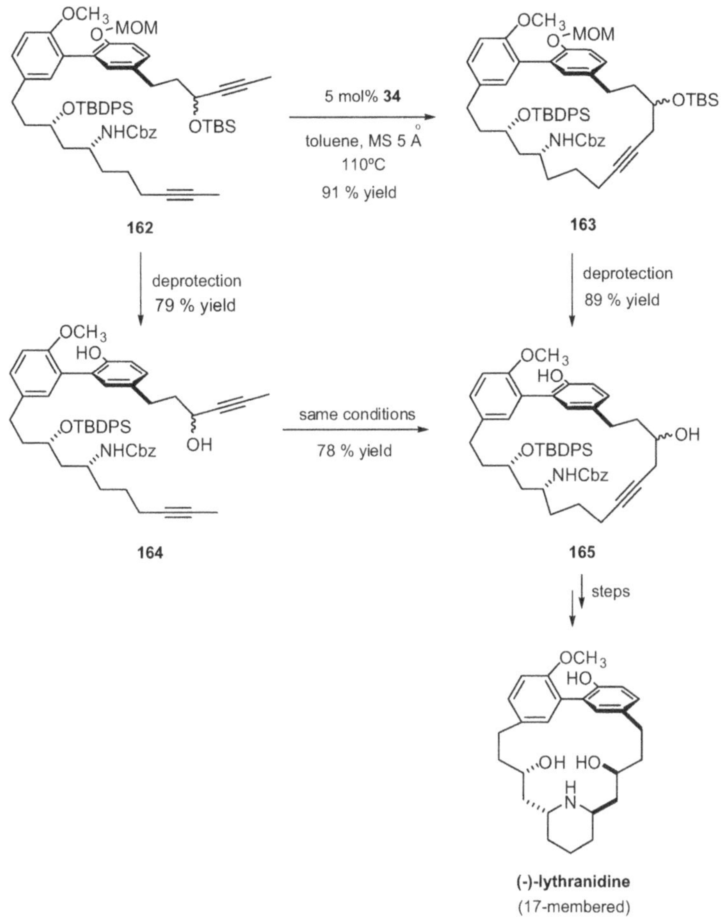

Scheme 43 Synthesis of (−)-lythranidine using RCAM methodology

substrates is in principle incompatible with alkylidyne catalysts. However, as we have shown previously in this chapter, the new molybdenum alkylidyne complex **34** (Fig. 4) endowed with triarylsilanolate ligands is fully compatible with many functional groups. The macrocyclization of diyne **162** through RCAM furnished the desired product **163** in 91% yield on a 1.4 g scale, and cleavage of the TBS ether and the phenolic MOM acetal gave the free propargyl alcohol **165**, an advanced intermediate in the synthesis of (−)-lythranidine. Interestingly the order of events could also be reversed and the free propargyl alcohol **164** successfully gave RCAM product **165** (Scheme 43) [168].

3.4 Macrocyclic Glycolipids

Macrocyclic structures that contain embedded carbohydrate residues can be defined as those where at least two bonds from a monosaccharide residue form part of the macrocycle [169]. Such structures occur in nature as resin glycosides or as other kinds of macrocyclic glycolipids.

Various resin glycosides and sugar-based macrodiolides such as tricolorins A and G and woodrosin I were synthesized by Fürstner's group some years ago using RCM and further hydrogenation [170]. Figure 14 shows the structures and RCM conditions used for the synthesis of tricolorins A and G, two natural products that are plant growth inhibitors and protect sugar cane against invasive weeds, as well as woodrosin I. The RCMs were carried out with the first-generation Ru-alkylidene complex **166** and proceeded in high yields.

More recently, the same group has disclosed the total synthesis of cytotoxic agents ipomoeassin B and E and gobienine A. Exposure of diene **167** to **G2** (10 mol %) in refluxing CH_2Cl_2 afforded the corresponding macrocycle in excellent yield as a mixture of isomers, which could be selectively hydrogenated/deprotected to give ipomoeassin E (Scheme 44a) [171, 172]. For the synthesis of gobienine A, the macrocyclization was performed on **168** using the ruthenium indenylidene complex

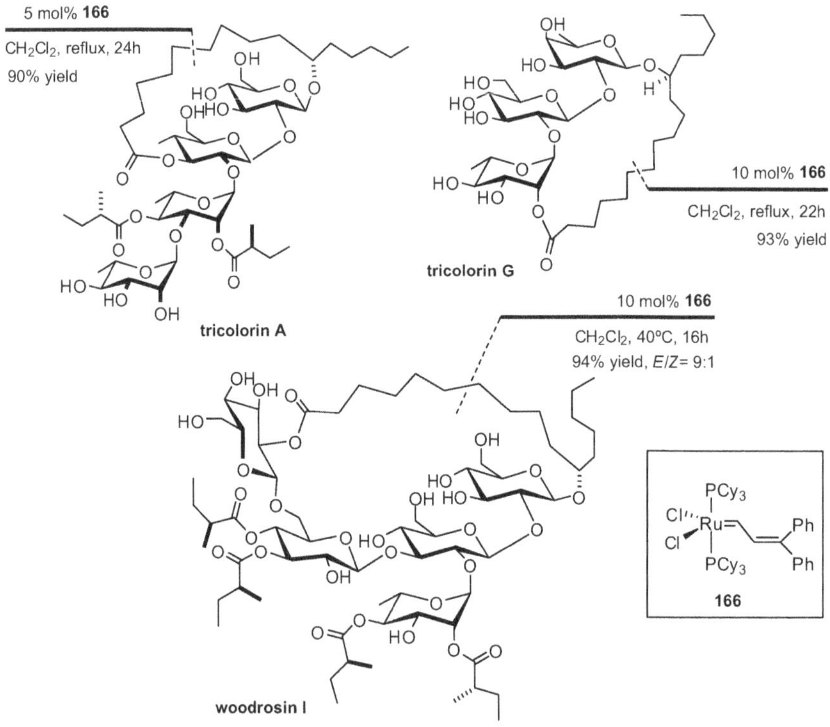

Fig. 14 Conditions of the RCM step for the construction of macrocyclic glycolipids

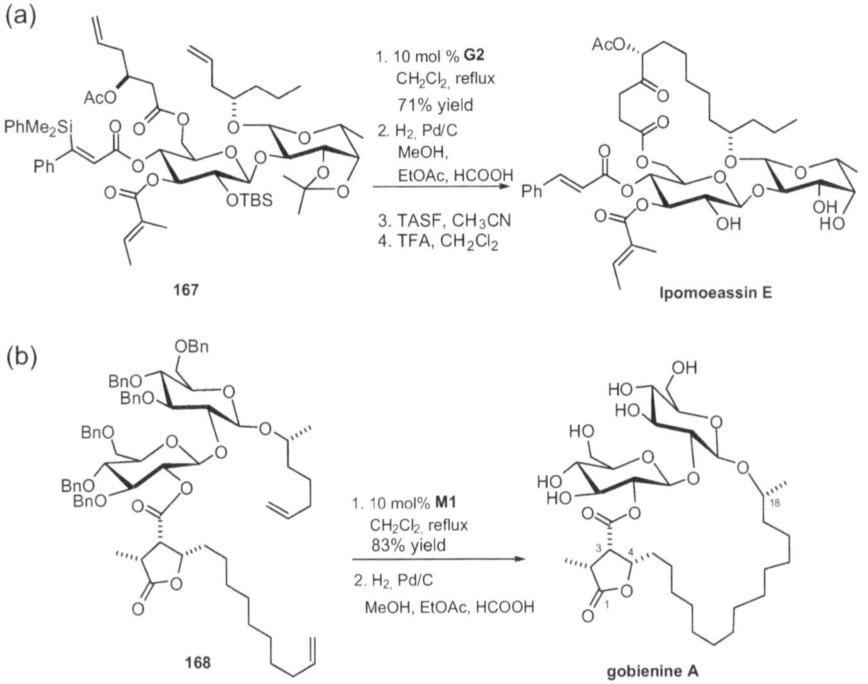

Scheme 44 Metathesis step for the synthesis of: (**a**) ipomoeassin B and E; (**b**) gobienine A

M1 as alternative to **G1**. RCM reactions proceeded smoothly, furnishing the desired product in 83% yield as an inseparable (*E*/*Z*) mixture. All peripheral benzyl ethers were hydrogenolytically cleaved while saturating the double bond over palladium on charcoal giving the target product (Scheme 44b) [173].

4 Nonnatural Product Synthesis

4.1 Synthetic Druggable Compounds

As we have seen in the examples highlighted in this chapter, macrocyclic scaffolds are commonly found in bioactive natural products and this has prompted the synthetic effort toward them and closely related analogs. In this section we will select some examples of synthesis of pharmaceutical drug-like macrocycles generally inspired from natural products but structurally differentiated from them [174].

4.1.1 Macrocyclic HCV Protease Inhibitors

One potential option for the treatment of hepatitis C virus (HCV) is the design of protease inhibitors. HCV serine protease NS3 represents an attractive drug target because it is not only essential for viral replication but also implicated in the viral evasion of the host immune response pathway through direct cleavage of key proteins in the human innate immune system. Among the most promising molecules with this activity, vaniprevir is a compound in late-phase clinical studies disclosed in 2010. The synthesis of this compound was published in 2011 using a RCM for the construction of macrocycle **170** among other alternatives. The optimization of the RCM route led to a procedure that gave good yields, albeit requiring high catalyst loadings and high-dilution conditions (<3 mM). The catalyst chosen was **M1** due to its easy access and relatively low cost. However, when catalyst loading was reduced or when the concentration of the reaction was increased, substantially lower yields were obtained [175]. Later on, a new synthesis of vaniprevir undertook the challenge of improving two aspects of the RCM reaction in order to maintain an economically viable process when scaling it up: the high-dilution conditions typically employed for RCM and the relatively high catalyst loading. The authors found that slow addition of the catalyst allowed loadings of less than 1 mol% (**HG2** was selected). The use of quinone additives, particularly 2,6-dichloroquinone, effectively decreased the formation of side products even under 20 mL/g diene dilution (Scheme 45) [176].

Through structure-based drug design and optimization, macrocyclic peptidomimetic molecules with increasing protease activity have been designed and

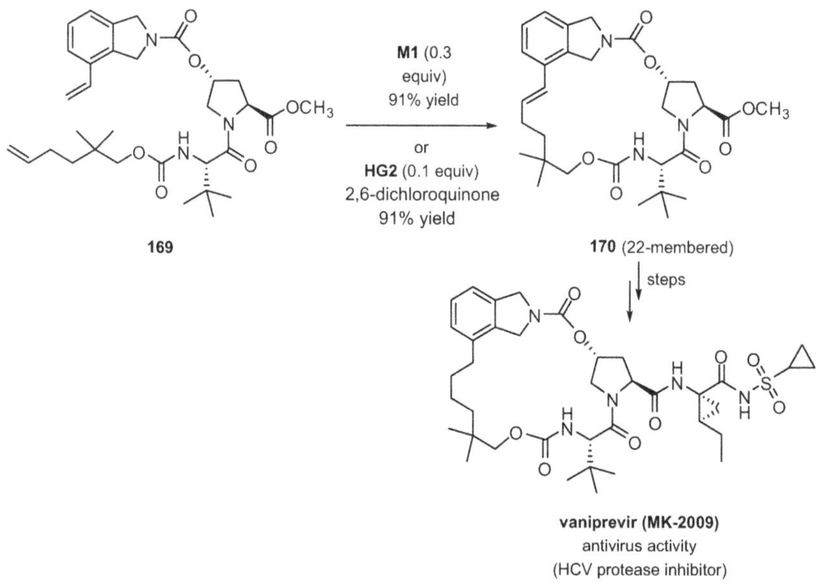

Scheme 45 Synthesis of hepatitis C virus (HCV) protease inhibitor vaniprevir

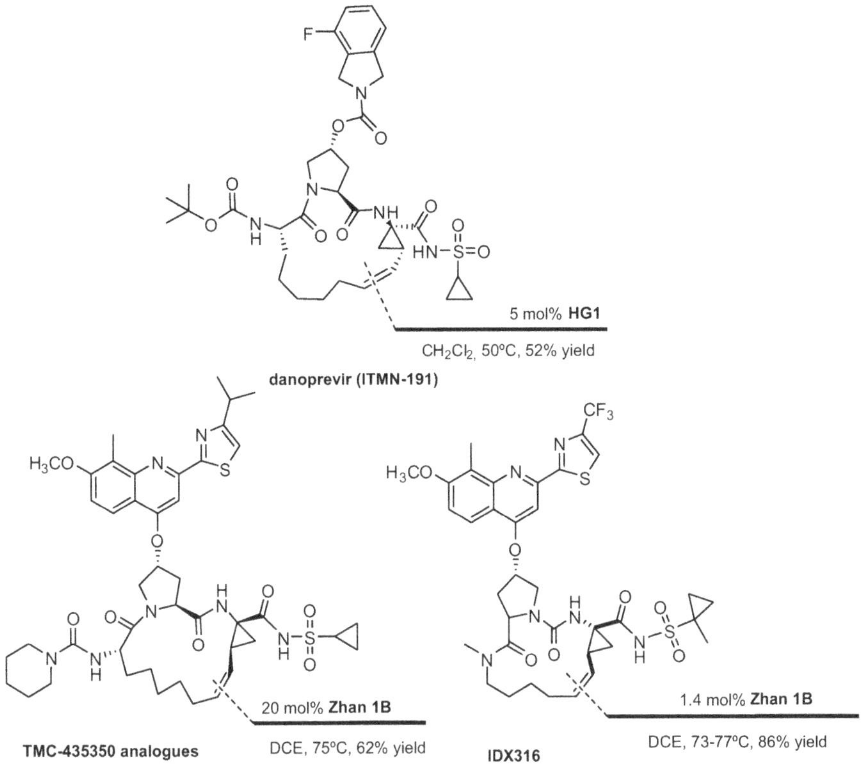

Fig. 15 Conditions for gram-scale RCM in the synthesis of HCV protease inhibitors

synthesized. There has been a great effort in finding scalable conditions for the RCM step, and in various cases extremely efficient procedures have been developed. Figure 15 shows the structures of several NS3 protease inhibitors: **danoprevir** [177], **TMC-435350**, and analogs **IDX316** [178] and **IDX320** [179].

Scheme 46 shows the synthesis of another compound of this group, a good example of how slight modifications in the substrate can lead to great improvements in the metathesis step. Thus, *N*-Boc protection was introduced in amide **171** in order to change the initiation site of the RCM reaction by interrupting the coordinative stabilization by the ester group. This allowed an increase in the reaction concentration to 0.1–0.2 M using **Grela** catalyst (0.1 mol%). The RCM reaction furnished the desired RCM product **172** in excellent 93% assay yield [180, 181].

4.1.2 Other Bioactive Compounds

Proteases are known to bind their substrates and inhibitors in a β-strand geometry. Thus, for other targeted proteases, cyclic peptidomimetics can be efficient inhibitors and several groups have designed and synthesized these kinds of macrocycles using RCM for the key cyclization. A group of macrocyclic transition-state

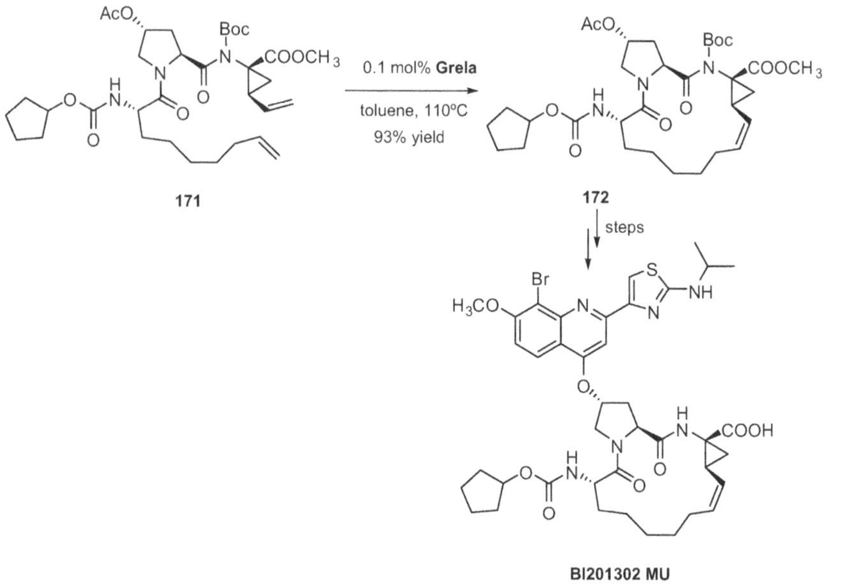

Scheme 46 Low catalyst loading in the RCM step for the synthesis of a HCV protease inhibitor

mimicking HIV-1 protease inhibitors were prepared and biologically evaluated. Their structure consisted of 14- and 15-membered macrolactams constructed by means of a RCM of the corresponding linear dienes. Figure 16 shows **173** as a representative example [182].

Abell's group reported the design and synthesis of macrocyclic *β*-strand templates **175** for use in the development of cysteine protease inhibitors. The key RCM of an appropriate peptide-based substrate **174** proceeded in principle with moderate yields (10 mol% of **G2** under trichlorethane reflux, yields from 20% to 50%), albeit high (*Z*)-selectivity [183]. Thus, they performed a study to optimize the reaction. It was found that under MW irradiation, chlorodicyclohexylborane addition as Lewis acid, and three sequential additions of **G2**, yields became nearly quantitative in most cases. The Lewis acid was added to avoid formation of ruthenium chelates with the carbonyl groups present in the substrates (Scheme 47) [184].

Insulin-regulated aminopeptidase (IRAP) inhibitors were disclosed recently based on angiotensin IV. A number of Ang IV-inspired macrocyclic derivatives were designed and synthesized. After attaching an azide to a resin, the authors performed classical solid-phase peptide chemistry, which gave the solid-supported diene. The attempted RCM of this diene was low yielding and thus they performed the metathesis step after cleaving the substrate from the resin. They selected **HG2** as the catalyst under MW heating and in the presence of benzoquinone (Scheme 48) [185].

Macrocyclic paclitaxel congeners were designed to mimic the bioactive conformation of paclitaxel. The design consisted of a rigidification of the baccatin moiety by connecting the C14 position and the *ortho* position of C3′*N*-benzoyl group with a

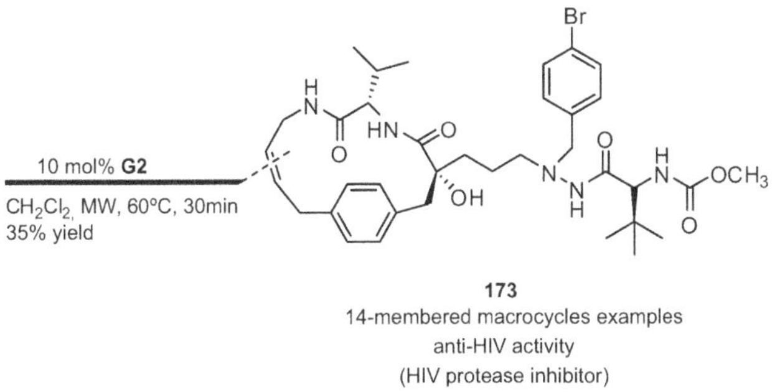

Fig. 16 Conditions of the RCM step in the synthesis of a HIV inhibitor

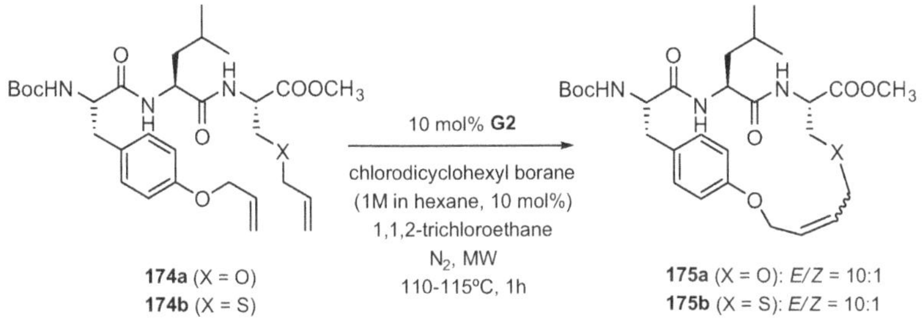

Scheme 47 (*Z*)-selective metathesis for the synthesis of a *β*-strand template

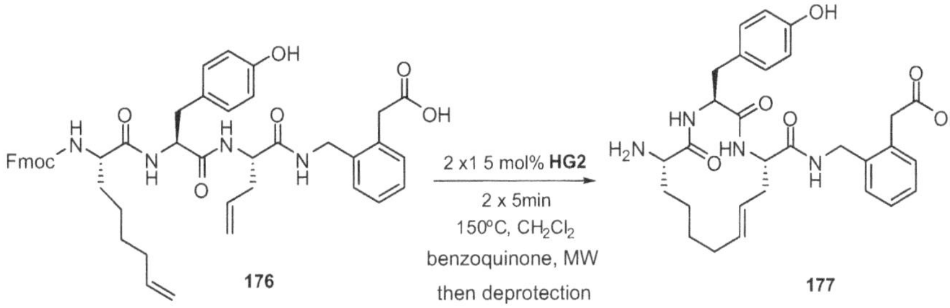

Scheme 48 Synthesis of angiotensin IV-inspired macrocyclic derivatives

short linker. Starting from an allyloxybaccatin III, they prepared the suitable paclitaxel-dienes **178**. The RCM using 20 mol% of **G1** gave the desired 15- and 16-membered macrocyclic taxoids in good yields. However, the RCM reaction to form the designed 14-membered macrocyclic taxoid from **178b** did not proceed as planned. Instead, the attempted RCM reaction led to the occurrence of Ru-catalyzed diene-coupling process, giving the corresponding 15-membered macrocyclic taxoid **181**, possibly through the formation of ruthenium hydride intermediate **180** (Scheme 49) [186].

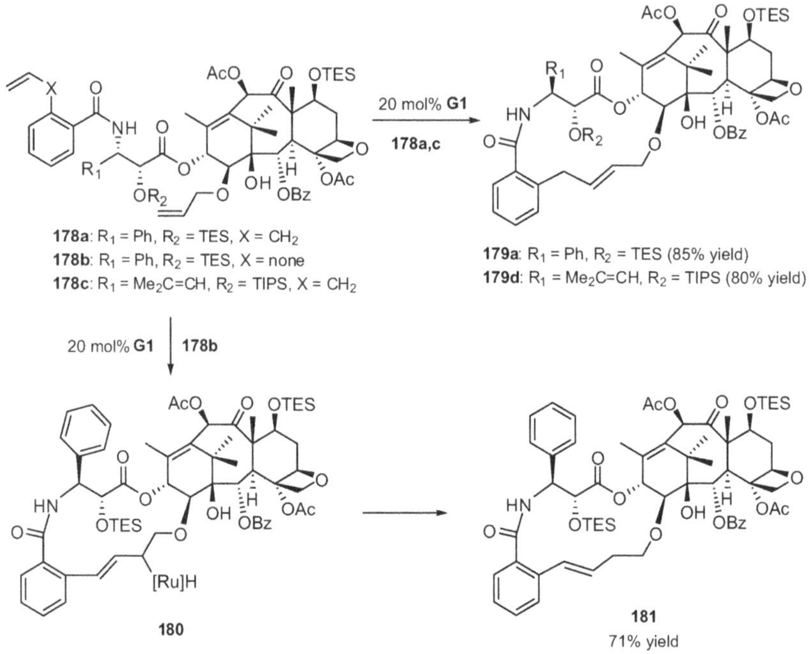

Scheme 49 Macrocyclic paclitaxel congeners obtained through RCM

4.1.3 Diversity-Oriented Synthesis of Macrocycles Using RCM

Diversity-oriented synthesis (DOS) is a powerful tool for the generation of diverse libraries generally based in a natural product structure (reviews [187, 188]). Creation of 13- to 18-membered macrolactam libraries with a high degree of diversity via RCM reaction was recently reported. After multiplying the number of starting diene isomers using a diastereoselective aldol reaction followed by an amide formation, the precursor dienes **182** were submitted to RCM macrocyclizaton under typical conditions. No significant differences were observed using **G1**, **G2**, or **HG2** catalysts in terms of conversions and stereoselectivity. This method provided a feasible and rapid entry to a complex macrocyclic library with 14,000 members by appropriately varying the nature and chain length of the alkenol fragment (Scheme 50) [189, 190].

4.2 Macrocycles for Supramolecular Chemistry

Not only macrocycles are prevalent in natural product structures and derived druggable molecules, but also they are crucial in materials science and supramolecular chemistry. Cyclophanes, catenanes, rotaxanes, structures able to host

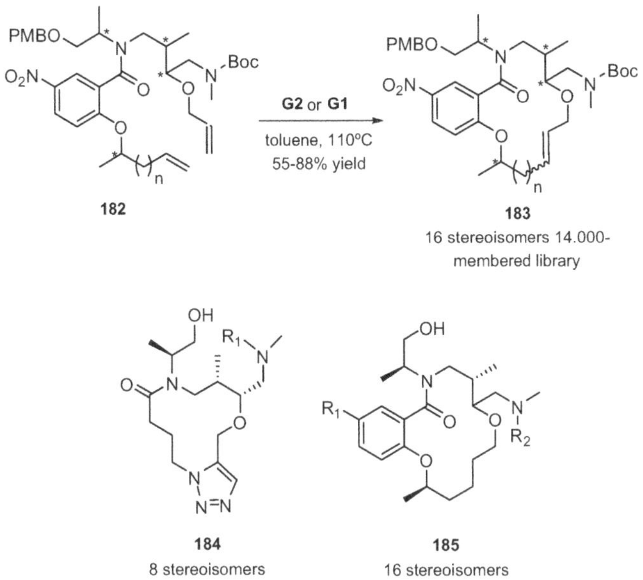

Scheme 50 Diversity-oriented synthesis for the creation of 13- to 18-membered macrolactam libraries via RCM

cations, or other organic molecules are among the most useful entities, and metathesis methodologies are among the synthetic ways to construct these targets. This section shows a selection of synthetic efforts in macrocyclizations toward heterocycles engaged in supramolecular or material chemistry. Some examples deal with molecules with potential application in medicinal chemistry. In general these are less functionalized products. However compatibility problems with the catalysts may arise in metal-templated reactions. Conformational control is sometimes difficult with less rigid substrates or if final products are sterically crowded and high dilution may not enable cyclization.

Rigidified or strained macrocycles are of particular interest as their shape-persistent structures can be exploited in medicinal chemistry and for well-defined carbon-based materials. Substrates with unfavorable conformations fail to react and give low yields or side-oligomerization products (see, for instance, [191]). In a recent synthesis of cyclophanes **187**, some conformational control was achieved in the substrate to favor cyclization using a RCM reaction catalyzed by **G1** (10 mol%). The conformational controlling element was a quinolinium salt that forced the reactive conformation by an intermolecular noncovalent interaction with the aromatic ring of substrate **186** (Scheme 51) [192].

Shape-persistent two-dimensional macrocycles and three-dimensional molecular cages have found applications as molecular wires, sensors, liquid crystals, or catalysts, among others. The metathesis methodology was successfully applied to the synthesis of heterosequenced carbazole-containing macrocycles consisting of up to three different building blocks as well as the construction of a 3-D shape-

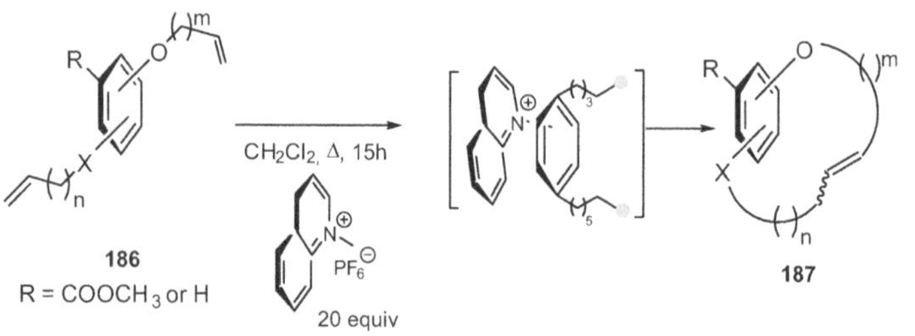

Scheme 51 RCM for the synthesis of cyclophanes with the aid of a conformational controlling quinolinium salt

persistent molecular cage **190**. For the metathesis step, **HG2** (20 mol%) was used as part of a tandem imine formation–macrocyclization reaction from **188** to **189** (Scheme 52) [193, 194].

Synthesis of structurally different bis(dihydrofuryl)cyclophane scaffolds, some of them containing 2-azetidinones, was described from carbonyl compounds. Selective RCM macrocyclization using 10 mol% **G2** was used to construct structures **191**. Mixtures of *syn*/*anti* or (*E*/*Z*) isomers sometimes were observed but not in the example showed in Fig. 17 in which the RCM was highly selective and efficient [195]. These macrocyclic bis-β-lactams are structurally similar to **192**, obtained by Ibrahim's group as mixtures of isomers under typical RCM conditions [196, 197].

Tris(2-aminoethyl)amine (TREN)-based ligands were prepared using a triple RCM over metal-templated substrate **193**. The choice of the catalyst proved to be difficult due to interaction with the central molybdenum. Fourteen Mo-, W-, and Ru-based catalysts were screened. Finally, tungsten-based MAP (monoaryloxide-pyrrolide) catalyst **194** effectively promoted the three intramolecular metatheses. After RCM the product mixture was hydrolyzed to generate TREN-based ligands **195** (Scheme 53) [198].

Olefin metathesis has been used for the construction of very large heterocycles designed for different purposes, like porphyrin supramolecules as membrane nanopores [199], macrocyclic hosts for fullerenes [200, 201], or large rings for the synthesis of catenanes [202, 203]. For the construction of these large macrocycles, **G1** catalyst under mild conditions is preferred avoiding polymerizations due to its low reactivity profile.

In particular, assembly of rotaxanes and catenanes is a difficult task where metathesis has found a new application. It is very convenient to use a template in order to construct the supramolecular entity. When using metathesis for the final ring closure, compatibility with template substrates has been a novel challenge. Scheme 54 shows the synthesis of pseudorotaxanes **199–200** due to Beer's group. Through the combination of halogen- and hydrogen-bonding noncovalent interactions, bromide and iodide anions templated the formation of assemblies between isophthalamide-containing macrocycle **196** and a bromo- or iodo-functionalized

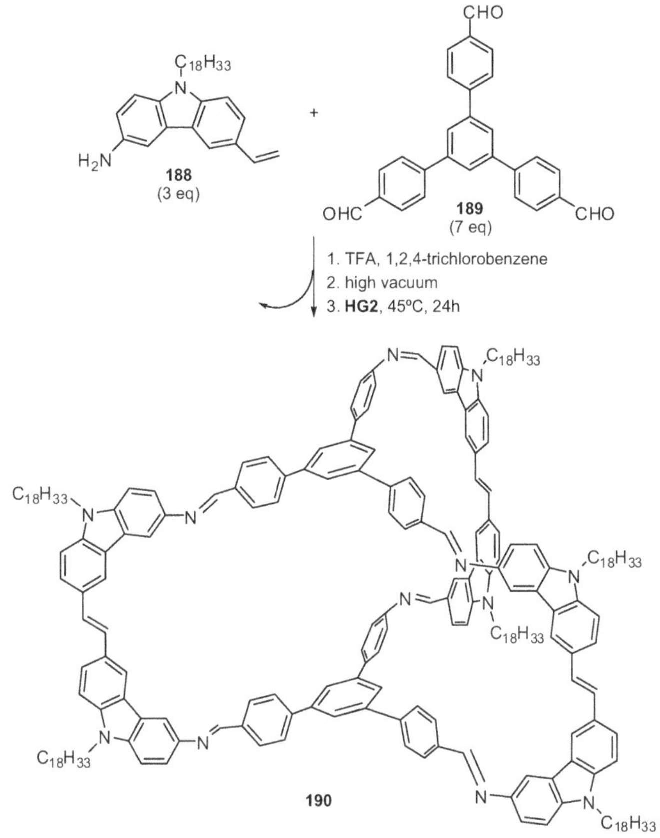

Scheme 52 Construction of a 3-D shape-persistent molecular cage using a tandem imine formation–macrocyclization metathesis process

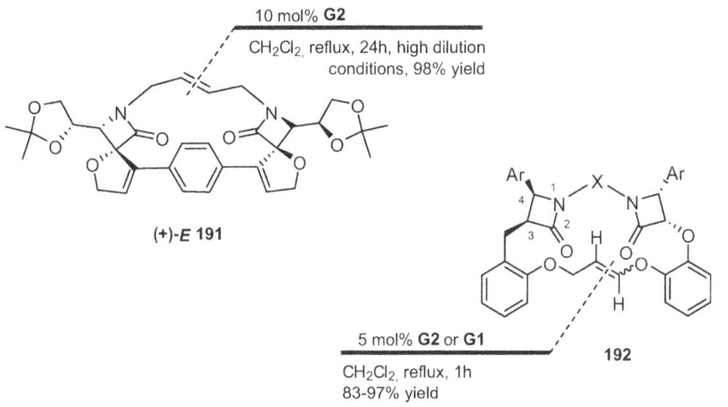

Fig. 17 Conditions of the RCM step for the synthesis of cyclophane scaffolds

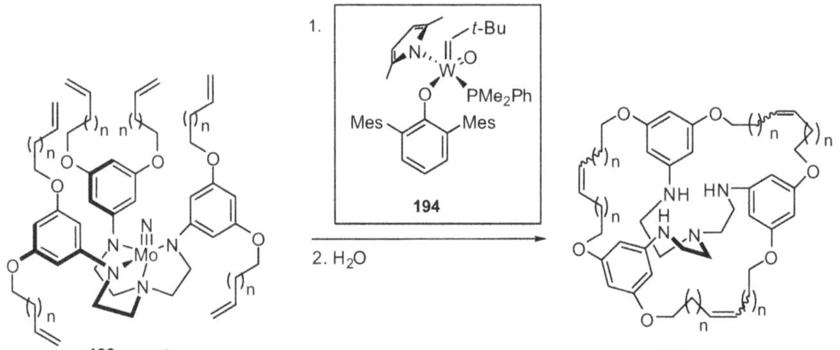

193a: n = 1
193b: n = 2
193c: n = 3

195a: n = 1
195b: n = 2
195c: n = 3

Scheme 53 RCM over metal-templated substrate **193** for the preparation of tris(2-aminoethyl) amine (TREN)-based ligands

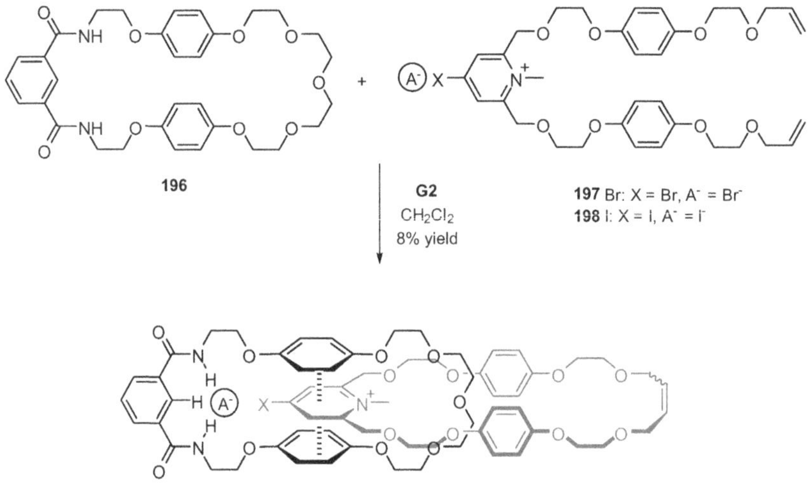

196

197 Br: X = Br, A⁻ = Br⁻
198 I: X = I, A⁻ = I⁻

199 Br: X = Br, A⁻ = Br⁻
200 I: X = I, A⁻ = I⁻

Scheme 54 Synthesis of pseudorotaxanes using RCM

pyridinium component **197/198**. RCM was developed using **G2** in CH_2Cl_2 (10 mol %) ([204]; for other examples, see [205–207]).

The synthesis of catenanes and rotaxanes using gold (I) or cobalt (III) as templates was disclosed by Leigh's group. Scheme 55 shows the RCM reaction of tridentate dianionic pyridine-2,6-dicarboxamido ligands **201**, each with two terminal alkene groups, coordinated to Co(III) in a mutually orthogonal arrangement. The macrocyclization catalyzed by **G1** under dilution conditions gave **202**, an interlocked molecular architecture ([208]; for a gold-catalyzed catenane, see [209]).

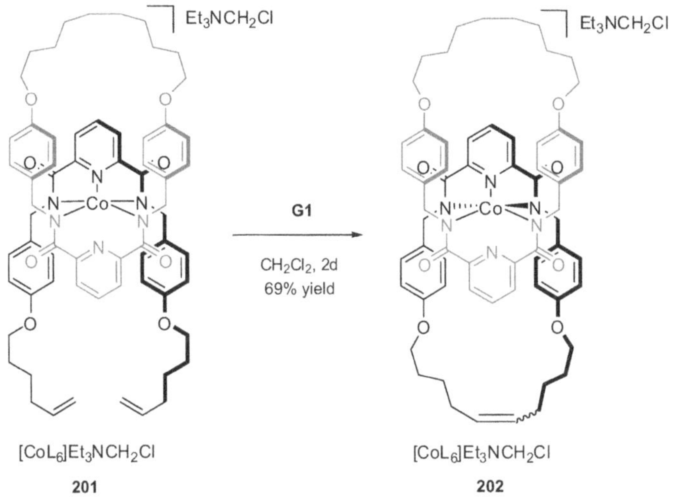

Scheme 55 Synthesis of a catenane from a cobalt (III) template substrate

RCAM is a powerful methodology to apply to the synthesis of large macrocyclic heterocycles, as mixtures of isomers are avoided and rigid structures are achieved. Zhang's group has utilized this strategy for the synthesis of shape-persistent structures able to interact with different molecules. Thus, an ethynylene-linked cage **205** was obtained via alkyne metathesis using the multidentate molybdenum catalysts **204** developed in the group (Scheme 56). This cage product was obtained in 56% isolated yield as the catalyst was able to convert large oligomers into the most thermodynamically stable macrocycle **205**. The combination of the conjugated system and the rigidity of the cage led to high binding affinity for fullerenes ([210, 211]; for review, see [212]). A C84 selective porphyrin macrocycle with an adaptable cavity was reported using a similar methodology for the synthesis [213].

Macrocyclic oligomers **208–209** possessing direction-defining ester linkages were synthesized via alkyne metathesis using molybdenum catalysts (Scheme 57) [214]. The authors had also developed a highly active and durable fumed silica-supported heterogeneous molybdenum(VI) catalyst, which was applied to ring-closing alkyne metathesis and cyclooligomerization reactions to give high yields of cyclooligomers of alkynes [215].

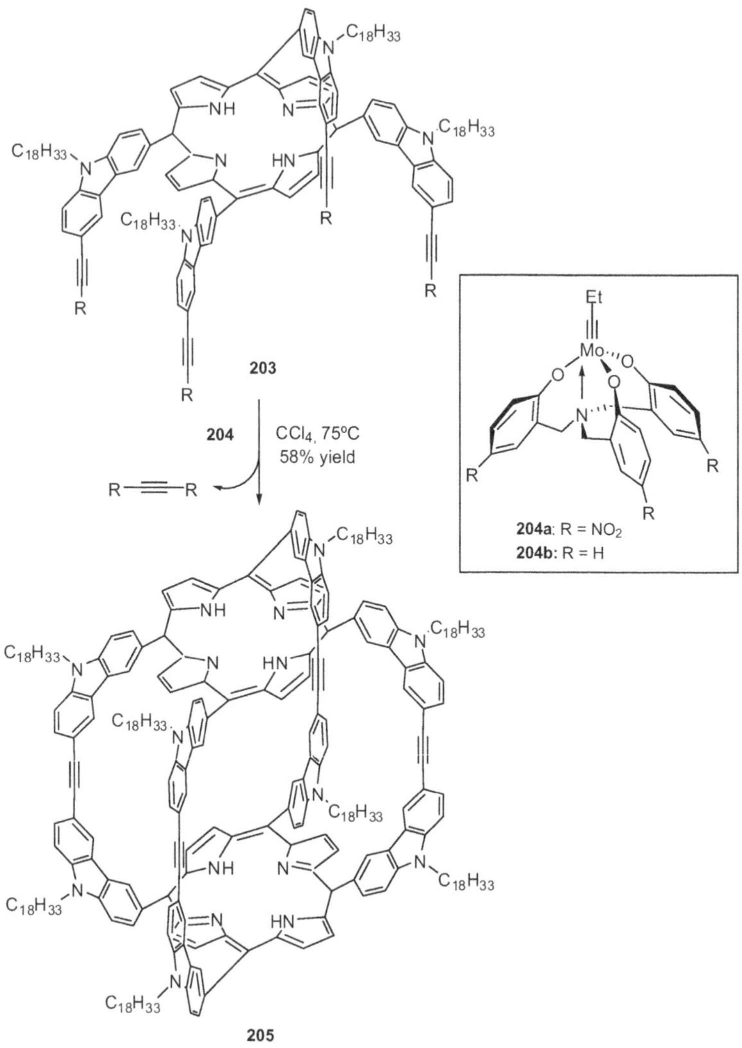

Scheme 56 Use of RCAM methodology for the synthesis of shape-persistent structure **205**

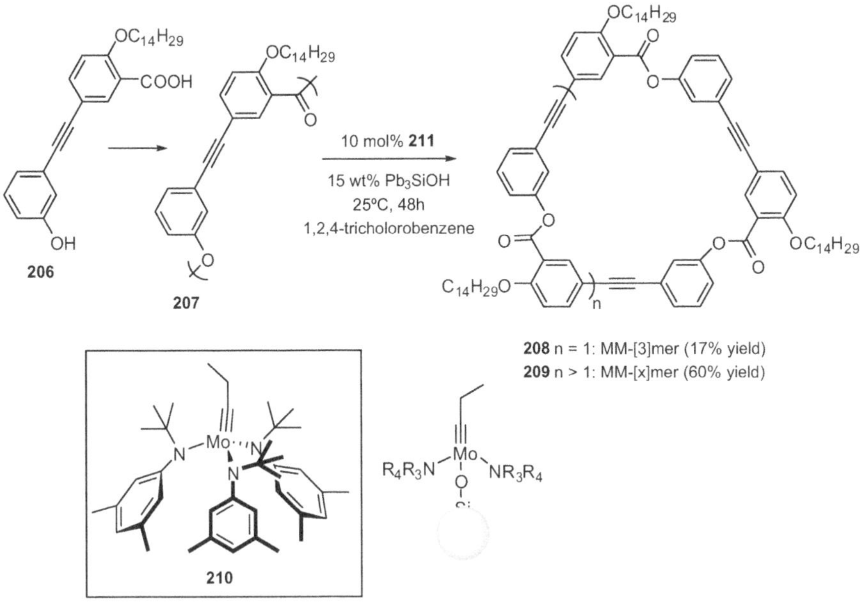

Scheme 57 Alkyne metathesis cyclooligomerization catalyzed by a silica-supported heterogeneous molybdenum (VI) catalyst

5 Conclusions and Outlook

The evolution of metathesis applied for the construction of large rings has been impressive in the past decade. Since the discovery of ruthenium and molybdenum catalysts, the metathesis reaction was applied to the construction of macrocycles with great success. However, the most recent years have accounted for the improvements in the limitations observed in early times. New catalysts are now able to produce kinetically favored (*Z*)-isomers controlling the side-oligomerization and double-bond isomerization reactions. The great development of RCAM makes this methodology a nice alternative to olefin metathesis as upon (*Z*)-selective hydrogenation of the formed cycloalkyne it gives access to (*Z*)-macrocycles. On the other hand, new catalysts both based on ruthenium and on molybdenum or tungsten are more group tolerant and work under milder conditions. This has allowed carrying out reactions with densely functionalized substrates bearing fragile moieties. Even free hydroxy groups are tolerated and sometimes the free alcohols react better than the protected derivatives, which are sterically hindered. It is interesting to note that the first introduced catalysts such as **G1** have still their utility when low reactivity profile is desired, and polymerization is a possible problem. Some regioselectivity problems could be solved using these first-generation catalysts, especially with polyenic substrates. In relation with metathesis of polyenes, new recent disclosed procedures use additional reagents such as quinones in combination with ruthenium carbenes in order to avoid undesired initiation of the metathesis with internal double bonds. Despite all these new findings, there is not a clear rule that indicates the best catalyst to use in each case. Subtle variations in the substrate structure and in the type of final product may lead to different outcomes with each type of carbene complexes. Overall a good amount of work is still very often necessary to find the best conditions for each particular synthesis, but in the end most efforts lead to successful metathesis processes in terms of yields and selectivity.

Regarding applications, a wide range of examples have been reported using metathesis in the context of natural macrocycle synthesis. Very complex molecules have been successfully prepared using as key step these methodologies. In addition, molecular structures designed for supramolecular chemistry or materials science have increased the applications of this reaction. Scaled-up procedures are paving the way for the use of metathesis in the pharmaceutical industry. Low catalyst loading procedures, anchored catalysts to solid matrices, and high concentration methods are being developed and used for multigram syntheses.

Some problems are still unresolved like the complete control of the (*E*/*Z*) ratio especially when the (*E*) isomer is desired in medium-sized rings, or the efficiency with conformationally unfavorable substrates, but it is clear that metathesis is a mature methodology for the synthesis of macrocycles and new exciting applications and findings will come soon.

References

1. Gradillas A, Pérez-Castells J (2010) In: Cossy J, Arseniyadis S, Meyer C (eds) Metathesis in natural product synthesis. Wiley, Weinheim
2. Prunet J, Rouge dos Santos A, Férézou JP (2008) In: Diederich F, Stang PJ, Tikwinski RR (eds) Modern supramolecular chemistry: strategies for macrocycle synthesis. Wiley, Weinheim
3. Mori M (2007) Adv Synth Catal 349:121–135
4. Schmidt B (2004) Eur J Org Chem 2004:1865–1880
5. Hoveyda AH, Gillingham DG, Van Veldhuizen JJ, Kataoka O, Garber SB, Kingsbury JS, Harrity JP (2004) Org Biomol Chem 2:8–23
6. Grubbs RH (2004) Tetrahedron 60:7117–7140
7. Prunet J (2003) Angew Chem Int Ed 42:2826–2830
8. Mulzer J, Öhler E (2004) Top Organomet Chem 13:269–366
9. Nicolaou KC, Bulger PG, Sarlah D (2005) Angew Chem Int Ed 44:4490–4527
10. Gaich T, Mulzer J (2005) Curr Top Med Chem 5:1473–1494
11. Van der Weghe J, Eustache J (2005) Curr Top Med Chem 5:1495–1519
12. Deiters A, Martin SF (2004) Chem Rev 104:2199–2238
13. Villemin D (1980) Tetrahedron Lett 21:1715–1718
14. Tsuji J, Hashiguchi S (1980) Tetrahedron Lett 21:2955–2958
15. Altmann KH (2004) Org Biomol Chem 2:2137–2152
16. Rivkin A, Cho YS, Gabarda AE, Yoshimura F, Danishefsky SJ (2004) J Nat Prod 67:139–143
17. Zhang R, Liu Z (2004) Curr Org Chem 8:267–290
18. Nicolaou KC, Wissinger N, Pastor J, Ninkovic S, Sarabia F, He Y, Vourloumis D, Li T, Giannakakou P, Hamel E (1997) Nature 390:268–272
19. Nishioka T, Iwabuchi Y, Irie H, Hatakeyama S (1998) Tetrahedron Lett 39:5597–5600
20. Schrock RR, Czekelius C (2007) Adv Synth Catal 349:55–77
21. Vougloukalakis GJ, Grubbs RH (2010) Chem Rev 110:1746–1787
22. Bieniek M, Michrowska A, Usanov DL, Grela K (2008) Chem Eur J 14:806–818

23. Conrad JC, Eelman MD, Duarte Silva J, Monfere S, Parnas H, Snelgrove JL, Fogg DE (2007) J Am Chem Soc 129:1024–1025
24. Jee JE, Cheong JL, Lim J, Chen C, Hong SH, Lee SS (2013) J Org Chem 78:3048–3056
25. Kadyrov R (2013) Chem Eur J 19:1002–1012
26. Raymond M, Holtz-Mulholland M, Collins SK (2014) Chem Eur J 20:1–6
27. Fürstner A, Dierkes T, Thiel OR, Blanda G (2001) Chem Eur J 7:5286–5298
28. Wu Y, Liao X, Wang R, Xie XS, de Brabander JK (2002) J Am Chem Soc 124:3245–3253
29. Chegondi R, Tan MM, Hanson PR (2011) J Org Chem 76:3909–3916
30. Wang Y, Jimenez M, Hansen AS, Raiber E-S, Schreiber SL, Young DW (2011) J Am Chem Soc 133:9196–9199
31. Wang Y, Jimenez M, Sheehan P, Zhong C, Hung AW, Tam CP, Young DW (2013) Org Lett 15:1218–1221
32. Ibrahem I, Yu M, Schrock RR, Hoveyda AH (2009) J Am Chem Soc 131:3844–3845
33. Nicolaou KC, He Y, Vourloumis D, Vallberg H, Roschangar F, Sarabia F, Ninkovic S, Yang Z, Trujillo JI (1997) J Am Chem Soc 119:7960–7973
34. Yu M, Wang C, Kyle AF, Jakubec P, Dixon DJ, Schrock RR, Hoveyda AH (2011) Nature 479:88–93
35. Wang C, Yu M, Kyle AF, Jakubec P, Dixon DJ, Schrock RR, Hoveyda AH (2013) Chem Eur J 19:2726–2740
36. Wang C, Haeffner F, Schrock RR, Hoveyda AH (2013) Angew Chem Int Ed 52:1939–1943
37. Torker S, Müller A, Chen P (2010) Angew Chem Int Ed 49:3762–3766
38. Endo K, Grubbs RH (2011) J Am Chem Soc 133:8525–8527
39. Khan RKM, O'Brien RV, Torker S, Li B, Hoveyda AH (2012) J Am Chem Soc 134: 12774–12779
40. Occhipinti G, Hansen FR, Törnroos KW, Jensen VR (2013) J Am Chem Soc 135:3331–3334
41. Marx VM, Herbert MB, Keitz BK, Grubbs RH (2013) J Am Chem Soc 135:94–97
42. Dang Y, Wang ZX, Wang X (2012) Organometallics 31:7222–7234
43. Persich P, Llaveria J, Lhermet R, Haro T, Stade R, Kondoh A, Fürstner A (2013) Chem Eur J 19:13047–13058
44. Wessjohann LA, Ruijter E, García-Rivera D, Bradt W (2005) Mol Divers 9:171–186
45. Wessjohann LA, Ruijter E (2004) Top Curr Chem 243:137–184
46. Rijkers DTS (2015) Synthesis of cyclic peptides and peptidomimetics by metathesis reactions. Top Heterocycl Chem. doi:10.1007/7081_2015_142
47. Compain P (2014) Synthesis of amine-containing heterocycles by metathesis reactions: recent advances and opportunities. Top Heterocycl Chem. doi:10.1007/7081_2014_139
48. Vincent G (2015) Synthesis of lactams by metathesis reactions. Top Heterocycl Chem. doi:10.1007/7081_2015_144
49. Katz L, Ashley GW (2005) Chem Rev 105:499–528
50. Yeung KS, Paterson I (2005) Chem Rev 105:4237–4313
51. McDaniel R, Welch M, Hutchinson CR (2005) Chem Rev 105:543–558
52. Fürstner A, Leitner A (2003) Angew Chem Int Ed 42:308–311
53. Caijo F, Tripoteau F, Bellec A, Crévisy C, Baslé O, Mauduit M, Briel O (2013) Catal Sci Technol 3:429–435
54. Lehr K, Fürstner A (2012) Tetrahedron 68:7695–7700
55. Sharma GVM, Reddy SV (2013) RSC Adv 3:21759–21762
56. Gersbach P, Jantsch A, Feyen F, Scherr N, Dangy JP, Pluschke G, Altmann K-H (2011) Chem Eur J 17:13017–13031
57. Kanada RM, Itoh D, Nagai M, Niijima J, Asai N, Mizui Y, Abe S, Kotake Y (2007) Angew Chem Int Ed 46:4350–4355
58. Sharma GVN, Reddy PS (2012) Eur J Org Chem 2012:2414–2421
59. Gallenkamp D, Fürstner A (2011) J Am Chem Soc 133:9232–9235
60. Gatti M, Vieille-Petit L, Luan X, Mariz R, Drinkel E, Linden A, Dorta R (2009) J Am Chem Soc 131:9498–9499

61. Gaul C, Njardarson JT, Shan D, Dorn DC, Wu KD, Tong WP, Huang XY, Moore MAS, Danishefsky SJ (2004) J Am Chem Soc 126:11326–11337
62. Glaus F, Altman KH (2012) Angew Chem Int Ed 51:3405–3409
63. Dai WM, Guan Y, Jin J (2005) Curr Med Chem 12:1947–1993
64. Rix D, Caijo F, Laurent I, Boeda F, Clavier H, Nolan SP, Mauduit M (2008) J Org Chem 73:4225–4228
65. Chevalley A, Prunet J, Mauduit M, Férézou JP (2012) Eur J Org Chem 2012:8265–8278
66. Xuan R, Oh HS, Lee Y, Kang HY (2008) J Org Chem 73:1459–1461
67. Winssinger N, Barluenga S (2007) Chem Commun 2007:22–36
68. Yang ZQ, Geng X, Solit D, Pratilas CA, Rosen N, Danishefsky SJ (2004) J Am Chem Soc 126:7881–7889
69. Geng X, Danishefsky SJ (2004) Org Lett 6:413–416
70. Day JEH, Sharp SY, Rowlands MG, Aherne W, Lewis W, Roe SM, Prodromou C, Pearl LH, Workman P, Moody CJ (2010) Chem Eur J 16:10366–10372
71. Barluenga S, Dakas PY, Fernandin Y, Meijer L, Winssinger N (2006) Angew Chem Int Ed 45:3951–3954
72. Bajwa N, Jennings MP (2012) In: Harmata M (ed) Strategies and tactics in organic synthesis, vol 8. Academic, Amsterdam, pp 153–169
73. Calo F, Richardson J, Barrett AGM (2009) Org Lett 11:4910–4913
74. Baird LJ, Timmer MSM, Teesdale-Spittle PH, Harvey JE (2009) J Org Chem 74:2271–2277
75. Barluenga S, Moulin E, Lopez P, Winssinger N (2005) Chem Eur J 11:4935–4952
76. Moulin E, Barluenga S, Winssinger N (2005) Org Lett 7:5637–5639
77. Moulin E, Zoete S, Barluenga S, Karplus M, Wissinger N (2005) J Am Chem Soc 127: 6999–7004
78. Martinez-Solorio D, Belmore K, Jennings MP (2011) J Org Chem 76:3898–3908
79. Yet L (2003) Chem Rev 103:4283–4306
80. Kang EJ, Cho EJ, Lee YE, Ji MK, Shin DM, Chung YK, Lee E (2004) J Am Chem Soc 126: 2680–2681
81. Kang EJ, Cho EJ, Ji MK, Lee YE, Shin DM, Choi SY, Chung YK, Kim JS, Kim HJ, Lee SG, Lah MS, Lee E (2005) J Org Chem 70:6321–6329
82. Soltani O, de Brabander JK (2005) Org Lett 7:2791–2793
83. Bolshakov S, Leighton JL (2005) Org Lett 7:3809–3812
84. Cheung LL, Marumoto S, Anderson CD, Rychnovsky SD (2008) Org Lett 10:3101–3104
85. Wang X, Porco JA (2003) J Am Chem Soc 125:6040–6041
86. Wang X, Bowman EJ, Bowman BJ, Porco JA (2004) Angew Chem Int Ed 43:3601–3605
87. Vintonyak VV, Maier ME (2007) Org Lett 9:655–658
88. Vintonyak VV, Maier ME (2007) Angew Chem Int Ed 46:5209–5211
89. Fürstner A, Bindl M, Jean L (2007) Angew Chem Int Ed 46:9275–9278
90. Bindl M, Jean L, Herrmann J, Mueller R, Fuerstner A (2009) Chem Eur J 15:12310–12319
91. Fouché M, Rooney L, Barrett AGM (2012) J Org Chem 77:3060–3070
92. Ramulu U, Ramesh D, Rajaram S, Reddy SP, Venkatesham K, Venkateswarlu Y (2012) Tetrahedron Asymmetry 23:117–123
93. Wadsworth AD, Furkert DP, Sperry J, Brimble MA (2012) Org Lett 14:5374–5377
94. Lisboa MP, Dudley GB (2013) Chem Eur J 19:16146–16168
95. Nicolaou KC, Sun YP, Guduru R, Banerji B, Chen DYK (2008) J Am Chem Soc 130: 3633–3644
96. Pawar AB, Prasad KR (2012) Chem Eur J 18:15202–15206
97. Pujari SA, Gowrisankar P, Kaliappan KP (2011) Chem Asian J 6:3137–3151
98. Florence GJ, Gardner NM, Paterson I (2008) Nat Prod Rep 25:342–375
99. Zhu W, Jiménez M, Jung W-H, Camarco DP, Balachandran R, Vogt A, Day BW, Curran DP (2010) J Am Chem Soc 132:9175–9187
100. Fürstner A, Nevado C, Waser M, Tremblay M, Chevrier C, Teplý F, Aïssa C, Moulin E, Müller O (2007) J Am Chem Soc 129:9150–9161

101. Gagnepain J, Moulin E, Fürstner A (2011) Chem Eur J 17:6964–6972
102. Moulin E, Nevado C, Gagnepain J, Kelter G, Fiebig HH, Fürstner A (2010) Tetrahedron 66:6421–6428
103. Hanson PR, Chegondi R, Nguyen J, Thomas CD, Waetzig JD, Whitehead A (2011) J Org Chem 76:4358–4370
104. Biswas K, Lin H, Njardarson JT, Chappell MD, Chou TC, Guan Y, Tong WP, He L, Horwitz SB, Danishefsky SJ (2002) J Am Chem Soc 124:9825–9832
105. Fürstner A, Bouchez LC, Funel JA, Liepins V, Porée FH, Gilmour R, Beaufils F, Laurich D, Tamiya M (2007) Angew Chem Int Ed 46:9265–9270
106. Hara A, Morimoto R, Iwasaki Y, Saitoh T, Ishikawa Y, Nishiyama S (2012) Angew Chem Int Ed 51:9877–9880
107. Cribiu R, Jaeger C, Nevado C (2009) Angew Chem Int Ed 48:8780–8783
108. Aïssa C, Riveiros R, Ragot J, Fürstner A (2003) J Am Chem Soc 125:15512–15520
109. Wu D, Li H, Jin J, Wu J, Dai WM (2011) Synlett 7:895–898
110. Li H, Wu J, Luo J, Dai WM (2010) Chem Eur J 16:11530–11534
111. Clark JS, Romiti F (2013) Angew Chem Int Ed 52:10072–10075
112. Jin J, Chen Y, Li Y, Wu J, Dai WM (2007) Org Lett 9:2585–2588
113. Va P, Roush WR (2006) J Am Chem Soc 128:15960–15961
114. Jung JH, Lee E (2009) Angew Chem Int Ed 48:5698–5700
115. Hwang MH, Han SJ, Lee DH (2013) Org Lett 15:3318–3321
116. Demonceau A, Dragutan I, Dragutan V, Le Gendre P (2012) Curr Org Synth 9:779–790
117. Xie J, Ma Y, Horne DA (2011) Tetrahedron 67:7485–7501
118. Li S, Chen Z, Xu Z, Ye T (2010) Chem Commun 46:4773–4775
119. Fang L, Yang J, Yang F (2010) Org Lett 12:3124–3127
120. Fuwa H, Yamagata N, Saito A, Sasaki M (2013) Org Lett 15:1630–1633
121. Fuwa H, Saito A, Sasaki M (2010) Angew Chem Int Ed 49:3041–3044
122. Hong SH, Sanders HP, Lee CW, Grubbs RH (2005) J Am Chem Soc 127:17160–17161
123. Díaz-Oltra S, Angulo-Pachon CA, Murga J, Carda M, Marco JA (2010) J Org Chem 75: 1775–1778
124. Wullschleger CW, Gertsch J, Altmann KH (2010) Org Lett 12:1120–1123
125. Wullschleger CW, Gertsch J, Altmann KH (2013) Chem Eur J 19:13105–13111
126. Zimmermann N, Pinard P, Carboni B, Gosselin P, Gaulon-Nourry C, Dujardin G, Collet S, Lebreton J, Mathé-Allainmat M (2013) Eur J Org Chem 12:2303–2315
127. Zhao Z, Taylor RE (2012) Org Lett 14:669–671
128. Ding F, Jennings MP (2008) J Org Chem 73:5965–5976
129. Hoye TR, Hu M (2003) J Am Chem Soc 125:9576–9577
130. Fürstner A, Fluegge S, Larionov O, Takahashi Y, Kubota T, Kobayashi J (2009) Chem Eur J 15:4011–4029
131. Fürstner A, Larionov O, Flügge S (2007) Angew Chem Int Ed 46:5545–5548
132. Valot G, Regens CS, O'Malley DP, Godineau E, Takikawa H, Fürstner A (2013) Angew Chem Int Ed 52:9534–9538
133. Brewitz L, Llaveria J, Yada A, Fürstner A (2013) Chem Eur J 19:4532–4537
134. Hickmann V, Kondoh A, Gabor B, Alcarazo M, Fürstner A (2011) J Am Chem Soc 133: 13471–13480
135. Altmann KH, Hofle G, Müller R, Mulzer J, Prantz K (2009) The epothilones: an outstanding family of anti-tumor agents-from soil to the clinic, vol 90. Springer, Vienna
136. Feyen F, Cachoux F, Gertsch J, Wartmann M, Altmann KH (2008) Acc Chem Res 41:21–31
137. Nicolaou KC (2005) J Med Chem 48:5613–5638
138. Mulzer J, Altmann KH, Höfle G, Müller R, Prantz K (2008) C R Chim 11:1336–1368
139. Schiess R, Gertsch J, Schweizer WB, Altmann KH (2011) Org Lett 13:1436–1439
140. Neuhaus CM, Liniger M, Stieger M, Altman KH (2013) Angew Chem Int Ed 52:5866–5870
141. Stivala CE, Gu Z, Smith LL, Zakarian A (2012) Org Lett 14:804–807
142. Fujiwara K, Suzuki Y, Koseki N, Aki Y, Kikuchi Y, Murata S, Yamamoto F, Kawamura M, Norikura T, Matsue H, Murai A, Kawai H, Suzuki T (2014) Angew Chem Int Ed 53:780–784

143. Sakamoto S, Sakazaki H, Hagiwara K, Kamada K, Ishii K, Noda T, Inoue M, Hirama M (2004) Angew Chem Int Ed 43:6505–6510
144. Stivala CE, Zakarian A (2008) J Am Chem Soc 130:3774–3776
145. Trost BM, Yang H, Dong G (2011) Chem Eur J 17:9789–9805
146. Trost BM, Yang H, Thiel OR, Frontier AJ, Brindle CS (2007) J Am Chem Soc 129: 2206–2207
147. Benson S, Collin M-P, Arlt A, Gabor B, Goddard R, Fürstner A (2011) Angew Chem Int Ed 50:8739–8744
148. Hassan H, Mutlak A (2010) Chem Commun 46:9100–9106
149. Evano G, Schaus JV, Panek JS (2004) Org Lett 6:525–528
150. Hayashi Y, Shoji M, Ishikawa H, Yamaguchi J, Tamaru T, Imai H, Nishigaya Y, Takabe K, Kakeya H, Osada H (2008) Angew Chem Int Ed 47:6657–6659
151. Del Valle DJ, Krische MJ (2013) J Am Chem Soc 135:10986–10989
152. Takada A, Uda K, Ohtani T, Tsukamoto S, Takahashi D, Toshima K (2013) J Antibiot 66:155–159
153. Fukuda H, Nishiyama Y, Nakamura S, Ohno Y, Eguchi T, Iwabuchi Y, Usui T, Kanoh N (2012) Chem Asian J 7:2872–2881
154. Bicchielli D, Borguet Y, Delaude L, Demonceau A, Dragutan I, Dragutan V, Hans M, Jossifov C, Nicks F, Willem Q (2012) Curr Org Synth 9:397–405
155. Smith AB, Mesaros EF, Meyer EA (2006) J Am Chem Soc 128:5292–5299
156. Magauer T, Martin HJ, Mulzer J (2010) Chem Eur J 16:507–519
157. Hoffmeister L, Persich P, Fürstner A (2014) Chem Eur J 20:4396–4402
158. Jasper C, Wittenberg R, Quitschalle M, Jakupovic J, Kirschning A (2005) Org Lett 7: 479–482
159. Nicolaou KC, Xu X (2006) Chem Comm 2006:600–602
160. Kwon MS, Sim SH, Chung YK, Lee E (2011) Tetrahedron 67:10179–10185
161. Toelle N, Weinstabl H, Gaich T, Mulzer J (2014) Angew Chem Int Ed 53:3859–3862
162. Jun JH, Javed S, Ndi CN, Hanson PR (2015) Synthesis of P-, S-, Si-, B-, and Se-heterocycles via ring-closing metathesis. Top Heterocycl Chem. doi:10.1007/7081_2015_162
163. Tannert R, Milroy LG, Ellinger B, Hu TS, Arndt HD, Waldmann H (2010) J Am Chem Soc 132:3063–3077
164. Tannert R, Hu TS, Arndt HD, Waldmann H (2009) Chem Commun 12:1493–1495
165. Aldrich LN, Berry CB, Bates BS, Konkol LC, So M, Lindsley CW (2013) Eur J Org Chem 2013:4215–4218
166. Amat M, Ballette R, Proto S, Pérez M, Bosch J (2013) Chem Commun 49:3149–3151
167. Luo SP, Guo LD, Gao LH, Li S, Huang PQ (2013) Chem Eur J 19:87–91
168. Gebauer K, Fürstner A (2014) Angew Chem Int Ed 53:6393–6396
169. Jarikote DV, Murphy PV (2010) Eur J Org Chem 2010:4959–4970
170. Fürstner A (2004) Eur J Org Chem 2004:943–958
171. Fürstner A, Nagano T (2007) J Am Chem Soc 129:1906–1907
172. Nagano T, Pospisil J, Chollet G, Schulthoff S, Hickmann V, Moulin E, Herrmann J, Mueller R, Fürstner A (2009) Chem Eur J 15:9697–9706
173. Kondoh A, Arlt A, Gabor B, Fürstner A (2013) Chem Eur J 19:7731–7738
174. Yu X, Sun D (2013) Molecules 18:6230–6268
175. Song ZJ, Tellers DM, Journet M, Kuethe JT, Lieberman D, Humphrey G, Zhang F, Peng Z, Waters MS, Zewge D, Nolting A, Zhao D, Reamer RA, Dormer PG, Belyk KM, Davies IW, Devine PN, Tschaen DM (2011) J Org Chem 76:7804–7815
176. Kong J, Chen C, Balsells-Padros J, Cao Y, Dunn RF, Dolman SJ, Janey J, Li H, Zacuto MJ (2012) J Org Chem 77:3820–3828
177. Jiang Y, Andrews SW, Condroski KR, Buckman B, Serebryany V, Wenglowsky S, Kennedy AL, Madduru MR, Wang B, Lyon M, Doherty GA, Woodard BT, Lemieux C, Do MG, Zhang H, Ballard J, Vigers G, Brandhuber BJ, Stengel P, Josey JA, Beigelman L, Blatt L, Seiwert SD (2014) J Med Chem 57:1753–1769

178. Kazmierski WM, Hamatake R, Duan M, Wright LL, Smith GK, Jarvest RL, Ji JJ, Cooper JP, Tallant MD, Crosby RM, Creech K, Wang A, Li X, Zhang S, Zhang YK, Liu Y, Ding CZ, Zhou Y, Plattner JJ, Baker SJ, Bu W, Liu L (2012) J Med Chem 55:3021–3026
179. Arumugasamy J, Arunachalam K, Bauer D, Becker A, Caillet CA, Glynn R, Latham GM, Lim J, Liu J, Mayes BA, Moussa A, Rosinovsky E, Salanson AE, Soret AF, Stewart A, Wang J, Wu X (2013) Org Process Res Dev 17:811–828
180. Wei X, Shu C, Haddad N, Zeng X, Patel ND, Tan Z, Liu J, Lee H, Shen S, Campbell S, Varsolona RJ, Busacca CA, Hossain A, Yee NK, Senanayake CH (2013) Org Lett 15: 1016–1019
181. Shu C, Zeng X, Hao MH, Wei X, Yee NK, Busacca CA, Han Z, Farina V, Senanayake CH (2008) Org Lett 10:1303–1306
182. de Rosa M, Unge J, Motwani H, Rosenquist A, Vrang L, Wallberg H, Larhed M (2014) J Med Chem 57:6444–6457
183. Abell AD, Jones MA, Coxon JM, Morton JD, Aitken SG, McNabb SB, Lee HYY, Mehrtens JM, Alexander NA, Stuart BG, Neffe AT, Bickerstaffe R (2009) Angew Chem Int Ed 48:1455–1458
184. Abell AD, Alexander NA, Aitken SG, Chen H, Coxon JM, Jones MA, McNabb SB (2009) J Org Chem 74:4354–4356
185. Andersson H, Demaegdt H, Johnsson A, Vauquelin G, Lindeberg G, Hallberg M, Erdelyi M, Karlen A, Hallberg A (2011) J Med Chem 54:3779–3792
186. Sun L, Geng X, Geney R, Li Y, Simmerling C, Li Z, Lauher JW, Xia S, Horwitz SB, Veith JM, Pera P, Bernacki RJ, Ojima O (2008) J Org Chem 73:9584–9593
187. Madsen CM, Clausen MH (2011) Eur J Org Chem 3107–3115
188. Schreiber SL (2000) Science 287:1964–1969
189. Dandapani S, Lowe JT, Comer E, Marcaurelle LA (2011) J Org Chem 76:8042–8048
190. Marcaurelle LA, Comer E, Dandapani S, Duvall JR, Gerard B, Kesavan S, Lee MD, Liu H, Lowe JT, Marie JC, Mulrooney CA, Pandya BA, Rowley A, Ryba TR, Suh BC, Wei J, Young DW, Akella LB, Ross NT, Zhang YL, Fass DM, Reis SA, Zhao WN, Haggarty SJ, Palmer M, Foley MA (2010) J Am Chem Soc 132:16962–16976
191. Kotha S, Chavan AS, Shaikh M (2012) J Org Chem 77:482–489
192. Bolduc P, Jacques A, Collins SK (2010) J Am Chem Soc 132:12790–12791
193. Okochi KD, Han GS, Aldridge IM, Liu Y, Zhang W (2013) Org Lett 15:4296–4299
194. Okochi KD, Jin Y, Zhang W (2013) Chem Commun 49:4418–4420
195. Alcaide B, Almendros P, Quirós MT, Lázaro C, Torres MR (2014) J Org Chem 79: 6244–6255
196. Ibrahim YA, Al-Azemi TF, Abd El-Halim MD (2010) J Org Chem 75:4508–4513
197. Ibrahim YA, Al-Azemi TF, Abd El-Halim MD, John E (2009) J Org Chem 74:4305–4310
198. Cain MF, Forrest WP, Peryshkov DV, Schrock RR, Müller P (2013) J Am Chem Soc 135: 15338–15341
199. Satake A, Yamamura M, Oda M, Kobuke Y (2008) J Am Chem Soc 130:6314–6315
200. Canevet D, Gallego M, Isla H, de Juan A, Perez EM, Martin N (2011) J Am Chem Soc 133: 3184–3190
201. de Juan A, Pouillon Y, Ruiz-Gonzalez L, Torres-Pardo A, Casado S, Martin N, Rubio A, Perez EM (2014) Angew Chem Int Ed 53:5394–5400
202. Niess F, Sauvage JP (2013) Chem Commun 49:10790–10792
203. Dasgupta S, Huang KW, Wu J (2012) Chem Commun 48:4821–4823
204. Gilday LC, Beer PD (2014) Chem Eur J 20:8379–8385
205. Evans NH, Serpell CH, Beer PD (2011) Chem Eur J 17:7734–7738
206. Mullen KM, Mercurio J, Serpell CJ, Beer PD (2009) Angew Chem Int Ed 48:4781–4784
207. Lee JJ, Baumes JM, Connell RD, Oliver AG, Smith BD (2011) Chem Commun 47: 7188–7190
208. Leigh DA, Lusby PJ, McBurney RT, Morelli A, Slawin AMZ, Thomson AR, Walker DB (2009) J Am Chem Soc 131:3762–3771

209. Goldup SM, Leigh DA, Lusby PJ, McBurney RT, Slawin AMZ (2008) Angew Chem Int Ed 47:6999–7003
210. Zhang C, Wang Q, Long H, Zhang W (2011) J Am Chem Soc 133:20995–21001
211. Lohrman J, Zhang CX, Zhang W, Ren SQ (2012) Chem Commun 48:8377–8379
212. Jin Y, Wang Q, Taynton P, Zhang W (2014) Acc Chem Res 47:1575–1586
213. Zhang CX, Long H, Zhang W (2012) Chem Commun 48:6172–6174
214. Scott WS, Moore JS (2012) J Am Chem Soc 134:9114–9117
215. Cho HM, Weissman H, Moore JS (2008) J Org Chem 73:4256–4258

Top Heterocycl Chem (2017) 47: 319–380
DOI: 10.1007/7081_2015_162

Published online: 30 October 2015

Synthesis of P-, S-, Si-, B-, and Se-Heterocycles via Ring-Closing Metathesis

Jung Ho Jun, Salim Javed, Cornelius N. Ndi, and Paul R. Hanson

Abstract Heterocyclic compounds are an important class of motif in natural products and have found various applications in agriculture, pharmaceutical agents, and material science. This chapter describes various synthetic strategies toward heterocycle compounds containing P, S, Si, B, and Se atoms utilizing RCM method. Herein, efficient and innovative synthetic methods ranging from small molecules to large rings for natural products in the use of RCM are discussed.

Keywords Boron · Grubbs/Hoveyda-Grubbs/Schrock catalyst · Heterocycles · Natural products · Phosphorus · Ring-Closing Metathesis · Selenium · Silicon · Sulfur

Contents

J.H. Jun, S. Javed, C.N. Ndi, and P.R. Hanson (✉)
Department of Chemistry, University of Kansas, 1251 Wescoe Hall Dr., Lawrence, KS 66045-7582, USA
e-mail: phanson@ku.edu

1 Introduction

Since the development of molybdenum and ruthenium metathesis catalysts in the early 1990s, olefin, enyne, and alkyne metathesis reactions have continued to advance as versatile tools for the formation of C–C bonds in small molecules and natural products [1–12]. These metathesis reactions have gained prominence over the last two decades due to the development of well-defined, stable metal carbene complexes that are mainly characterized by tunable reactivity, thermal stability, and functional group compatibility.

Among several known metathesis catalysts, seminal advances in the field were made on the strength of pioneering work with the classical Schrock catalyst (**S**) [13, 14], the Grubbs first-generation (**G1**) [15–17] and second-generation (**G2**) catalysts [18], as well as the Hoveyda–Grubbs first-generation catalyst (**HG1**) [19] and second-generation (**HG2**) catalysts [20, 21]. The Schrock catalyst (**S**) [13, 14] is a molybdenum (IV)-based alkylidene complex discovered in 1990. The Ru(II)-carbenoid complex known as the Grubbs first-generation catalyst (**G1**) [15–17] was discovered in 1995 and is generally regarded as more functional group tolerant than the Mo-based counterparts. In 1999, the Grubbs second-generation catalyst (**G2**) [18] was developed and has generally been shown to possess higher activity. In the same year, Hoveyda and coworkers reported the Hoveyda–Grubbs first-generation catalyst (**HG1**) [19], whereby the phosphine ligand in the **G1** catalyst was switched to a benzylidene ligand containing a chelating *ortho*-isopropoxy group attached to a benzene ring. In 2000, the Hoveyda–Grubbs second-generation catalyst (**HG2**) [20, 21] was developed employing a phosphine-free structure.

Among several attractive features in olefin metathesis is the employment of nonfunctional olefins as substrates and transfer of highly functionalized fragments to generate complex synthetic architecture [22–27]. The metathesis reaction involves exchange between metal-alkylidene species and alkenes and encompasses ring-closing metathesis (RCM), cross metathesis (CM), ring-closing enyne metathesis (RCEYM), alkyne metathesis, ring-opening metathesis polymerization (ROMP), as well as methods employing tandem combinations of these processes. Among these various metathesis methods, RCM has a profound influence on the synthesis of both carbocycles and heterocycles.

The applications of RCM in the synthesis of heterocycles in materials, small molecules, and natural products have grown tremendously in the last two decades. With the advent of new metathesis catalysts, previous reactivity issues involving strongly coordinating heteroatoms are being resolved, resulting in a pronounced use of RCM in the construction of heterocycles. In this regard, the use of RCM to form phosphorus and sulfur heterocycles has emerged as an important area and, in 2004, was comprehensively reviewed with respect to methods development and biological utility [28]. In 2010, comprehensive reviews appeared that highlighted RCM methods to form phosphorus, sulfur, and silicon heterocycles for use in natural product chemistry (P, S, and Si) [29–31]. In addition, in 2012 an excellent review covering use of silicon tethers in natural product chemistry was reported [32]. In related work, the use of RCM in the synthesis of boron heterocycles was a major subtopic covered in reviews reported in 2009 [33] and 2012 [34]. It is therefore the aim of this chapter to highlight recent, representative examples in the use of RCM for the synthesis of P-, S-, Si-, B-, and Se-containing heterocycles in the context of their broader utility.

1.1 Bioactive Heterocyclic Compounds

Heterocycles containing P, S, Si, B, and Se atoms have been shown to possess a broad array of bioactive properties with a few representative examples shown in Fig. 1. The unique heterocyclic properties P-, S-, Si-, B-, and Se-heterocycles render them very attractive for use in medicinal chemistry. In terms of phosphorus-based heterocycles, ifosfamide (**1a**) and trofosfamide (**1b**) are oxazaphosphorine analogs used as alkylating agents in the treatment of cancer and possess immunomodulating activity by halting or slowing down the growth of cancer cells [35, 36]. Cyclic glycophostone analogs (**1c**, cyclic phosphonamidates) were synthesized as biologically relevant sugars and tested for neuraminidase B inhibiting activity [37–45]. Cyclophostin (**1d**) [46–49] and the cyclipostins (**1e**, **1f**) [48–50] are structurally unique and very potent naturally occurring organophosphate-based inhibitors of acetyl cholinesterase (AChE) and hormone-sensitive lipase (HSL), respectively. These analogs that share a common core unit and diverge only by the chain length of the phosphate ester moiety offer a prospective opportunity to develop inhibitors for a wide range of lipase targets.

Heterocycles containing S, Si, B, and Se nuclei have also shown promising activity and include ampiroxicam, a COX-2 inhibitor that is one of an array of bioactive sultams (Fig. 1) [51]. Ampiroxicam (**1g**) is a prodrug of piroxicam (**1h**), a

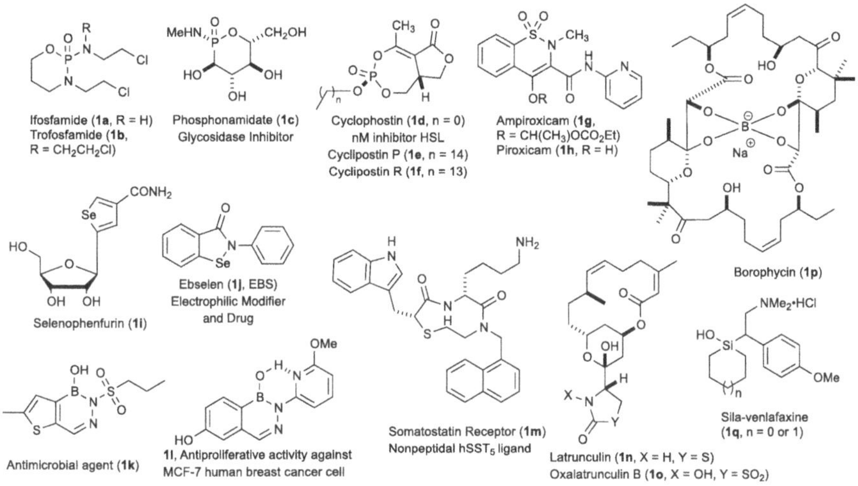

Fig. 1 Representative bioactive P-, S-, Si-, B-, and Se-containing heterocycles

non-steroidal anti-inflammatory drug (NSAID), and is used to relieve the painful symptoms of arthritis [52]. Sulfur-containing β-turn mimetics (**1m**) as medium-sized heterocycle compounds were studied as $hSST_5$ ligands for somatostatin [53]. Selenophenfurin (**1i**) is a selenium-containing inosine 5′-monophosphate dehydrogenase (IMPDH) inhibitor analog [54]. Ebselen (**1j**, EBS) is another selenium-containing heterocycle that has been investigated as a potential cure for reperfusion injury and stroke, tinnitus, and bipolar disorder [55]. It also serves as a scavenger for hydrogen peroxide and hydroperoxide.

In a recent review [56], boron-containing compounds were shown to have therapeutic potency against a variety of biological targets (Fig. 1). Antibacterial activities of diazaborine compounds (**1k**) have been observed against *Proteus*, *Klebsiella*, and *Salmonella* with the minimum inhibitory concentrations (MICs) of 1.25 μg/mL [57]. In addition, diazaborine compounds (**1l**) containing an intramolecular hydrogen bond have demonstrated estrogen-like conformations [58], possessing anti-proliferative activity against MCF-7 human breast cancer cells with an IC_{50} value of 5 μM [59]. Borophycin (**1p**) is a potent boron-containing macrocyclic cytotoxin that was isolated from the cyanobacterium *Nostoc Spongiaeforme* and the marine cyanobacterium *Nostoc linckia* [60–63], while sila-venlafaxine (**1q**), a sila-analog of serotonin, is in clinical use as an antidepressant and has promising therapeutic benefits in the treatment of nerve system disorders [64].

Medium-sized heterocyclic sulfide lactams have also been studied as β-turn mimetics, as seen by the non-peptidal $hSST_5$ ligand for somatostatin (**1m**) (Fig. 1) [53]. Macrocycles in these classes are also prevalent, including the latrunculin (**1n**) family of marine natural products [65, 66], possessing a macrocyclic lactone ring of 14 or 16 carbon atoms and a 2-thiazolidinone moiety, which has been studied as a novel agent for targeting cancer [67, 68] and the inhibition of actin polymerization [69]. Oxalatrunculin B (**1o**) [70, 71], a latrunculin analog, possesses a novel and highly oxidized 2-thiazolidinone ring and has shown actin polymerization inhibition properties, as well as antifungal and anticancer activity. Collectively, these representative examples are testament to the broad bioactivity seen in these unique heterocyclic classes.

2 P-Containing Heterocycles

Phosphorus compounds are widely distributed in naturally occurring biological systems; this ubiquity has made them ideal targets for a variety of pharmaceutical [46, 72–81] and agricultural [82] applications [83–85]. In particular, phosphorus heterocycles have been shown to impart a myriad of biological activities including antitumor (compound **2a**, **2b**) and antiproliferative effects (compound **2c**) [86–89] as well as inhibitory effects against a variety of enzymes [inhibition of endothelin-converting enzyme (ECE) (compound **2d**) [38, 44, 45] and imidazoleglycerol phosphate dehydrogenase (compound **2e**) to name a few] (Fig. 2) [90]. In addition,

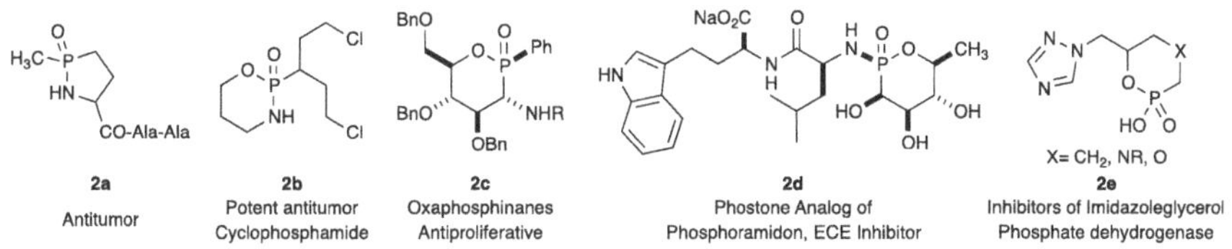

Fig. 2 Representative *P*-heterocycles

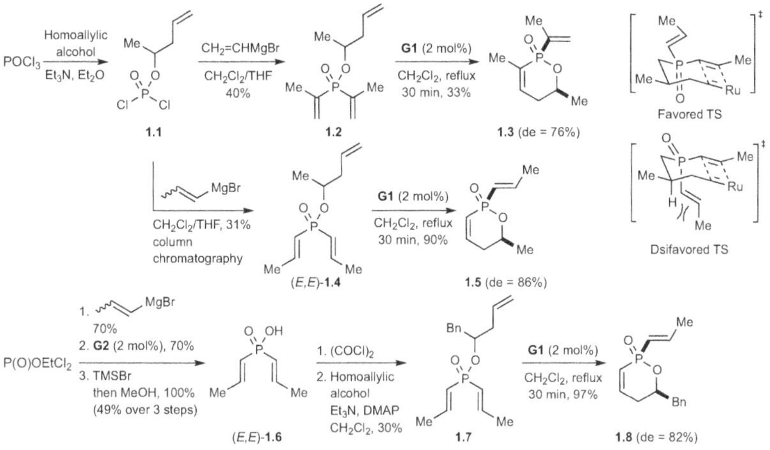

Scheme 1 Synthesis of *P*-stereogenic cyclic phosphinates utilizing RCM

the ability of phosphorus compounds to possess varied valency and metal-binding modes has enhanced their utility in asymmetric catalysis [91–97]. The diverse biological and chemical profile of phosphorus compounds highlighted in Figs. 1 and 2, along with more cited in this chapter, has motivated efforts in the field to develop new strategies for the synthesis of phosphorus-containing compounds with both synthetic and biological utility. Among several approaches, use of RCM as a synthetic route for the construction of phosphorus-containing heterocycles is a versatile method that has been implemented in the last decade [28, 29]. This section aims to highlight recent uses of RCM to generate a diverse range of cyclized phosphorus heterocycles.

In 2005, Gouverneur and coworkers reported a versatile and efficient synthetic method for the preparation of *P*-stereogenic phosphinates utilizing diastereoselective RCM as the key step (Scheme 1) [98], representing the first examples of diastereoselective RCM of prochiral trienic phosphinates (for the formation of *P*-stereogenic heterocycles via diastereotopic differentiation of pseudo-C_2-symmetric phosphonamides and phosphonates using RCM, see [99]). The synthesis of cyclic phosphinate **1.3** and **1.5** commenced using a three-step process employing initial coupling of a homoallylic alcohol with $POCl_3$, followed by double addition of an alkenyl Grignard reagent, and RCM. Reaction of $POCl_3$ with a homoallylic alcohol furnished dichlorophosphinate **1.1**. Subsequent bis-addition of vinylmagnesium bromide or isopropenylmagnesium bromide to **1.1** provided diene **1.2** and **1.4**, respectively, in moderate yields. RCM of **1.2** and **1.4** in the presence of the **G1** catalyst (2 mol%) in CH_2Cl_2 afforded the desired cyclic phosphinates **1.3** and **1.5** with diastereomeric excesses of 76% and 86%, accordingly. The construction of the major *cis* isomer was rationalized by the favored transition state shown in Scheme 1, whereby both substituents are placed in an equatorial position to avoid an unfavorable steric interaction.

The authors also developed an elegant approach to the substituted 5,6-dihydro-1,2-oxaphosphinine 2-oxide **1.8** that utilized olefin isomerization with the **G2** catalyst to avoid the inconvenience of having to purify three geometrical isomers (*EE*/*EZ*/*ZZ*) [98]. Thus, divinyl phosphinic acid (*E*,*E*)-**1.6** was prepared through a three-step reaction employing with ethyl dichlorophosphinate with propenylmagnesium bromide, *cis–trans* isomerization of the alkene using the **G2** catalyst, and sequential addition of TMSBr and methanolysis using MeOH. It is worthy to note that *cis–trans* isomerization of an alkene using RCM catalyst has been relatively underexplored in synthesis in spite of its high value [100, 101]. Coupling with activated phosphinyl chloride generated from treatment of (*E*,*E*)-**1.6** with an excess amount of oxalyl chloride, followed by the addition of homoallylic alcohol, afforded RCM precursor **1.7**. Ring-closing metathesis of **1.7** was performed in the presence of the **G1** catalyst (2 mol%) in refluxing CH_2Cl_2 to furnish cyclic phosphinate **1.8** in 97% yield and 82% de.

In 2005, Hanson and coworkers reported the facile and stereoselective synthesis of novel *P*-sugar analogs applying an RCM strategy (Scheme 2) [102, 103]. This work augments efforts in the field aimed at the development of synthetic strategies to functional and structural carbohydrate mimetics that can potentially serve as biological probes and therapeutic agents [104–108]. The strategy started with construction of the key *P*-chiral allylphostone building block **2.1** derived via a diastereoselective addition pathway (not shown) [102, 103]. With the diene intermediate **2.1** in hand, RCM using the **G1** catalyst (3 mol%) in CH_2Cl_2 was carried out to generate phostone **2.2** in 95% yield. Dihydroxylation of **2.2** with OsO_4/NMO in the presence of citric acid furnished diol **2.3**, which was converted into the

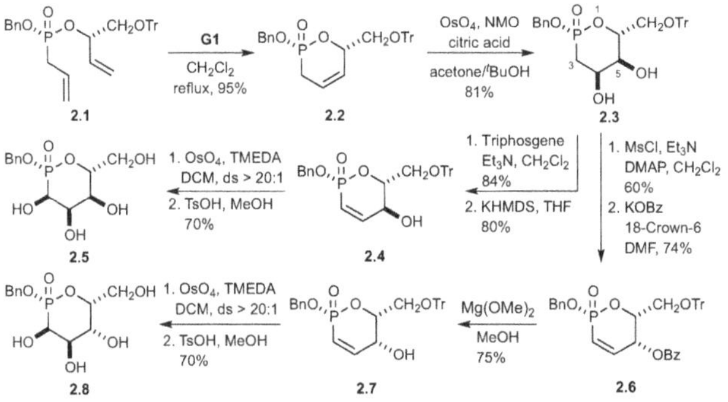

Scheme 2 Preparation of *P*-sugar analogs applying an RCM strategy

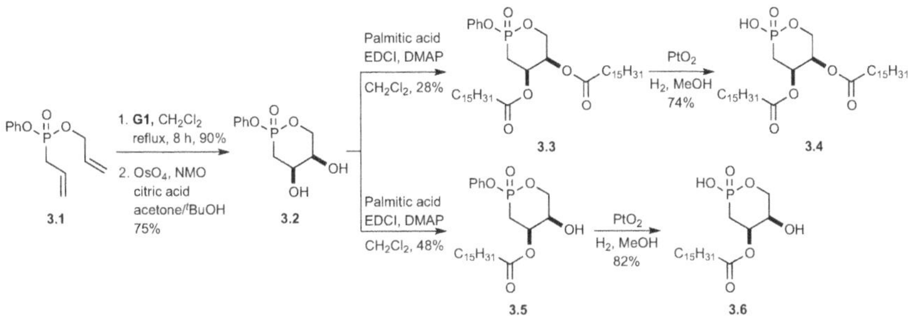

Scheme 3 Synthesis of cyclic phosphonate analogs and free cyclic phosphonic acids

corresponding carbonate, and transformed into vinyl phosphonate **2.4** in good yield upon treatment with KHMDS. Subsequent dihydroxylation using Donohoe conditions [109], and removal of the trityl-group using TsOH in MeOH, afforded the fully deprotected tetraol **2.5** in 70% yield. Alternatively, a synthetic route to the *P*-sugar diastereomer **2.8** was developed and is outlined in Scheme 2. In this route, vinyl phosphonate **2.7** with a *syn* stereochemical relationship between the C(5) and C(6) substituents was successfully accessed by conversion of diol **2.3** into a bis-mesylate with elimination and S_N2 substitution reaction occurring via treatment with BzOK in DMF. Deprotection of the benzoate in **2.6**, followed by dihydroxylation and deprotection of trityl-group, afforded the tetraol *P*-sugar **2.8** as a single diastereoisomer in 70% yield.

In 2006, Prestwich and coworkers described the synthesis of acyloxy-substituted 6-membered cyclic phosphonate analogs of (lyso)phosphatidic acid using RCM (Scheme 3) [110]. Lysophosphatidic acid (LPA) and phosphatidic acid (PA) are major constituents of cell membranes that possess interesting metabolic and cell-signaling properties, prompting analog synthesis to probe their role in receptor-mediated LP signaling pathways [111–113]. LPA has been found to play a crucial role in brain development [114, 115] and cell invasion in cancer [116]. These pleiotropic actions of LPA inspired the authors to search for metabolically stabilized agonists and antagonists for LPA receptors using RCM. The bis-acylated PA analog **3.3** and mono-acylated LPA analog **3.5** were prepared using an RCM/dihydroxylation/EDCI coupling strategy. Ring-closing methathesis of phosphonate **3.1** with the **G1** catalyst (3 mol%) in refluxing CH_2Cl_2 furnished the unsaturated 6-membered cyclic phosphonate, which was converted to dihydroxylated intermediate **3.2** in moderate yield. Treatment of diol **3.2** with palmitic acid in CH_2Cl_2 using EDCI furnished two desired products, the bis-acylated PA analog **3.3** and mono-acylated LPA analog **3.5**. Reductive cleavage of the phenyl phosphonates **3.3** and **3.5** with 1 atm H_2/PtO_2 in MeOH afforded the free cyclic phosphonic acids **3.4** and **3.6**, respectively.

In 2008, Montchamp and coworkers reported the synthesis of 1-(benzyloxy)-2,5-dihydrophosphole 1-oxide (**4.4**) (Scheme 4) [117]. In this work, the authors used a novel catalytic system for the direct allylation of *H*-phosphinic acids with allylic alcohols employing a Pd catalyst. The versatility of these allylated products was

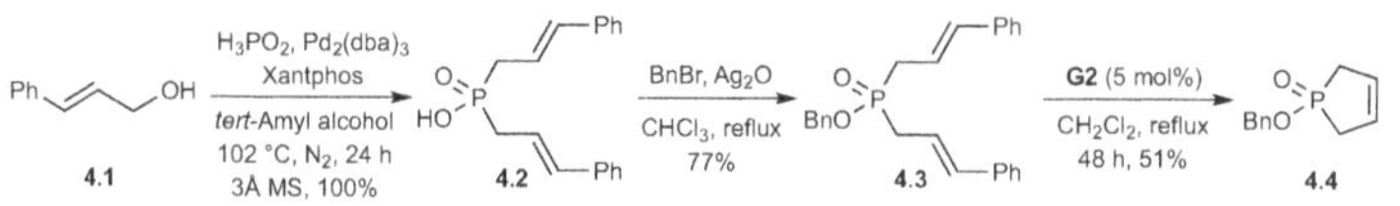

Scheme 4 Synthesis of 1-(benzyloxy)-2,5-dihydrophosphole 1-oxide (**4.4**) via allylation and RCM

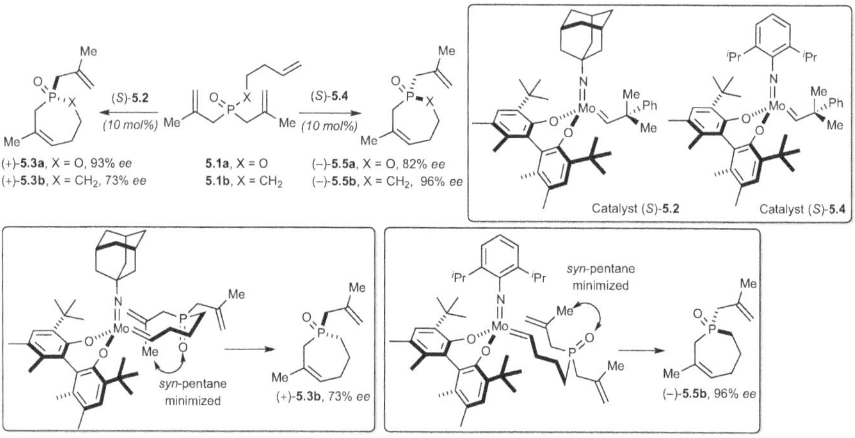

Scheme 5 Enantioselective olefin metathesis reactions toward *P*-stereogenic heterocycles

highlighted by further subjection to a range of reactions including RCM. The synthesis of phospholenes using RCM is sufficiently documented in the literature [118, 119] with the caveat that all the RCM precursors are terminal olefins. In particular, the acyclic phosphinate **4.3**, possessing internal alkenes, was subjected to an unprecedented RCM reaction using the **G2** catalyst to furnish phospholene **4.4** in modest yield and slow reaction time.

In 2009, Hoveyda and Gouverneur reported the elegant synthesis of enantiomerically enriched 5-, 6-, and 7-membered *P*-stereogenic heterocycles prepared by the desymmetrization of prochiral phosphinates and phosphine oxides using novel enantioselective Mo-catalyzed olefin metathesis reactions (Scheme 5) [120]. Chiral, non-racemic phosphines, phosphine oxides, and their derivatives have been implemented in widespread applications as ligands for metal catalysis or as catalysts in organic synthesis [84, 121–124]. Historically, these chiral phosphines have been synthesized by methods based on resolutions of racemic mixtures or the use of chiral auxiliaries. Interestingly, the use of chiral Mo-based metathesis catalysts (*S*)-**5.2** and (*S*)-**5.4**, containing the same chiral, non-racemic diol ligands, but with different achiral imido ligands, led to opposite enantiomers of the 7-membered heterocyclic phosphinate **5.3a** and phosphine oxide **5.5b** with good to excellent enantioselectivity (up to 96% *ee*). The result of this opposite stereoinduction was explained using different reactive alkylidene geometrical isomers, whereby the *syn* isomer was favored for catalyst (*S*)-**5.2** because of the less hindered adamantyl

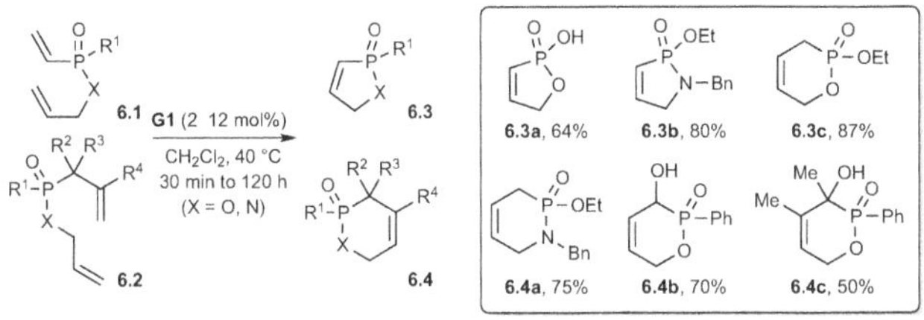

Scheme 6 Construction of oxaphospholene and oxaphosphinene heterocycles via RCM

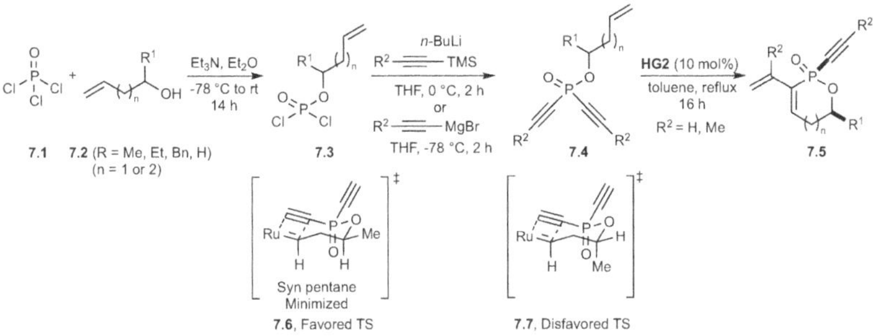

Scheme 7 Metathesis approach to *P*-stereogenic heterocyclic compounds

group, and the *anti* isomer was favored with catalyst (*S*)-**5.4** due to steric repulsion between the isopropyl and the alkylidene groups (Scheme 5).

In 2010, Virieux and coworkers reported the synthesis of oxaphospholene and oxaphosphinene heterocycles via RCM of unsymmetric and unsaturated functional precursors (Scheme 6) [125]. The starting vinyl- and allyl-phosphonates **6.1** and **6.2**, respectively, were readily synthesized by the Arbuzov reaction and subjected to RCM with the **G1** catalyst in refluxing dichloromethane. The 5- and 6-membered *P*-heterocycles **6.3a-c** and **6.4a-c** were obtained in yields ranging from 50 to 87%. The 5-membered oxaphospholene ester **6.3** ($R^1 = OEt$, $X = O$) was not stable and easily hydrolyzed by air moisture to its acid form **6.3a** as observed in previously reported literature with similar molecules [126]. It is also highlighted that the Lewis basicity of both the phosphoryl and hydroxyl groups in hydroxyoxaphosphinanes (precursors to **6.4b** and **6.4c**) did not affect the RCM and the products were obtained in acceptable yields.

In 2010, Gouverneur and coworkers described efforts to use diastereoselective ring-closing enyne metathesis (RCEYM) with *P*-containing ene-diynes for the diastereoselective synthesis of *P*-stereogenic heterocycles (Scheme 7) [127]. To yield a single regioisomer upon RCM, a scope of ene-diynes **7.4** were considered. Dichlorophosphinate intermediate **7.3** was furnished by the *O*-alkylation of chiral unsaturated alcohols **7.2** to phosphorus oxychloride (**7.1**). Double nucleophile addition of either trimethylsilylethynyl lithium or ethynyl magnesium bromide to **7.3** provided the generation of desired dichlorophosphinate intermediates **7.4**. To investigate the feasibility of diastereoselective metathesis to afford alkynyl

Table 1 Optimization studies for diastereoselective RCEYM

Entry	Cat. (mol%)	Atmosphere	Solvent	Conv. (%)	Dr
1	**G2** (2)	Ar	CH_2Cl_2	34	3:1
2	**G2** (2)	Ethylene	CH_2Cl_2	40	2:1
3	**G2** (5)	Ar	CH_2Cl_2	75	3:1
4	**G2** (5)	Ethylene	CH_2Cl_2	73	3:1
5	**G2** (10)	Ar	CH_2Cl_2	90	4:1
6	**G1** (10)	Ar	Toluene	45	3:1
7	**G2** (10)	Ar	Toluene	>98	10:1
8	***HG2 (10)***	***Ar***	***Toluene***	***>98***	***18:1***

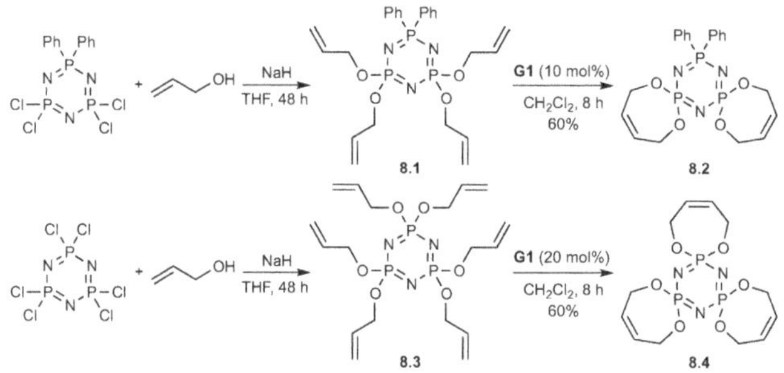

Scheme 8 Terminal alkene-derived trispirocyclic phosphazene compounds

substituted *P*-stereogenic heterocycle **7.5** ($R^1 = Me$, $R^2 = H$), several screening conditions were applied (Table 1). When the reaction was performed using 2 mol % of the **G2** catalyst in CH_2Cl_2, the desired product was formed in low yield and modest diastereoselectivity (entry 1). Reaction under different atmosphere (Ar or ethylene) did not lead to any improvement of conversion. With increment of catalyst loading to 10 mol%, conversion to the desired product was fully increased in either CH_2Cl_2 or toluene. Furthermore, diastereoselectivity was improved when 10 mol% of the **HG2** catalyst was employed in toluene to afford heterocyclic compound **7.5** in high yield (>98%) and diastereoisomeric excess (18:1) (entry 8). Furthermore, the desired Ru-carbene intermediate transition state **7.6** ($R^1 = Me$, $R^2 = H$) displayed both anomeric stabilization induced by donation of the endocyclic oxygen into the σ* orbital of the exocyclic P=O bond and a favorable *syn* pentane minimization by having two small groups (H atoms) in the axial positions. Both 6- and 7-membered *P*-stereogenic heterocycles were synthesized in good yields and selectivity by employing the aforementioned diastereoselective RCEYM. The authors also determined that the nature of the solvent might modulate catalyst activity or enantioselectivity for asymmetric metathesis reactions [128–135].

In 2011, Elias and coworkers reported the synthesis of terminal alkene-derived cyclic phosphazenes (Scheme 8) [136]. Initially, the first reaction of *gem*-diphenyl substituted cyclophosphazene was performed with allyl alcohol to generate the

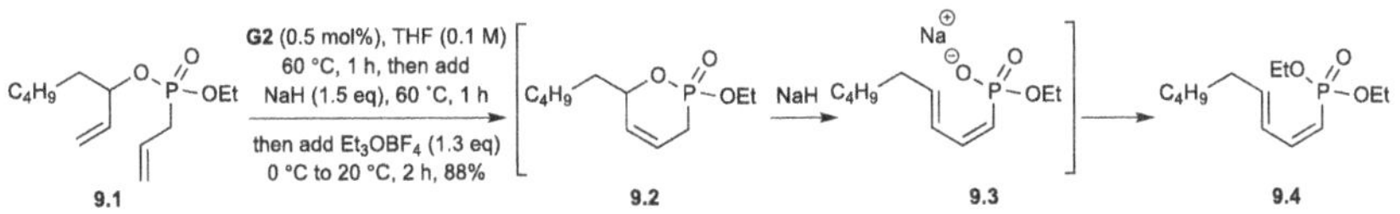

Scheme 9 Construction of conjugated dienyl phosphonates via sequential RCM/ring-opening/alkylation

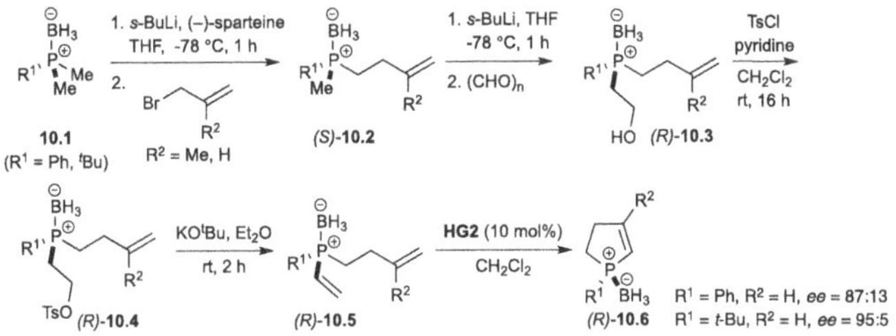

Scheme 10 Synthesis of *P*-stereogenic phospholene boranes

tetrakis(allyloxy) derivative **8.1** which was converted to **8.2** through RCM. *Hexakis* (allyloxy)cyclotriphosphazene **8.4** was furnished in a similar manner. Noticeably, the use of 10 mol% of the **G1** catalyst in RCM of hexaene **8.3** led to the formation of the corresponding monospirocyclic compound, while treatment of **8.3** with 20 mol % of the **G1** catalyst led to the exclusive generation of trispirocyclic product **8.4**.

In 2013, Schmidt and coworkers reported a highly diastereoselective, one-pot synthesis of conjugated dienyl phosphonates from allylphosphonates utilizing an RCM/ring-opening/alkylation sequence (Scheme 9) [137]. Ring-closing metathesis of allylphosphonate **9.1** in the presence of the **G2** catalyst (0.5 mol%) furnished **9.2**, followed by sequential addition of NaH (1.5 equiv.) and Meerwein's reagent [Et_3OBF_4] to afford the alkylated (1*Z*,3*E*)-configured diene **9.4** in situ via ring-opening product **9.3** in good yield.

In 2013, O'Brien and coworkers described the first examples of the asymmetric synthesis of *P*-stereogenic phospholene boranes **10.6** (Scheme 10) [138]. Using the Evans protocol [139], initial intermediate phosphine boranes **10.2** were prepared from lithiation of phosphine boranes **10.1** using *s*-BuLi and (−)-sparteine suspended in THF at −78°C, followed by reaction with methylallyl bromide or allyl bromide. With enantioenriched allylated phosphine boranes (*S*)-**10.2** in hand, lithiation via BuLi in THF at −78°C followed by the addition of paraformaldehyde, generated hydroxyl phosphine borane (*R*)-**10.3**. Subsequent tosylation and elimination produced diene (*R*)-**10.5**. With no prior reported examples of the carbocyclic vinyl phospholenes (*R*)-**10.6**, several conditions were investigated to find optimal RCM reaction with the **HG2** catalyst. When $R^2 = H$, use of 1 mol% of the **HG2** catalyst worked effectively; however, for more sterically hindered alkene ($R^2 = Me$), higher catalyst loading (10 mol%) was required for cyclization.

In 2013, Matsuda and coworkers reported the synthesis of novel phosphine oxide compounds from phosphorus derivatives (Scheme 11) [140]. While RCM has been

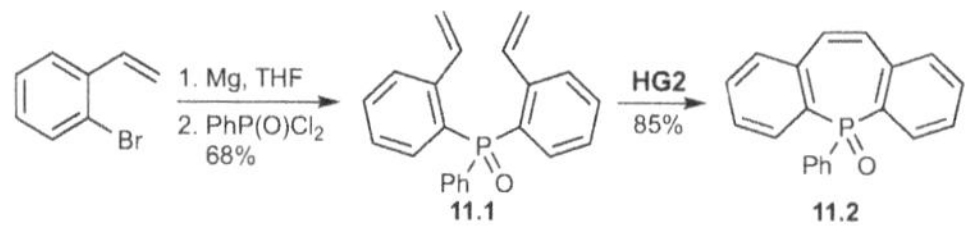

Scheme 11 Preparation of novel phosphine oxide compounds

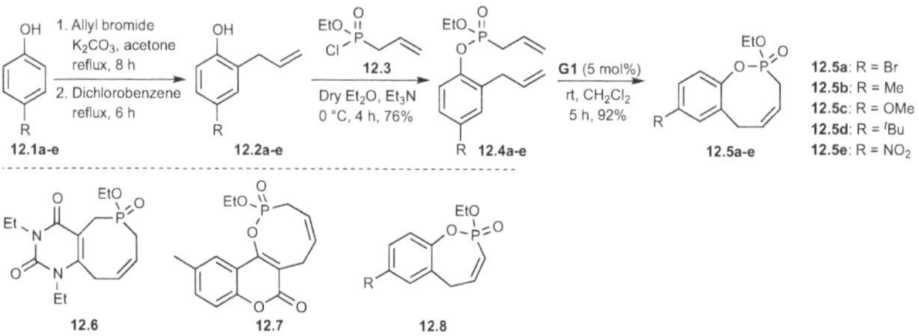

Scheme 12 Synthetic route of phosphorus-containing heterocycles by RCM

widely applied for the synthesis of carbo- and heterocyclic alkenes with different ring sizes, examples for the synthesis of fully unsaturated or aromatic compounds via RCM are relatively limited. Treatment of (2-vinylphenyl)magnesium bromide and phenylphosphonic dichloride led to the generation of phosphorus-tethered diene **11.1**. Terminal olefin **11.1** was consumed for the RCM reaction to give dibenzo[*b*,*f*]phosphepine oxide **11.2** in 85% yield. Matsuda suggested that these novel phosphine oxides have potential use as ligands in catalysis [141, 142].

In 2014, Majumdar and coworkers reported an innovative synthetic route to medium ring benzo-fused *P*-heterocycles using a sequential Claisen rearrangement–RCM sequence (Scheme 12) [143]. The synthesis of benzo-fused *P*-heterocycles is relatively unexplored, which inspired the authors to target the 2-ethoxy-2,5-dihydrobenzo[*f*][1,2]oxaphosphepine 2-oxide and (*Z*)-2-ethoxy-3,6-dihydro-2*H*-benzo[*g*][1,2]oxaphosphocine 2-oxide classes of compounds. To generate *C*-allylated phenols **12.2a–e**, *O*-allylated derivatives were used for the thermal Claisen rearrangement. Phosphonate derivatives **12.4a–e** were prepared by using coupling reactions of *C*-allylated phenols **12.2a–e** and allylphosphonochloridate **12.3**. Unsymmetric alkenyl derivatives **12.4a–e** were treated with the **G1** catalyst for the formations of derivatives **12.5a–e**. In an analogous manner, compounds **12.6**, **12.7**, and **12.8** were synthesized utilizing similar conditions.

Hanson and coworkers utilized two novel enantiomeric bicyclo[4.3.1]-phosphate scaffolds (*S*,*S*,*S*$_P$)-**13.6** and (*R*,*R*,*R*$_P$)-**13.6** [144–146] to afford key polyol building blocks in the synthesis of a number of different natural products including (−)-tetrahydrolipstatin (**13.1**) in 2010 [147], salicylihalamide A in 2011 (**13.3**, formal synthesis) [148], dolabelide C (**13.2**) in 2011 [149, 150], and (+)-strictifolione (**13.4**) in 2014 [151], as well as (6*R*)-6[(*E*, 4*R*, 6*R*)-4,6-dihydroxy-10-phenyl-1-decenyl]-5,6-dihydro-2H-2-pyrone (**13.5**) [151] in 2014 (Scheme 13). The bicyclic phosphate **13.6** was synthesized by a phosphate tether-mediated

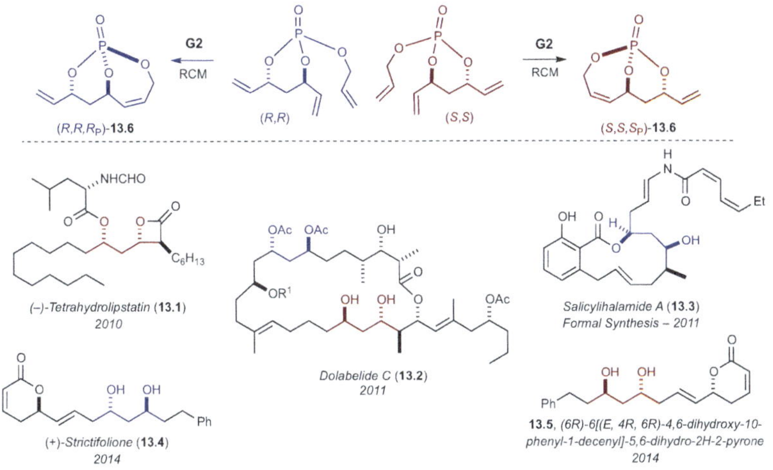

Scheme 13 Natural products prepared from phosphate tether-mediated methods

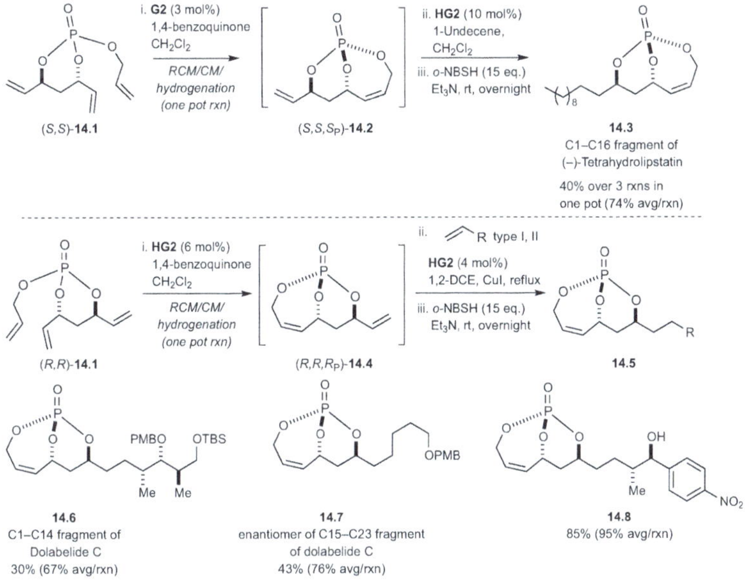

Scheme 14 Synthesis of fragments of (−)-tetrahydrolipstatin and dolabelide C

desymmetrization of the C_2-symmetric 1,3-*anti* diol diene via RCM reaction of the corresponding triene in the presence of the **G2** catalyst. Using the inherent properties of these bicyclic templates [144, 152], a number of synthetically useful transformations enabled the synthesis of the aforementioned natural products.

In 2010, Hanson and coworkers reported a modular, one-pot, sequential protocol involving RCM, CM, and chemoselective hydrogenation for the synthesis of the C1–C15 fragments **14.3** en route to (−)-tetrahydrolipstatin (Scheme 14) [147]. This methodology involved RCM, CM, and chemoselective diimide hydrogenation with the processes occurring without the isolation of any of the intermediates. Initially the process involved use of the **G2** catalyst for RCM and use of the **HG2** catalyst for CM. Optimization showed that RCM could be performed on the starting triene (*S*, *S*)-**14.1** in the presence of the **HG2** catalyst and benzoquinone additive to afford the 8-membered bicyclo[4.3.1]-phosphate intermediate (S,S,S_P)-**14.2**. Following the completion of the RCM reaction, the solvent was evaporated, and the olefin CM partner (type 1) in 1,2-dichloroethane or methylene chloride was added, along with an additional 4 mol% of the **HG2** catalyst. After the completion of CM, chemoselective diimide hydrogenation of the exocyclic olefin was performed utilizing *o*-NBSH and Et_3N to afford the hydrogenated bicyclic phosphate **14.3** [C1–C15 fragment of (–)-tetrahydrolipstatin] in good yield.

In 2012 [153], Hanson and coworkers further elaborated this protocol to include a broader array of CM partners (both type I and II). Formation of **14.5** was achieved via a one-pot, sequential process under optimized condition using the **HG2** catalyst as the sole RCM/CM catalyst. Benzoquinone was added to scavenge ruthenium hydride generated in situ. With careful selection of both the triene and olefin partners, a number of different polyol subunits (**14.6**, **14.7**, and **14.8** are a few representative examples) were synthesized including the C1–C14 and C15–C23 fragments of dolabelide C **14.6** and **14.7**, respectively, in good yields over three reactions in one-pot.

In 2013, Hanson and coworkers explored the role of stereochemistry within a phosphate tether coupling partner, where a number of factors were probed, including product ring size, in the synthesis of 1,3 *anti*-diol-containing subunits (Scheme 15)

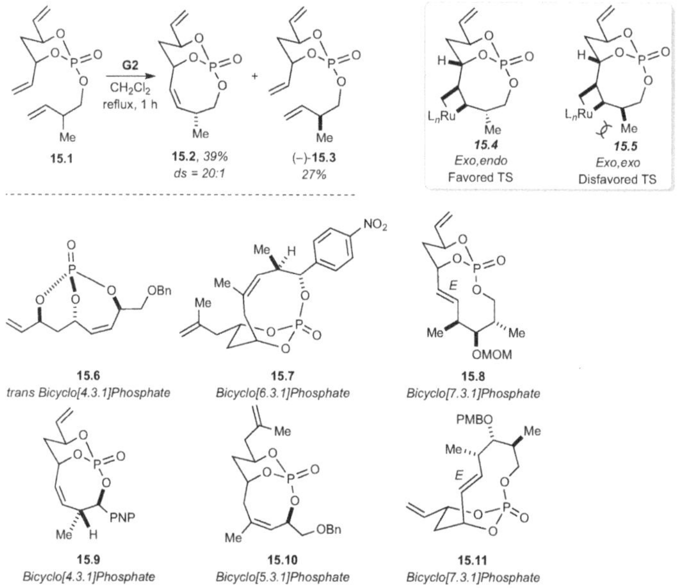

Scheme 15 Application of phosphate tether into the synthesis of 1,3 anti-diol-containing subunits

[154]. Postulated metallocyclobutane models were proposed to gain insight into the observed stereochemical outcomes of the heterocyclic phosphate products obtained from the RCM reactions using the **G2** catalyst. Of notable significance was the effect of allylic stereochemistry in the formation of the 8-membered bicyclo[5.3.1]-phosphate. The allylic effect was demonstrated in the double diastereotopic differentiation of triene **15.1** using RCM to afford exclusively the bicyclo[5.3.1]-phosphate diastereomer **15.2**, accompanied by unreacted diastereomeric triene (−)-**15.3**. The orientation of the allylic methyl group in the *trans,anti*-**15.2** RCM product was shown by X-ray crystallographic analysis to be *endo*. A plausible metallocyclobutane intermediate *exo,endo*-**15.4** was proposed to support the obtained RCM product, while the *exo* allylic methyl group in **15.5** prevented the RCM event from occurring. A number of bicycle phosphates (**15.6–15.11**) of different ring sizes were also synthesized. In the production of the 10-membered bicyclo[7.3.1]-phosphates **15.8** and **15.11**, only the *E*-configured alkene stereoisomer was observed.

In 2014, Hanson and coworkers reported a modular, "pot economical" total synthesis of (+)-strictifolione (**16.3**) and the related natural product, (*6R*)-6 [(*E*, *4R*, *6R*)-4,6-dihydroxy-10-phenyl-1-decenyl]-5,6-dihydro-2H-2-pyrone (**16.6**) (Scheme 16) [151]. In this one-pot, sequential process, the synthesis of **16.3** and **16.6** involved subjection of the pseudo-C_2-symmetric trienes (*R,R*)-**16.1** and (*S,S*)-**16.4** to the **HG2** catalyst (6 mol%) in refluxing CH_2Cl_2, followed by solvent removal and addition of the CM partner (*cis*-stilbene or 4-phenyl-1-butene), which led to smooth CM reactions, and final chemoselective diimide hydrogenation of the exocyclic double bond to yield the bicyclic phosphates, **16.2** and **16.5** in 52% and 54% yield, respectively, over 3 reactions in one-pot. The resulting bicyclic phosphates **16.2** and **16.5** were further elaborated to the final natural products **16.3** and the related natural product **16.6** using a one-pot, sequential Pd-catalyzed allylic transposition/$LiAlH_4$ tether removal, followed by a final CM. Overall, **16.3** and the related natural product **16.6** were produced from triene phosphate triesters (*R,R*)-**16.1** and (*S,S*)-**16.4** using consecutive phosphate tether-mediated, one-pot,

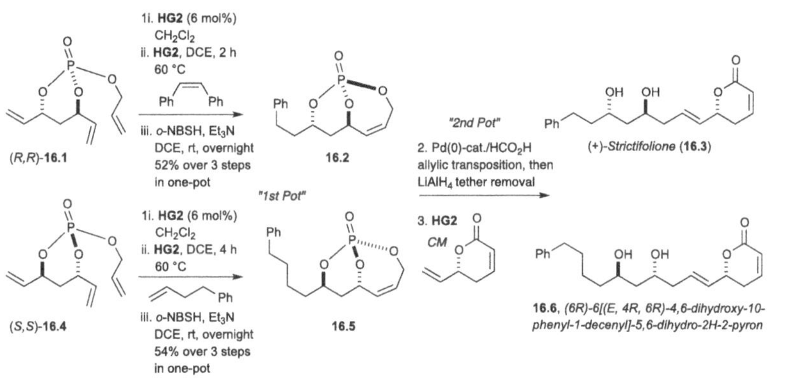

Scheme 16 Consecutive phosphate tether-mediated, one-pot, sequential protocols, followed by CM

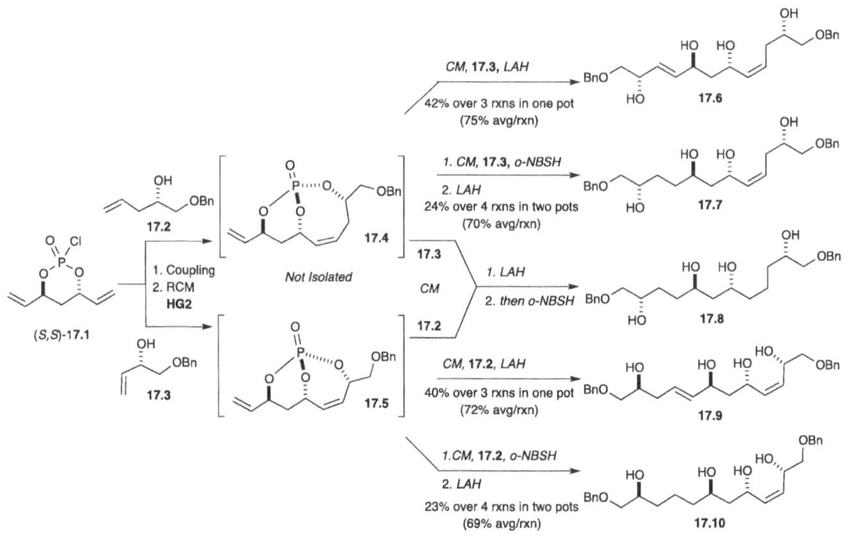

Scheme 17 Phosphate tether-mediated, one-pot, sequential protocols

sequential protocols, followed by CM, representing an efficient, scalable, and library-amenable approach to both.

In 2014, Hanson and coworkers reported the usage of a modular, phosphate tether-mediated, one-pot, sequential protocol for the divergent synthesis of polyol subunits (Scheme 17) [155]. In this work, one-pot sequences comprised of RCM/CM/LAH, or RCM/CM/H_2 – and LAH, as well as RCM/CM/LAH – followed by global hydrogenation, were employed to generate different polyol subunits. Starting with three readily prepared compounds and utilizing the aforementioned procedures, the authors were able to synthesize five different stereochemically rich polyols fragments. Work in this area was initiated from the pseudo-C_2-symmetric phosphoryl chloride (*S*,*S*)-**17.1**, which was coupled with the olefinic alcohols, **17.2** and **17.3**, to yield the corresponding trienes that were isolated and subjected to RCM using 6 mol% of the **HG2** catalyst to generate the bicyclic phosphates, **17.4** and **17.5**, respectively. Chemoselective CM of the exocyclic double bond of **17.4** with olefinic alcohol **17.2**, followed by tether removal (LAH), resulted in the construction of tetraol **17.6** in 42% yield over three reactions in one pot with an average yield of 75% per reaction. Tetraol **17.6** possesses both *E*- and *Z*-configured olefin geometries with the latter governed by the phosphate tether in the RCM event.

In a similar fashion, tetraol **17.7** was constructed with CM between **17.4** and **17.2**, followed by a chemoselective diimide-mediated hydrogenation of the exocyclic double bond. The phosphate tether was next removed in a separate pot (LAH) yielding tetraol **17.7** in 24% yield over four reactions in two pots (70% avg yield per reaction). In a third sequence of reactions, the RCM/CM/LAH sequence, combined with a diimide-mediated global hydrogenation of both double bonds, yielded tetraol

17.8 in 34% over four reactions in two pots. The process was repeated with the in situ-generated bicyclic phosphate **17.5** performing similar protocols to furnish two additional tetraols, **17.9** and **17.10**, in good overall yields.

3 *S*-Containing Heterocycles

Sulfur-containing compounds and sulfur heterocycles (*S*-heterocycles) in particular have been extensively studied and shown to possess broad bioactivity (Fig. 3) [156–159]. This importance is underscored in a 2012 survey of the 200 top selling drugs, of which 18 contain *S*-heterocyclic subunits [160]. Some representative examples of bioactive sultams are highlighted in Fig. 3 and include the γ-sultam compound S-2474, which was synthesized as an anti-arthritic agent and has shown potent inhibitory effects on both cyclooxygenase (COX)-2 and 5-lipoxygenase (5-LO) [161]. A number of substituted 2,3-dihydrobenzo[d]isothiazole 1,1-dioxide sultam analogs were tested in combination therapy and were shown to interrupt proliferation of HIV [162]. In addition, a benzopyridothiadiazepine has also been shown to possess anti-proliferative activity toward the murine L1210 leukemia cell line with an IC_{50} value of 0.11 μM [163]. Finally, potential biologically active 1,2-benzothiazins were synthesized and shown to possess antimicrobial activity against *Escherichia coli* [164].

Ring-closing metathesis represents one method for the synthesis of both carbocycles and heterocycles [10, 11, 165–169]. The continual development of new metathesis catalysts has prompted the emergence of new RCM methods for the synthesis of a wide array of *S*-heterocycles; this section highlights select examples.

Sultones, the internal esters of hydroxyl sulfonic acids, are very useful heterocycles as key intermediates for natural product synthesis. Many methodologies have been developed for sultone synthesis, including the intramolecular Diels–Alder reaction [170], Rh-catalyzed C–H insertion [171], Pd-catalyzed intramolecular coupling reactions [172, 173], as well as several other methods [174, 175]. In 2005 and 2006 [176–178], Cossy and coworkers exploited a sulfonate tether method they reported in 2003 for the facile total synthesis of the antihelminthic

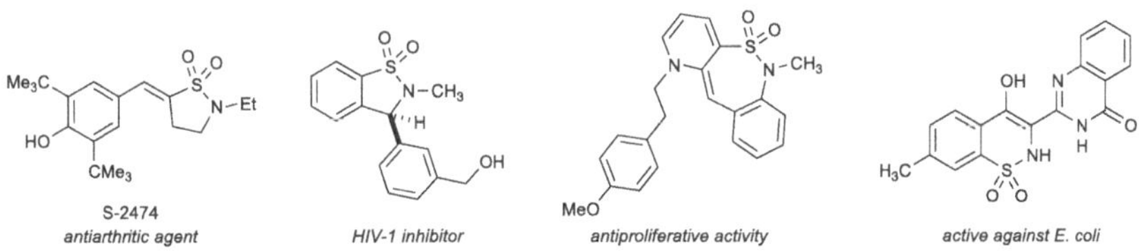

Fig. 3 Representative *S*-heterocycles

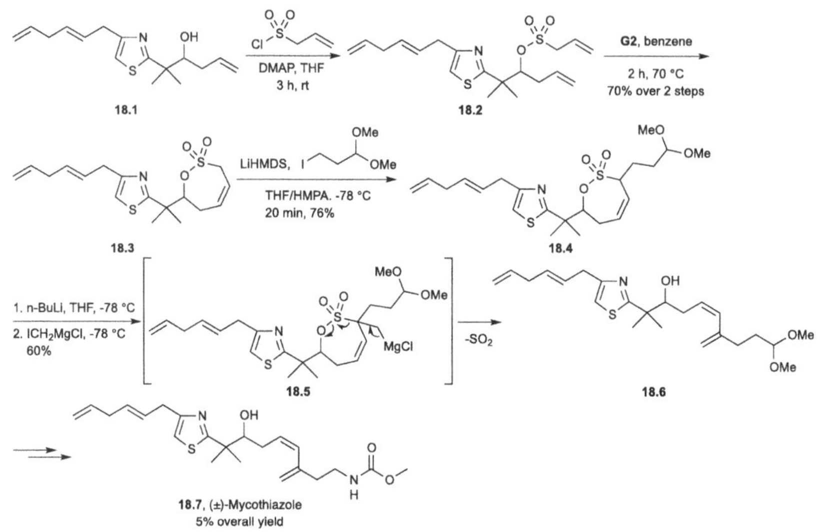

Scheme 18 Total synthesis of (±)-mycothiazole using sulfonate tether

[179] and antitumor[1] agent (±)-mycothiazole **18.7** (Scheme 18) [180]. In this work, a sulfonate tether was used for both selective and efficient C–C bond formation, as well as providing additional synthetic handles for further functionalization. To elaborate the conjugated (Z)-dienol moiety of mycothiazole **18.7** from secondary allylic alcohol **18.1**, preparation of intermediate unsaturated sultone **18.3** was necessary. To construct the RCM precursor **18.2**, the sterically hindered allylic alcohol **18.1** was treated with allylsulfonyl chloride in the presence of DMAP in THF. RCM of the unsaturated sulfonate **18.2**, in the presence of the **G2** catalyst (15 mol%) with heating benzene, provided the key intermediate unsaturated sultone **18.3** in 70% yield over two steps. Alkylation with readily available 1,1-dimethoxy-3-iodopropane as an electrophile with LiHMDS and HMPA at −78°C afforded the mono-alkylated sultone **18.4** in 76% yield. Deprotonation with *n*-butyllithium, and sequential addition of carbenoid (ICH_2MgCl), led to the generation of the transient species **18.5**, which via β-elimination afforded the conjugated (Z)-dienol **18.6** in 60% yield. Further elaboration of dienol **18.6** achieved the total synthesis of (±)-mycothiazole **18.7** in 5% overall yield.

In 2008, Hanson and coworkers reported the synthesis of chiral, non-racemic 6- and 7-membered δ-sultams and ε-sultams, respectively, and their derivatives using RCM as a key reaction (Scheme 19) [181]. Starting with the Boc-protected allyl sulfonamide **19.1**, a Mitsunobu reaction was performed with chiral allyl alcohol **19.2** in THF to produce the RCM precursor **19.3** in 78% yield. Subsequent RCM, using the **G2** catalyst in refluxing toluene, furnished the 6-membered sultam **19.4** in

[1] The results of the National Cancer Institute Human Tumor Cell Line Screen mean graph can be consulted on the Internet at http://dtp.nci.nih.gov (NCS number 647640).

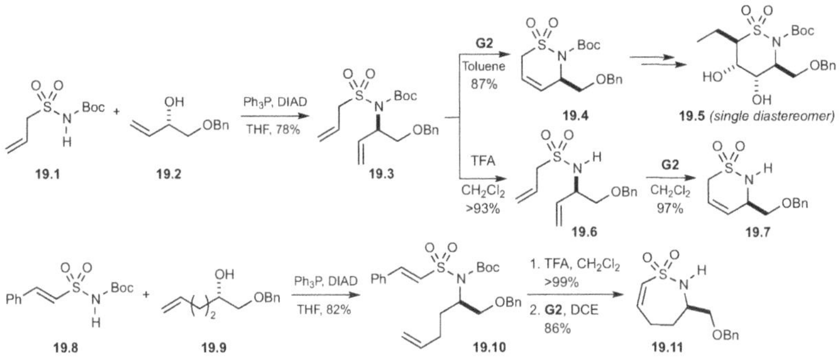

Scheme 19 Synthesis of sultam via RCM

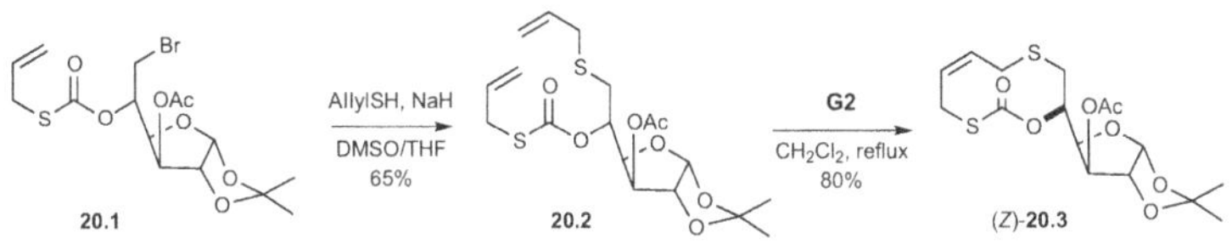

Scheme 20 Synthesis of the *Z*-configured 10-membered macrocyclic thiolcarbonate **20.3**

87% yield. This sultam scaffold was further diversified using various regio- and diastereoslective reactions to generate substituted sultams such as **19.5**. It was found that Boc-removal in **19.3** using TFA in CH_2Cl_2 at 0°C, followed by RCM in refluxing CH_2Cl_2, generated the sultam **19.7** in higher yield (97%). Presumably, the lower yields and longer reaction times in the RCM of **19.3** are the result of increased A1,2-strain between the CH_2OBn group and vicinally substituted Boc group in the TS leading to the desired transformation. In an analogous manner, formation of the 7-membered vinylic sultam **19.11** started by employing Mitsunobu alkylation of vinyl sulfonamide **19.8** with the chiral, non-racemic alcohol **19.9** to provide diene RCM precursor **19.10**. Boc-deprotection, followed by RCM, using the **G2** catalyst in CH_2Cl_2 furnished the 7-membered sultam **19.11** in good yield.

In 2009, Benazza and coworkers reported the stereoselective synthesis of the *Z*-configured 10-membered macrocyclic thiolcarbonate glycol conjugates **20.3**, utilizing *vic*-diol cyclic thiolcarbonate haloalkylation (VHA) and RCM as the key reactions (Scheme 20) [182]. The starting bromide-containing compound **20.1**, possessing a D-gluco-configuration, was treated with allylthiol in the presence of NaH base to produce diene **20.2** in 65% yield. Diene **20.2** was successfully cyclized using RCM with the **G2** catalyst, to furnish the *Z*-configured 10-membered thiolcarbonate **20.3** in 80% yield.

In 2009, Wang and coworkers reported the synthesis of ferrocene-tagged ruthenium carbene catalysts **21.2** and **21.3** that can be immobilized in an ionic liquid (Scheme 21) [183]. These catalysts overcome some difficulties that many homogeneous catalysts have, such as recycling and removing residual catalysts from the products. As illustrated in Scheme 21, reaction of ferrocene-tagged ligand **21.1** with

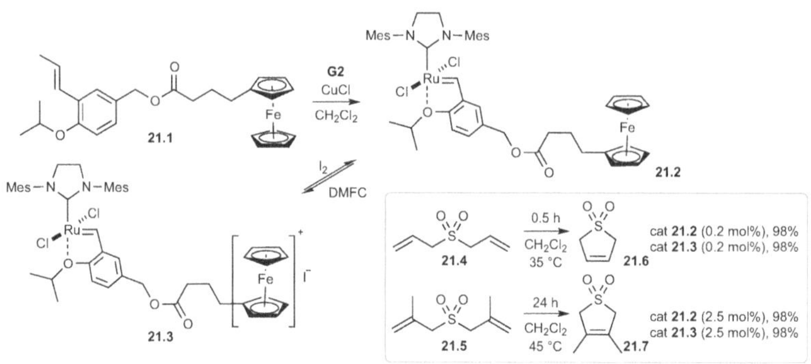

Scheme 21 RCM using ferrocene-tagged ruthenium carbene catalysts

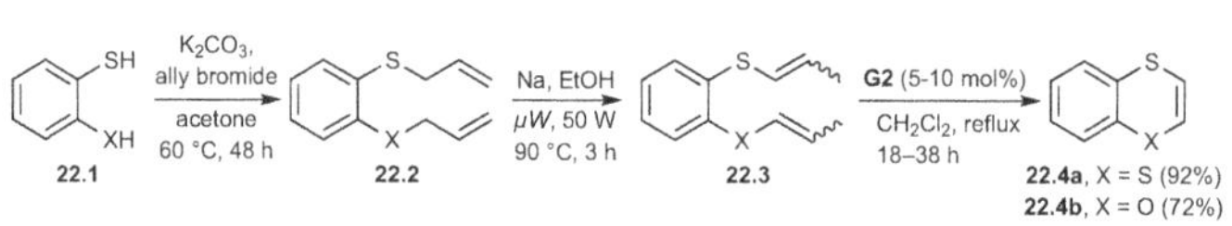

Scheme 22 Synthetic routes for 1,4-heteroatom containing benzo-fused heterocycles

the **G2** catalyst in the presence of CuCl provided the ferrocene-tagged ruthenium complex **21.2** in 68% yield. Furthermore, I_2 was employed as an oxidant to furnish the cationic ferrocene-tagged ruthenium complex **21.3** efficiently in quantitative yield; this complex was shown to be soluble in an ionic liquid. A number of dienes including the sulfur-containing substrates **21.4** and **21.5** were investigated in order to explore the scope of these catalysts. In all reported cases, catalysts **21.2** and **21.3** produced the desired products in very high yields.

In 2009, Otterlo and coworkers reported a strategic method to afford the 1,4-heteroatom-containing benzo-fused heterocycles **22.4** (Scheme 22) [184]. Some benzo-fused, unsaturated heterocycles have been investigated as potential medical scaffolds for pharmaceutical needs [185–192]. Treatment of the bis-allyl intermediates **22.2** with excess amounts of NaOEt using microwave reaction conditions readily provided the isomerized 2-propenyl intermediates **22.3**. The desired 6-membered heterocyclic benzo[b][1,4]dithiine (**22.4a**, X = S) and benzo[b][1,4]oxathiine (**22.4b**, X = O) were generated by the application of RCM using the **G2** catalyst under refluxing conditions.

In 2009, Bolm and coworkers developed the synthesis of heterocyclic sulfoximines using ring-closing enyne metathesis (RCEYM) (Scheme 23) [193]. Sulfoximines are stable compounds possessing versatile features that offer additional opportunities for substitution. In these systems (**23.7**), the nature of the substituent on the nitrogen atom attenuates basicity at the nitrogen and acidity at the CH_2 carbon. These weakly basic systems can coordinate with a metal ion, and in addition, the free sulfoximines ($R^1 = H$) are capable of phosphorylation in vivo [194, 195]. The unique hydrophilic sulfoximine group is very attractive for medicinal chemistry [196]. Compound **23.6** can be generated from *N*-TMS-protected *S*-

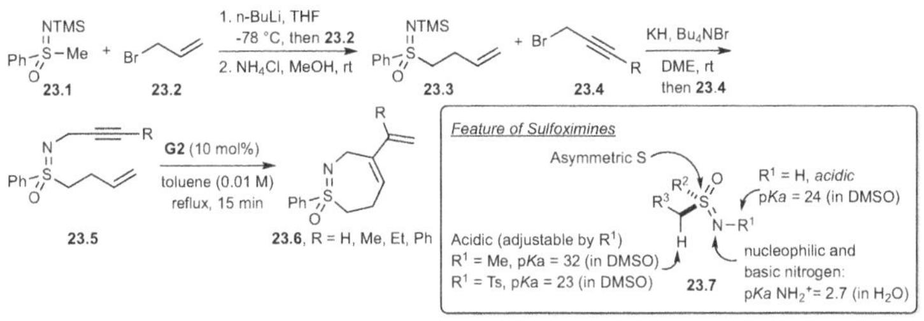

Scheme 23 Construction of heterocyclic sulfoximines by RCM

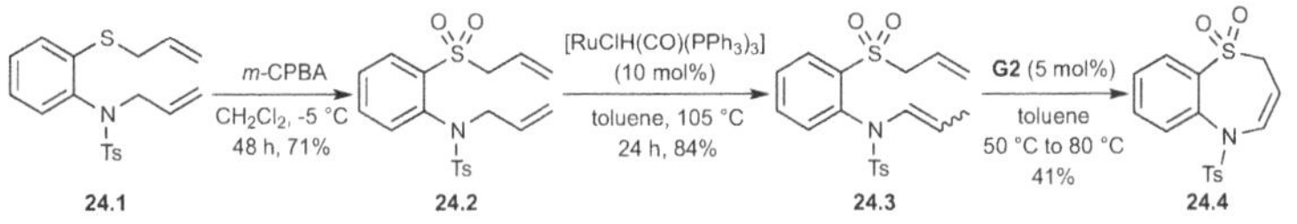

Scheme 24 Route to generate 1,5-benzothiazepine **24.4**

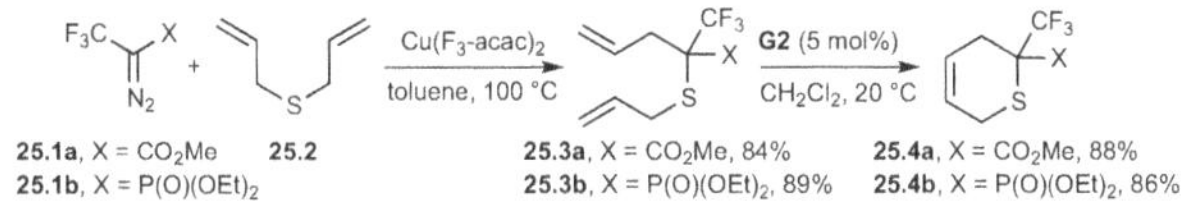

Scheme 25 Synthesis of trifluoromethyl substituted sulfur-containing heterocycles

methyl-*S*-phenyl sulfoximine **23.1**. Deprotonation of **23.1** with *n*-BuLi and treatment with allyl bromide led to the generation of alkenyl phenyl sulfoximine **23.3**, which was coupled with bromoalkyne **23.4** to afford the doubly unsaturated sulfoximine intermediates **23.5**. Application of RCM using the **G2** catalyst in toluene, furnished the desired cyclized products **23.6** in moderate to good yields [R = H (65%), Me (70%), Et (79%), and Ph (45%)].

In 2011, Otterlo and coworkers also reported the generation of 7-membered benzo-fused heterocyclic compounds containing two heteroatoms, which were employed for the synthesis of several bioactive compounds in medicinal studies (Scheme 24) [197]. In this regard, the bis-olefin substrate **24.1** was oxidized to sulfone **24.2** by the addition of *m*-CPBA at −5°C. Selective alkene isomerization on **24.2** occurred only at the *N*-allyl group to furnish intermediate **24.3** in 84% yield. In this work, Kuźnik showed that allyl sulfones can be isomerized to the thermodynamically more stable internal vinyl form [198], but in these examples, only the *N*-allyl group was effectively isomerized to **24.3** by the Ru catalyst. Subsequent treatment of **24.3** with the **G2** catalyst afforded the protected-2,5-dihydrobenzo[*b*][1,4]thiazepine 1,1-dioxide **24.4** in moderate yield.

In 2011, Osipov and coworkers reported a novel efficient method for the synthesis of trifluoromethyl-substituted sulfur-containing heterocycles **25.4a** and **25.4b** via [2,3]-sigmatropic rearrangement and RCM (Scheme 25) [199]. α-Trifluoromethyl-substituted α-diazocarboxylate **25.1a** and α-diazophosphonate

25.1b are distinctive building blocks for the installation of trifluoromethyl, carboxylic, and phosphonic functionalities into organic molecules. The carefully selected catalyst Cu(F_3-acac)$_2$ was used for the reaction of **25.1a** and **25.1b** with diallylsulfide **25.2**, whereby heating in toluene at 100°C for 2–3 h generated the corresponding terminal olefin containing acyclic sulfides **25.3a** and **25.3b** in good yield. These two dienyl sulfides were easily transformed into the corresponding 6-membered fluoro-containing heterocycles **25.4a** and **25.4b** via RCM using the **G2** catalyst in CH_2Cl_2 at room temperature in 88% and 86% yield, respectively.

In 2012, Otterlo and coworkers reported the synthesis of 2-allyl-3,4-dihydro-2*H*-1,4-benzothiazines from the corresponding diene precursors [200]. The two products isolated were dihydro-2*H*-1,6-benzothiazocine **26.2** and the by-product allyl-substituted benzo[1,4]thiazine **26.3** in nearly the same yield as shown in Scheme 26. Interestingly, RCM reaction in the presence of the **G2** catalyst did not occur when substrates **26.1a–c** (Y = O, S, NH) were used. Furthermore, it was necessary to add exceedingly large amounts of the **G2** catalyst (5 × 10 mol% over 24 h). It was also recognized that the tosyl group was important, as the starting material **26.1c** failed to provide the desired product. To investigate if the oxidation state of the sulfur atom was crucial, sulfoxide **26.4** was generated through *m*-CPBA oxidation. Subsequent RCM of **26.4** afforded the cyclized product **26.5** in high yield (95%) using only a catalytic amount of the **G2** catalyst.

In 2012, Zhao and coworkers reported the route to generate chiral benzo-fused *N*,*S*-heterocycles via Ir-catalyzed allylation and RCM (Scheme 27) [201]. Previously,

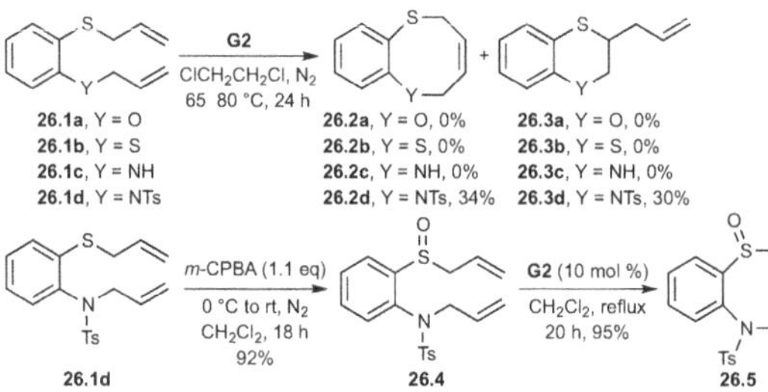

Scheme 26 Synthesis of benzothiazines utilizing RCM

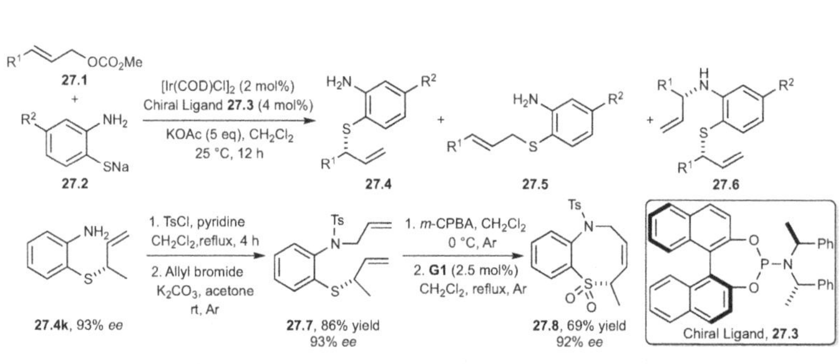

Scheme 27 Synthetic route of the enantio-enriched benzofused *N*,*S*-heterocycle **27.8**

Table 2 Enantioselective allylic alkylation of sodium 2-aminobenzenethiolate with various methyl allylic carbonates utilizing Ir catalyst

Entry	R^1	R^2	Yield of **27.4**	**27.4/27.5/27.6**	*ee* (%)
1	Ph	H	**27.4a**, 83	96/4/trace	96
2	3-$MeOC_6H_4$	H	**27.4b**, 74	84/8/8	96
3	4-$MeOC_6H_4$	H	**27.4c**, 70	90/10/trace	96
4	4-MeC_6H_4	H	**27.4d**, 67	93/7/trace	96
5	4-ClC_6H_4	H	**27.4e**, 58	94/6/trace	97
6	4-BrC_6H_4	H	**27.4f**, 67	94/6/trace	93
7	3-$CF_3C_6H_4$	H	**27.4g**, 75	94/6/trace	97
8	$BnCH_2$	H	**27.4h**, 84	90/10/trace	94
9	*n*-Pr	H	**27.4i**, 75	96/4/trace	94
10	Et	H	**27.4j**, 76	97/3/trace	81
11	Me	H	**27.4k**, 54	98/2/trace	94
12	Ph	Cl	**27.4l**, 58	86/14/trace	94
13	4-BrC_6H_4	Cl	**27.4m**, 52	85/15/trace	93
14	*n*-Pr	Cl	**27.4n**, 56	90/10/trace	89

Hartwig [202], You [203, 204], and Zhao [205, 206] had demonstrated stereospecific installation of stereogenic C–S bond formation catalyzed via Ir-complexes; however, application of sodium 2-aminobenzenethiolate in the Ir-catalyzed allylation was void in the literature. To generate benzo-fused *N,S*-heterocycles, the authors optimized reaction conditions by screening solvents, base, and temperature. These efforts revealed that phosphoramidite ligand **27.3** in the presence of KOAc in CH_2Cl_2 were efficient conditions for the Ir-catalyzed allylic alkylation. The presence of electron donating groups or electron withdrawing groups on the phenyl ring (entries **27.4a–g** in Table 2) afforded products in good to high yields. Allylation of aliphatic substrates **27.4h–k** gave excellent regioselectivity and enantioselectivities in moderate to good yields.

A method for the synthesis of an 8-membered benzo-fused *N,S*-heterocycle was described starting with substrate **27.4k**. Intermediate **27.7** was produced by the protection and alkylation with allyl bromide in 86% yield. Subsequent oxidation with *m*-CPBA, followed by RCM in the presence of the **G1** catalyst, afforded the benzo-fused *N,S*-heterocyclic compound **27.8** in 62% yield with 92% *ee*.

In 2012, Rao and coworkers reported the synthesis of 13-membered sulfur-containing heterocycles utilizing RCM reaction (Scheme 28) [207]. Allyl protected 2-hydroxybenzaldehyde **28.1** was reduced to provide benzyl alcohol **28.2**. Treatment of **28.2** with thionyl chloride afforded the allyl-protected benzyl chloride **28.3** in 86% yield. Generation of sulfone **28.4** was accomplished by thio-ether formation using Na_2S in basic and refluxing conditions, followed by oxidation with oxone in MeOH. Heterocyclic compounds **28.5** and **28.6** were afforded through RCM as an inseparable mixture of *E*/*Z* isomers (52:48), utilizing the **G1** catalyst (10 mol%) in CH_2Cl_2.

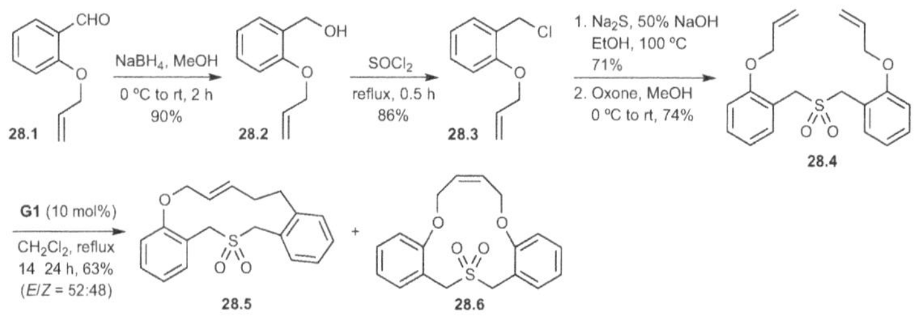

Scheme 28 Synthesis of 13-membered sulfur-containing heterocycles

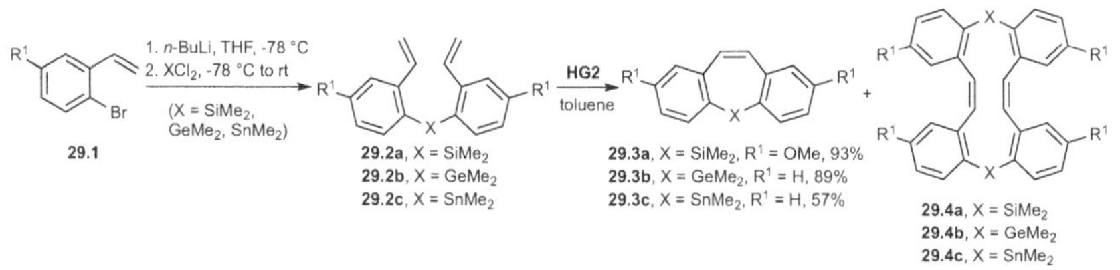

Scheme 29 Formation of dibenzoheteropines

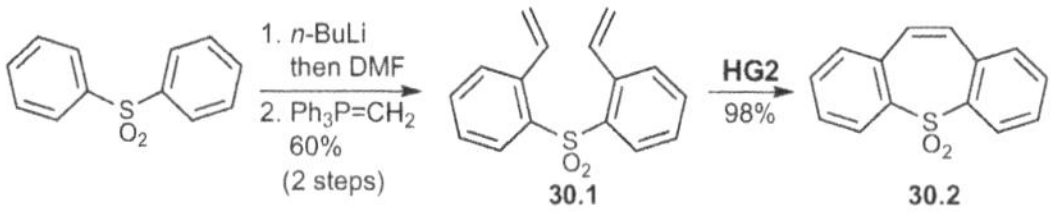

Scheme 30 Application of RCM to generate dibenzo[*b*,*f*]thiepine 5,5-dioxide

In 2013, Sato and coworkers reported dibenzoheteropines-containing group 13–16 elements (Si, Ge, Sn, B, N, P, O, and S) via RCM of the corresponding heteroatom-tethered dienes (Scheme 29) [140]. Thus, the bis-(2-vinylphenyl)silane, bis-(2-vinylphenyl)germane, and bis-(2-vinylphenyl)stannane **29.2a–c** were prepared through the reaction of the corresponding dialkyl substrates and 2 equiv. of (2-vinylphenyl)lithium that was generated from 2-bromostyrene **29.1** and *n*-BuLi. When intermediates **29.2a–c** were subjected to the standard reaction conditions (**HG2**, toluene, 0.1 M, 5 mol%, 100°C), the dimerized 14-membered heterocycles **29.4a–c** were produced in 19–40% yield, in addition to smaller amounts of the desired products **29.3a–c**. To minimize the generation of the self-dimerization product **29.4a–c**, intermediates **29.2a–c** were employed to perform RCM using the **HG2** catalyst (0.002 M) in toluene to afford heteropines **29.3a–c** in moderate to high yields (57–93%).

In 2013, Matsuda and coworkers reported the RCM of a bis-(2-vinylphenyl) sulfone (Scheme 30) [140]. Examples of aromatic or fully unsaturated compounds synthesized through RCM are relatively limited [9, 208]. Heteroatom-substituted cyclopentadienes acquire a variety of intriguing properties, and several synthetic methods have been developed. Synthesis of the bis-(2-vinylphenyl)sulfone **30.1** was performed via *ortho*-formylation of diphenyl sulfone and subsequent Wittig

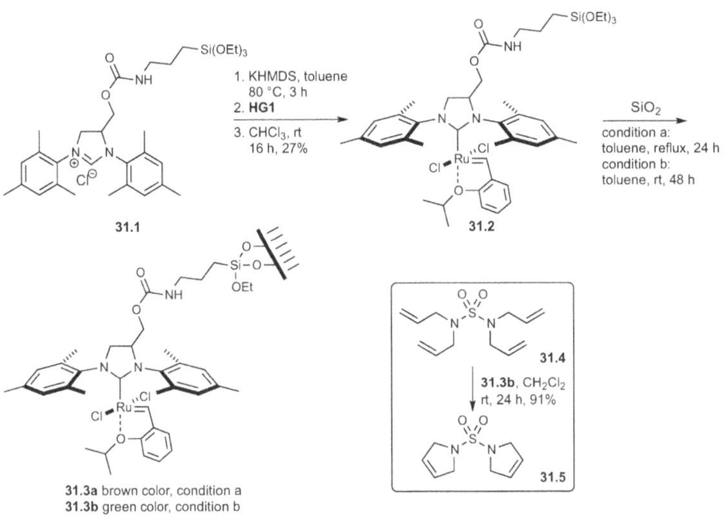

Scheme 31 Application of silica-immobilized NHC-ruthenium complexes for RCM

methylenation. Ring-closing metathesis of **30.1** at 100°C in toluene (0.1 M) in the presence of the **HG2** catalyst (5 mol%) afforded dibenzo[*b*,*f*]thiepine 5,5-dioxide **30.2** in 98% yield.

In 2013, Pleixats and coworkers reported application of silica-immobilized *N*-heterocyclic carbene (NHC)-ruthenium complexes for RCM (Scheme 31) [209]. Efforts in the field have studied various modifications on the metal environment of RCM catalysts in order to alter their stability, reactivity, selectivity, and reusability [210–212]. In this regard, the authors immobilized Ru-alkylidene complexes onto insoluble silica. Other benefits for catalyst immobilization are ease of separation of catalyst from the reaction mixture and lower metal contamination in the final products. Treatment of silylated carbamate **31.1** with KHMDS provided the in situ-generated free carbene, which was subsequently treated with the **HG1** catalyst to furnish Ru-complex **31.2** that was attached to a silica surface via *ortho*-silicate exchange. Complexes **31.3a** and **31.3b** were prepared using different grafting conditions to pre-prepared silica. RCM reactions with complex **31.3b** proved to be very selective, avoiding the generation of cycloisomerization by-products. Double RCM of the acyclic tetraene sulfamide **31.4** afforded the bis-cyclized sulfamide **31.5** in high yield (91%).

In 2014, Abbaz and coworkers reported the synthesis of cyclic sulfamide compounds linked to tetrathiafulvalene (TTF) and reported their antibacterial activity (Scheme 32) [213]. The authors surmised that since TTF derivatives are planar, these compounds could be used as chemical probes for the intercalation of DNA or to potentially modulate various metabolic pathways. After the synthesis of symmetric sulfamide **32.1** from the condensation reaction of 2 equiv. of amine with sulfuryl chloride, diallylation of **32.1** with allyl bromide at refluxing condition afforded symmetric diallyl sulfamide **32.2** in good yield. The 7-membered cyclic

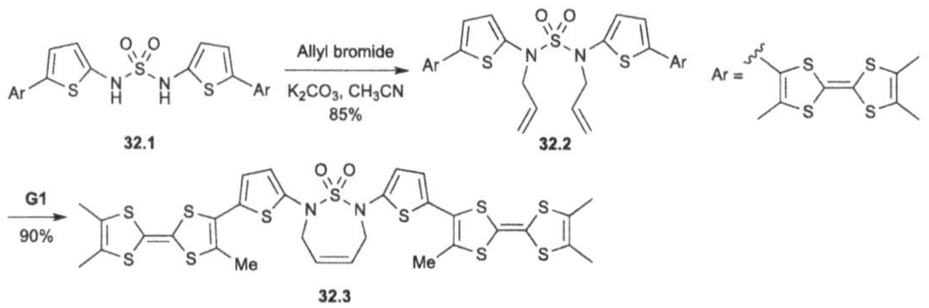

Scheme 32 7-Membered cyclic sulfamide linked to TTF

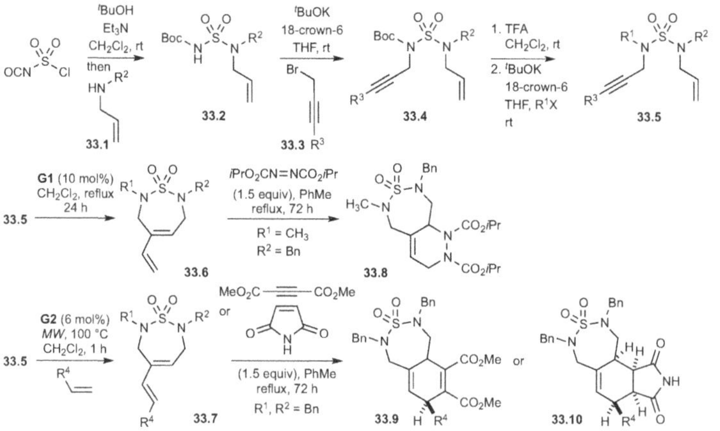

Scheme 33 Synthesis of polycyclic sulfonamides via RCEYM/Diels–Alder reaction

sulfamide **32.3** was obtained in high yield using the **G1** catalyst. Through this synthetic strategy, analogs of 7- to 10-membered cyclic sulfamides were synthesized.

In 2014, Brown and coworkers reported an efficient, one-pot protocol for the synthesis of polycyclic sulfonamides through use of an enyne RCM-Diels–Alder reaction sequence (Scheme 33) [214]. In previous related examples, both the Hanson and Brown groups utilized RCM as a method to build cyclic sulfonamides [215–217] and sulfamides [218–221] and to construct cyclic sulfamoyl carbamates and ureas in 2005 [222]. One of the benefits of enyne RCM is the potential for manipulation of the 1,3-diene products, via cycloaddition reactions, such as the Diels–Alder reaction to build bi- and tricyclic structures. The enyne metathesis substrates were prepared from chlorosulfonyl isocyanate (CSI) coupling with ^{t}BuOH, followed by addition of *N*-allylamine **33.1**, to furnish sulfamide **33.2**. *N*-Alkylation with 1-bromo-2-butyne, Boc cleavage in TFA, and *N*-alkylation of the deprotected sulfamides furnished sulfamides **33.5** ready for subsequent enyne RCM reactions. Treatment of enyne **33.5** with the **G1** catalyst using CH_2Cl_2 reflux conditions yielded the products **33.6**. Microwave-mediated RCM of **33.5** with 2–3 equiv. of a mono-substituted alkene and the **G2** catalyst afforded sulfamides **33.7** successfully with excellent levels of *E*-selectivity ($\geq$15:1 by ^{1}H NMR). Thermal

Diels–Alder reactions of enyne **33.6** with electron deficient dienophiles such as di-iso-propylazodicarboxylate (DIAD), dimethyl acetylenedicarboxylate (DMAD), and maleimide at reflux in toluene (72 h) were utilized to construct the desired polycyclic sulfamide products **33.8**, **33.9**, and **33.10** in yields of 65–86%.

In the aforementioned study, Brown and coworkers were able to prepare a series of bicyclic and tricyclic sulfamides with high diastereoselectivity employing a Yb $(OTf)_3$-promoted Diels–Alder cycloaddition reaction furnishing the *endo*-adduct as the only stereoisomer. To improve yields and avoid degradation of the metathesis catalyst from extended heating times, Lewis acids were screened. $Yb(OTf)_3$ was effective in increasing the yields in the cycloaddition of this one-pot, sequential metathesis Diels–Alder reaction (Table 3). Enyne RCM in toluene and addition of the diverse dienophiles (DIAD, DMAD, maleimide, and *N*-phenylmaleimide) and $Yb(OTf)_3$ were effective to generate the desired products with reduced reaction time. An excess of the dienophile (1.5 equiv.) was necessary for the reaction completion during the Diels–Alder step.

In 2014, Debnath and coworkers reported the synthesis of 7-membered sultones with different carbo- and heterocyclic structures through RCM (Scheme 34)

Table 3 Sequential RCEYM/Diels–Alder reactions for the formation of bicyclic and tricyclic sulfamides

Entry	Substrate	Reaction condition	Product	Yield (%)
1	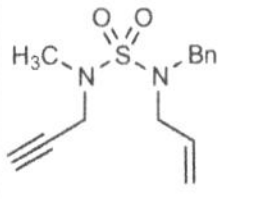	**G1** (10 mol%), PhMe, 50°C, 12 h then DIAD (1.5 equiv.), $Yb(OTf)_3$ (1 equiv.), reflux, 12 h	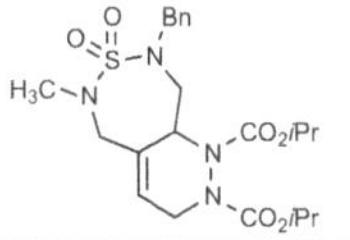	78
2	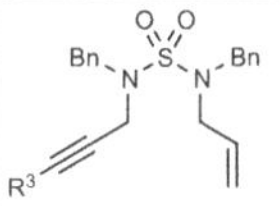	**G1** (10 mol%) or **G2** (3 mol%), PhMe, 50°C, 12 h then DMAD (1.5 equiv.), $Yb(OTf)_3$ (1 equiv.), reflux, 12 h	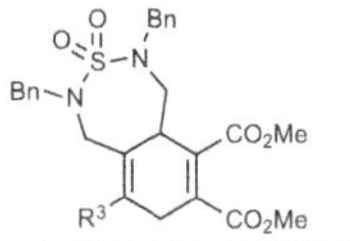	$R^3 = H$, 86 $R^3 = Me$, 79
3	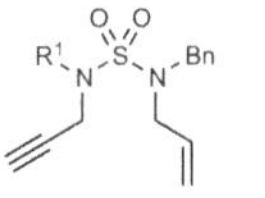	**G1** (10 mol%), PhMe, 50°C, 12 h then maleimide (1.5 equiv.), Yb $(OTf)_3$ (1 equiv.), reflux, 12 h	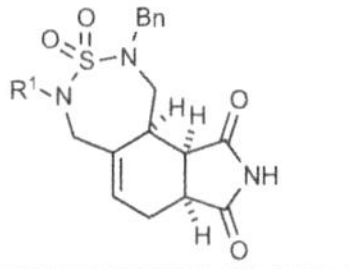	$R^1 = Me$, 63 $R^1 = Bn$, 86
4	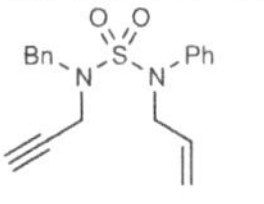	**G1** (10 mol%), PhMe, 50°C, 12 h then *N*-phenylmaleimide (1.5 equiv.), $Yb(OTf)_3$ (1 equiv.), reflux, 12 h	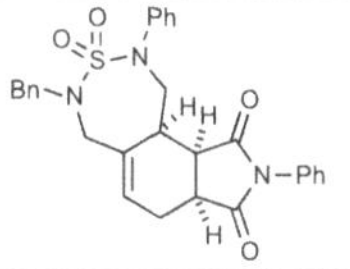	79

Scheme 34 Synthesis of 7-membered ring-fused sultones

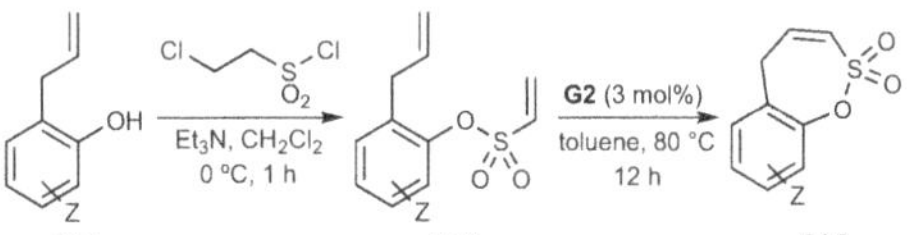

34.1 34.2 34.3

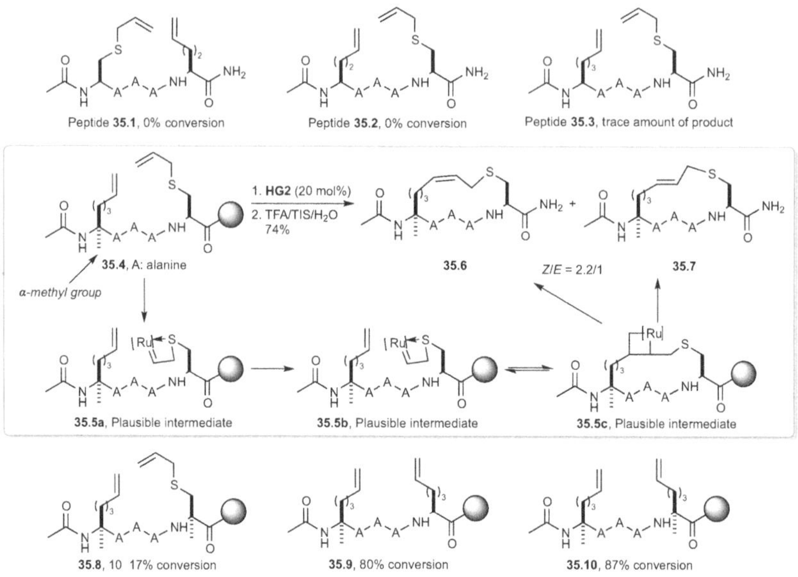

Scheme 35 Effect of α-methylation on metathesis efficiency

[223]. *Ortho*-allylphenol derivatives **34.1** as starting substrates were prepared via Claisen rearrangement of the corresponding allyl ether derivatives. Sulfonylation of **34.1** with 2-chloroethylsulfonyl chloride was carried out in the presence of Et_3N to produce the vinyl sulfonates **34.2** in good yields (81–95%). RCM with the **G1** catalyst (5 mol%) gave no conversion to products. Switching to the **G2** catalyst (3 mol%) in refluxing toluene with **34.2** afforded the ring-fused sultones **34.3** in moderate to good yields (70–80%).

In 2014, Li and coworkers reported the influence of α-methylation for the generation of stapled peptides with RCM (Scheme 35) [224]. In this study, pentapeptides **35.1–35.10**, which are precursors to thioether macrocyclic stapled peptides, were synthesized with three alanine residues placed at the internal residues, and unnatural amino acids bearing olefins installed at each terminus. Subsequent RCM of Ac-*S*-allyl-C-AAA-m*S*4-NH_2 **35.1** with the **HG2** catalyst produced no conversion to the macrocyclic thioethers as observed with LC–MS. The peptide Ac-m*S*4-AAA-*S*-allyl-C-NH_2 **35.2**, whereby the amino acids at the N- and C-terminus were switched, also did not provide detectable product. In addition, RCM of Ac-m*S*5-AAA-*S*-allyl-C-NH_2 **35.3**, possessing one more additional atom in the tether, only produced <5% of the desired product. However, further studies of Ac-*S*5-AAA-*S*-allyl-C-NH_2 **35.4** containing α-methylation demonstrated that RCM with the **HG2** catalyst smoothly generated compounds **35.6** and **35.7** in 74% yield with moderate *Z*/*E* selectivity (2.2:1). The authors proposed that the selectivity could be explained by the formation of plausible intermediates **35.5a–c**, which led to generation of the favored product **35.6**. Moreover, the presence of α-methyl substitution was considered critical for optimal RCM. When these

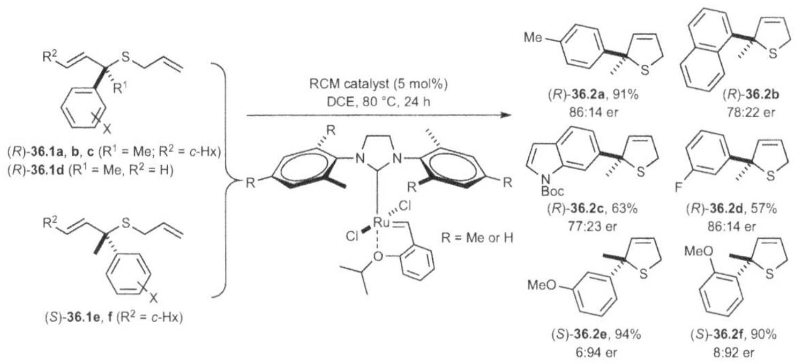

Scheme 36 RCM of allylsulfides to construct enantiomerically enriched 2,5-dihydrothiophenes

unnatural amino acids were switched as in peptides **35.9** and **35.10**, RCM using the **HG2** catalyst gave high conversion to cyclized product.

In 2014, Clayden and coworkers reported the synthesis of 2,5-dihydrothiophenes containing sulfur-bearing quaternary centers (Scheme 36) [225]. Dihydrothiophenes are important analogs in the research of biologically active compounds, natural products, and material application [226–231]; however, until this work, development of their stereoselective synthesis had received little attention [226, 227]. In this regard, the authors focused on the construction of enantiomerically enriched 2,5-dihydrothiophenes. After sequential stereospecific rearrangement and allylation, the key *S*-allylated sulfide intermediates **36.1** were produced for subsequent RCM. Regardless of the presence of the sulfur atom and the adjacent quaternary center, RCM successfully occurred to afford dihydrothiophenes **36.2** in good to high yields.

4 Si-Containing Heterocycles

The application of silicon tethers in a variety of chemical transformations is an area that is continually growing and has been the subject of a number of reviews (for reviews on Si disposable tethers, see [232, 233], for reviews on silicon tethered ring-closing metathesis, see [234–238]. In 2010, Evans highlighted work on natural products synthesized using temporary silicon-tethered ring-closing metathesis (TST-RCM) [30, 31]. In addition, a comprehensive 2012 review by Čusak reported the use of TST-RCM in method development and the re-functionalization of silicon en route to natural products synthesis [32]. In this section, we highlight selected examples and recent RCM work to Si-heterocycles.

In 1992, Grubbs and Fu reported the first TST-RCM for the synthesis of achiral 1,4-diol (Fig. 4a) [239]. In this method, a bis-allylic siloxane was treated with the **S** catalyst to furnish the 7-membered heterocyclic siloxane intermediate (not isolated), which was subsequently subjected to TBAF to afford the (*Z*)-configured

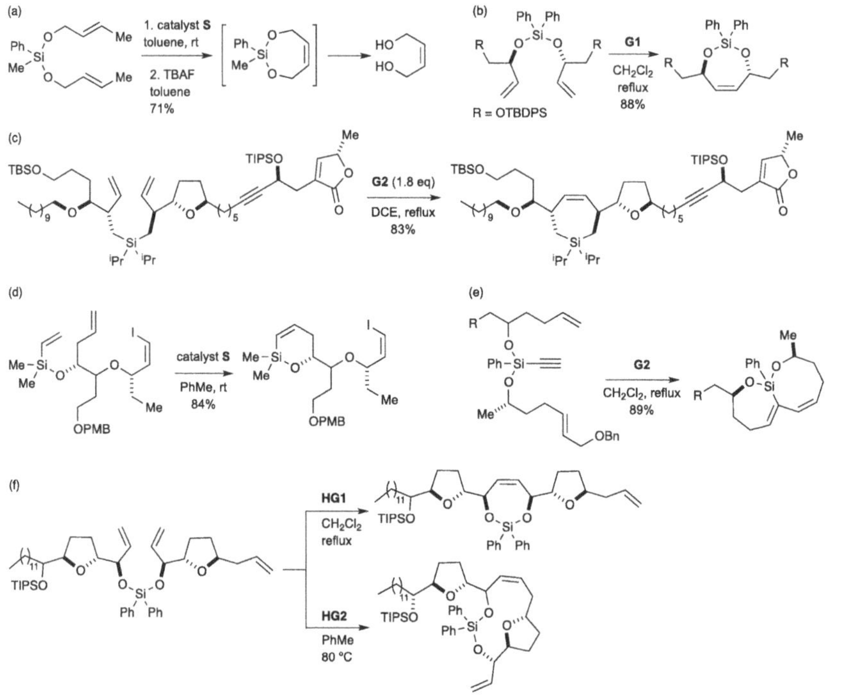

Fig. 4 Selected examples of TST-RCM prior to 2006. (**a**) Grubbs [239], (*Z*)-but-2-ene-1,4-diol. (**b**) Evans [240], D-Altritol. (**c**) Evans [30], (−)-Mucocin. (**d**) Denmark [241], (+)-Brasilenyne. (**e**) Lee [242], C1–C21 fragment of Tartrolon B. (**f**) Hoye [243], (+)-Gigantecin and (+)-14-Deoxy-9-oxygigantecin

1,4-diol in good overall yield. In 1998, Evans and coworkers reported the first application of TST-RCM in the synthesis of C_2-symmetric diols (Fig. 4b) [240] and highlighted the greater stability of the diphenyl-substituted silaketals when compared to their silane counterparts. Evans and Murthy demonstrated the usefulness of this method by subjecting the heterocyclic silane to *syn*-dihydroxylation and in two additional steps completed the total synthesis of D-altritol. This work subsequently culminated in the total synthesis of (−)-mucocin in 2003 via a TST-RCM strategy involving a non-symmetrical silaketal intermediate (Fig. 4c) [244]. Denmark [241] (Fig. 4d) and Lee [242] also had significant contributions in the advancement of TST-RCM that are highlighted in Fig. 4e (more recent contributions are discussed later in this chapter). In 2006, Hoye and coworkers further broadened the synthetic application of the TST-RCM reaction with the total synthesis of (+)-gigantecin and (+)-14-deoxy-9-oxgigantecin (Fig. 4f) ([243], for additional TST RCM work by Hoye and coworkers, see [245]; related work through 2008 was reviewed by Evans in the aforementioned chapter [30].

In 2008, Harvey and coworkers reported the synthesis of the C12–C24 fragment of Peloruside A, utilizing a TST diastereomer-discriminating RCM as a key step to form the C16–C17 (Z)-configured alkene (Scheme 37) [246]. This work by Harvey

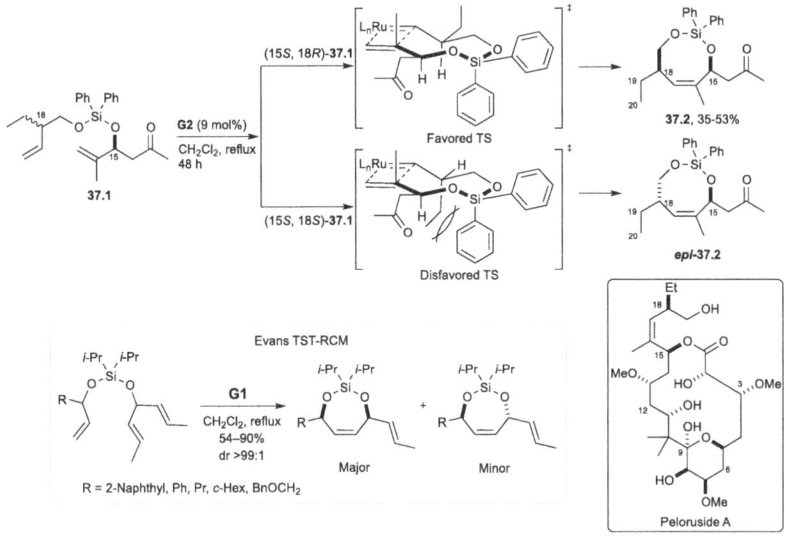

Scheme 37 Synthesis of C12–C24 fragment of Peloruside A utilizing RCM

involved 8-membered silicon heterocycles and built upon prior work by Evans [30] demonstrating long-range asymmetric induction using temporary silicon-tethered RCM (TST-RCM) in 7-membered Si-heterocycles (Scheme 37) [31]. Peloruside A displays synergy and enhanced antimitotic action when combined with taxoid-site drugs [247, 248]. After exploring several reactions, the optimized conditions involved slow addition of diene **37.1** (mixture of diastereomers) to a solution of the **G2** catalyst (9 mol%) in CH_2Cl_2 at refluxing temperature to yield the 8-membered siloxane diastereomer **37.2** in good yield. In addition, the unreacted diastereomeric triene was recovered along with the product of cross metathesis, providing evidence that a kinetic resolution was in play in the RCM event. The observed stereoselectivity could be explained by the favored transition state in which both the ethyl group and the substituent at C15 reside in the pseudo-equatorial position as elaborated in Scheme 37.

In 2008, Otterlo and coworkers reported the synthesis of 9- and 10-membered silicon-containing benzo-fused heterocycles (Scheme 38) [249]. Benzo-fused heterocycles with silicon incorporated into the backbone represent privileged medicinal scaffolds for use in drug discovery [250–254]. In the initial study, compound **38.1** was subjected to RCM utilizing the **G2** catalyst to afford the 9-membered benzoxazolinone **38.2** in good yield (78%). With this promising result, attempts were made to synthesize the corresponding benzodioxasilonines **38.4** and **38.6** via RCM. However, this RCM strategy was unsuccessful providing no product in the case of **38.3**, and only 22% yield for RCM of **38.5**. Efforts were then directed toward the synthesis of larger benzo-fused rings containing silicon. RCM precursor **38.7** was subjected to the **G2** catalyst to afford the 10-membered benzo-fused ring **38.8**, which has previously been challenging to synthesize via RCM [255].

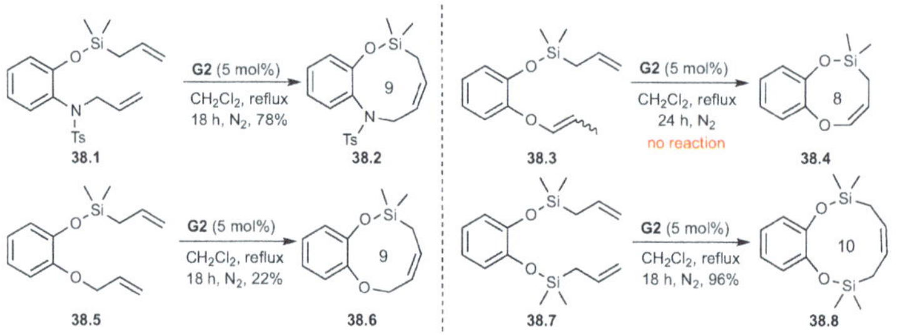

Scheme 38 Synthesis of 8-, 9-, and 10-membered benzo-fused Si heterocycles

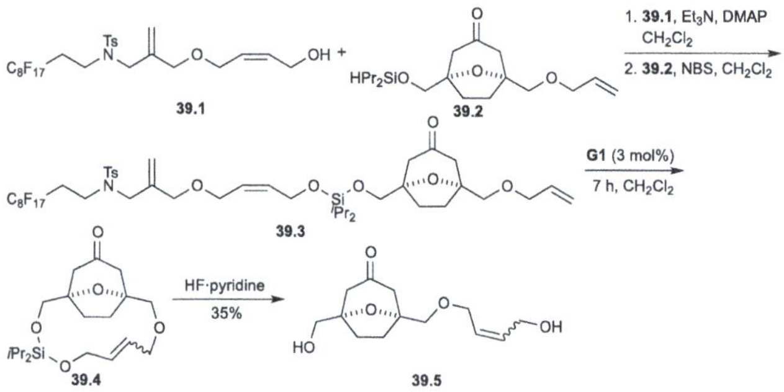

Scheme 39 Synthesis of unsymmetric silaketal-tethered macrocycles

In 2008, Nelson and coworkers reported an efficient and chemoselective method for the synthesis of unsymmetric silaketal-tethered macrocycles from the corresponding silyl ethers (Scheme 39) [256]. In the design of the RCM precursor, a fluorous tag was incorporated to facilitate purification for application in parallel synthesis. The fluorous-tagged RCM precursor **39.3** was synthesized by a modified procedure involving inverse addition of alcohol **39.1** to a solution of the NBS-activated silyl ether **39.2**. Alkene **39.3** was next subjected to RCM utilizing the **G1** catalyst to afford the 13-membered silaketal macrocycle **39.4**. Following the RCM reaction, the tether was removed to afford diol **39.5** as an *E*:*Z* mixture (55:45) in 35% yield over two steps. This method highlighted TST-RCM for the synthesis of both small cycles and macrocycles in which the alkene was remote from the silicon tether.

In 2008, Brown and coworkers employed an efficient TST-RCM to couple two THF-containing fragments as a key step for the asymmetric total synthesis of *cis*-sylvaticin **40.4** (Scheme 40) [257]. The RCM precursor **40.3** was obtained by the sequential addition of **40.1** and **40.2** to dichlorodiisopropylsilane. The resulting siloxane **40.3** was cyclized via RCM utilizing the **G2** catalyst to furnish the 7-membered unsaturated siloxane heterocycle, which was subsequently hydrogenated to afford siloxane **40.4** in 59% yield over three steps. Siloxane **40.4** was converted to the final natural product **40.5** in four steps with an overall yield of

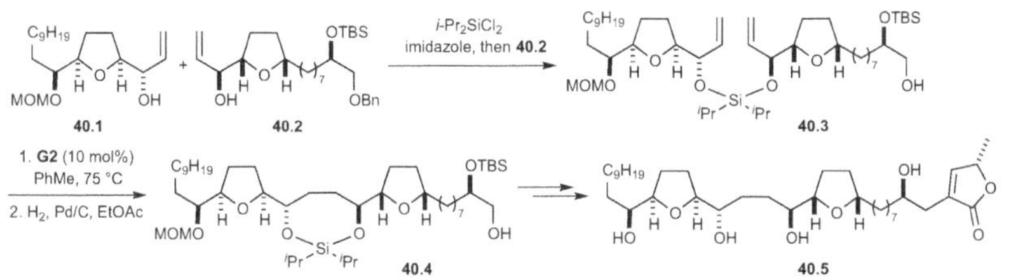

Scheme 40 Synthesis of *cis*-sylvaticin through TST-RCM coupling

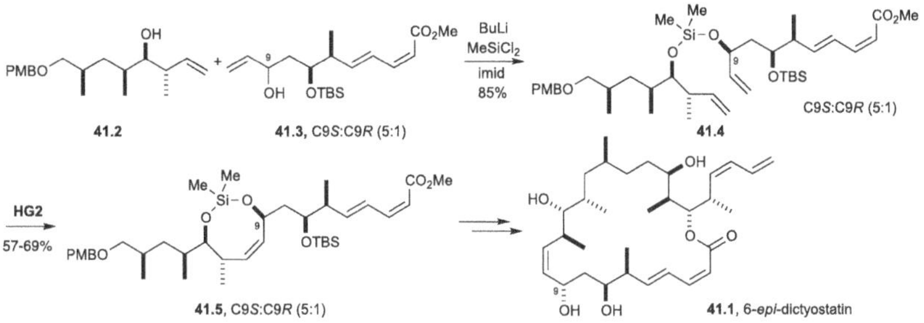

Scheme 41 Scale-up synthesis of 6-epi-dictyostatin using TST-RCM

7.8%. Employing this route, Brown and coworkers were able to synthesize hundred milligram quantities of *cis*-sylvaticin in an overall 21 linear-step sequence.

In 2008, Curran and coworkers reported a scale-up synthesis of the potent antitumor agent 6-*epi*-dictyostatin **41.1** by utilizing a TST-RCM method to form the pivotal *Z*-configured olefinic bond of siloxane **41.5** (Scheme 41) [258]. Among other analogs, 6-*epi*-dictyostatin was identified by structure activity studies as an active epimer of dictyostatin and was selected for further evaluation in animals. En route to the southern portion of 6-*epi*-dictyostatin, intermediates **41.2** and **41.3** were coupled utilizing dimethyldichlorosilane to afford the RCM precursor **41.4** in 85% yield. The resulting silyl acetal **41.4** was then subjected to RCM using the **HG2** catalyst to afford an inseparable mixture of the 8-membered siloxanes **41.5** in 57–69% yield. The siloxane was then deprotected with dichloroacetic acid, the diastereomers were separated, and the desired diol was further converted to 6-*epi*-dictyostatin **41.1** (30 mg).

In 2008, Vilarrasa and coworkers utilized TST-RCM to generate a challenging trisubstituted *E*-olefin en route to the total synthesis of Amphidinolide X **42.4** (Scheme 42) [259]. Amphidinolide X is one of several cytotoxic macrolides isolated from *Amphidinium dinoflagellates* [260, 261] and is remarkable for its intriguing nonsymmetrical diolide structure as well as an *E*-trisubstituted olefin. Unfortunately, initial efforts to employ RCM as a macrocyclization reaction were unsuccessful, yielding exclusively the undesired *Z*-trisubstituted olefin. As a result, an alternative strategy involving the use of a TST was then attempted. In this regard, diene **42.1** was subjected to the RCM utilizing the **S** catalyst, affording

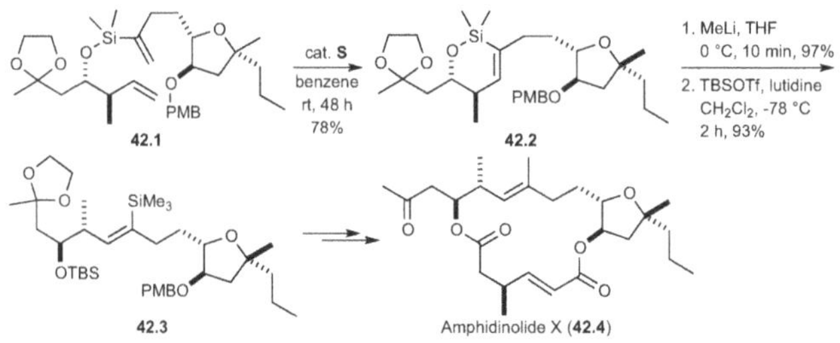

Scheme 42 Synthesis of Amphidinolide X utilizing TST-RCM of trisubstituted E-olefin

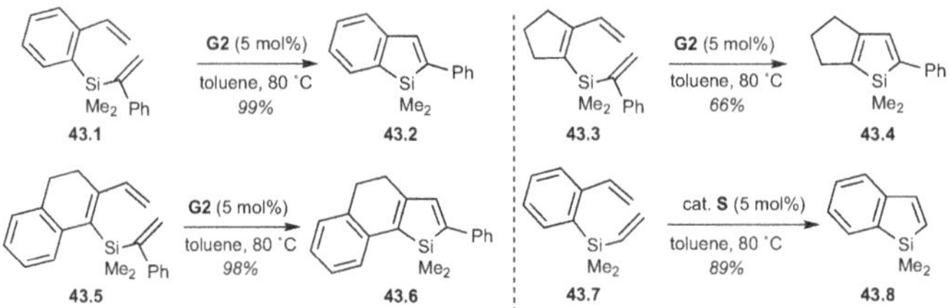

Scheme 43 Synthesis of 2, 3, 4-trisubstituted siloles by RCM reaction

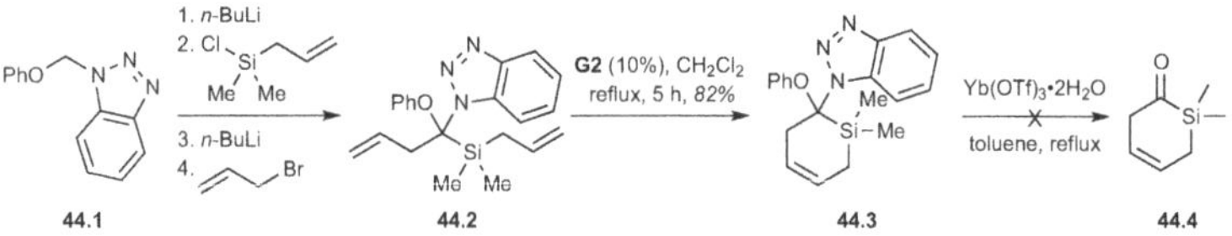

Scheme 44 Synthesis of unsaturated cyclic tetrahydrosiline **44.3** via RCM reaction

the 6-membered cyclic siloxane **42.2** in good yield (78%) and with the desired olefinic geometry. Subsequent tether cleavage, followed by TBS-protection, furnished vinylsilane **42.3**, which was converted to the final natural product **42.4**.

In 2008, Murakami and coworker reported the synthesis of densely functionalized silole skeletons through RCM (Scheme 43) [262]. Silole analogs can function as electron-transporting materials in OLEDs and have potential utility in organic photonics and electronics. To this goal, Murakami and coworkers sought to explore RCM for the synthesis of these highly π-conjugated silole skeletons. Initial studies were conducted to screen catalysts (**G1**, **G2**, and **S**) and a variety of substituted silanes. In this work, use of the **G2** catalyst afforded the highest yield for the 2-aryl-1-silaindenes **43.2**, while catalyst **S** furnished 2-*H*-1-silaindenes **43.8** in good yield. Various aryl-substituted silatrienes were also efficiently converted to the corresponding silole skeletons (**43.4** and **43.6**) via RCM by utilizing the **G2** catalyst. Overall, this RCM method provided an efficient route for the synthesis of fused and densely functionalized silole skeletons for applications in organic photonics and electronics.

In 2009, Portella and coworkers reported the synthesis of unsaturated cyclic tetrahydrosilines **44.3** via RCM (Scheme 44) [263]. It was well known that analogs

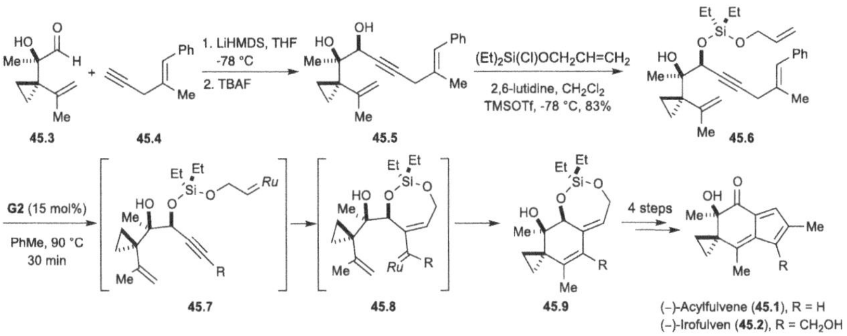

Scheme 45 Enantioselective total synthesis of (−)-acylfulvene and (−)-irofulven via TST-RCEYM

of acylsilane **44.4** are useful synthetic intermediates with very interesting reactivity profiles [264–271]. The RCM precursor **44.2** was obtained from triazole **44.1** through deprotonations and coupling reactions. Intermediate **44.2** was subjected to RCM employing the **G2** catalyst to afford the 6-membered tetrahydrosiline **44.3** in good yield (82%). Several attempts to convert the resulting cyclic tetrahydrosiline **44.3** to the corresponding unsaturated cyclic acylsilane **44.4** derivative were unsuccessful. Nonetheless, RCM has been shown to be an effective method for the synthesis of benzotriazolyl-substituted silacyclic compounds.

In 2009, Movassaghi and coworkers illustrated the enantioselective total synthesis of (−)-acylfulvene (**45.1**) and (−)-irofulven (**45.2**) featuring a TST enyne-RCM cascade reaction in a challenging setting (Scheme 45) [272]. After synthesizing aldehyde **45.3** in sufficient quantity, addition of alkyne **45.4** to aldehyde **45.3**, followed by desilylation, afforded diol intermediate **45.5**. Selective diethylallyloxysilane tether formation was carried out using allyldiethylsilyl chloride with the secondary hydroxyl group of diol **45.5** to provide trienyne **45.6**. Through the plausible mechanism of the RCEYM reaction with trienyne and desilylation, intermediate **45.6** was successfully converted to the metathesis product **45.9** in good yield (79%). Initial metathesis occurs at the terminal olefin **45.4** using the **G2** catalyst to provide the ruthenium alkylidene **45.7**, which enabled intramolecular olefin metathesis to furnish compound **45.9** via the final cyclization of the ruthenium alkylidene **45.8**. The desired cyclic silane intermediate **45.9** was converted to complete the synthesis of the final products **45.1** and **45.2** through a tandem RCM/dehydrogenation protocol for the required ring construction. In another report, the authors described a similar synthetic method using the **G2** catalyst at refluxing toluene to furnish analogs of 7,6-bicycle intermediate **45.9** (R = ^{i}Bu, CH_2OPMB) in high yields [273].

In 2009, Lee and coworkers reported the application of Si-tethered tandem RCEYM strategy for the construction of key intermediate **46.4** in the formal total synthesis of (−)-cochleamycin A (Scheme 46) [274]. This work was built upon prior work in this area by Lee demonstrating the synthetic benefits of dienyne TST-RCM with both symmetrical and unsymmetrical silanes for the stereoselective

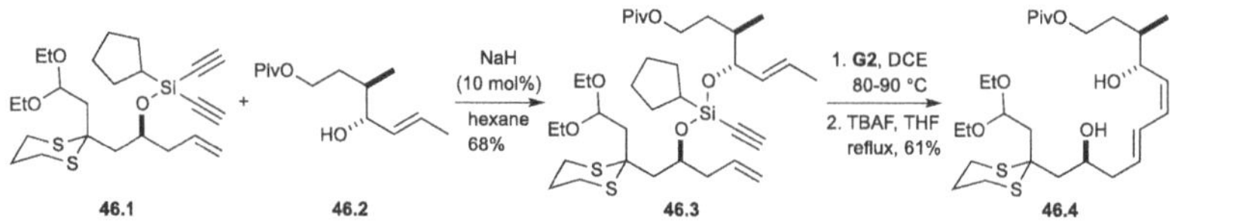

Scheme 46 Application of Si-tethered tandem RCEYM for the synthesis of (−)-cochleamycin A

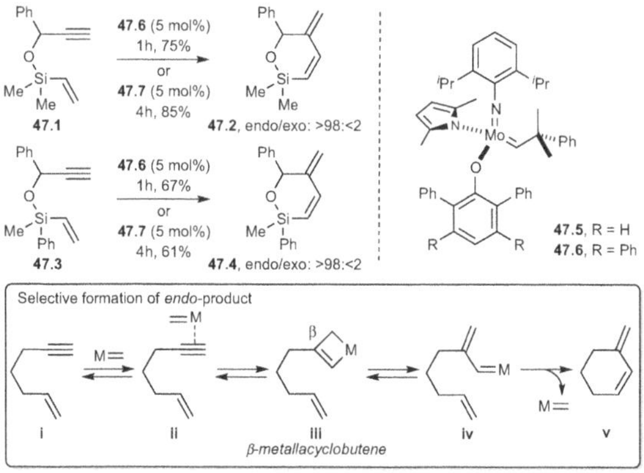

Scheme 47 Synthesis of *endo*-cyclic unsaturated siloxanes through RCEYM

synthesis of 1,3-butadienes subunits [242, 275–279] and applied in the synthesis of the C1–C21 fragment of tartrolon B [242].

Toward the synthesis of (−)-cochleamycin A, silyl ether **46.1** and alcohol **46.2** were coupled to furnish the desired silaketal **46.3** in 68% yield. Silaketal **46.3** was designed to control the *E*/*Z* stereochemistry of 1,3-diene. The eloquent design of the silaketal **46.3** allowed for the productive RCM initiation to occur at the more accessible, unsubstituted terminal olefin [275, 276]. Bis(alkoxyl)silane **46.3** was treated with the **G2** catalyst to afford a bicyclic heterocycle, which was subsequently subjected to a TBAF-mediated protodesilylation to furnish stereo-defined dienediol **46.4** in 61% yield over two steps.

In 2009, Hoveyda and coworkers reported an *endo*-selective RCEYM with the newly developed stereogenic-at-Mo monoalkoxide and monoaryloxide Mo catalysts **47.5** and **47.6** (Scheme 47) [280]. This method represented the first examples of enantioselective RCEYM and facilitated the selective access to *endo* products via the *β*-metallacyclobutene intermediate **iii** as summarized in Scheme 47. The sterically congested vinylsilanes **47.1** and **47.3** were among the substrates screened to explore the reactivity of the novel catalysts **47.5** and **47.6**. Subjecting vinylsilanes **47.1** and **47.3** to RCM yielded the corresponding *endo*-cyclic unsaturated siloxanes in good yields and selectivity.

In 2010, Kobayashi and coworkers described an elegant TST intramolecular RCM study leading to the selective formation of 8-membered *E*- and *Z*-products in differentially substituted silicon-tethered dienes (Scheme 48) [281]. This work

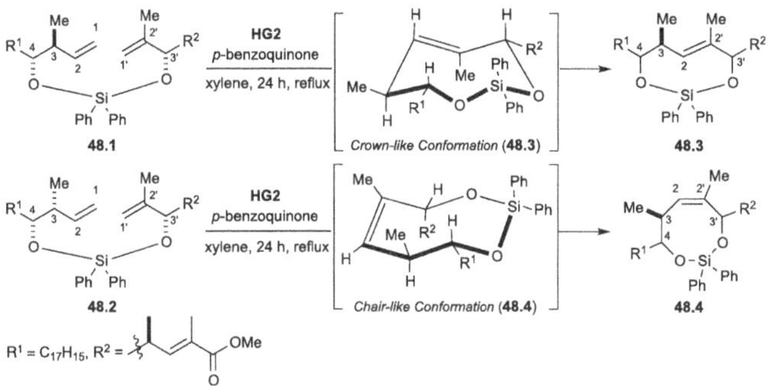

Scheme 48 Intramolecular TST-RCM for the synthesis of 8-membered *E*- and *Z*-olefin

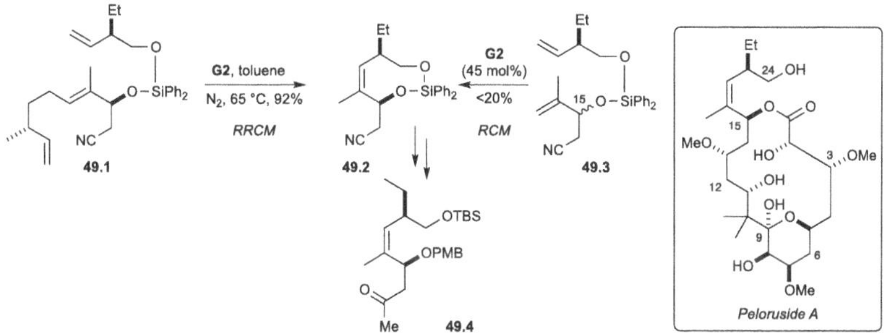

Scheme 49 Total synthesis of Peloruside A via RRCM

substantiated and built upon the aforementioned prior studies by Evans [244] and Harvey [246]. Trienes **48.1** and **48.2** were subjected to RCM employing the **HG2** catalyst (10–20 mol%) to furnish the *E*-configured siloxane **48.3** (exists in a crown-like conformation) and the *Z*-configured siloxane **48.4** (exists in a chair-like conformation), respectively. Based on results from the Nuclear Overhauser Effect (NOE) experiments, the newly formed C–C double bond in **48.3** was assigned the *E*-configuration. The authors reasoned that the factors facilitating the formation of the unusual *E*-configured, 8-membered ring were mainly the *cis* relationship of R^1 and R^2 substituents, the *anti* stereochemistry of homoallylic alcohol, the presence of C2′–Me, and the presence of a bulky substituent at C4′. This study represents an example of the RCM reaction in which a Si-tether reduces the ring strain energy in the transition state thereby enabling the formation of the *E*-olefin functionality.

In 2010, Hoye and coworkers reported the total synthesis of Peloruside A through relay ring-closing metathesis cyclization (RRCM) reactions (Scheme 49) [282, 283]. Synthetic challenges of the hindered *Z*-alkene in **49.4** were resolved in the course of the synthesis of the key intermediate **49.2**. Initially, RCM reaction was employed for the construction of *Z*-alkene in **49.2** from the corresponding diene **49.3**, which was an epimeric mixture at C15. The formation of this diene gave poor yield (<20%) using 45 mol% of the **G2** catalyst. In contrast, RRCM reaction of triene substrate **49.1** using the **G2** catalyst was performed efficiently to afford the corresponding 8-membered silacycle **49.2** in 92% yield. Intermediate **49.2** was

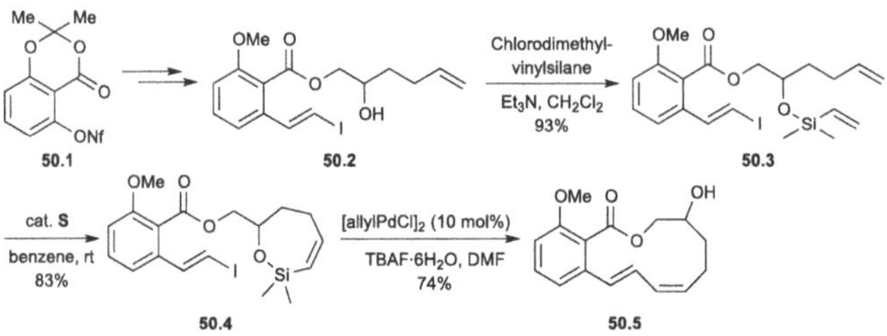

Scheme 50 Synthesis of polyunsaturated macrolactones via TST-RCM/intramolecular cross-coupling

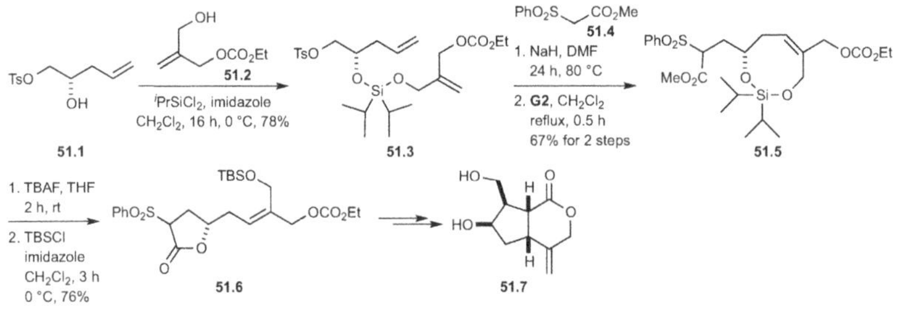

Scheme 51 Total synthesis of an iridoid applying a TST-RCM strategy

further converted to the desired final product, Peloruside A, via the sterically hindered *Z*-configured alkene containing ester **49.4**.

In 2010, Denmark and coworkers reported a general strategy for the synthesis of polyunsaturated macrolactones by employing a sequential TST-RCM/intramolecular Hiyama cross-coupling [284] reaction (Scheme 50) [285]. Commonly, it is known that construction of medium- and large-sized macrocycles is challenging to accomplish due to entropy and enthalpic factors [286–289]. Efforts to develop new synthetic methods to construct analogs of polyunsaturated macrocycles are shown in Scheme 50. Highly functionalized macrocycles [290–294] possessing a conjugated diene or triene moiety and other functional groups within the rings are considered important in early phase drug discovery [295–298] and supramolecular chemistry [299]. Thus, vinyl iodo alcohol **50.2** was generated from the aryl nonaflate **50.1**, which was prepared from the corresponding phenol [300]. The vinyl iodo alcohol **50.2** was next silylated with chlorodimethylvinylsilane to furnish the silyl ether RCM precursor **50.3** in good yield. Intermediate **50.3** was next treated with the **S** catalyst to afford siloxane **50.4** in 83% yield. Subsequent Hiyama coupling via addition of siloxane **50.4** in DMF (0.1 M) into a solution of $[allylPdCl]_2$ (10 mol%) and TBAF furnished the benzo-fused macrocyclic lactone **50.5** in good yield.

In 2011, Suh and coworkers employed the TST-RCM strategy in the first total synthesis of 6-hydroxy-7-(hydroxymethyl)-4-methylenehexahydrocyclopenta[*c*]pyran-1(3*H*)-one **51.7**, a member of the iridoid family of natural monoterpenoid products (Scheme 51) [301]. Iridoid analogs are structurally characterized by a *cis*-

Scheme 52 TST-RCM/vinylogous Mukaiyama aldol sequence for the synthesis of TMC-151C

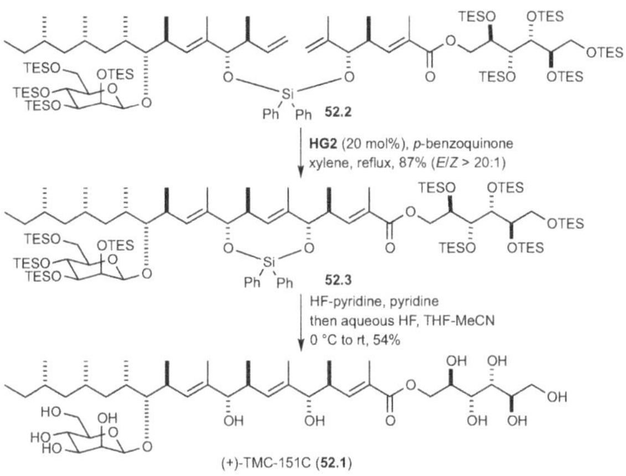

fused cyclopenta[*c*]pyran ring system. To install the γ-lactone in intermediate **51.6**, Si-tethered alkoxysilane **51.3** was prepared from hydroxyl tosylate **51.1** and allyl carbonate **51.2** in good yield. Anion formation of benzenesulfonyl acetate **51.4** and alkylation, followed by RCM using the **G2** catalyst, furnished the 8-membered cyclic bis-alkoxysilane **51.5** in 67% yield over two steps. Removal of the Si-tether with TBAF in THF led the impromptu cyclization of the hydroxyl ester and generated the γ-lactone intermediate, which was next protected with TBSCl to provide allylic carbonate **51.6** in good yield (76%). The allylic carbonate **51.6** was subsequently converted to the desired lactone iridoid **51.7**.

In 2011, Kobayashi and coworkers addressed the total synthesis of (+)-TMC-151C (**52.1**) by exploiting a TST-RCM method and a vinylogous Mukaiyama aldol reaction (Scheme 52) [302]. Initial attempts to the synthesis of (+)-TMC-151C (**52.1**) involved three iterative vinylogous Mukaiyama aldol reactions (VMAR); however, the final VMAR was not successful to deliver **52.1**. As an alternative route, building on their previous result (Scheme 48), silylene acetal **52.2**, was prepared using $Et_2NPh_2SiCl/Et_3N/DMAP$ to link the west and east fragments while setting up the molecule for a TST-RCM. Ring-closing metathesis of **52.2** employing the **HG2** catalyst (20 mol%) in the presence of *p*-benzoquinone in refluxing xylene provided the *E*-selective silicon-tethered olefin **52.3** in 87% yield with high selectivity ($E/Z > 20{:}1$). Deprotection of **52.3** with HF · pyridine and aqueous HF successfully led to the total synthesis of (+)-TMC-151C (**52.1**) in 54% yield.

In 2011, Lee and coworkers reported the synthesis of medium-sized cyclic allylsilyl ether derivatives via coupling and TST-RCM reaction in order to investigate the regioselective allylic transposition of the C–O bond using a rhenium catalyst (Scheme 53) [303]. Allylic transposition and ring contraction with medium-sized rings containing ring strain is expected to be regioselective due to the following two primary reasons: (1) first, the internal double bond has a *cis* configuration in most cases of medium-sized rings, and it is a major driving force to

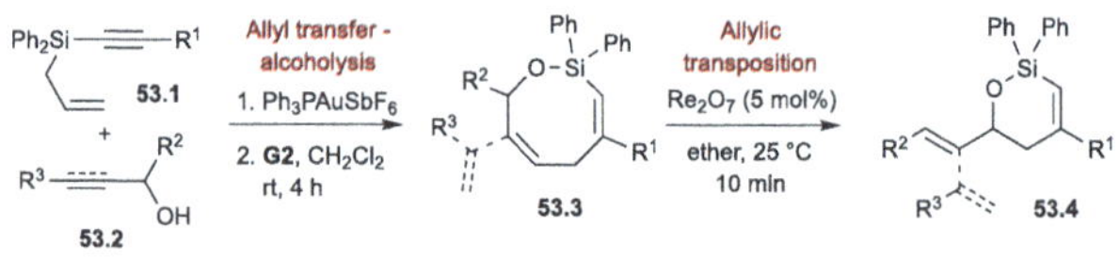

Scheme 53 Study of ring contraction employing Si-tethered-RCM and regioselective allylic transposition

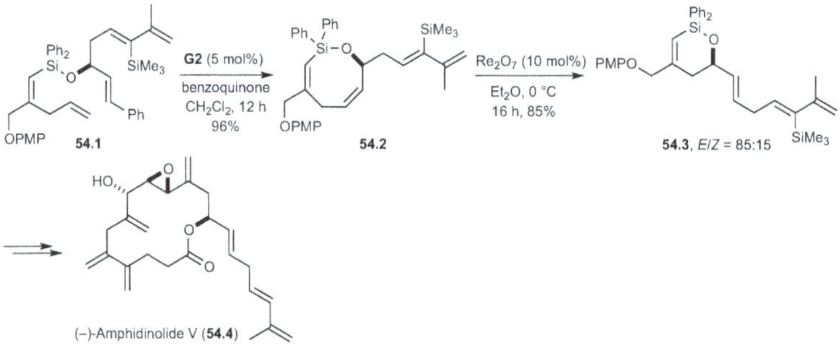

Scheme 54 Construction of amphidinolide V segment, via Si-tethered-RCM and allylic 1,3-transposition

convert to a *trans*-double bond after the ring contraction, and (2) second, medium-sized rings are generally more strained than their smaller complements. In this regard, favorable ring contraction of 8-membered rings to the corresponding 6-membered rings will relieve the ring strain. Synthesis of 8-membered rings such as compound **53.3** was started from the reaction of silylalkyne **53.1** and allyl alcohol or propargylic alcohol **53.2**. An intramolecular allyl transfer reaction afforded silyl ether with *Z*-stereochemistry, which was converted to 8-membered siloxacycle **53.3** via RCM using the **G2** catalyst (5 mol%) in moderate to high yields (47–96%). Silicon-directed ring contraction was carried out by treatment of **53.3** with a catalytic amount of Re_2O_7, affording siloxene **53.4** in moderate to good yield (49–87%). Chirality transfer during allylic transposition was examined with chiral substrates **53.3** obtained from chiral alcohol **53.2** using chiral HPLC and Mosher ester analysis to provide products with 76–79% *ee*.

In 2013, Lee and coworkers utilized this ring contraction through an allylic transposition strategy to convert siloxene intermediate **54.3** to a key fragment for the total synthesis of (−)-amphidinolide V (**54.4**) (Scheme 54) [304], a macro-lactone characterized by its arrayed alkene moieties, two of which are 1,3-dienes. The authors envisioned that these 1,3-diene subunits could be efficiently constructed by enyne metathesis. Thus, RCM of silyl ether **54.1** using the **G2** catalyst, in the presence of benzoquinone (20 mol%), yielded the allylic 1,3-transposition precursor **54.2** in 96% yield. In this RCM, the trimethylsilyl substituent of the 1,3-diene moiety was strategically installed to prevent RCM with the 1,3-diene and ultimately enabled RCM to take place selectively with the styryl group. The resulting siloxane **54.2** was next treated with Re_2O_7 (10 mol%) at

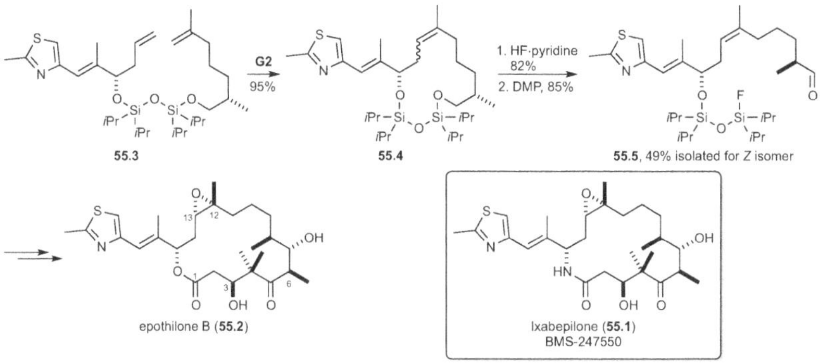

Scheme 55 Bis-siloxane-tether in the construction of analogs of ixabepilone

0°C to facilitate 1,3-allylic transposition generating siloxene **54.3** as a mixture of *E*/*Z* isomers (85:15) in 85% yield. Final manipulations resulted in the asymmetric total synthesis of amphidinolide V, which was achieved in 22 steps, highlighted by TST ring-closing enyne and diene metathesis, as well as the allylic transposition of silyl ethers to install 1,3- and 1,5-diene motifs.

In 2012, Lin and coworkers utilized a bis-siloxane tether in the construction of analogs of ixabepilone (**55.1**), an anti-breast cancer medication known as azaepothilone B and (−)-epothilone B (**55.2**) (Scheme 55) [305]. In this synthesis, selective installation of the trisubstituted *Z*-configured olefin at C12–C13 was deemed crucial en route to epothilone B (**55.2**). Toward this goal, RCM of the bis-siloxane-tethered substrate **55.3**, constructed through the use of $ClSi(iPr)_2OSi(iPr)_2Cl$ with two corresponding alcohol segments, was performed using the **G2** catalyst to furnish the cyclized bis-siloxane-containing macrocycle **55.4** in 95% yield with a selectivity of 1.7/1 favoring the (*Z*) isomer. The least hindered Si–O bond was selectively cleaved by HF·pyridine generating $FSi(iPr)_2OSi(iPr)_2$ [FSOS] group, which was next oxidized using the Dess–Martin periodinane to provide aldehyde **55.5** in 40% yield over three steps.

In 2013, Kumar and coworkers utilized a TST-RCM for the synthesis of umuravumbolide (**56.4a**) and hyptolide (**56.4b**), naturally occurring pharmacologically active α,β-unsaturated δ-lactones (Scheme 56) [306]. For the synthesis of these bioactive natural products, intermediates **56.1a** and **56.1b** were cyclized through TST-RCM using the **G2** or **HG2** catalysts in heating toluene, which furnished the corresponding cyclic intermediates **56.2a** and **56.2b** in good yield. Removal of protecting groups and construction of the unsaturated lactone ring led to the completion of the syntheses of umuravumbolide (**56.4a**) and hyptolide (**56.4b**).

In 2013, Rao and coworkers reported the total synthesis of pectinolide C (**57.6**) in 14 steps (11.45%), employing TST-RCM, Jacobsen resolution, and Wadsworth–Emmons olefination as the key reactions (Scheme 57) [307]. Structurally, analogs of pectinolide possess an electrophilic α,β-unsaturated δ-lactone ring, a side chain with a *Z*-configured olefin, as well as both acetyl and hydroxyl functional groups.

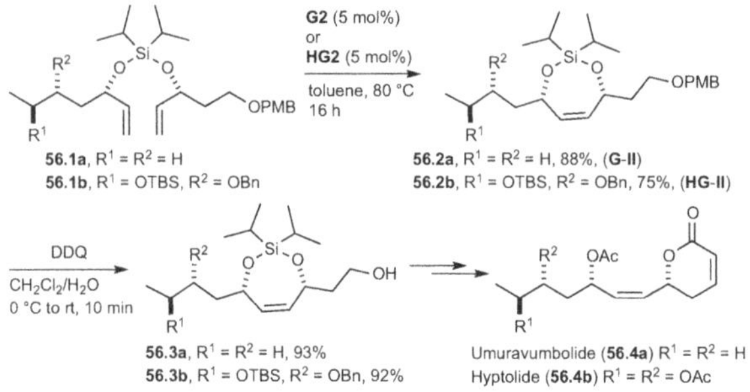

Scheme 56 Utilization of a TST-RCM for the synthesis of umuravumbolide

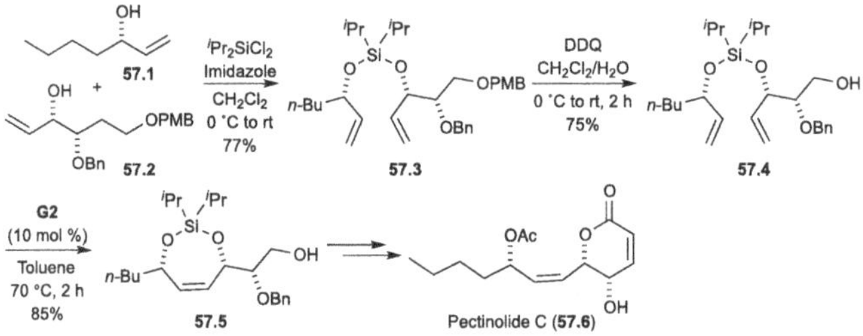

Scheme 57 Synthesis of pectinolide C employing TST-RCM reaction

Toward this goal, the authors first synthesized the 7-membered silicon heterocycle **57.5** through the use of TST-RCM utilizing the **G2** catalyst (10 mol%). The unsymmetrical bis-alkoxylsilane **57.3** was synthesized by the sequential reaction of dichlorodiisopropylsilane with the two key allylic alcohols **57.1** and **57.2**. The PMB group was selectively removed using DDQ in CH_2Cl_2/H_2O to generate alcohol **57.4**. The resulting alcohol **57.4** was cyclized through RCM utilizing the **G2** (10 mol%) catalyst at 70°C to afford the 7-membered silicon heterocycle **57.5** in 85% yield. The stereochemistry of the *Z*-configured double bond was confirmed by ^{1}H NMR spectrum study (coupling constant, $J = 11.6$ Hz). The resulting silicon heterocycle was further converted to the final natural product pectinolide C (**57.6**) in five steps.

In 2014, Quinn and coworkers reported the synthesis of the C_2-symmetric non-adjacent bis(tetrahydrofuran) core **58.4** of *cis*-sylvaticin **58.5** with 24% overall yield from (2*R*,3*S*)-1,2-epoxy-4-pentene-3-ol **58.1** in seven steps (Scheme 58) [308]. In this work, the authors were able to synthesize the bis-epoxide siloxane **58.3** by assembling the 1,4-diol unit using TST-RCM. Epoxide **58.1** was subjected to dichlorodiisopropylsilane and 2,6-lutidine to obtain substrate the siloxane diene **58.2** in 74% yield. The siloxane diene **58.2** was next subjected to a one-pot RCM/H_2 sequence in the presence of the **HG2** catalyst to yield the bis-epoxy siloxane **58.3** in gram quantities and excellent yield (90% over two steps). The authors further investigated use of less-active metathesis catalysts, including the **G1** catalyst, but found they were all less effective at catalyzing the RCM of **58.2**. In all cases, high catalyst loadings and longer reaction times were required for completion of

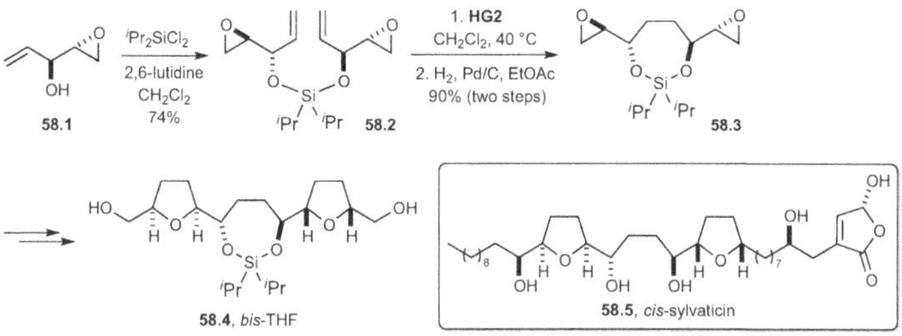

Scheme 58 TST RCM en route to C_2-symmetric bis-THF core **58.4** of *cis*-sylvaticin **58.5**

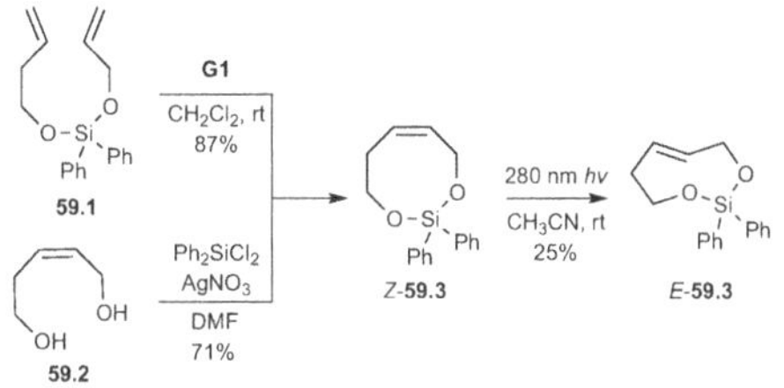

Scheme 59 A TST-RCM strategy for the formation of the 8-membered cyclic *Z*-configured siloxane **59.3** and the photochemical isomerization to the *E*-configured siloxane **59.3**

reaction. Further elaboration of the siloxane yielded the core intermediate **58.4** en route to *cis*-sylvaticin **58.5**. Currently, efforts to desymmetrize **58.4**, as needed for completion of **58.5,** are underway.

In 2014, Tomooka and coworkers utilized a TST-RCM strategy and a subsequent isomerization reaction in the synthesis of planar chiral heterocyclic alkenes for applications in asymmetric synthesis without the need for an external chiral element [309]. The 8-membered *Z*-**59.3** was prepared using the Evans TST-RCM method [240]. The RCM precursor **59.1** was subjected to RCM using the **G1** catalyst to afford the corresponding 8-membered Si-heterocycle **59.3** in good yield (Scheme 59). Alternatively, **59.3** was also synthesized by the direct silylation of the *Z*-configured alkene **59.2** under dilute conditions, albeit in slightly lower yields. Siloxane *Z*-**59.3** was subsequently subjected to a photochemical isomerization reaction utilizing 280 nm UV light in the presence of dimethyl isophthalate (sensitizer) to afford a 25:75 mixture of the *E*/*Z* isomers. The *E*-configured siloxane *E*-**59.3** was successfully isolated and employed as substrate in epoxidation, Diels–Alder, and cycloaddition reactions.

5 B-Containing Heterocycles

Boron plays a very important role in the field of synthetic organic chemistry [310–312]. Compounds containing B–N bonds have emerged with renewed interest over the last two decades due to their isoelectronic properties with C–C bonds found in

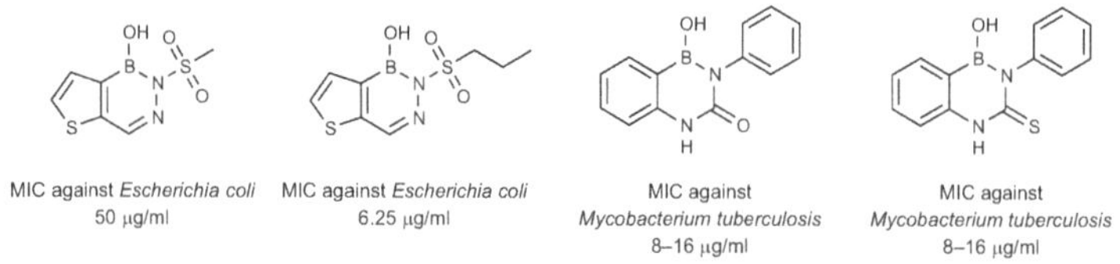

Fig. 5 Therapeutic potential of boron-containing heterocycles

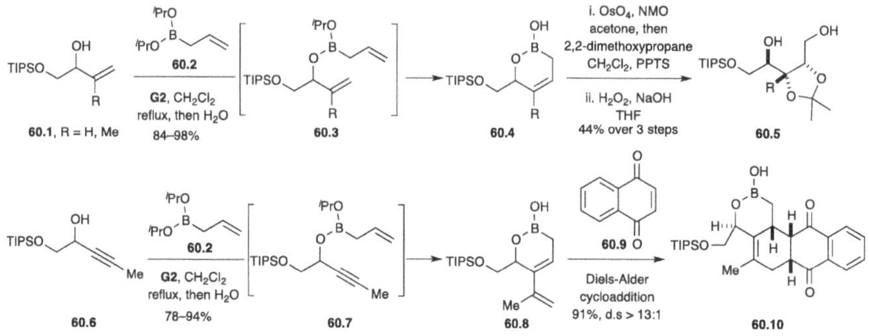

Scheme 60 RCM/ Diels–Alder cycloaddition strategy for multiple ring installation

sp^3, sp^2 (olefins), and sp^2 (aromatics) skeletons [313]. In particular, their unique electronic structure and ability to form covalent bonds with carbon have engendered medicinal and biomedical interest in boron heterocycles [314]. In addition, applications in optoelectronic materials [315], as well as their use as novel organic synthons, has inspired new RCM methods for the synthesis of boron heterocycles.

Boron heterocycles have shown promising bioactivity (Fig. 5) [56–58, 316–318]. In this regard, 2-*N*-sulfonylalkyl-thienodiazaborine has shown activity against *E. coli* with MIC values of 6.25–50 μg/mL. Furthermore, various analogs of diazaborine heterocycles have demonstrated promising activity against *Mycobacterium tuberculosis*, including 2,4,1-benzo[*e*] diazaborines possessing an MIC value of 8–16 μg/mL, which is superior to the known drug Pyrazinamide (~200 μg/mL) on the market.

In 2002, Schreiber and coworkers reported a diversity-oriented approach for the synthesis of complex molecules having multiple rings, stereocenters, and unsaturated units, in a few steps with moderate to good yields from the corresponding boronic ester **60.2** and allylic or propargylic alcohols **60.1** or **60.6** (Scheme 60) [319]. In this annulation procedure, the authors carried out transesterification of unsaturated boronic ester **60.2** with the corresponding alcohol **60.1** or **60.6** followed by RCM utilizing the **G2** catalyst to provide cyclic allylboronic acid **60.4** and cyclic dienylic boronic acid **60.8**. The resulting cyclic allylboronic acid **60.4** was efficiently converted to xylitol derivative **60.5** through a diastereoselective dihydroxylation, protection, and oxidation sequence with an overall 44% yield. The cyclic dienylic boronic acid **60.8** was also further converted to the corresponding tetracyclic boronic acid **60.10** through Diels–Alder cycloaddition with 13:1 diastereoselectivity and 91% yield.

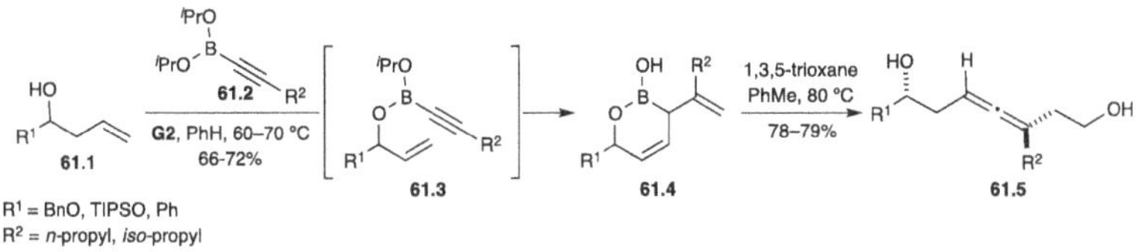

Scheme 61 RCEYM-hydroxalkylation en route to the trisubstituted allene **61.5**

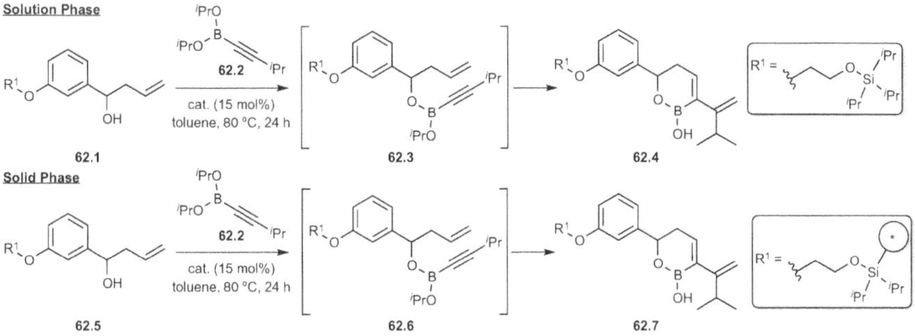

Entry	Phase	Substrate	Boronate	Catalyst	Product	Conversion
1	solution phase	**62.1**	**62.2**	**HG2**	**62.4**	54%
2	solution phase	**62.1**	**62.2**	**G2**	**62.4**	69%
3	solid phase	**62.5**	**62.2**	**HG2**	**62.7**	54%
4	solid phase	**62.5**	**62.2**	**G2**	**62.7**	36%

Scheme 62 Observations of catalytic ability for solution- and solid-phase boronic ester annulations

Based on their previous annulation strategy [319, 320], Schreiber and coworkers reported the *trans*-esterification reaction of homoallylic alcohols **61.1** and electron poor alkynyl boronic esters **61.2**, followed by RCEYM using the **G2** catalyst (11–15 mol%), to generate the functionalized 6-membered heterocyclic dialkenyl-boronic acids **61.4** in moderate to good yield (Scheme 61) [321]. These annulated products were readily transformed to structurally diverse small molecules with good yields and high diastereoselectivity through oxidation and allene-forming hydroxyalkylation reactions. The dialkenyl boronic acids **61.4**, obtained after an aqueous washing of the impure product, were directly hydroxymethylated in the presence of 1,3,5-trioxane to afford the trisubstituted allene **61.5** in 78–79% yield with 6:1 diastereoselectivity.

In 2005, Schreiber and coworkers reported the crossover in the relative efficiency of two commonly utilized metathesis catalysts (**G2** and **HG2**) in solution and solid phase for the synthesis of heterocyclic dialkenyl boronic esters [322]. The boronic esters **62.4** and **62.7** were generated from the corresponding homoallylic alcohol **62.1** and **62.5**, respectively, and alkynylboronic esters **62.2** through RCEYM using either the **G2** or **HG2** catalysts (Scheme 62). In solution phase, use of the **G2** catalyst resulted in greater degree of conversion of starting substrate compared to use of the **HG2** catalyst. Interestingly, this order was reversed in the

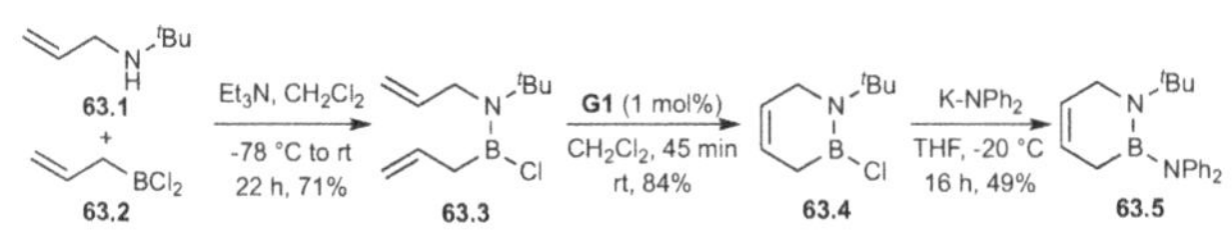

Scheme 63 Synthesis of azaborine compounds through RCM

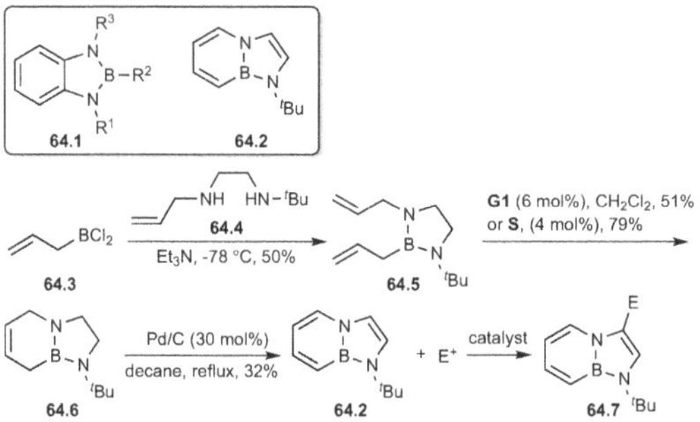

Scheme 64 Synthetic routes for "BN-fused" indole compounds using the Schrock catalyst

solid phase variant. On further investigation, it was observed that crossover in the ability of both ruthenium pre-catalysts in RCEYM was independent of the solid supported homoallylic alcohol **62.5** and the alkynylboronic esters **62.2**.

In 2008, Liu and coworkers addressed the synthesis and characterization of 1,2-azaborine **63.5** (Scheme 63) [34, 323–325], using the RCM precursor **63.3**, which was synthesized by coupling of allylboron dichloride **63.2** and *tert*-butylallylamine **63.1** in CH_2Cl_2 at $-78°C$. The resulting diene **63.3** was subjected to RCM using the **G1** catalyst (1 mol%) in CH_2Cl_2, which furnished the 6-membered BN-heterocycle **63.4** in 84% yield. The Cl atom of **63.4** was substituted by diphenylamide through simple nucleophilic attack of K-NPh_2 to afford the desired azaborine heterocycle **63.5** in moderate yield.

In 2010, Liu and coworkers reported the synthesis of the first example of a "BN-fused" indole **64.2** (Scheme 64) [326]. BN indoles can efficiently undergo electrophilic aromatic substitution (EAS) reactions, such as bromination, Mannich reaction, Michael addition, and Friedel-Crafts acylation to generate various substituted BN indole analogs **64.7**. Before 2010, the only BN-substituted indoles reported in the literature [327–329] were phenylenediamine-type compounds having an external unit as shown in compound **64.1**. The RCM precursor **64.5** was synthesized via the condensation of *N*-*t*-Bu-*N′*-allylethylenediamine **64.4** with allylboron dichloride **64.3**. The diene **64.5** was subjected to RCM utilizing 6 mol% of the **G1** catalyst to furnish the bicyclic BN-substituted indole precursor **64.6** in 51% yield. A significantly improved yield (79%) for RCM was observed when the **S** catalyst (4 mol%) was employed for the formation of the cyclized intermediate **64.6**. Dehydrogenation of intermediate **64.6** was achieved using Pd/C in refluxing decane to obtain **64.2** in 32% yield.

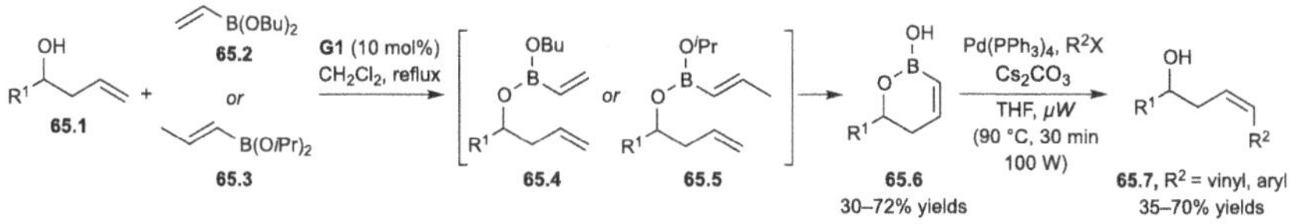

Scheme 65 Synthesis of cyclic alkenyl boronic half acids by *trans*-esterification and RCM

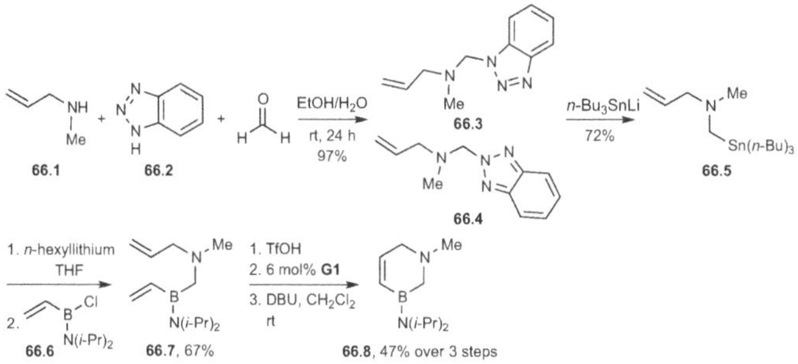

Scheme 66 Synthesis of 1,3-dihydro-1,3-azaborine compounds

In 2010, McNulty and coworkers reported the synthesis of cyclic alkenyl boronic half acids **65.6** utilizing transesterification and RCM as the key reactions (Scheme 65) [330]. These boronic half acids can serve as substrates in various metal coupling reactions for the formation of new C–C bonds [319, 321, 331, 332]. The cyclic alkenyl boronic acids **65.6** were obtained through RCM (using 10 mol% of the **G1** catalyst) of the in situ-generated intermediates **65.4** or **65.5**. Intermediates **65.4** or **65.5** were synthesized by the *trans*-esterification of homoallylic alcohols **65.1** and corresponding vinyl boronic dibutyl ester **65.2** or propenyl boronic diisopropyl ester **65.3**. The cyclic alkenyl boronic half acids **65.6** can be effectively coupled with vinyl and arylhalides in a Suzuki–Miyaura coupling reaction to afford the corresponding stereoselective dienes **65.7** in moderate to good yields (35–70%).

In 2011, Liu and coworkers reported the synthesis of 1,3-dihydro-1,3-azaborine **66.8**, which is an isostere of benzene, using RCM with the **G1** catalyst (Scheme 66) [333]. Toward the goal of synthesizing 1,3-azaborines, the authors carried out a one-pot, three-component coupling reaction of *N*-methylallylamine **66.1**, formaldehyde and 1,2,3-benzotriazole **66.2**, which resulted in 4:1 mixture of intermediates **66.3** and **66.4** in high yield (97%). Intermediate **66.5** was produced in 72% yield by treating a mixture of **66.3** and **66.4** with n-Bu_3SnH in the presence of LDA. Allyl amine **66.5** was converted to the desired coupling product **66.7** in 67% yield through lithium–tin exchange of **66.5** with *n*-hexyllithium, followed by the addition of electrophile **66.6**. Ring-closing metathesis of **66.7** using either the **S** or **G1** catalysts were unsuccessful due to the degradation of both RCM catalysts by the relatively nucleophilic amine group in **66.7**. However, it is known that the **G1** catalyst is compatible with dienes possessing ammonium salt [334]. Therefore,

Scheme 67 Synthesis of 1,2-azaborines and their applications

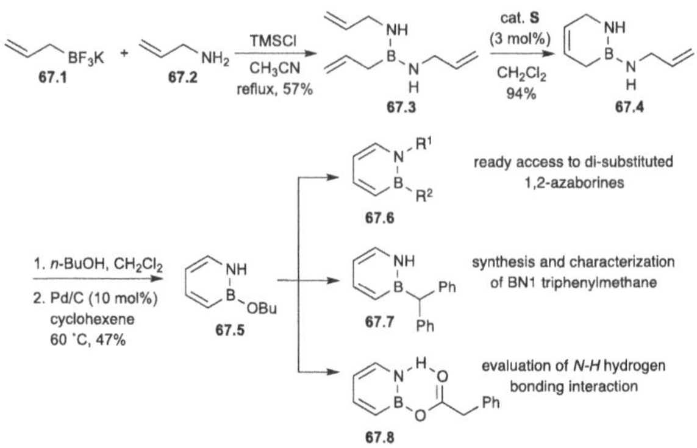

after screening a number of Brønsted acids, triflic acid was found to provide suitable protection for the nitrogen lone pair in **66.7**. Subsequent RCM of protonated **66.7**, followed by deprotonation employing DBU, afforded the desired heterocycle **66.8** in 47% overall yield over three steps from **66.7**. The 1,3-azaborine analogs are thermally stable aromatic boron heterocycles, which can undergo electrophilic aromatic substitution reactions and nucleophilic substitution reaction at the boron atom.

In 2013, Liu and coworkers reported the protecting group-free synthesis of 1,2-azaborine **67.5** (Scheme 67) [335]. In spite of the advances in azaborine chemistry by a number of research groups, significant synthetic challenges still remain to unleash the full potential of azaborine heterocycles [33, 325, 336–352]. In this work, allylboron dichloride was synthesized in situ from the commercially available potassium allyltrifluoroborate. Allyltrifluoroborate was further treated with allylamine, which furnished the diene intermediate **67.3** in 57% yield. Ring-closing metathesis of diene intermediate **67.3** using the **S** catalyst (3 mol%) afforded boron heterocycle **67.4** in 94% yield. Boron heterocycle **67.4** was further converted to the desired compound **67.5** in 47% yield by substituting the allylamino fragment of boron with *n*-butanol and subsequent oxidation using catalytic amounts of Pd/C. This protocol provided access to gram quantities of 1,2-azaborine synthon **67.5** in four steps (25% overall yield). The 1,2-azaborine obtained was further diversified to the corresponding di-substituted 1,2-azaborine **67.6**, BN1 triphenylmethane **67.7**, and the BN isostere of phenyl phenylacetate **67.8**. The authors also reported the use of similar BN compounds for hydrogen storage [353].

In 2013, Matsuda and coworkers performed RCM of boron-tethered dienes to generate dibenzo[*b*,*f*]borepines (Scheme 68) [140]. To the best of their knowledge, it was the first examples of dibenzo[*b*,*f*]borepines with a tetracoordinate boron center. After the generation of tris(2-vinylphenyl)borane **68.1** as described below, treatment of **68.1** with dibenzoylmethane and 8-quinoline furnished boron-tethered dienes **68.2** and **68.4**, respectively. Next, RCM was performed in the presence of the **HG2** catalyst at 100°C in toluene to generate the spirocyclic tetracoordinate dibenzoborepines **68.3** and **68.5**, respectively, in high yield.

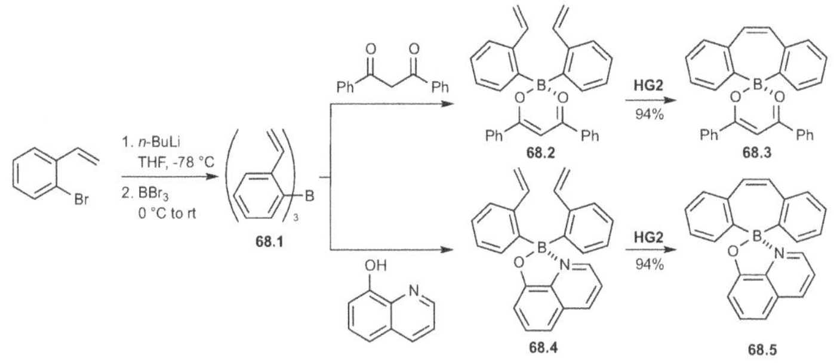

Scheme 68 Synthesis of spirocyclic dibenzoborepines from boron-tethered dienes via RCM

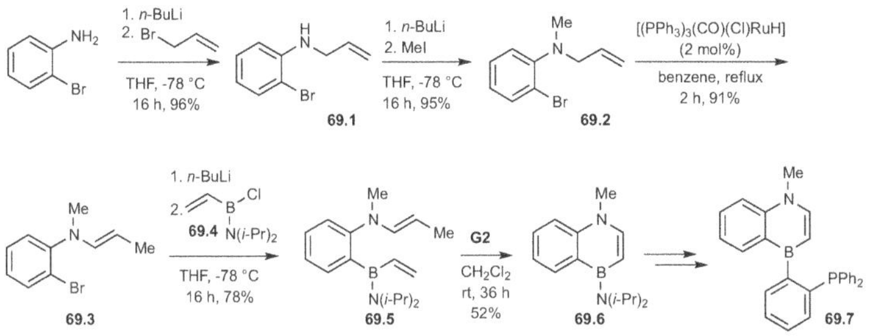

Scheme 69 Construction of monobenzofused 1,4-azaborines using RCM

In 2014, Liu and coworkers described the synthesis of boron-substituted, monobenzofused 1,4-azaborines using RCM as a key synthetic strategy (Scheme 69) [354]. Later, this azaborine **69.6** was consumed for the synthesis of azaborine phosphine ligand **69.7** for use in the assembly of a Pd(0) complex as a hydroboration catalyst. Synthesis of monobenzofused 1,4-azaborine **69.6** was effectively achieved through the route illustrated in Scheme 69. *N*-Allylation followed by *N*-methylation of 2-bromoaniline generated *N*-allyl-*N*-methyl-2-bromoaniline **69.2** in high yield. Alkene isomerization through the use of $[(PPh_3)_3(CO)(Cl)RuH]$ (2 mol%) furnished *N*-vinyl aniline **69.3**. Lithiation of **69.3** and quenching with vinylboron chloride **69.4** generated intermediate **69.5**, which was subsequently cyclized by RCM using the **G2** catalyst to afford the desired product **69.6** in 52% yield.

In 2014, Ashe and coworker reported the synthesis of BN-naphthalene **70.5** and [6,6]-fused-ring 1,2-azaborines **70.6** via RCM in three steps with 14% overall yield (Scheme 70) [355]. To the best of their knowledge, it was the first experimental work on BN-naphthalene systems. Diallylboron chloride **70.2** was initially produced by treating allyltributylstannane **70.1** with BCl_3 in pentane at −78°C. Without purification, diallylboron **70.2** was treated with diallylamine, followed by addition of triethylamine, and the resulting mildly air-sensitive intermediate **70.3** was obtained by distillation. Intermediate **70.3** was employed in RCM utilizing the **G1** (3 mol%) catalyst at 25°C followed by reflux, which resulted in the

Scheme 70 Synthesis of BN-naphthalene and tetralin using the **G1** catalyst

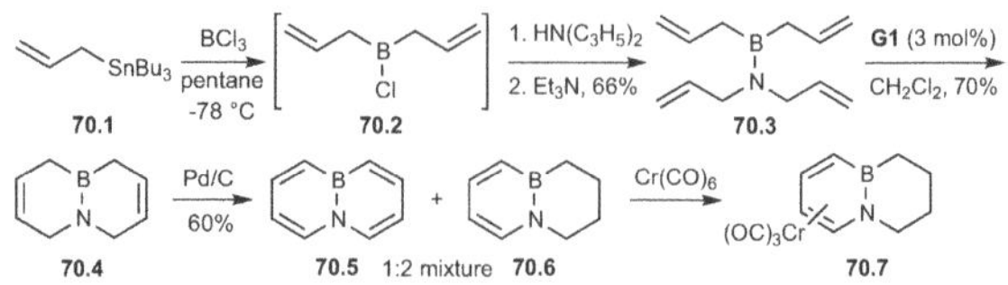

generation of the doubly cyclized [6,6]-fused-ring 1,2-azaborine **70.4** in 70% yield. No other side products were detected in the RCM reaction of intermediate **70.3**. 1,2-azaborine **70.4** was subjected to catalytic dehydrogenation in excess cyclohexene for 18 h resulting in a 1:2 mixture of BN-naphthalene **70.5** and BN-tetralin **70.6**. The authors further converted the tetrahydro BN-naphthalene compound **70.6** to the Cr complex **70.7** in order to explore its coordination chemistry.

6 Se-Containing Heterocycles

Berzelius discovered selenium in 1817 [356], and in 1847 the first organoselenium compound ethylselenol was synthesized by Wöhler and Siemens [357]. Despite the similarities between selenium compounds and sulfur- and oxygen-containing compounds, less progress has been made in advancing selenium chemistry when compared to sulfur and oxygen [358–362]. This is in part due to the toxicity and instability of selenium-containing compounds [363–365]. In this small section, two literature examples for the synthesis of selenium heterocycles via RCM are discussed.

In 2009, Koketsu and coworkers reported the synthesis of selenium-containing bicyclic β-lactams utilizing RCM as a key reaction (Scheme 71) [366]. Selenium was incorporated at the C4 position of the azetidinone unit by the treatment of 4-acetoxyazetidinone **71.3** with selenating reagents **71.2a**, **b** to furnish intermediates **71.4** in good yield. *N*-alkylation of **71.4** was successfully achieved with alkene bromides to afford the *N*-alkylated intermediates **71.5a–e**. Initial attempts to cyclize intermediates **71.5** via RCM using either the **G1** or **S** catalysts were unsuccessful. Utilizing 5 mol% of the **G2** catalyst in refluxing toluene was found to be the optimal reaction condition to afford the bicyclic β-lactams, selenazepines, selenazocines, and selenazonine **71.6a–e** in good yields.

In 2010, Koketsu and coworkers focused on RCEYM of β-lactam skeletons having propargylseleno moieties at the C4 position (Scheme 72) [367]. The TSE-protected azetidin-2-one **72.1** was selected as a starting material. After several attempts for the selective removal of TSE group of **72.1**, and subsequent in situ alkylation of the selenolate anion, treatment of **72.1** with TBAF (2.5 equiv.), followed by alkylation with numerous propargyl halides in DMF, readily furnished the key intermediate **72.2**. Treatment with propargyl chloride was found to be superior to that with propargyl bromide. Subsequent RCEYM in the presence of the **G2** catalyst afforded the desired 1,3-selenazepines **72.3b–d** correspondingly in

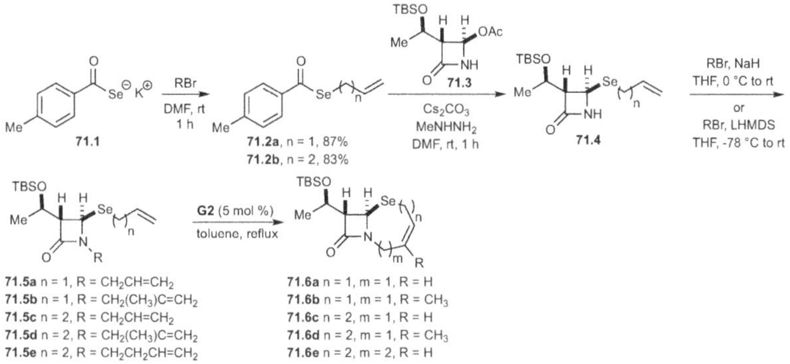

Scheme 71 Synthesis of Se-containing bicyclic β-lactams employing RCM

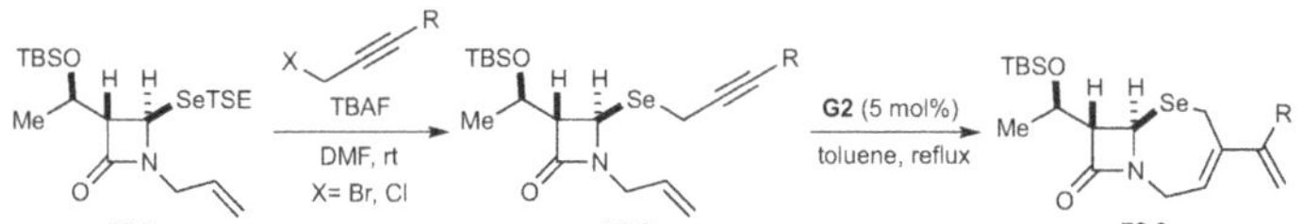

Entry	Substrate	Time	Product		Yield (%)
1	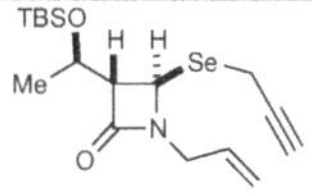	4 days	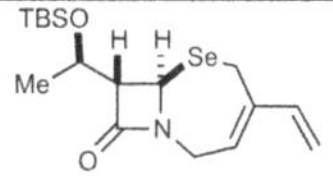	**72.3a**	0
2	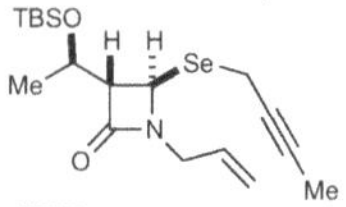	2 days	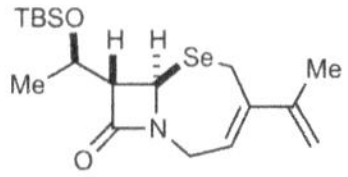	**72.3b**	44
3	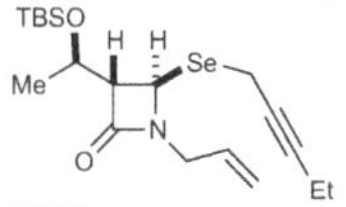	1 day	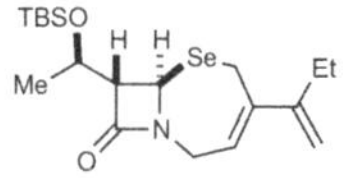	**72.3c**	46
4	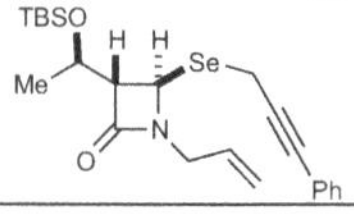	1 day	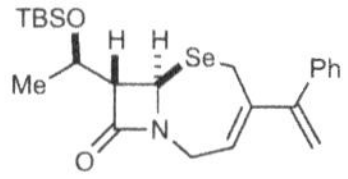	**72.3d**	59

Scheme 72 Construction of β-lactam skeletons utilizing RCEYM

good yield, except for substrate **72.3a**. It is well known that the synthesis of higher-membered selenium heterocycles is very difficult [368]. To the best of knowledge, there are very few reports for 7-membered selenium-containing heterocycles (1,3-selenazepines) [366, 369–371] and one report for the synthesis of 8-membered selenium heterocycles [372]. Koketsu also reported the synthesis of selenium-β-lactams (3-selena-1-dethiacephems and selenazepines) via iodocyclization [373].

7 Conclusion

In conclusion, the purpose of this chapter has been to highlight the importance of RCM in the synthesis of P-, S-, Si-, B-, and Se-containing heterocycles. The olefin metathesis reaction has been known since the 1960s, and its application has been accelerated to generate the functionalized rings present in natural products and other biologically active compounds. From the 1990s, the progress in the development of RCM catalysts and metathesis reactions has guided the explosion of research in this area. RCM has been applied to the synthesis of many heterocyclic compounds and natural products, and frequently, this technological advance provides a unique entry to substrates much quicker than classic approaches. We hope that this work motivates researchers to use this versatile synthetic approach innovatively for the generation of new heterocyclic compounds that were not regarded in traditional ways. Furthermore, newer advances in catalyst development will lead the advance of unique P-, S-, Si-, B-, and Se-containing heterocycles exhibiting biological and synthetic potential and their application in medicinal chemistry and reagent development.

Acknowledgment The authors would like to thank the National Institute of General Medical Sciences (NIH R01 GM077309) and the State of Kansas for their generous support of our program.

References

1. Grubbs RH (2003) Handbook of metathesis. Wiley-VCH, Weinheim, pp 1–3
2. Diver ST, Giessert AJ (2004) Chem Rev 104:1317–1382
3. Deiters A, Martin SF (2004) Chem Rev 104:2199–2238
4. Nicolaou KC, Bulger PG, Sarlah D (2005) Angew Chem Int Ed 44:4490–4527
5. Parenty A, Moreau X, Campagne JM (2006) Chem Rev 106:911–939
6. Hoveyda AH, Zhugralin AR (2007) Nature 450:243–251
7. Kotha S, Lahiri K (2007) Synlett 2767–2784
8. Grubbs RH (2008) Handbook of metathesis: catalyst development. Wiley-VCH, Weinheim
9. van Otterlo WAL, de Koning CB (2009) Chem Rev 109:3743–3782
10. Monfette S, Fogg DE (2009) Chem Rev 109:3783–3816
11. Nolan SP, Clavier H (2010) Chem Soc Rev 39:3305–3316
12. Prunet J (2011) Eur J Org Chem 3634–3647
13. Schrock RR, Murdzek JS, Bazan GC, Robbins J, Dimare M, Oregan M (1990) J Am Chem Soc 112:3875–3886
14. Bazan GC, Oskam JH, Cho HN, Park LY, Schrock RR (1991) J Am Chem Soc 113: 6899–6907
15. Schwab P, France MB, Ziller JW, Grubbs RH (1995) Angew Chem Int Ed Engl 34: 2039–2041
16. Wu Z, Nguyen ST, Grubbs RH, Ziller JW (1995) J Am Chem Soc 117:5503–5511
17. Schwab P, Grubbs RH, Ziller JW (1996) J Am Chem Soc 118:100–110
18. Scholl M, Ding S, Lee CW, Grubbs RH (1999) Org Lett 1:953–956
19. Kingsbury JS, Harrity JPA, Bonitatebus PJ, Hoveyda AH (1999) J Am Chem Soc 121: 791–799

20. Garber SB, Kingsbury JS, Gray BL, Hoveyda AH (2000) J Am Chem Soc 122:8168–8179
21. Gessler S, Randl S, Blechert S (2000) Tetrahedron Lett 41:9973–9976
22. Yet L (2000) Chem Rev 100:2963–3007
23. Chatterjee AK, Choi TL, Sanders DP, Grubbs RH (2003) J Am Chem Soc 125:11360–11370
24. Connon SJ, Blechert S (2003) Angew Chem Int Ed 42:1900–1923
25. Grandillas A, Perez-Castells J (2006) Angew Chem 118:6232–6247
26. Gradillas A, Perez-Castells J (2006) Angew Chem Int Ed 45:6086–6101
27. Cossy J, Arseniyadis S, Meyer C, Grubbs RH (2010) Metathesis in natural product synthesis: strategies, substrates and catalysis. Wiley-VCH, Weinheim
28. McReynolds MD, Dougherty JM, Hanson PR (2004) Chem Rev 104:2239–2258
29. Thomas CD, Hanson PR (2010) Phosphorus and sulfur heterocycles via ring-closing metathesis: application in natural product synthesis. In: Cossy J, Areniyadis S, Meyer C (eds) Metathesis in natural product synthesis. Wiley-VCH, Weinheim, pp 129–144
30. Evans PA, Cui J, Buffone GP (2003) Angew Chem Int Ed 42:1734–1737
31. Evans PA (2010) Temporary silicon-tethered ring-closing metathesis reactions in natural product synthesis. In: Cossy J, Areniyadis S, Meyer C (eds) Metathesis in natural product synthesis. Wiley-VCH, Weinheim, pp 225–259
32. Čusak A (2012) Chem Eur J 18:5800–5824
33. Ashe AJ (2009) Organometallics 28:4236–4248
34. Campbell PG, Marwitz AJV, Liu SY (2012) Angew Chem Int Ed 51:6074–6092
35. Brassfield HA, Jacobson RA, Verkade JG (1975) J Am Chem Soc 97:4143–4144
36. Reichardt P, Pink D, Tilgner J, Kretzschmar A, Thuss-Patience PC, Dorken B (2002) Onkologie 25:541–546
37. Thiem J, Günther M, Paulsen H, Kopf J (1977) Chem Ber 110:3190–3200
38. Harvey TC, Simiand C, Weiler L, Withers SG (1997) J Org Chem 62:6722–6725
39. Wróblewski AE (1986) Tetrahedron 42:3595–3606
40. Molin H, Noren JO, Claesson A (1989) Carbohydr Res 194:209–221
41. Darrow JW, Drueckhammer DG (1994) J Org Chem 59:2976–2985
42. Hanessian S, Galeotti N, Rosen P, Oliva G, Babu S (1994) Bioorg Med Chem Lett 4: 2763–2768
43. Darrow JW, Drueckhammer DG (1996) Bioorg Med Chem 4:1341–1348
44. Hanessian S, Rogel O (1999) Bioorg Med Chem Lett 9:2441–2446
45. Hanessian S, Rogel O (2000) J Org Chem 65:2667–2674
46. Engel R (1992) Handbook of organo-phosphorus chemistry. Marcel Dekker, New York
47. Kurokawa T, Suzuki K, Hayaoka T, Nagakawa T, Izawa T, Kobayashi M, Harada M (1993) J Antibiot 46:1315–1318
48. Malla RK, Bandyopadhyay S, Spilling CD, Dutta S, Dupureur CM (2011) Org Lett 13: 3094–3097
49. Point V, Malla RK, Diomande S, Martin BP, Delorme V, Carriere F, Canaan S, Rath NP, Spilling CD, Cavalier JF (2012) J Med Chem 55:10204–10219
50. Vertesy L, Beck B, Bronstrup M, Ehrlick K, Kurz M, Müller G, Schummer D, Seibert G (2002) J Antibiot 55:480–494
51. Carty TJ, Marfat A, Moore PF, Falkner FC, Twomey TM, Weissman A (1993) Agents Actions 39:157–165
52. Weintraub M, Jacox RF, Angevine CD, Atwater EC (1977) J Rheumatol 4:393–404
53. Souers AJ, Virgilio AA, Rosenquist A, Fenuik W, Ellman JA (1999) J Am Chem Soc 121: 1817–1825
54. Franchetti P, Cappellacci L, AbuSheikha G, Jayaram HN, Gurudutt VV, Sint T, Schneider BP, Jones WD, Goldstein BM, Perra G, DeMontis A, Loi AG, LaColla P, Grifantini M (1997) J Med Chem 40:1731–1737
55. Wilson AJ, Kerns JK, Callahan JF, Moody CJ (2013) J Med Chem 56:7463–7476
56. Baker SJ, Ding CZ, Akama T, Zhang YK, Hernandez V, Xia Y (2009) Future Med Chem 1: 1275–1288

57. Grassberger MA, Turnowsky F, Hildebrandt J (1984) J Med Chem 27:947–953
58. Robinson PD, Groziak MP (1999) Acta Crystallogr C55:1701–1704
59. Groziak MP (2001) Am J Ther 8:321–328
60. Hemscheidt T, Puglisi MP, Larsen LK, Patterson GML, Moore RE, Rios JL, Clardy J (1994) J Org Chem 59:3467–3471
61. Banker R, Carmeli S (1998) J Nat Prod 61:1248–1251
62. Dembitsky VM, Smoum R, Al-Quntar AA, Abu AH, Pergament I, Srebnik M (2002) Plant Science 163:931–942
63. Ramachandran PV (2013) Future Med Chem 5:611–612
64. Daiss JO, Burschka C, Mills JS, Montana JG, Showell GA, Warneck JBH, Tacke R (2006) Organometallics 25:1188–1198
65. Kelly M, Sammaai T (2002) Family Podospongiidae De Laubenfels. In: Hooper NA, Van Soest RWM (eds) Systema Porifera: a guide to classification of sponges, vol 1. Kluwer, New York, pp 698–699
66. Hoye TR, Ayyad SN, Eklov BM, Hashish NE, Shier WT, El Sayed KA, Hamann MT (2002) J Am Chem Soc 124:7405–7410
67. Yamazaki D, Kurisu S, Takenawa T (2005) Cancer Sci 96:379–386
68. El Sayed KA, Youssef DT, Marchetti DJ (2006) J Nat Prod 69:219–223
69. Spector I, Shochet NR, Kashman Y, Groweiss A (1983) Science 219:493–495
70. Ahmed SA, Odde S, Daga PR, Bowling JJ, Mesbah MK, Youssef DT, Khalifa SI, Doerksen RJ, Hamann MT (2007) Org Lett 9:4773–4776
71. Bowling JJ, Daga PR, Odde S, Ahmed SA, Mesbah MK, Youssef DT, Kudrimoti S, Khalifa S, Doerksen RJ, Hamann MT (2008) Planta Medica 74:341–342
72. Ludeman SM, Zon G (1975) J Med Chem 18:1251–1253
73. Ludeman SM, Boyd VL, Regan JB, Gallo KA, Zon G, Ishii K (1986) J Med Chem 29: 716–727
74. Kafarski P, Lejczak B (1991) Phosphorus Sulfur Silicon Relat Elem 63:193–215
75. Mader MM, Bartlett PA (1997) Chem Rev 97:1281–1301
76. Franz JE, Mao MK, Sikorski JA (1997) Glyphosate: a unique global herbicide. American Chemical Society, Washington, DC
77. Seto H, Kuzuyama T (1999) Nat Prod Rep 16:580–596
78. Quin LD (2000) A guide to organophosphorus chemistry. Wiley-Interscience, New York
79. Kafarski P, Lejczak B (2001) Curr Med Chem Anti-Cancer Agents 1:301–312
80. Gholivand K, Dorosti N, Shariatinia Z, Ghaziany F, Sarikhani S, Mirshahi M (2011) Med Chem Res 20:1287–1293
81. Elliott TS, Slowey A, Ye YL, Conway SJ (2012) Med Chem Comm 3:735–751
82. Déath NJ, Klostermyer J. WO 2001021626
83. Carion L, Jacquard C, Sainte-Catherine O, Loiseau S, Filippini D, Hirlemann MH, Volle JN, Virieux D, Lecouvey M, Pirat JL, Bakalara N (2012) J Med Chem 55:2196–2211
84. Dubrovina NV, Borner A (2004) Angew Chem Int Ed 43:5883–5886
85. Point V, Malla RK, Diomande S, Martin BP, Delorme V, Carriere F, Canaan S, Rath NP, Spilling CD, Cavalier JF (2012) J Med Chem 55:10204–10219
86. Nachev I (1989) J Chem Soc Perkin Trans 1:125–131
87. Clarion L, Jacquard C, Sainte-Catherine O, Loiseau S, Filippini D, Hirlemann MH, Volle JN, Virieux D, Lecouvey M, Pirat JL, Bakalara N (2012) J Med Chem 55:2196–2211
88. Colvin OM (1999) Current Pharmaceutical Design 5:555–560
89. Maynard-Faure P, Gonser C, Vaime V, Bouchu D (1998) Tetrahedron Lett 39:2315–2318
90. Hoffmann MG, Bauer K, Bieringer H, Rosinger C, Lindell SD (1997) DE Patent 19542305
91. Bayardon J, Cavazzini M, Maillard D, Pozzi G, Quici S, Sinou D (2003) Tetrahedron-Asym 14:2215–2224
92. Castillon S, Claver C, Diaz Y (2005) Chem Soc Rev 34:702–713
93. Chelucci G, Orru G, Pinna GA (2003) Tetrahedron 59:9471–9515

94. Claver C, Fernandez E, Gillon A, Heslop K, Hyett DJ, Martorell A, Orpen AG, Pringle PG (2000) Chem Commun 961–962
95. Claver C, Fernandez E, Gillon A, Heslop K, Hyett DJ, Martorell A, Orpen AG, Pringle PG (2000) Chem Commun 1447
96. Vlahovic S, Schadel N, Tussetschlager S, Laschat S (2013) Eur J Org Chem 1580–1590
97. Wassenaar J, Reek JNH (2011) Org Biomol Chem 9:1704–1713
98. Dunne KS, Bisaro F, Odell B, Paris JM, Gouverneur V (2005) J Org Chem 70:10803–10809
99. Stoianova DS, Hanson PR (2000) Org Lett 2:1769–1772
100. Bilhou JL, Basset JM, Mutin R, Graydon WF (1976) J Chem Soc Chem Commun 970–971
101. Kumar VG, Kummel K (1983) J Polym Sci Part A: Polym Chem 21:1183–1193
102. Stoianova DS, Whitehead A, Hanson PR (2005) J Org Chem 70:5880–5889
103. Stoianova DS, Hanson PR (2001) Org Lett 3:3285–3288
104. Gruner SAW, Locardi E, Lohof E, Kessler H (2002) Chem Rev 102:491–514
105. Lillelund VH, Jensen HH, Liang X, Bols M (2002) Chem Rev 102:515–553
106. Postema MHD, Piper JL, Betts RL (2005) Synlett 1345–1358
107. Plumet J, Gomez AM, Lopez JC (2007) Mini Rev Org Chem 4:201–216
108. Cipolla L, La Ferla B, Airoldi C, Zona C, Orsato A, Shaikh N, Russo L, Nicotra F (2010) Future Med Chem 2:587–599
109. Donohoe TJ (2002) Synlett 1223–1232
110. Zhang H, Tsukuhara R, Tigyi G, Prestwich GD (2006) J Org Chem 71:6061–6066
111. van Meer G (2005) EMBO J 24:3159–3165
112. Fukushima N, Ishii I, Contos JJ, Weiner JA, Chun J (2001) Annu Rev Pharmacol Toxicol 41: 507–534
113. Tanyi J, Croetzer D, Wolf J, Yu S, Hasegawa Y, Lahad J, Cheng KW, Umezu-Goto M, Prestwich GD, Morris AJ, Newman RA, Felix EA, Lapis R, Mills GB (2006) In: Feng L, Prestwich GD (eds) Functional lipidomics. CRC/Taylor & Francis, New York, pp 101–123
114. Contos JJ, Ishii I, Chun J (2000) Mol Pharmacol 58:1188–1196
115. Contos JJ, Chun J (2001) Gene 267:243–253
116. Hama K, Aoki J, Fukaya M, Kishi Y, Sakai T, Suzuki R (2004) J Biol Chem 279: 17634–17639
117. Coudray L, Bravo-Altamirano K, Montchamp JL (2008) Org Lett 10:1123–1126
118. Bujard M, Gouverneur V, Mioskowski C (1999) J Org Chem 64:2219–2231
119. Briot A, Bujard M, Gouverneur V, Nolan SP, Mioskowsky C (2000) Org Lett 2:1517–1519
120. Harvey JS, Malcolmson SJ, Dunne KS, Meek SJ, Thompson AL, Schrock RR, Hoveyda AH, Gouverneur V (2009) Angew Chem Int Ed 48:762–766
121. Pietrusiewicz KM, Zablocka M (1994) Chem Rev 94:1375–1411
122. Jiang XB, Minnaard AJ, Hessen B, Feringa BL, Duchateau ALL, Andrien JGO, Boogers JAF, de Vries JG (2003) Org Lett 5:1503–1506
123. Teichert JF, Feringa BL (2010) Angew Chem Int Ed 49:2486–2528
124. van Leeuwen PWNM, Kamer PCJ, Claver C, Pàmies O, Diéguez M (2011) Chem Rev 111: 2077–2118
125. Fourgeaud P, Midrier C, Vors JP, Volle JN, Pirat JL, Virieux D (2010) Tetrahedron 66: 758–764
126. Machida Y, Saito I (1979) J Org Chem 44:865–866
127. Harvey JS, Giuffredi GT, Gouverneur V (2010) Org Lett 12:1236–1239
128. Samojlowicz C, Bieniek M, Zarecki A, Kadyrov R, Grela K (2008) Chem Commun 6282–6284
129. Funk TW, Berlin JM, Grubbs RH (2006) J Am Chem Soc 128:1840–1846
130. Grandbois A, Collins SK (2008) Chem Eur J 14:9323–9329
131. Fujimura O, Grubbs RH (1998) J Org Chem 63:824–832
132. Sattely ES, Cortez GA, Moebius DC, Schrock RR, Hoveyda AH (2005) J Am Chem Soc 127: 8526–8533
133. Fujimura O, Grubbs RH (1996) J Am Chem Soc 118:2499–2500

134. Lee AL, Malcolmson SJ, Puglisi A, Schrock RR, Hoveyda AH (2006) J Am Chem Soc 128:5153–5157
135. Rost D, Porta M, Gessler S, Blechert S (2008) Tetrahedron Lett 49:5968–5971
136. Kumar D, Singh N, Keshav K, Elias AJ (2011) Inorg Chem 50:250–260
137. Schmidt B, Kunz O (2013) Org Lett 15:4470–4473
138. Wu X, O'Brien P, Ellwood S, Secci F, Kelly B (2013) Org Lett 15:192–195
139. Muci AR, Campos KR, Evans DA (1995) J Am Chem Soc 117:9075–9076
140. Matsuda T, Sato S (2013) J Org Chem 78:3329–3335
141. Roggen M, Carreira EM (2012) Angew Chem Int Ed Engl 51:8652–8655
142. Mariz R, Briceno A, Dorta R, Dorta R (2008) Organometallics 27:6605–6613
143. Majumdar KC, Nandi RK, Ganai S (2014) Tetrahedron Lett 55:1247–1250
144. Whitehead A, McReynolds MD, Moore JD, Hanson PR (2005) Org Lett 7:3375–3378
145. Waetzig JD, Hanson PR (2006) Org Lett 8:1673–1676
146. Waetzig JD, Hanson PR (2008) Org Lett 10:109–112
147. Venukadasula PKM, Chegondi R, Maitra S, Hanson PR (2010) Org Lett 12:1556–1559
148. Chegondi R, Tan MML, Hanson PR (2011) J Org Chem 76:3909–3916
149. Whitehead A, Waetzig JD, Thomas CD, Hanson PR (2008) Org Lett 10:1421–1424
150. Hanson PR, Chegondi R, Nguyen J, Thomas CD, Waetzig JD, Whitehead A (2011) J Org Chem 76:4358–4370
151. Jayasinghe S, Venukadasula PKM, Hanson PR (2014) Org Lett 16:122–125
152. Whitehead A, McParland JP, Hanson PR (2006) Org Lett 8:5025–5028
153. Venukadasula PKM, Chegondi R, Suryn GM, Hanson PR (2012) Org Lett 14:2634–2637
154. Chegondi R, Maitra S, Markley JL, Hanson PR (2013) Chem Eur J 19:8088–8093
155. Hanson PR, Jayasinghe S, Maitra S, Ndi CN, Chegondi R (2014) Beilstein J Org Chem 10:2332–2337
156. Drews J (2000) Science 287:1960–1964
157. Scozzafava A, Owa T, Mastrolorenzo A, Supuran CT (2003) Curr Med Chem 10:925–953
158. Hanessian S, Sailes H, Therrien E (2003) Tetrahedron 59:7047–7056
159. Bouras N, Mathieu F, Sabaou N, Lebrihi A (2006) J Appl Microbiol 100:390–397
160. University of Arizona. Department of Chemistry and Biochemistry. Top 200 Brand Name Drugs by US Retail Sales in 2012. http://cbc.arizona.edu/njardarson/group/top-pharmaceuticals-poster. Accessed 12 Dec 2014
161. Inagaki M, Tsuri T, Jyoyama H, Ono T, Yamada K, Kobayashi M, Hori Y, Arimura A, Yasui K, Ohno K, Kakudo S, Koizumi K, Suzuki R, Kato M, Kawai S, Matsumoto S (2000) J Med Chem 43:2040–2048
162. Baker DC, Jiang B (2002) US Patent No. 6353112 B1
163. Lebegue N, Gallet S, Flouquet N, Carato P, Pfeiffer B, Renard P, Leonce S, Pierre A, Chavatte P, Berthelot P (2005) J Med Chem 48:7363–7373
164. Zia-ur-Rehman M, Choudary JA, Ahmad S, Siddiqui HL (2006) Chem Pharm Bull 54: 1175–1178
165. Donohoe TJ, Bower JF, Chan LKM (2012) Org Biomol Chem 10:1322–1328
166. Vougioukalakis GC, Grubbs RH (2010) Chem Rev 110:1746–1787
167. Lozano-Vila AM, Monsaert S, Bajek A, Verpoort F (2010) Chem Rev 110:4865–4909
168. Samojłowicz C, Bieniek M, Grela K (2009) Chem Rev 109:3708–3742
169. Bieniek M, Michrowska A, Usanov DL, Grela K (2008) Chem Eur J 14:806–818
170. Plietker B, Seng D, Frohlich P, Metz P (2000) Tetrahedron 56:873–879
171. Wolckenhauer SA, Devlin AS, Du Bois J (2007) Org Lett 9:4363–4366
172. Majumdar KC, Mondal S, Ghosh D (2009) Tetrahedron Lett 50:4781–478
173. Majumdar KC, Chattopadhyay B, Sinha B (2008) Synthesis 23:3857–3863
174. Mondal S (2012) Chem Rev 112:5339–5355
175. Morimoto Y, Kurihara H, Kinoshita T (2000) Chem Commun 189–190
176. Le Flohic A, Meyer C, Cossy J (2005) Org Lett 7:339–342
177. Le Flohic A, Meyer C, Cossy J (2006) Tetrahedron 62:9017–9037

178. Cossy J, Arseniyadis S (2012) Modern tools for the synthesis of complex bioactive molecules. Wiley, New Jersey
179. Crews P, Kakou Y, Quiñoà E (1988) J Am Chem Soc 110:4365–4368
180. Le Flohic A, Meyer C, Cossy J, Desmurs JR, Galland JC (2003) Synlett 667–670
181. Jiménez-Hopkins M, Hanson PR (2008) Org Lett 10:2223–2226
182. Benazza M, Kanso R, Demailly G (2009) Tetrahedron 65:8885–8890
183. Liu GY, He HY, Wang JH (2009) Adv Synth Catal 351:1610–1620
184. Morgans GL, Ngidi EL, Madeley LG, Khanye SD, Michael JP, de Koning CB, van Otterlo WAL (2009) Tetrahedron 65:10650–10659
185. Thiéry V, Coudert G, Bizot-Espiard J-G, Pfeiffer B, Renard P, Lindenbaum A, Guillaumet G (2001) J Med Chem 44:3904–3914
186. Charton I, Mamai A, Bennejean C, Renard P, Howell HE, Guardiola-Lemaître B, Delagrange P, Morgan PJ, Viaud M-C, Guillaumet G (2000) Bioorg Med Chem 8:105–114
187. Mamai A, Bennejean C, Renard P, Delagrange P, Guardiola-Lemaître B, Howell HE, Viaud M-C, Guillaumet G (1999) Pharm Pharmacol Commun 5:199–206
188. Charton I, Mamai A, Bennejean C, Renard P, Delagrange P, Morgan PJ, Howell HE, Gourdel-Martin ME, Viaud M-C, Guillaumet G (2000) Pharm Pharmacol Commun 6:49–60
189. Rybczynski PJ, Zeck RE, Combs DW, Turchi I, Burris TP, Xu JZ, Yang M, Demarest KT (2003) Bioorg Med Chem Lett 13:2359–2362
190. Bergman J, Rosenzweig-Lipson S, Spealman RD (1995) J Pharmacol Exp Ther 273:40–48
191. Spealman RD (1996) J Pharmacol Exp Ther 278:1128–1137
192. Armenise D, Trapani G, Arrivo V, Laraspata E, Morlacchi F (2000) J Heterocycl Chem 37: 1611–1616
193. Fuger B, Bolm C (2009) Synlett 1601–1604
194. Reggelin M, Zur C (2000) Synthesis 1–64
195. Lucking U (2013) Angew Chem Int Ed 52:9399–9408
196. Roeb E, Purucker E, Gartung C, Geier A, Jansen B, Winograd R, Matern S (2003) Scand J Gastroentero 38:878–885
197. Yadav DB, Morgans GL, Aderibigbe BA, Madeley LG, Fernandes MA, Michael JP, de Koning CB, van Otterlo WAL (2011) Tetrahedron 67:2991–2997
198. Kuźnik N, Krompiec S, Bieg T, Baj S, Skutil K, Chrobok A (2003) J Organomet Chem 665: 167–175
199. Vorobyeva DV, Mailyan AK, Peregudov AS, Karimova NM, Vasilyeva TP, Bushmarinov IS, Bruneau C, Dixneuf PH, Osipov SN (2011) Tetrahedron 67:3524–3532
200. Morgans GL, Yadav DB, Fernandes MA, de Koning CB, Michael JP, van Otterlo WAL (2012) Tetrahedron Lett 53:2384–2387
201. Gao N, Guo XW, Zheng SC, Yang WK, Zhao XM (2012) Tetrahedron 68:9413–9418
202. Ueda M, Hartwig JF (2010) Org Lett 12:92–94
203. Xu QL, Liu WB, Dai LX, You SL (2010) J Org Chem 75:4615–4618
204. Xu QL, Dai LX, You SL (2010) Org Lett 12:800–803
205. Zheng SC, Gao N, Liu W, Liu DG, Zhao XM, Cohen T (2010) Org Lett 12:4454–4457
206. Gao N, Zheng SC, Yang WK, Zhao XM (2011) Org Lett 13:1514–1516
207. Rao HP, Rafi S, Kumar PR, Guravaiah C, Muthanna N (2012) Tetrahedron Lett 53: 6877–6880
208. Donohoe TJ, Orr AJ, Bingham M (2006) Angew Chem Int Ed 45:2664–2670
209. Monge-Marcet A, Pleixats R, Cattoen X, Man MWC (2013) Tetrahedron 69:341–348
210. Tzur E, Szadkowska A, Ben-Asuly A, Makal A, Goldberg I, Wozniak K, Grela K, Lemcoff NG (2010) Chem Eur J 16:8726–8737
211. Kress S, Blechert S (2012) Chem Soc Rev 41:4389–4408
212. Hoveyda AH, Malcolmson SJ, Meek SJ, Zhugralin AR (2000) Angew Chem Int Ed 49:34–43
213. Abbaz T, Bendjeddou A, Gouasmia A, Bouchouk D, Boualleg C, Kaouachi N, Inguimbert N, Villemin D (2014) Lett Org Chem 11:59–63

214. Hill-Cousins JT, Salim SS, Bakar YM, Bellingham RK, Light ME, Brown RCD (2014) Tetrahedron 70:3700–3706
215. Hanson PR, Probst DA, Robinson RE, Yau M (1999) Tetrahedron Lett 40:4761–4764
216. Brown RCD, Castro JL, Moriggi JD (2000) Tetrahedron Lett 41:3681–3685
217. Moriggi JD, Brown LJ, Castro JL, Brown RCD (2004) Org Biomol Chem 2:835–844
218. Dougherty JM, Probst DA, Robinson RE, Moore JD, Klein TA, Snelgrove KA, Hanson PR (2000) Tetrahedron 56:9781–9790
219. Jun JH, Dougherty JM, Jimenez MD, Hanson PR (2003) Tetrahedron 59:8901–8912
220. Salim SS, Bellingham RK, Satcharoen V, Brown RCD (2003) Org Lett 5:3403–3406
221. Salim SS, Bellingham RK, Brown RCD (2004) Eur J Org Chem 800–806
222. Dougherty JM, Jimenez M, Hanson PR (2005) Tetrahedron 61:6218–6230
223. Mondal S, Debnath S (2014) Tetrahedron Lett 55:1577–1580
224. Zhang Q, Shi X, Jiang Y, Li Z (2014) Tetrahedron 70:7621–7626
225. Mingat G, McDouall JJW, Clayden J (2014) Chem Commun 50:6754–6757
226. Morita N, Krause N (2006) Angew Chem Int Ed 45:1897–1899
227. Benetti S, De Risi C, Pollini GP, Zanirato V (2012) Chem Rev 112:2129–2163
228. Tang J, Xu DQ, Xia AB, Wang YF, Jiang JR, Luo SP, Xu ZY (2010) Adv Synth Catal 352: 2121–2126
229. Dotsenko VV, Krivokolysko SG, Chernega AN, Litvinov VP (2007) Russ Chem Bull 56: 1431–1436
230. Samet AV, Shestopalov AM, Nesterov VN, Semenov VV (1997) Synthesis 623–625
231. Shestopalov AM, Bogomolova OP, Litvinov VP (1991) Synthesis 277–279
232. Gauthier DR Jr, Zandi KS, Shea KJ (1998) Tetrahedron 54:2289–2338
233. Van de Weghe P, Bisseret P, Blanchard N, Eustache J (2006) J Organomet Chem 691: 5078–5108
234. Bols M, Skrydstrup T (1995) Chem Rev 95:1253–1277
235. Fensterbank L, Malacria M, Sieburth SM (1997) Synthesis 813–854
236. Marciniec B, Pietraszuk C (2003) Curr Org Chem 7:691–735
237. Brown RCD, Satcharoen V (2006) Heterocycles 70:705–736
238. Bracegirdle S, Anderson EA (2010) Chem Soc Rev 39:4114–4129
239. Fu GC, Grubbs RH (1992) J Am Chem Soc 114:5426–5427
240. Evans PA, Murthy VS (1998) J Org Chem 63:6768–6769
241. Denmark SE, Yang S-M (2004) J Am Chem Soc 126:12432–12440
242. Kim YJ, Lee D (2006) Org Lett 8:5219–5222
243. Hoye TR, Eklov BM, Jeon J, Khoroosi M (2006) Org Lett 8:3383–3386
244. Evans PA, Cui J, Gharpure SJ, Polosukhin A, Zhang H-R (2003) J Am Chem Soc 125: 14702–14703
245. Hoye TR, Promo MA (1999) Tetrahedron Lett 40:1429–1432
246. Casey EM, Teesdale-Spittle P, Harvey JE (2008) Tetrahedron Lett 49:7021–7023
247. Gaitanos TN, Buey RM, Díaz JF, Northcote PT, Teesdale-Spittle P, Andreu JM, Miller JH (2004) Cancer Res 64:5063–5067
248. Wilmes A, Bargh K, Kelly C, Northcote PT, Miller JH (2007) Mol Pharm 4:269–280
249. Scalzullo SM, Islam RU, Morgans GL, Michael JP, van Otterlo WAL (2008) Tetrahedron Lett 49:7403–7405
250. Rousseau G, Blanco L (2006) Tetrahedron 62:7951–7993
251. Ando W (1996) Bull Chem Soc Jpn 69:1–16
252. Mignani S, Damour D, Doble A, Labaudinière R, Malleron JL, Piot O, Gueremy C (1993) Bioorg Med Chem Lett 3:1913–1918
253. Aoyama T, Sato Y, Suzuki T, Shirai H (1978) J Organomet Chem 153:193–207
254. Sato Y, Fukami Y, Shirai H (1974) J Organomet Chem 78:75–81
255. Chang D, Grubbs RH (1997) Tetrahedron Lett 38:4757–4760
256. Cordier C, Morton D, Leach S, Woodhall T, O'Leary-Steele C, Warriner S, Nelson A (2008) Org Biomol Chem 6:1734–1737

257. Brown LJ, Spurr IB, Kemp SC, Camp NP, Gibson KR, Brown RCD (2008) Org Lett 10: 2489–2492
258. Eiseman JL, Bai L, Jung WH, Moura-Letts G, Day BW, Curran DP (2008) J Med Chem 51: 6650–6653
259. Rodriguez-Escrich C, Urpi F, Vilarrasa J (2008) Org Lett 10:5191–5194
260. Kobayashi J, Kubota T (2007) J Nat Prod 70:451–460
261. Kobayashi J, Tsuda M (2004) Nat Prod Rep 21:77–93
262. Matsuda T, Yamaguchi Y, Murakami M (2008) Synlett 4:561–564
263. Hammaecher C, Bouillon JP, Portella C (2009) Tetrahedron 65:5527–5534
264. Brook AG, Mauris RJ (1957) J Am Chem Soc 79:971–973
265. Ricci A, Degl'Innocenti A (1989) Synthesis 647–660
266. Page PCB, Klair SS, Rosenthal S (1990) Chem Soc Rev 19:147–195
267. Cirillo PF, Panek JS (1992) Org Prep Proced Int 24:555–582
268. Najera C, Yus M (1995) Org Prep Proced Int 27:385–456
269. Bonini BF, Comes-Franchini M, Fochi M, Mazzanti G, Ricci AJ (1997) Chim Ital 127: 619–628
270. Bonini BF, Comes-Franchini M, Fochi M, Mazzanti G, Ricci A (1998) J Organomet Chem 567:181–189
271. Page PCB, McKenzie MJ, Klair SS, Rosenthal S (1998) In: Patai S, Rappoport Z (eds) The chemistry of organosilicon compounds, vol 2. Wiley, Chichester, 1599 p, Part 2
272. Siegel DS, Piizzi G, Piersanti G, Movassaghi M (2009) J Org Chem 74:9292–9304
273. Movassaghi M, Piizzi G, Siegel DS, Piersanti G (2009) Tetrahedron Lett 50:5489–5492
274. Mukherjee S, Lee D (2009) Org Lett 11:2916–2919
275. Maifeld SV, Miller RL, Lee D (2004) J Am Chem Soc 126:12228–12229
276. Maifeld SV, Lee D (2005) Chem Eur J 11:6118–6126
277. Grimm JB, Otte RD, Lee D (2005) J Organomet Chem 690:5508–5516
278. Hansen EC, Lee D (2006) Acc Chem Res 39:509–519
279. Kim YJ, Grimm JB, Lee D (2007) Tetrahedron Lett 48:7961–7964
280. Lee YJ, Schrock RR, Hoveyda AH (2009) J Am Chem Soc 131:10652–10661
281. Matsui R, Seto K, Fujita K, Suzuki T, Nakazaki A, Kobayashi S (2010) Angew Chem 122: 10266–10271
282. Hoye TR, Jeon J, Kopel LC, Ryba TD, Tennakoon MA, Wang YN (2010) Angew Chem Int Ed 49:6151–6155
283. Hoye TR, Jeffrey CS, Tennakoon MA, Wang J, Zhao H (2004) J Am Chem Soc 126: 10210–10211
284. Hatanaka Y, Hiyama T (1988) J Org Chem 53:918–920
285. Denmark SE, Muhuhi JM (2010) J Am Chem Soc 132:11768–11778
286. Langer P, Freiberg W (2004) Chem Rev 104:4125–4149
287. Rousseau G (1995) Tetrahedron 51:2777–2849
288. Mandolini L (1986) Adv Phys Org Chem 22:1–111
289. Illuminati G, Mandolini L (1981) Acc Chem Res 14:95–102
290. Kim JW, Shin-Ya K, Furihata K, Hayakawa Y, Seto H (1999) J Org Chem 64:153–155
291. Hayakawa Y, Tomikawa T, Shin-ya K, Arao N, Nagai K, Suzuki K, Furihata K (2003) J Antibiot 56:905–908
292. Winssinger N, Barluenga S (2007) Chem Commun 22–36
293. Delmotte P, Delmotte-Plaquee J (1953) Nature 171:344
294. Ayer WA, Lee SP, Tsuneda A, Hiratsuka Y (1980) Can J Microbiol 26:766–773
295. Driggers EM, Hale SP, Lee J, Terrett NK (2008) Nat Rev Drug Discovery 7:608–624
296. Harvey AL (2008) Drug Discovery Today 13:894–901
297. Butler MS (2008) Nat Prod Rep 25:475–516
298. Butler MS (2005) Nat Prod Rep 22:162–195
299. Wessjohann LA, Rivera DG, Vercillo OE (2009) Chem Rev 109:796–814
300. Furstner A, Nagano T, Muller C, Seidel G, Muller O (2007) Chem Eur J 13:1452–1462

301. Lee S, Paek SM, Yun H, Kim NJ, Suh YG (2011) Org Lett 13:3344–3347
302. Matsui R, Seto K, Sato Y, Suzuki T, Nakazaki A, Kobayashi S (2011) Angew Chem Int Ed 50:680–683
303. Volchkov I, Park S, Lee D (2011) Org Lett 13:3530–3533
304. Volchkov I, Lee D (2013) J Am Chem Soc 135:5324–5327
305. Wang J, Sun BF, Cui K, Lin GQ (2012) Org Lett 14:6354–6357
306. Chowdhury PS, Kumar P (2013) Eur J Org Chem 4586–4593
307. Kumar TVG, Suresh SK, Babu SJ, Rao JM (2013) Tetrahedron Lett 54:1397–1400
308. Quinn KJ, Gage KE, Stanners CM, Hayes CO, Shanley KE, Couvertier SM (2014) Tetrahedron Lett 55:376–378
309. Tomooka K, Miyasaka S, Motomura S, Igawa K (2014) Chem Eur J 20:7598–7602
310. Lipscomb WN (1977) Angew Chem 89:685–696
311. Brown HC (1980) Angew Chem 92:675–683
312. Suzuki A (2011) Angew Chem 123:6854–6869
313. Abbey ER, Liu SY (2013) Org Biomol Chem 11:2060–2069
314. Baker SJ, Tomsho JW, Benkovic SJ (2011) Chem Soc Rev 40:4279–4285
315. Hudson ZM, Wang SN (2011) Dalton T 40:7805–7816
316. Gronowit S, Dahlgren T, Namtvedt J, Roos C, Sjoberg B, Forsgren U (1971) Acta Pharmaceutica Suecica 8:377–390
317. Gronowit S, Forsgren U, Namtvedt J, Sjoberg B, Roos C, Rosen G, Dahlgren T (1971) Acta Pharmaceutica Suecica 8:623–638
318. Davis MC, Franzblau SG, Martin AR (1998) Bioorg Med Chem Lett 8:843-846
319. Micalizio GC, Schreiber SL (2002) Angew Chem Int Ed 41:152–154
320. Micalizio GC, Schreiber SL (2002) Angew Chem 114:160–162
321. Micalizio GC, Schreiber SL (2002) Angew Chem Int Ed 41:3272–3276
322. Brittain DEA, Gray BL, Schreiber SL (2005) Chem Eur J 11:5086–5093
323. Abbey ER, Zakharov LN, Liu SY (2008) J Am Chem Soc 130:7250–7252
324. Abby ER, Liu SY (2013) Org Biomol Chem 11:2060–2069
325. Marwitz AJV, Matus MH, Akharov LN, Dixon DA, Liu SY (2009) Angew Chem Int Ed 48:973–977
326. Abbey ER, Zakharov LN, Liu SY (2010) J Am Chem Soc 132:16340–16342
327. Ulmschneider D, Goubeau J (1957) Chemische Berichte-Recueil 90:2733–2738
328. Goubeau J, Schneider H (1964) Annalen Der Chemie-Justus Liebig 675:1–9
329. Weber L (2008) Coord Chem Rev 252:1–31
330. McNulty L, Kohlbacher K, Borin K, Dodd B, Bishop J, Fuller L, Wright Z (2010) J Org Chem 75:6001–6004
331. Jernelius JA, Schrock RR, Hoveyda AH (2004) Tetrahedron 60:7345–7351
332. Hansen EC, Lee D (2006) J Am Chem Soc 128:8142–8143
333. Xu S, Zakharov LN, Liu SY (2011) J Am Chem Soc 133:20152–20155
334. Fu GC, Nguyen SBT, Grubbs RH (1993) J Am Chem Soc 115:9856–9857
335. Abbey ER, Lamm AN, Baggett AW, Zakharov LN, Liu SY (2013) J Am Chem Soc 135:12908–12913
336. Ashe AJ, Fang X (2000) Org Lett 2:2089–2091
337. Pan J, Kampf JW, Ashe AJ III (2009) J Organomet Chem 694:1036–1040
338. Pan J, Kampf JW, Ashe AJ III (2006) Organometallics 25:197–202
339. Pan J, Kampf JW, Ashe AJ III (2009) Organometallics 28:506–511
340. Emslie DJ, Piers WE, Parvez M (2003) Angew Chem Int Ed 42:1252–1255
341. Bosdet MJ, Piers WE, Sorensen TS, Parvez M (2007) Angew Chem Int Ed 46:4940–4943
342. Jaska CA, Emslie DJ, Bosdet MJ, Piers WE, Sorensen TS, Parvez M (2006) J Am Chem Soc 128:10885–10896
343. Jaska CA, Piers WE, McDonald R, Parvez M (2007) J Org Chem 72:5234–5243
344. Bosdet MJ, Jaska CA, Piers WE, Sorensen TS, Parvez M (2007) Org Lett 9:1395–1398

345. Lepeltier M, Lukoyanova O, Jacobson A, Jeeva S, Perepichka DF (2010) Chem Commun 46:7007–7009
346. Taniguchi T, Yamaguchi S (2010) Organometallics 29:5732–5735
347. Hatakeyama T, Hashimoto S, Seki S, Nakamura M (2011) J Am Chem Soc 133:18614–18617
348. Hatakeyama T, Hashimoto S, Oba T, Nakamura M (2012) J Am Chem Soc 134:19600–19603
349. Wang XY, Lin HR, Lei T, Yang DC, Zhuang FD, Wang JY, Yuan SC, Pei J (2013) Angew Chem Int Ed 52:3117–3120
350. Lu JS, Ko SB, Walters NR, Kang Y, Sauriol F, Wang S (2013) Angew Chem Int Ed 52: 4544–4548
351. Pan J, Kampf JW, Ashe AJ III (2004) Organometallics 23:5626–5629
352. Marwitz AJV, Abbey ER, Jenkins JT, Zakharov LN, Liu SY (2007) Org Lett 9:4905–4908
353. Campbell PG, Zakharov LN, Grant DJ, Dixon DA, Liu SY (2010) J Am Chem Soc 132: 3289–3291
354. Xu SM, Haeffner F, Li B, Zakharov LN, Liu SY (2014) Angew Chem Int Ed 53:6795–6799
355. Rohr AD, Kampf JW, Ashe AJ (2014) Organometallics 33:1318–1321
356. Berzelius JJ, Afhandl F (1818) Kemi Mineral 6:42
357. Wöhler F, Siemens C (1847) Ann Chem 61:360–362
358. Ninomiya M, Garud DR, Koketsu M (2011) Coord Chem Rev 255:2968–2990
359. Garud DR, Makimura M, Koketsu M (2011) New J Chem 35:581–586
360. Garud DR, Toyoda Y, Koketsu M (2009) Tetrahedron Lett 50:3035–3037
361. Garud DR, Tanahashi N, Ninomiya M, Koketsu M (2009) Tetrahedron 65:4775–4780
362. Koketsu M, Yamamura Y, Ishihara H (2006) Synthesis-Stuttgart 2738–2742
363. Garud DR, Koketsu M, Ishihara H (2007) Molecules 12:504–535
364. Litvinov VP, Dyachenko VD (1997) Russ Chem Rev 66:923–951
365. Terazawa R, Garud DR, Hamada N, Fujita Y, Itoh T, Nozawa Y, Nakane K, Deguchi T, Koketsu M, Ito M (2010) Bioorg Med Chem 18:7001–7008
366. Garud DR, Garud DD, Koketsu M (2009) Org Biomol Chem 7:2591–2598
367. Bankar DB, Koketsu M (2010) Eur J Org Chem 14:2742–2745
368. Hua G, Slawin MZA, Randall AMR, Cordes BD, Crawford L, Bühl M, Woollins JD (2013) Chem Commun 49:2619–2621
369. Garud DR, Ando H, Kawai Y, Ishihara H, Koketsu M (2007) Org Lett 9:4455–4458
370. Sommen GL, Linden A, Heimgartner H (2005) Tetrahedron Lett 46:6723–6725
371. Nurbaev KI, Zakhidov KA, Oripov EO, Smiev RA, Shakhidoyatov KM (1996) Chem Abstr 126:47303
372. Clive DLJ, Cheng H (2001) Chem Commun 7:605–606
373. Garud DR, Koketsu M (2008) Org Lett 10:3319–3322

Top Heterocycl Chem (2017) 47: 381–382
DOI: 10.1007/7081_2016_203

Published online: 10 July 2016

Erratum to: Synthesis of Cyclic Peptides and Peptidomimetics by Metathesis Reactions

Dirk T.S. Rijkers

Erratum to: Top Heterocycl Chem
DOI: 10.1007/7081_2015_142

In Scheme 10 (b), 3D structure was originally missing.

The missing 3D structure has been included now.

The updated online version of the chapter can be found under
DOI 10.1007/7081_2015_142

D.T.S. Rijkers (✉)
Department of Pharmaceutical Sciences, Faculty of Science, Medicinal Chemistry and Chemical Biology, Utrecht Institute for Pharmaceutical Sciences (UIPS), Utrecht University, P.O. Box 80082, 3508 Utrecht, The Netherlands
e-mail: D.T.S.Rijkers@uu.nl

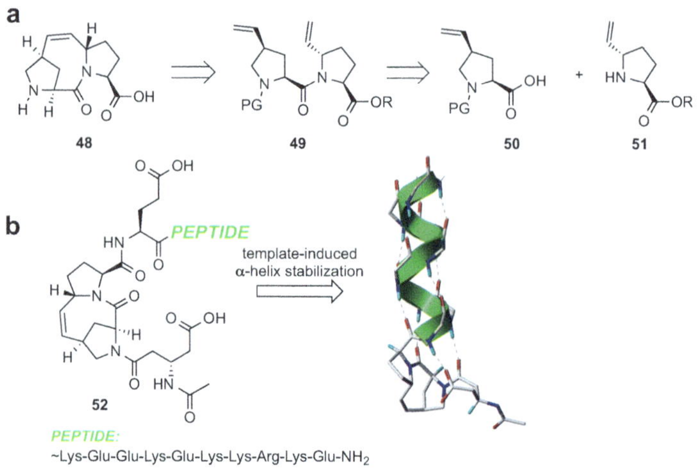

Scheme 10 (**a**) Retrosynthesis of tricyclic diproline scaffold **48**, (**b**) model of an N-capped α-helical peptide. Based on Angew Chem Int Ed (2013) 52:9539–9543. © Professor H. G. Schmalz, Universität zu Köln, Germany

Index

The manufacturer's authorised representative in the EU is Springer Nature Customer Service Centre GmbH, Europaplatz 3, 69115 Heidelberg, Germany. If you have any concerns regarding our products, please contact ProductSafety@springernature.com

Printed and bound by CPI Group (UK) Ltd, Croydon, CR0 4YY
30/09/2026
02225652-0016